advanced
classical field theory

advanced classical field theory

giovanni giachetta
university of camerino, italy

luigi mangiarotti
university of camerino, italy

gennadi sardanashvily
moscow state university, russia

World Scientific

NEW JERSEY · LONDON · SINGAPORE · BEIJING · SHANGHAI · HONG KONG · TAIPEI · CHENNAI

Published by

World Scientific Publishing Co. Pte. Ltd.
5 Toh Tuck Link, Singapore 596224
USA office: 27 Warren Street, Suite 401-402, Hackensack, NJ 07601
UK office: 57 Shelton Street, Covent Garden, London WC2H 9HE

British Library Cataloguing-in-Publication Data
A catalogue record for this book is available from the British Library.

ADVANCED CLASSICAL FIELD THEORY

Copyright © 2009 by World Scientific Publishing Co. Pte. Ltd.

All rights reserved. This book, or parts thereof, may not be reproduced in any form or by any means, electronic or mechanical, including photocopying, recording or any information storage and retrieval system now known or to be invented, without written permission from the Publisher.

For photocopying of material in this volume, please pay a copying fee through the Copyright Clearance Center, Inc., 222 Rosewood Drive, Danvers, MA 01923, USA. In this case permission to photocopy is not required from the publisher.

ISBN-13 978-981-283-895-7
ISBN-10 981-283-895-3

Printed in Singapore.

Preface

Contemporary quantum field theory is mainly developed as quantization of classical fields. Classical field theory thus is a necessary step towards quantum field theory. This book provides an exhaust mathematical foundation of Lagrangian classical field theory and its BRST extension for the purpose of quantization.

Lagrangian theory of Grassmann-graded (even and odd) fields on fibre bundles and graded manifolds is presented in the book in a very general setting. It is adequately formulated in geometric and algebraic topological terms of the jet manifolds and the variational bicomplex. The main ingredients in this formulation are cohomology of the variational bicomplex, the global first variational formula, variational symmetries and supersymmetries, the first Noether theorem, Noether identities, the direct and inverse second Noether theorems, and gauge symmetries.

Degenerate Lagrangian field theories are comprehensively investigated. The hierarchies of their non-trivial reducible Noether identities and gauge symmetries are described in homology terms. The relevant direct and inverse second Noether theorems are formulated in a very general setting.

The study of degeneracy of Lagrangian field theory straightforwardly leads to its BRST extension by Grassmann-graded antifields and ghosts which constitute the chain and cochain complexes of non-trivial Noether identities and gauge symmetries. In particular, a gauge operator is prolonged to a nilpotent BRST operator, and an original field Lagrangian is extended to a non-trivial solution of the classical master equation of Lagrangian BRST theory. This is a preliminary step towards quantization of classical Lagrangian field theory in terms of functional integrals.

The basic field theories, including gauge theory on principal bundles, gravitation theory on natural bundles, theory of spinor fields and topological field theory, are presented in the book in a very complete way.

Our book addresses to a wide audience of theoreticians and mathematical physicists, and aims to be a guide to advanced differential geometric and algebraic topological methods in field theory.

With respect to mathematical prerequisites, the reader is expected to be familiar with the basics of differential geometry of fibre bundles. We have tried to give the necessary mathematical background, thus making the exposition self-contained. For the sake of convenience of the reader, several relevant mathematical topics are compiled in Appendixes.

Contents

Preface   v

Introduction   1

1. Differential calculus on fibre bundles   5
   1.1 Geometry of fibre bundles 5
       1.1.1 Manifold morphisms 6
       1.1.2 Fibred manifolds and fibre bundles 7
       1.1.3 Vector and affine bundles 12
       1.1.4 Vector fields, distributions and foliations 18
       1.1.5 Exterior and tangent-valued forms 21
   1.2 Jet manifolds 26
   1.3 Connections on fibre bundles 29
       1.3.1 Connections as tangent-valued forms 30
       1.3.2 Connections as jet bundle sections 32
       1.3.3 Curvature and torsion 34
       1.3.4 Linear connections 36
       1.3.5 Affine connections 38
       1.3.6 Flat connections 39
       1.3.7 Second order connections 41
   1.4 Composite bundles 42
   1.5 Higher order jet manifolds 46
   1.6 Differential operators and equations 51
   1.7 Infinite order jet formalism 54

2. Lagrangian field theory on fibre bundles   61
   2.1 Variational bicomplex 61

2.2	Lagrangian symmetries	66
2.3	Gauge symmetries	70
2.4	First order Lagrangian field theory	73
	2.4.1 Cartan and Hamilton–De Donder equations	75
	2.4.2 Lagrangian conservation laws	78
	2.4.3 Gauge conservation laws. Superpotential	80
	2.4.4 Non-regular quadratic Lagrangians	83
	2.4.5 Reduced second order Lagrangians	87
	2.4.6 Background fields	88
	2.4.7 Variation Euler–Lagrange equation. Jacobi fields	90
2.5	Appendix. Cohomology of the variational bicomplex	92

3. Grassmann-graded Lagrangian field theory — 99

3.1	Grassmann-graded algebraic calculus	99
3.2	Grassmann-graded differential calculus	104
3.3	Geometry of graded manifolds	107
3.4	Grassmann-graded variational bicomplex	115
3.5	Lagrangian theory of even and odd fields	120
3.6	Appendix. Cohomology of the Grassmann-graded variational bicomplex	125

4. Lagrangian BRST theory — 129

4.1	Noether identities. The Koszul–Tate complex	130
4.2	Second Noether theorems in a general setting	140
4.3	BRST operator	147
4.4	BRST extended Lagrangian field theory	150
4.5	Appendix. Noether identities of differential operators	154

5. Gauge theory on principal bundles — 165

5.1	Geometry of Lie groups	165
5.2	Bundles with structure groups	169
5.3	Principal bundles	171
5.4	Principal connections. Gauge fields	175
5.5	Canonical principal connection	179
5.6	Gauge transformations	181
5.7	Geometry of associated bundles. Matter fields	184
5.8	Yang–Mills gauge theory	188
	5.8.1 Gauge field Lagrangian	188

		5.8.2	Conservation laws	190
		5.8.3	BRST extension	192
		5.8.4	Matter field Lagrangian	194
	5.9		Yang–Mills supergauge theory	196
	5.10		Reduced structure. Higgs fields	198
		5.10.1	Reduction of a structure group	198
		5.10.2	Reduced subbundles	200
		5.10.3	Reducible principal connections	202
		5.10.4	Associated bundles. Matter and Higgs fields	203
		5.10.5	Matter field Lagrangian	207
	5.11		Appendix. Non-linear realization of Lie algebras	211

6. Gravitation theory on natural bundles — 215

	6.1	Natural bundles	215
	6.2	Linear world connections	219
	6.3	Lorentz reduced structure. Gravitational field	223
	6.4	Space-time structure	228
	6.5	Gauge gravitation theory	232
	6.6	Energy-momentum conservation law	236
	6.7	Appendix. Affine world connections	238

7. Spinor fields — 243

	7.1	Clifford algebras and Dirac spinors	243
	7.2	Dirac spinor structure	246
	7.3	Universal spinor structure	252
	7.4	Dirac fermion fields	258

8. Topological field theories — 263

	8.1		Topological characteristics of principal connections	263
		8.1.1	Characteristic classes of principal connections	264
		8.1.2	Flat principal connections	266
		8.1.3	Chern classes of unitary principal connections	270
		8.1.4	Characteristic classes of world connections	274
	8.2		Chern–Simons topological field theory	278
	8.3		Topological BF theory	283
	8.4		Lagrangian theory of submanifolds	286

9. Covariant Hamiltonian field theory — 293

	9.1	Polysymplectic Hamiltonian formalism	293
	9.2	Associated Hamiltonian and Lagrangian systems	298
	9.3	Hamiltonian conservation laws	304
	9.4	Quadratic Lagrangian and Hamiltonian systems	306
	9.5	Example. Yang–Mills gauge theory	313
	9.6	Variation Hamilton equations. Jacobi fields	316

10. Appendixes 319

    10.1 Commutative algebra 319
    10.2 Differential operators on modules 324
    10.3 Homology and cohomology of complexes 327
    10.4 Cohomology of groups 330
    10.5 Cohomology of Lie algebras 333
    10.6 Differential calculus over a commutative ring 334
    10.7 Sheaf cohomology 337
    10.8 Local-ringed spaces 346
    10.9 Cohomology of smooth manifolds 348
    10.10 Leafwise and fibrewise cohomology 354

Bibliography   359

Index   369

Introduction

Contemporary quantum field theory is mainly developed as quantization of classical fields. In particular, a generating functional of Green functions in perturbative quantum field theory depends on an action functional of classical fields. In contrast with quantum field theory, classical field theory can be formulated in a strict mathematical way.

Observable classical fields are an electromagnetic field, Dirac spinor fields and a gravitational field on a world real smooth manifold. Their dynamic equations are Euler–Lagrange equations derived from a Lagrangian. One also considers classical non-Abelian gauge fields and Higgs fields. Basing on these models, we study Lagrangian theory of classical Grassmann-graded (even and odd) fields on an arbitrary smooth manifold in a very general setting. Geometry of principal bundles is known to provide the adequate mathematical formulation of classical gauge theory. Generalizing this formulation, we define even classical fields as sections of smooth fibre bundles and, accordingly, develop their Lagrangian theory as Lagrangian theory on fibre bundles.

Note that, treating classical field theory, we are in the category of finite-dimensional smooth real manifolds, which are Hausdorff, second-countable and paracompact. Let X be such a manifold. If classical fields form a projective $C^\infty(X)$-module of finite rank, their representation by sections of a fibre bundle follows from the well-known Serre–Swan theorem.

Lagrangian theory on fibre bundles is adequately formulated in algebraic terms of the variational bicomplex of exterior forms on jet manifolds [3; 17; 50]. This formulation is straightforwardly extended to Lagrangian theory of even and odd fields by means of the Grassmann-graded variational bicomplex [9; 14; 59]. Cohomology of this bicomplex provides the global first variational formula for Lagrangians and Euler–Lagrange operators, the first

Noether theorem and conservation laws in a general case of supersymmetries depending on derivatives of fields of any order.

Note that there are different descriptions of odd fields on graded manifolds [27; 118] and supermanifolds [29; 45]. Both graded manifolds and supermanifolds are described in terms of sheaves of graded commutative algebras [10]. However, graded manifolds are characterized by sheaves on smooth manifolds, while supermanifolds are constructed by gluing of sheaves on supervector spaces. Treating odd fields on a smooth manifold X, we follow the Serre–Swan theorem generalized to graded manifolds [14]. It states that, if a Grassmann $C^\infty(X)$-algebra is an exterior algebra of some projective $C^\infty(X)$-module of finite rank, it is isomorphic to the algebra of graded functions on a graded manifold whose body is X.

Quantization of Lagrangian field theory essentially depends on its degeneracy characterized by a family of non-trivial reducible Noether identities [9; 15; 63]. A problem is that any Euler–Lagrange operator satisfies Noether identities which therefore must be separated into the trivial and non-trivial ones. These Noether identities can obey first-stage Noether identities, which in turn are subject to the second-stage ones, and so on. If certain conditions hold, this hierarchy of Noether identities is described by the exact Koszul–Tate chain complex of antifields possessing the boundary operator whose nilpotentness is equivalent to all non-trivial Noether and higher-stage Noether identities [14; 15].

The inverse second Noether theorem formulated in homology terms associates to this Koszul–Tate complex the cochain sequence of ghosts with the ascent operator, called the gauge operator, whose components are non-trivial gauge and higher-stage gauge symmetries of Lagrangian field theory [15]. These gauge symmetries are parameterized by odd and even ghosts so that k-stage gauge symmetries act on $(k-1)$-stage ghosts.

It should be emphasized that the gauge operator unlike the Koszul–Tate one is not nilpotent, unless gauge symmetries are Abelian. Gauge symmetries are said to be algebraically closed if this gauge operator admits a nilpotent extension where k-stage gauge symmetries are extended to k-stage BRST (Becchi–Rouet–Stora–Tyitin) transformations acting both on $(k-1)$-stage and k-stage ghosts [61]. This nilpotent extension is called the BRST operator. If the BRST operator exists, the cochain sequence of ghosts is brought into the BRST complex.

The Koszul–Tate and BRST complexes provide a BRST extension of original Lagrangian field theory. This extension exemplifies so called field-antifield theory whose Lagrangians are required to satisfy a certain condition, called the classical master equation. An original Lagrangian is extended to a proper solution of the master equation if the BRST operator exists [15]. This extended Lagrangian, dependent on original fields, ghosts and antifields, is a first step towards quantization of classical field theory in terms of functional integrals [9; 63].

The basic field theories, including gauge theory on principal bundles (Chapter 5), gravitation theory on natural bundles (Chapter 6), theory of spinor fields (Chapter 7) and topological field theory (Chapter 8) are presented in the book in a complete way.

The reader also can find a number of original topics, including: general theory of connections (Section 1.3), geometry of composite bundles (Section 1.4), infinite-order jet formalism (Section 1.7), generalized symmetries (Section 2.2), Grassmann-graded Lagrangian field theory (Section 3.5), second Noether theorems in a general setting (Section 4.2), the BRST complex (Section 4.3), classical Higgs field theory (Section 5.10), gauge theory of gravity as a Higgs field (Section 6.5), gauge energy-momentum conservation laws (Section 6.6), composite spinor bundles (Section 7.3), global Chern–Simons topological field theory (Section 8.2), topological BF (background field) theory (Section 8.3), covariant Hamiltonian field theory (Chapter 9).

For the sake of convenience of the reader, several relevant mathematical topics are compiled in Chapter 10.

Chapter 1

Differential calculus on fibre bundles

This Chapter summarizes the relevant material on fibre bundles, jet manifolds and connections which find application in classical field theory. The material is presented in a fairly informal way. It is tacitly assumed that the reader has some familiarity with the basics of differential geometry [69; 92; 147; 164].

1.1 Geometry of fibre bundles

Throughout the book, all morphisms are smooth (i.e. of class C^∞) and manifolds are smooth real and finite-dimensional. A smooth real *manifold* is customarily assumed to be Hausdorff and second-countable (i.e., it has a countable base for topology). Consequently, it is a locally compact space which is a union of a countable number of compact subsets, a separable space (i.e., it has a countable dense subset), a paracompact and completely regular space. Being paracompact, a smooth manifold admits a partition of unity by smooth real functions (see Remark 10.7.4). One can also show that, given two disjoint closed subsets N and N' of a smooth manifold X, there exists a smooth function f on X such that $f|_N = 0$ and $f|_{N'} = 1$. Unless otherwise stated, manifolds are assumed to be connected (and, consequently, arcwise connected). We follow the notion of a manifold without boundary.

The standard symbols $\otimes$, $\vee$, and $\wedge$ stand for the tensor, symmetric, and exterior products, respectively. The interior product (contraction) is denoted by $\rfloor$. By ∂^A_B are meant the partial derivatives with respect to coordinates with indices ^B_A.

If Z is a manifold, we denote by
$$\pi_Z : TZ \to Z, \qquad \pi_Z^* : T^*Z \to Z$$
its tangent and cotangent bundles, respectively. Given coordinates (z^α) on Z, they are equipped with the *holonomic coordinates*
$$(z^\lambda, \dot{z}^\lambda), \qquad \dot{z}'^\lambda = \frac{\partial z'^\lambda}{\partial z^\mu} \dot{z}^\mu,$$
$$(z^\lambda, \dot{z}_\lambda), \qquad \dot{z}'_\lambda = \frac{\partial z'^\mu}{\partial z^\lambda} \dot{z}_\mu,$$
with respect to the *holonomic frames* $\{\partial_\lambda\}$ and *coframes* $\{dz^\lambda\}$ in the tangent and cotangent spaces to Z, respectively. Any manifold morphism $f : Z \to Z'$ yields the *tangent morphism*
$$Tf : TZ \to TZ', \qquad \dot{z}'^\lambda \circ Tf = \frac{\partial f^\lambda}{\partial x^\mu} \dot{z}^\mu.$$
The symbol $C^\infty(Z)$ stands for the ring of smooth real functions on a manifold Z.

1.1.1 Manifold morphisms

Let us consider manifold morphisms of maximal rank. They are immersions (in particular, imbeddings) and submersions. An injective immersion is a submanifold, and a surjective submersion is a fibred manifold (in particular, a fibre bundle).

Given manifolds M and N, by the *rank of a morphism* $f : M \to N$ at a point $p \in M$ is meant the rank of the linear morphism
$$T_p f : T_p M \to T_{f(p)} N.$$
For instance, if f is of maximal rank at $p \in M$, then $T_p f$ is injective when $\dim M \leq \dim N$ and surjective when $\dim N \leq \dim M$. In this case, f is called an *immersion* and a *submersion* at a point $p \in M$, respectively.

Since $p \to \mathrm{rank}_p f$ is a lower semicontinuous function, then the morphism $T_p f$ is of maximal rank on an open neighbourhood of p, too. It follows from the inverse function theorem that:
- if f is an immersion at p, then it is locally injective around p.
- if f is a submersion at p, it is locally surjective around p.

If f is both an immersion and a submersion, it is called a *local diffeomorphism* at p. In this case, there exists an open neighbourhood U of p such that
$$f : U \to f(U)$$

1.1. Geometry of fibre bundles

is a diffeomorphism onto an open set $f(U) \subset N$.

A manifold morphism f is called the immersion (resp. submersion) if it is an immersion (resp. submersion) at all points of M. A submersion is necessarily an *open map*, i.e., it sends open subsets of M onto open subsets of N. If an immersion f is open (i.e., f is a homeomorphism onto $f(M)$ equipped with the relative topology from N), it is called the *imbedding*.

A pair (M, f) is called a *submanifold* of N if f is an injective immersion. A submanifold (M, f) is an *imbedded submanifold* if f is an imbedding. For the sake of simplicity, we usually identify (M, f) with $f(M)$. If $M \subset N$, its natural injection is denoted by $i_M : M \to N$.

There are the following criteria for a submanifold to be imbedded.

Theorem 1.1.1. *Let (M, f) be a submanifold of N.*

(i) The map f is an imbedding if and only if, for each point $p \in M$, there exists a (cubic) coordinate chart (V, ψ) of N centered at $f(p)$ so that $f(M) \cap V$ consists of all points of V with coordinates $(x^1, \ldots, x^m, 0, \ldots, 0)$.

(ii) Suppose that $f : M \to N$ is a proper map, i.e., the pre-images of compact sets are compact. Then (M, f) is a closed imbedded submanifold of N. In particular, this occurs if M is a compact manifold.

(iii) If $\dim M = \dim N$, then (M, f) is an open imbedded submanifold of N.

1.1.2 Fibred manifolds and fibre bundles

A triple

$$\pi : Y \to X, \qquad \dim X = n > 0, \tag{1.1.1}$$

is called a *fibred manifold* if a manifold morphism π is a surjective submersion, i.e., the tangent morphism

$$T\pi : TY \to TX$$

is a surjection. One says that Y is a *total space* of a fibred manifold (1.1.1), X is its *base*, π is a *fibration*, and $Y_x = \pi^{-1}(x)$ is a *fibre* over $x \in X$.

Any fibre is an imbedded submanifold of Y of dimension $\dim Y - \dim X$. Unless otherwise stated, we assume that

$$\dim Y \neq \dim X,$$

i.e., fibred manifolds with discrete fibres are not considered.

Theorem 1.1.2. *A surjection (1.1.1) is a fibred manifold if and only if a manifold Y admits an atlas of coordinate charts $(U_Y; x^\lambda, y^i)$ such that (x^λ)*

are coordinates on $\pi(U_Y) \subset X$ and coordinate transition functions read
$$x'^\lambda = f^\lambda(x^\mu), \qquad y'^i = f^i(x^\mu, y^j).$$
These coordinates are called fibred coordinates compatible with a fibration π.

By a *local section* of a surjection (1.1.1) is meant an injection $s : U \to Y$ of an open subset $U \subset X$ such that $\pi \circ s = \mathrm{Id}\, U$, i.e., a section sends any point $x \in X$ into the fibre Y_x over this point. A local section also is defined over any subset $N \in X$ as the restriction to N of a local section over an open set containing N. If $U = X$, one calls s the *global section*. Hereafter, by a *section* is meant both a global section and a local section (over an open subset).

Theorem 1.1.3. *A surjection π (1.1.1) is a fibred manifold if and only if, for each point $y \in Y$, there exists a local section s of $\pi : Y \to X$ passing through y.*

The range $s(U)$ of a local section $s : U \to Y$ of a fibred manifold $Y \to X$ is an imbedded submanifold of Y. It also is a *closed map*, which sends closed subsets of U onto closed subsets of Y. If s is a global section, then $s(X)$ is a closed imbedded submanifold of Y. Global sections of a fibred manifold need not exist.

Theorem 1.1.4. *Let $Y \to X$ be a fibred manifold whose fibres are diffeomorphic to $\mathbb{R}^m$. Any its section over a closed imbedded submanifold (e.g., a point) of X is extended to a global section [147]. In particular, such a fibred manifold always has a global section.*

Given fibred coordinates $(U_Y; x^\lambda, y^i)$, a section s of a fibred manifold $Y \to X$ is represented by collections of local functions $\{s^i = y^i \circ s\}$ on $\pi(U_Y)$.

A fibred manifold $Y \to X$ is called a *fibre bundle* if admits a fibred coordinate atlas $\{(\pi^{-1}(U_\xi); x^\lambda, y^i)\}$ over a cover $\{\pi^{-1}(U_\iota)\}$ of Y which is the inverse image of a cover $\mathfrak{U} = \{U_\xi\}$ is a cover of X. In this case, there exists a manifold V, called a *typical fibre*, such that Y is locally diffeomorphic to the splittings
$$\psi_\xi : \pi^{-1}(U_\xi) \to U_\xi \times V, \qquad (1.1.2)$$
glued together by means of *transition functions*
$$\varrho_{\xi\zeta} = \psi_\xi \circ \psi_\zeta^{-1} : U_\xi \cap U_\zeta \times V \to U_\xi \cap U_\zeta \times V \qquad (1.1.3)$$

1.1. Geometry of fibre bundles

on overlaps $U_\xi \cap U_\zeta$. Transition functions $\varrho_{\xi\zeta}$ fulfil the *cocycle condition*

$$\varrho_{\xi\zeta} \circ \varrho_{\zeta\iota} = \varrho_{\xi\iota} \qquad (1.1.4)$$

on all overlaps $U_\xi \cap U_\zeta \cap U_\iota$. Restricted to a point $x \in X$, trivialization morphisms ψ_ξ (1.1.2) and transition functions $\varrho_{\xi\zeta}$ (1.1.3) define diffeomorphisms of fibres

$$\psi_\xi(x): Y_x \to V, \qquad x \in U_\xi, \qquad (1.1.5)$$

$$\varrho_{\xi\zeta}(x): V \to V, \qquad x \in U_\xi \cap U_\zeta. \qquad (1.1.6)$$

Trivialization charts (U_ξ, ψ_ξ) together with transition functions $\varrho_{\xi\zeta}$ (1.1.3) constitute a *bundle atlas*

$$\Psi = \{(U_\xi, \psi_\xi), \varrho_{\xi\zeta}\} \qquad (1.1.7)$$

of a fibre bundle $Y \to X$. Two bundle atlases are said to be *equivalent* if their union also is a bundle atlas, i.e., there exist transition functions between trivialization charts of different atlases.

A fibre bundle $Y \to X$ is uniquely defined by a bundle atlas. Given an atlas Ψ (1.1.7), there is a unique manifold structure on Y for which $\pi: Y \to X$ is a fibre bundle with the typical fibre V and the bundle atlas Ψ. All atlases of a fibre bundle are equivalent.

Remark 1.1.1. The notion of a fibre bundle introduced above is the notion of a *smooth locally trivial fibre bundle*. In a general setting, a *continuous fibre bundle* is defined as a continuous surjective submersion of topological spaces $Y \to X$. A continuous map $\pi: Y \to X$ is called a *submersion* if, for any point $y \in Y$, there exists an open neighborhood U of the point $\pi(y)$ and a right inverse $\sigma: U \to Y$ of π such that $\sigma \circ \pi(y) = y$, i.e., there exists a local section of π. The notion of a *locally trivial continuous fibre bundle* is a repetition of that of a smooth fibre bundle, where trivialization morphisms ψ_ξ and transition functions $\varrho_{\xi\zeta}$ are continuous.

We have the following useful criteria for a fibred manifold to be a fibre bundle.

Theorem 1.1.5. *If a fibration $\pi: Y \to X$ is a proper map, then $Y \to X$ is a fibre bundle. In particular, a fibred manifold with a compact total space is a fibre bundle.*

Theorem 1.1.6. *A fibred manifold whose fibres are diffeomorphic either to a compact manifold or $\mathbb{R}^r$ is a fibre bundle [115].*

A comprehensive relation between fibred manifolds and fibre bundles is given in Remark 1.3.2. It involves the notion of an Ehresmann connection.

Unless otherwise stated, we restrict our consideration to fibre bundles. Without a loss of generality, we further assume that a cover $\mathfrak{U}$ for a bundle atlas of $Y \to X$ also is a cover for a manifold atlas of the base X. Then, given a bundle atlas Ψ (1.1.7), a fibre bundle Y is provided with the associated *bundle coordinates*

$$x^\lambda(y) = (x^\lambda \circ \pi)(y), \qquad y^i(y) = (y^i \circ \psi_\xi)(y), \qquad y \in \pi^{-1}(U_\xi),$$

where x^λ are coordinates on $U_\xi \subset X$ and y^i, called *fibre coordinates*, are coordinates on a typical fibre V.

The forthcoming Theorems 1.1.7 – 1.1.9 describe the particular covers which one can choose for a bundle atlas. Throughout the book, only *proper covers* of manifolds are considered, i.e., $U_\xi \neq U_\zeta$ if $\zeta \neq \xi$. Recall that a cover $\mathfrak{U}'$ is a refinement of a cover $\mathfrak{U}$ if, for each $U' \in \mathfrak{U}'$, there exists $U \in \mathfrak{U}$ such that $U' \subset U$. Of course, if a fibre bundle $Y \to X$ has a bundle atlas over a cover $\mathfrak{U}$ of X, it admits a bundle atlas over any refinement of $\mathfrak{U}$.

A fibred manifold $Y \to X$ is called trivial if Y is diffeomorphic to the product $X \times V$. Different trivializations of $Y \to X$ differ from each other in surjections $Y \to V$.

Theorem 1.1.7. *Any fibre bundle over a contractible base is trivial.*

However, a fibred manifold over a contractible base need not be trivial, even its fibres are mutually diffeomorphic.

It follows from Theorem 1.1.7 that any cover of a base X consisting of *domains* (i.e., contractible open subsets) is a bundle cover.

Theorem 1.1.8. *Every fibre bundle $Y \to X$ admits a bundle atlas over a countable cover $\mathfrak{U}$ of X where each member U_ξ of $\mathfrak{U}$ is a domain whose closure $\overline{U}_\xi$ is compact [69].*

If a base X is compact, there is a bundle atlas of Y over a finite cover of X which obeys the condition of Theorem 1.1.8.

Theorem 1.1.9. *Every fibre bundle $Y \to X$ admits a bundle atlas over a finite cover $\mathfrak{U}$ of X, but its members need not be contractible and connected.*

Proof. Let Ψ be a bundle atlas of $Y \to X$ over a cover $\mathfrak{U}$ of X. For any cover $\mathfrak{U}$ of a manifold X, there exists its refinement $\{U_{ij}\}$, where $j \in \mathbb{N}$ and i runs through a finite set such that $U_{ij} \cap U_{ik} = \emptyset$, $j \neq k$. Let $\{(U_{ij}, \psi_{ij})\}$

1.1. Geometry of fibre bundles

be the corresponding bundle atlas of a fibre bundle $Y \to X$. Then Y has the finite bundle atlas

$$U_i = \bigcup_j U_{ij}, \qquad \psi_i(x) = \psi_{ij}(x), \qquad x \in U_{ij} \subset U_i.$$

$\square$

Morphisms of fibre bundles, by definition, are *fibrewise morphisms*, sending a fibre to a fibre. Namely, a *bundle morphism* of a fibre bundle $\pi : Y \to X$ to a fibre bundle $\pi' : Y' \to X'$ is defined as a pair (Φ, f) of manifold morphisms which form a commutative diagram

$$\begin{array}{ccc} Y & \xrightarrow{\Phi} & Y' \\ \pi \downarrow & & \downarrow \pi' \\ X & \xrightarrow{f} & X' \end{array}, \qquad \pi' \circ \Phi = f \circ \pi.$$

Bundle injections and surjections are called *bundle monomorphisms and epimorphisms*, respectively. A bundle diffeomorphism is called a *bundle isomorphism*, or a *bundle automorphism* if it is an isomorphism to itself. For the sake of brevity, a bundle morphism over $f = \mathrm{Id}\, X$ is often said to be a bundle morphism over X, and is denoted by $Y \xrightarrow[X]{} Y'$. In particular, a bundle automorphism over X is called a *vertical automorphism*.

A bundle monomorphism $\Phi : Y \to Y'$ over X is called a *subbundle* of a fibre bundle $Y' \to X$ if $\Phi(Y)$ is a submanifold of Y'. There is the following useful criterion for an image and an inverse image of a bundle morphism to be subbundles.

Theorem 1.1.10. *Let $\Phi : Y \to Y'$ be a bundle morphism over X. Given a global section s of the fibre bundle $Y' \to X$ such that $s(X) \subset \Phi(Y)$, by the kernel of a bundle morphism Φ with respect to a section s is meant the inverse image*

$$\mathrm{Ker}\,_s \Phi = \Phi^{-1}(s(X))$$

of $s(X)$ by Φ. If $\Phi : Y \to Y'$ is a bundle morphism of constant rank over X, then $\Phi(Y)$ and $\mathrm{Ker}\,_s \Phi$ are subbundles of Y' and Y, respectively.

In conclusion, let us describe the following standard constructions of new fibre bundles from the old ones.

• Given a fibre bundle $\pi : Y \to X$ and a manifold morphism $f : X' \to X$, the *pull-back* of Y by f is called the manifold

$$f^*Y = \{(x', y) \in X' \times Y \,:\, \pi(y) = f(x')\} \tag{1.1.8}$$

together with the natural projection $(x',y) \to x'$. It is a fibre bundle over X' such that the fibre of f^*Y over a point $x' \in X'$ is that of Y over the point $f(x') \in X$. There is the canonical bundle morphism

$$f_Y : f^*Y \ni (x',y)|_{\pi(y)=f(x')} \xrightarrow{f} y \in Y. \qquad (1.1.9)$$

Any section s of a fibre bundle $Y \to X$ yields the *pull-back section*

$$f^*s(x') = (x', s(f(x')))$$

of $f^*Y \to X'$.

• If $X' \subset X$ is a submanifold of X and $i_{X'}$ is the corresponding natural injection, then the pull-back bundle

$$i_{X'}^*Y = Y|_{X'}$$

is called the *restriction* of a fibre bundle Y to the submanifold $X' \subset X$. If X' is an imbedded submanifold, any section of the pull-back bundle

$$Y|_{X'} \to X'$$

is the restriction to X' of some section of $Y \to X$.

• Let $\pi : Y \to X$ and $\pi' : Y' \to X$ be fibre bundles over the same base X. Their *bundle product* $Y \times_X Y'$ over X is defined as the pull-back

$$Y \underset{X}{\times} Y' = \pi^*Y' \quad \text{or} \quad Y \underset{X}{\times} Y' = \pi'^*Y$$

together with its natural surjection onto X. Fibres of the bundle product $Y \times Y'$ are the Cartesian products $Y_x \times Y'_x$ of fibres of fibre bundles Y and Y'.

1.1.3 Vector and affine bundles

A *vector bundle* is a fibre bundle $Y \to X$ such that:

• its typical fibre V and all the fibres $Y_x = \pi^{-1}(x)$, $x \in X$, are real finite-dimensional vector spaces;

• there is a bundle atlas Ψ (1.1.7) of $Y \to X$ whose trivialization morphisms ψ_ξ (1.1.5) and transition functions $\varrho_{\xi\zeta}$ (1.1.6) are linear isomorphisms.

Accordingly, a vector bundle is provided with *linear bundle coordinates* (y^i) possessing linear transition functions $y'^i = A^i_j(x)y^j$. We have

$$y = y^i e_i(\pi(y)) = y^i \psi_\xi(\pi(y))^{-1}(e_i), \qquad \pi(y) \in U_\xi, \qquad (1.1.10)$$

where $\{e_i\}$ is a fixed basis for the typical fibre V of Y, and $\{e_i(x)\}$ are the fibre bases (or the *frames*) for the fibres Y_x of Y associated to the bundle atlas Ψ.

1.1. Geometry of fibre bundles

By virtue of Theorem 1.1.4, any vector bundle has a global section, e.g., the canonical global *zero-valued section* $\widehat{0}(x) = 0$. Global sections of a vector bundle $Y \to X$ constitute a projective $C^\infty(X)$-module $Y(X)$ of finite rank. It is called the *structure module of a vector bundle*.

Theorem 1.1.11. *Let a vector bundle $Y \to X$ admit $m = \dim V$ nowhere vanishing global sections s_i which are linearly independent, i.e., $\underset{m}{\wedge} s_i \neq 0$. Then Y is trivial.*

Proof. Values of these sections define the frames $\{s_i(x)\}$ for all fibres V_x, $x \in X$. Linear fibre coordinates (y^i) with respect to these frames form a bundle coordinate atlas with identity transition functions of fibre coordinates. □

Theorem 10.9.2 and Serre–Swan Theorem 10.9.3 state the categorial equivalence between the vector bundles over a smooth manifold X and projective $C^\infty(X)$-modules of finite rank. Therefore, the differential calculus (including linear differential operators, linear connections) on vector bundles can be algebraically formulated as the differential calculus on these modules. We however follow fibre bundle formalism because classical field theory is not restricted to vector bundles.

By a morphism of vector bundles is meant a *linear bundle morphism*, which is a linear fibrewise map whose restriction to each fibre is a linear map.

Given a linear bundle morphism $\Phi : Y' \to Y$ of vector bundles over X, its *kernel* $\operatorname{Ker} \Phi$ is defined as the inverse image $\Phi^{-1}(\widehat{0}(X))$ of the canonical zero-valued section $\widehat{0}(X)$ of Y. By virtue of Theorem 1.1.10, if Φ is of constant rank, its kernel and its range are vector subbundles of the vector bundles Y' and Y, respectively. For instance, monomorphisms and epimorphisms of vector bundles fulfil this condition.

There are the following particular constructions of new vector bundles from the old ones.

- Let $Y \to X$ be a vector bundle with a typical fibre V. By $Y^* \to X$ is denoted the *dual vector bundle* with the typical fibre V^* dual of V. The *interior product* of Y and Y^* is defined as a fibred morphism

$$\rfloor : Y \underset{X}{\otimes} Y^* \longrightarrow X \times \mathbb{R}.$$

- Let $Y \to X$ and $Y' \to X$ be vector bundles with typical fibres V and V', respectively. Their *Whitney sum* $Y \underset{X}{\oplus} Y'$ is a vector bundle over X with the typical fibre $V \oplus V'$.

- Let $Y \to X$ and $Y' \to X$ be vector bundles with typical fibres V and V', respectively. Their *tensor product* $Y \underset{X}{\otimes} Y'$ is a vector bundle over X with the typical fibre $V \otimes V'$. Similarly, the *exterior product* of vector bundles $Y \underset{X}{\wedge} Y'$ is defined. The exterior product

$$\wedge Y = X \times \mathbb{R} \underset{X}{\oplus} Y \underset{X}{\oplus} \overset{2}{\wedge} Y \underset{X}{\oplus} \cdots \overset{k}{\wedge} Y, \qquad k = \dim Y - \dim X, \qquad (1.1.11)$$

is called the *exterior bundle*.

Remark 1.1.2. Given vector bundles Y and Y' over the same base X, every linear bundle morphism

$$\Phi : Y_x \ni \{e_i(x)\} \to \{\Phi_i^k(x) e'_k(x)\} \in Y'_x$$

over X defines a global section

$$\Phi : x \to \Phi_i^k(x) e^i(x) \otimes e'_k(x)$$

of the tensor product $Y \otimes Y'^*$, and *vice versa*.

A sequence

$$Y' \xrightarrow{i} Y \xrightarrow{j} Y''$$

of vector bundles over the same base X is called *exact* at Y if $\operatorname{Ker} j = \operatorname{Im} i$. A sequence of vector bundles

$$0 \to Y' \xrightarrow{i} Y \xrightarrow{j} Y'' \to 0 \qquad (1.1.12)$$

over X is said to be a *short exact sequence* if it is exact at all terms Y', Y, and Y''. This means that i is a bundle monomorphism, j is a bundle epimorphism, and $\operatorname{Ker} j = \operatorname{Im} i$. Then Y'' is the *factor bundle* Y/Y' whose structure module is the quotient $Y(X)/Y'(X)$ of the structure modules of Y and Y'. Given an exact sequence of vector bundles (1.1.12), there is the exact sequence of their duals

$$0 \to Y''^* \xrightarrow{j^*} Y^* \xrightarrow{i^*} Y'^* \to 0.$$

One says that an exact sequence (1.1.12) is *split* if there exists a bundle monomorphism $s : Y'' \to Y$ such that $j \circ s = \operatorname{Id} Y''$ or, equivalently,

$$Y = i(Y') \oplus s(Y'') = Y' \oplus Y''.$$

Theorem 1.1.12. *Every exact sequence of vector bundles (1.1.12) is split* [80].

1.1. Geometry of fibre bundles

This theorem is a corollary of the Serre–Swan Theorem 10.9.3 and Theorem 10.1.3.

The tangent bundle TZ and the cotangent bundle T^*Z of a manifold Z exemplify vector bundles.

Remark 1.1.3. Given an atlas $\Psi_Z = \{(U_\iota, \phi_\iota)\}$ of a manifold Z, the tangent bundle is provided with the *holonomic bundle atlas*

$$\Psi_T = \{(U_\iota, \psi_\iota = T\phi_\iota)\}, \qquad (1.1.13)$$

where $T\phi_\iota$ is the tangent morphism to ϕ_ι. The associated linear bundle coordinates are holonomic (or *induced*) coordinates $(\dot{z}^\lambda)$ with respect to the holonomic frames $\{\partial_\lambda\}$ in tangent spaces T_zZ.

The tensor product of tangent and cotangent bundles

$$T = (\overset{m}{\otimes} TZ) \otimes (\overset{k}{\otimes} T^*Z), \qquad m, k \in \mathbb{N}, \qquad (1.1.14)$$

is called a *tensor bundle*, provided with holonomic bundle coordinates $\dot{x}^{\alpha_1 \cdots \alpha_m}_{\beta_1 \cdots \beta_k}$ possessing transition functions

$$\dot{x}'^{\alpha_1 \cdots \alpha_m}_{\beta_1 \cdots \beta_k} = \frac{\partial x'^{\alpha_1}}{\partial x^{\mu_1}} \cdots \frac{\partial x'^{\alpha_m}}{\partial x^{\mu_m}} \frac{\partial x^{\nu_1}}{\partial x'^{\beta_1}} \cdots \frac{\partial x^{\nu_k}}{\partial x'^{\beta_k}} \dot{x}^{\mu_1 \cdots \mu_m}_{\nu_1 \cdots \nu_k}.$$

Let $\pi_Y : TY \to Y$ be the tangent bundle of a fibre bundle $\pi : Y \to X$. Given bundle coordinates (x^λ, y^i) on Y, it is equipped with the holonomic coordinates $(x^\lambda, y^i, \dot{x}^\lambda, \dot{y}^i)$. The tangent bundle $TY \to Y$ has the subbundle $VY = \text{Ker}\,(T\pi)$, which consists of the vectors tangent to fibres of Y. It is called the *vertical tangent bundle* of Y and is provided with the holonomic coordinates $(x^\lambda, y^i, \dot{y}^i)$ with respect to the vertical frames $\{\partial_i\}$. Every bundle morphism $\Phi : Y \to Y'$ yields the linear bundle morphism over Φ of the vertical tangent bundles

$$V\Phi : VY \to VY', \qquad \dot{y}'^i \circ V\Phi = \frac{\partial \Phi^i}{\partial y^j} \dot{y}^j. \qquad (1.1.15)$$

It is called the *vertical tangent morphism*.

In many important cases, the vertical tangent bundle $VY \to Y$ of a fibre bundle $Y \to X$ is trivial, and is isomorphic to the bundle product

$$VY = Y \underset{X}{\times} \overline{Y} \qquad (1.1.16)$$

where $\overline{Y} \to X$ is some vector bundle. It follows that VY can be provided with bundle coordinates $(x^\lambda, y^i, \overline{y}^i)$ such that transition functions of coordinates $\overline{y}^i$ are independent of coordinates y^i. One calls (1.1.16) the *vertical* splitting. For instance, every vector bundle $Y \to X$ admits the *canonical vertical splitting*

$$VY = Y \underset{X}{\oplus} Y. \qquad (1.1.17)$$

The *vertical cotangent bundle* $V^*Y \to Y$ of a fibre bundle $Y \to X$ is defined as the dual of the vertical tangent bundle $VY \to Y$. It is not a subbundle of the cotangent bundle T^*Y, but there is the canonical surjection

$$\zeta : T^*Y \ni \dot{x}_\lambda dx^\lambda + \dot{y}_i dy^i \to \dot{y}_i \overline{d}y^i \in V^*Y, \qquad (1.1.18)$$

where $\{\overline{d}y^i\}$, possessing transition functions

$$\overline{d}y'^i = \frac{\partial y'^i}{\partial y^j} \overline{d}y^j,$$

are the duals of the holonomic frames $\{\partial_i\}$ of the vertical tangent bundle VY.

For any fibre bundle Y, there exist the exact sequences of vector bundles

$$0 \to VY \longrightarrow TY \xrightarrow{\pi_T} Y \underset{X}{\times} TX \to 0, \qquad (1.1.19)$$

$$0 \to Y \underset{X}{\times} T^*X \to T^*Y \to V^*Y \to 0. \qquad (1.1.20)$$

Their splitting, by definition, is a connection on $Y \to X$.

For the sake of simplicity, we agree to denote the pull-backs

$$Y \underset{X}{\times} TX, \qquad Y \underset{X}{\times} T^*X$$

by TX and T^*X, respectively.

Let $\overline{\pi} : \overline{Y} \to X$ be a vector bundle with a typical fibre $\overline{V}$. An *affine bundle* modelled over the vector bundle $\overline{Y} \to X$ is a fibre bundle $\pi : Y \to X$ whose typical fibre V is an affine space modelled over $\overline{V}$ such that the following conditions hold.

• All the fibres Y_x of Y are affine spaces modelled over the corresponding fibres $\overline{Y}_x$ of the vector bundle $\overline{Y}$.

• There is an affine bundle atlas

$$\Psi = \{(U_\alpha, \psi_\chi), \varrho_{\chi\zeta}\}$$

of $Y \to X$ whose local trivializations morphisms ψ_χ (1.1.5) and transition functions $\varrho_{\chi\zeta}$ (1.1.6) are affine isomorphisms.

Dealing with affine bundles, we use only *affine bundle coordinates* (y^i) associated to an affine bundle atlas Ψ. There are the bundle morphisms

$$Y \underset{X}{\times} \overline{Y} \underset{X}{\to} Y, \qquad (y^i, \overline{y}^i) \to y^i + \overline{y}^i,$$

$$Y \underset{X}{\times} Y \underset{X}{\to} \overline{Y}, \qquad (y^i, y'^i) \to y^i - y'^i,$$

where $(\overline{y}^i)$ are linear coordinates on the vector bundle $\overline{Y}$.

By virtue of Theorem 1.1.4, affine bundles have global sections, but in contrast with vector bundles, there is no canonical global section of an affine bundle. Let $\pi : Y \to X$ be an affine bundle. Every global section s of an affine bundle $Y \to X$ modelled over a vector bundle $\overline{Y} \to X$ yields the bundle morphisms

$$Y \ni y \to y - s(\pi(y)) \in \overline{Y}, \qquad (1.1.21)$$

$$\overline{Y} \ni \overline{y} \to s(\pi(y)) + \overline{y} \in Y. \qquad (1.1.22)$$

In particular, every vector bundle Y has a natural structure of an affine bundle due to the morphisms (1.1.22) where $s = \widehat{0}$ is the canonical zero-valued section of Y. For instance, the tangent bundle TX of a manifold X is naturally an affine bundle ATX called the *affine tangent bundle*.

Theorem 1.1.13. *Any affine bundle $Y \to X$ admits bundle coordinates $(x^\lambda, \widetilde{y}^i)$ possessing linear transition functions $\widetilde{y}'^i = A^i_j(x)\widetilde{y}^j$ (see Example 5.10.2).*

Proof. Let s be a global section of $Y \to X$. Given fibre coordinates $y \to (y^i)$, let us consider the fibre coordinates

$$y \to (\widetilde{y}^i = y^i - s^i(\pi(y))).$$

Due to the morphism (1.1.21), they possess linear transition functions. $\square$

One can define the *Whitney sum* $\overline{Y}' \oplus Y$ of a vector bundle $\overline{Y}' \to X$ and an affine bundle $Y \to X$ modelled over a vector bundle $\overline{Y} \to X$. This is an affine bundle modelled over the Whitney sum of vector bundles $\overline{Y}' \oplus \overline{Y}$.

By a morphism of affine bundles is meant a bundle morphism $\Phi : Y \to Y'$ whose restriction to each fibre of Y is an affine map. It is called an *affine bundle morphism*. Every affine bundle morphism $\Phi : Y \to Y'$ of an affine bundle Y modelled over a vector bundle $\overline{Y}$ to an affine bundle Y' modelled over a vector bundle $\overline{Y}'$ yields an unique linear bundle morphism

$$\overline{\Phi} : \overline{Y} \to \overline{Y}', \qquad \overline{y}'^i \circ \overline{\Phi} = \frac{\partial \Phi^i}{\partial y^j} \overline{y}^j, \qquad (1.1.23)$$

called the *linear derivative* of Φ.

Similarly to vector bundles, if $\Phi : Y \to Y'$ is an affine morphism of affine bundles of constant rank, then $\Phi(Y)$ and $\operatorname{Ker} \Phi$ are affine subbundles of Y' and Y, respectively.

Every affine bundle $Y \to X$ modelled over a vector bundle $\overline{Y} \to X$ admits the *canonical vertical splitting*

$$VY = Y \underset{X}{\times} \overline{Y}. \qquad (1.1.24)$$

Note that Theorems 1.1.8 and 1.1.9 on a particular cover for bundle atlases remain true in the case of linear and affine atlases of vector and affine bundles.

1.1.4 Vector fields, distributions and foliations

Vector fields on a manifold Z are global sections of the tangent bundle $TZ \to Z$.

The set $\mathcal{T}(Z)$ of vector fields on Z is both a $C^\infty(Z)$-module and a real Lie algebra with respect to the *Lie bracket*

$$u = u^\lambda \partial_\lambda, \qquad v = v^\lambda \partial_\lambda,$$
$$[v, u] = (v^\lambda \partial_\lambda u^\mu - u^\lambda \partial_\lambda v^\mu) \partial_\mu.$$

Given a vector field u on X, a *curve*

$$c : \mathbb{R} \supset () \to Z$$

in Z is said to be an *integral curve* of u if $Tc = u(c)$. Every vector field u on a manifold Z can be seen as an *infinitesimal generator* of a local one-parameter group of diffeomorphisms (a *flow*), and *vice versa* [92]. One-dimensional orbits of this group are integral curves of u. A vector field is called *complete* if its flow is a one-parameter group of diffeomorphisms of Z. For instance, every vector field on a compact manifold is complete.

A vector field u on a fibre bundle $Y \to X$ is called *projectable* if it projects onto a vector field on X, i.e., there exists a vector field τ on X such that

$$\tau \circ \pi = T\pi \circ u.$$

A projectable vector field takes the coordinate form

$$u = u^\lambda(x^\mu)\partial_\lambda + u^i(x^\mu, y^j)\partial_i, \qquad \tau = u^\lambda \partial_\lambda. \qquad (1.1.25)$$

Its flow is a local one-parameter group of automorphisms of $Y \to X$ over a local one-parameter group of diffeomorphisms of X whose generator is τ. A projectable vector field is called *vertical* if its projection onto X vanishes, i.e., if it lives in the vertical tangent bundle VY.

A vector field $\tau = \tau^\lambda \partial_\lambda$ on a base X of a fibre bundle $Y \to X$ gives rise to a vector field on Y by means of a connection on this fibre bundle (see the formula (1.3.6)). Nevertheless, every tensor bundle (1.1.14) admits the *canonical lift* of vector fields

$$\widetilde{\tau} = \tau^\mu \partial_\mu + [\partial_\nu \tau^{\alpha_1} \dot{x}^{\nu \alpha_2 \cdots \alpha_m}_{\beta_1 \cdots \beta_k} + \ldots - \partial_{\beta_1} \tau^\nu \dot{x}^{\alpha_1 \cdots \alpha_m}_{\nu \beta_2 \cdots \beta_k} - \ldots]\dot{\partial}^{\beta_1 \cdots \beta_k}_{\alpha_1 \cdots \alpha_m}, \qquad (1.1.26)$$

1.1. Geometry of fibre bundles

where we employ the compact notation

$$\dot\partial_\lambda = \frac{\partial}{\partial \dot x^\lambda}. \tag{1.1.27}$$

This lift is functorial, i.e., it is an $\mathbb{R}$-linear monomorphism of the Lie algebra $\mathcal{T}(X)$ of vector fields on X to the Lie algebra $\mathcal{T}(Y)$ of vector fields on Y (see Section 6.1). In particular, we have the functorial lift

$$\widetilde\tau = \tau^\mu \partial_\mu + \partial_\nu \tau^\alpha \dot x^\nu \frac{\partial}{\partial \dot x^\alpha} \tag{1.1.28}$$

of vector fields on X onto the tangent bundle TX and their functorial lift

$$\widetilde\tau = \tau^\mu \partial_\mu - \partial_\beta \tau^\nu \dot x_\nu \frac{\partial}{\partial \dot x_\beta} \tag{1.1.29}$$

onto the cotangent bundle T^*X.

A fibre bundle admitting functorial lift of vector fields on its base is called the natural bundle [94; 153] (see Section 6.1).

A subbundle **T** of the tangent bundle TZ of a manifold Z is called a regular *distribution* (or, simply, a distribution). A vector field u on Z is said to be *subordinate* to a distribution **T** if it lives in **T**. A distribution **T** is called *involutive* if the Lie bracket of **T**-subordinate vector fields also is subordinate to **T**.

A subbundle of the cotangent bundle T^*Z of Z is called a *codistribution* **T*** on a manifold Z. For instance, the *annihilator* Ann **T** of a distribution **T** is a codistribution whose fibre over $z \in Z$ consists of covectors $w \in T_z^*$ such that $v \rfloor w = 0$ for all $v \in \mathbf{T}_z$.

Theorem 1.1.14. *Let* **T** *be a distribution and* Ann **T** *its annihilator. Let* $\wedge$Ann **T**(Z) *be the ideal of the exterior algebra* $\mathcal{O}^*(Z)$ *which is generated by sections of* Ann **T** $\to Z$. *A distribution* **T** *is involutive if and only if the ideal* $\wedge$Ann **T**(Z) *is a differential ideal [164], i.e.,*

$$d(\wedge \text{Ann}\,\mathbf{T}(Z)) \subset \wedge \text{Ann}\,\mathbf{T}(Z).$$

The following local coordinates can be associated to an involutive distribution [164].

Theorem 1.1.15. *Let* **T** *be an involutive r-dimensional distribution on a manifold Z, $\dim Z = k$. Every point $z \in Z$ has an open neighborhood U which is a domain of an adapted coordinate chart $(z^1, \ldots, z^k)$ such that, restricted to U, the distribution* **T** *and its annihilator* Ann **T** *are spanned by the local vector fields $\partial/\partial z^1, \cdots, \partial/\partial z^r$ and the one-forms $dz^{r+1}, \ldots, dz^k$, respectively.*

A connected submanifold N of a manifold Z is called an *integral manifold* of a distribution $\mathbf{T}$ on Z if $TN \subset \mathbf{T}$. Unless otherwise stated, by an integral manifold is meant an integral manifold of dimension of $\mathbf{T}$. An integral manifold is called *maximal* if no other integral manifold contains it. The following is the classical theorem of Frobenius.

Theorem 1.1.16. *Let $\mathbf{T}$ be an involutive distribution on a manifold Z. For any $z \in Z$, there exists a unique maximal integral manifold of $\mathbf{T}$ through z, and any integral manifold through z is its open subset.*

Maximal integral manifolds of an involutive distribution on a manifold Z are assembled into a regular foliation $\mathcal{F}$ of Z. A regular r-dimensional *foliation* (or, simply, a foliation) $\mathcal{F}$ of a k-dimensional manifold Z is defined as a partition of Z into connected r-dimensional submanifolds (the *leaves* of a foliation) F_ι, $\iota \in I$, which possesses the following properties [126; 151]. A foliated manifold $(Z, \mathcal{F})$ admits an adapted coordinate atlas

$$\{(U_\xi; z^\lambda; z^i)\}, \quad \lambda = 1, \ldots, n - r, \quad i = 1, \ldots, r, \qquad (1.1.30)$$

such that transition functions of coordinates z^λ are independent of the remaining coordinates z^i and, for each leaf F of a foliation $\mathcal{F}$, the connected components of $F \cap U_\xi$ are given by the equations $z^\lambda =$const. These connected components and coordinates (z^i) on them make up a coordinate atlas of a leaf F. It follows that tangent spaces to leaves of a foliation $\mathcal{F}$ constitute an involutive distribution $T\mathcal{F}$ on Z, called the *tangent bundle to the foliation* $\mathcal{F}$. The factor bundle

$$V\mathcal{F} = TZ/T\mathcal{F},$$

called the *normal bundle* to $\mathcal{F}$, has transition functions independent of coordinates z^i. Let $T\mathcal{F}^* \to Z$ denote the dual of $T\mathcal{F} \to Z$. There are the exact sequences

$$0 \to T\mathcal{F} \xrightarrow{i_\mathcal{F}} TX \longrightarrow V\mathcal{F} \to 0, \qquad (1.1.31)$$

$$0 \to \operatorname{Ann} T\mathcal{F} \longrightarrow T^*X \xrightarrow{i_\mathcal{F}^*} T\mathcal{F}^* \to 0 \qquad (1.1.32)$$

of vector bundles over Z.

It should be emphasized that leaves of a foliation need not be closed or imbedded submanifolds. Every leaf has an open *tubular neighborhood* U, i.e., if $z \in U$, then a leaf through z also belongs to U.

A pair $(Z, \mathcal{F})$ where $\mathcal{F}$ is a foliation of Z is called a *foliated manifold*. For instance, any submersion $f : Z \to M$ yields a foliation

$$\mathcal{F} = \{F_p = f^{-1}(p)\}_{p \in f(Z)}$$

1.1. Geometry of fibre bundles

of Z indexed by elements of $f(Z)$, which is an open submanifold of M, i.e., $Z \to f(Z)$ is a fibred manifold. Leaves of this foliation are closed imbedded submanifolds. Such a foliation is called *simple*. It is a fibred manifold over $f(Z)$. Any (regular) foliation is locally simple.

1.1.5 Exterior and tangent-valued forms

An *exterior r-form* on a manifold Z is a section

$$\phi = \frac{1}{r!}\phi_{\lambda_1...\lambda_r}dz^{\lambda_1} \wedge \cdots \wedge dz^{\lambda_r}$$

of the exterior product $\overset{r}{\wedge} T^*Z \to Z$, where

$$dz^{\lambda_1} \wedge \cdots \wedge dz^{\lambda_r} = \frac{1}{r!}\epsilon^{\lambda_1...\lambda_r}{}_{\mu_1...\mu_r}dx^{\mu_1} \otimes \cdots \otimes dx^{\mu_r},$$

$$\epsilon^{...\lambda_i...\lambda_j...}{}_{...\mu_p...\mu_k...} = -\epsilon^{...\lambda_j...\lambda_i...}{}_{...\mu_p...\mu_k...} = -\epsilon^{...\lambda_i...\lambda_j...}{}_{...\mu_k...\mu_p...},$$

$$\epsilon^{\lambda_1...\lambda_r}{}_{\lambda_1...\lambda_r} = 1.$$

Sometimes, it is convenient to write

$$\phi = \phi'_{\lambda_1...\lambda_r}dz^{\lambda_1} \wedge \cdots \wedge dz^{\lambda_r}$$

without the coefficient $1/r!$.

Let $\mathcal{O}^r(Z)$ denote the vector space of exterior r-forms on a manifold Z. By definition, $\mathcal{O}^0(Z) = C^\infty(Z)$ is the ring of smooth real functions on Z. All exterior forms on Z constitute the $\mathbb{N}$-graded commutative algebra $\mathcal{O}^*(Z)$ of global sections of the exterior bundle $\wedge T^*Z$ (1.1.11) endowed with the *exterior product*

$$\phi = \frac{1}{r!}\phi_{\lambda_1...\lambda_r}dz^{\lambda_1} \wedge \cdots \wedge dz^{\lambda_r}, \qquad \sigma = \frac{1}{s!}\sigma_{\mu_1...\mu_s}dz^{\mu_1} \wedge \cdots \wedge dz^{\mu_s},$$

$$\phi \wedge \sigma = \frac{1}{r!s!}\phi_{\nu_1...\nu_r}\sigma_{\nu_{r+1}...\nu_{r+s}}dz^{\nu_1} \wedge \cdots \wedge dz^{\nu_{r+s}} =$$

$$\frac{1}{r!s!(r+s)!}\epsilon^{\nu_1...\nu_{r+s}}{}_{\alpha_1...\alpha_{r+s}}\phi_{\nu_1...\nu_r}\sigma_{\nu_{r+1}...\nu_{r+s}}dz^{\alpha_1} \wedge \cdots \wedge dz^{\alpha_{r+s}},$$

such that

$$\phi \wedge \sigma = (-1)^{|\phi||\sigma|}\sigma \wedge \phi,$$

where the symbol $|\phi|$ stands for the form degree. The algebra $\mathcal{O}^*(Z)$ also is provided with the *exterior differential*

$$d\phi = dz^\mu \wedge \partial_\mu \phi = \frac{1}{r!}\partial_\mu \phi_{\lambda_1...\lambda_r}dz^\mu \wedge dz^{\lambda_1} \wedge \cdots \wedge dz^{\lambda_r}$$

which obeys the relations
$$d \circ d = 0, \qquad d(\phi \wedge \sigma) = d(\phi) \wedge \sigma + (-1)^{|\phi|}\phi \wedge d(\sigma).$$
The exterior differential d makes $\mathcal{O}^*(Z)$ into a differential graded algebra which is the minimal Chevalley–Eilenberg differential calculus $\mathcal{O}^*\mathcal{A}$ over the real ring $\mathcal{A} = C^\infty(Z)$. Its de Rham complex is (10.9.12).

Given a manifold morphism $f : Z \to Z'$, any exterior k-form ϕ on Z' yields the *pull-back exterior form* $f^*\phi$ on Z given by the condition
$$f^*\phi(v^1,\ldots,v^k)(z) = \phi(Tf(v^1),\ldots,Tf(v^k))(f(z))$$
for an arbitrary collection of tangent vectors $v^1,\cdots,v^k \in T_zZ$. We have the relations
$$f^*(\phi \wedge \sigma) = f^*\phi \wedge f^*\sigma,$$
$$df^*\phi = f^*(d\phi).$$

In particular, given a fibre bundle $\pi : Y \to X$, the pull-back onto Y of exterior forms on X by π provides the monomorphism of graded commutative algebras $\mathcal{O}^*(X) \to \mathcal{O}^*(Y)$. Elements of its range $\pi^*\mathcal{O}^*(X)$ are called *basic forms*. Exterior forms
$$\phi : Y \to \overset{r}{\wedge} T^*X,$$
$$\phi = \frac{1}{r!}\phi_{\lambda_1\ldots\lambda_r}dx^{\lambda_1} \wedge \cdots \wedge dx^{\lambda_r},$$
on Y such that $u\rfloor\phi = 0$ for an arbitrary vertical vector field u on Y are said to be *horizontal forms*. Horizontal forms of degree $n = \dim X$ are called *densities*. We use for them the compact notation
$$L = \frac{1}{n!}L_{\mu_1\ldots\mu_n}dx^{\mu_1} \wedge \cdots \wedge dx^{\mu_n} = \mathcal{L}\omega, \qquad \mathcal{L} = L_{1\ldots n},$$
$$\omega = dx^1 \wedge \cdots \wedge dx^n = \frac{1}{n!}\epsilon_{\mu_1\ldots\mu_n}dx^{\mu_1} \wedge \cdots \wedge dx^{\mu_n}, \qquad (1.1.33)$$
$$\omega_\lambda = \partial_\lambda\rfloor\omega, \qquad \omega_{\mu\lambda} = \partial_\mu\rfloor\partial_\lambda\rfloor\omega,$$
where ϵ is the skew-symmetric *Levi–Civita symbol* with the component $\epsilon_{\mu_1\ldots\mu_n} = 1$.

The *interior product* (or *contraction*) of a vector field u and an exterior r-form ϕ on a manifold Z is given by the coordinate expression
$$u\rfloor\phi = \sum_{k=1}^r \frac{(-1)^{k-1}}{r!}u^{\lambda_k}\phi_{\lambda_1\ldots\lambda_k\ldots\lambda_r}dz^{\lambda_1} \wedge \cdots \wedge \widehat{dz}^{\lambda_k} \wedge \cdots \wedge dz^{\lambda_r} =$$
$$\frac{1}{(r-1)!}u^\mu\phi_{\mu\alpha_2\ldots\alpha_r}dz^{\alpha_2} \wedge \cdots \wedge dz^{\alpha_r},$$

where the caret ̂ denotes omission. It obeys the relations
$$\phi(u_1,\ldots,u_r) = u_r \rfloor \cdots u_1 \rfloor \phi,$$
$$u \rfloor (\phi \wedge \sigma) = u \rfloor \phi \wedge \sigma + (-1)^{|\phi|} \phi \wedge u \rfloor \sigma.$$

The *Lie derivative* of an exterior form ϕ along a vector field u is
$$\mathbf{L}_u \phi = u \rfloor d\phi + d(u \rfloor \phi),$$
$$\mathbf{L}_u(\phi \wedge \sigma) = \mathbf{L}_u \phi \wedge \sigma + \phi \wedge \mathbf{L}_u \sigma.$$

It is a derivation of the graded algebra $\mathcal{O}^*(Z)$ such that
$$\mathbf{L}_u \circ \mathbf{L}_{u'} - \mathbf{L}_{u'} \circ \mathbf{L}_u = \mathbf{L}_{[u,u']}.$$

In particular, if f is a function, then
$$\mathbf{L}_u f = u(f) = u \rfloor df.$$

An exterior form ϕ is invariant under a local one-parameter group of diffeomorphisms $G(t)$ of Z (i.e., $G(t)^*\phi = \phi$) if and only if its Lie derivative along the infinitesimal generator u of this group vanishes, i.e.,
$$\mathbf{L}_u \phi = 0.$$

A *tangent-valued r-form* on a manifold Z is a section
$$\phi = \frac{1}{r!} \phi^\mu_{\lambda_1\ldots\lambda_r} dz^{\lambda_1} \wedge \cdots \wedge dz^{\lambda_r} \otimes \partial_\mu \tag{1.1.34}$$

of the tensor bundle
$$\overset{r}{\wedge} T^*Z \otimes TZ \to Z.$$

Remark 1.1.4. There is one-to-one correspondence between the tangent-valued one-forms ϕ on a manifold Z and the linear bundle endomorphisms
$$\widehat{\phi} : TZ \to TZ, \quad \widehat{\phi} : T_z Z \ni v \to v \rfloor \phi(z) \in T_z Z, \tag{1.1.35}$$
$$\widehat{\phi}^* : T^*Z \to T^*Z, \quad \widehat{\phi}^* : T_z^* Z \ni v^* \to \phi(z) \rfloor v^* \in T_z^* Z, \tag{1.1.36}$$

over Z (see Remark 1.1.2). For instance, the *canonical tangent-valued one-form*
$$\theta_Z = dz^\lambda \otimes \partial_\lambda \tag{1.1.37}$$

on Z corresponds to the identity morphisms (1.1.35) and (1.1.36).

Remark 1.1.5. Let $Z = TX$, and let TTX be the tangent bundle of TX. There is the bundle endomorphism
$$J(\partial_\lambda) = \dot{\partial}_\lambda, \quad J(\dot{\partial}_\lambda) = 0 \tag{1.1.38}$$

of TTX over X. It corresponds to the canonical tangent-valued form
$$\theta_J = dx^\lambda \otimes \dot{\partial}_\lambda \tag{1.1.39}$$

on the tangent bundle TX. It is readily observed that $J \circ J = 0$.

The space $\mathcal{O}^*(Z) \otimes \mathcal{T}(Z)$ of tangent-valued forms is provided with the Frölicher–Nijenhuis bracket

$$[\ ,\ ]_{\mathrm{FN}} : \mathcal{O}^r(Z) \otimes \mathcal{T}(Z) \times \mathcal{O}^s(Z) \otimes \mathcal{T}(Z) \to \mathcal{O}^{r+s}(Z) \otimes \mathcal{T}(Z),$$
$$[\alpha \otimes u,\ \beta \otimes v]_{\mathrm{FN}} = (\alpha \wedge \beta) \otimes [u,v] + (\alpha \wedge \mathbf{L}_u \beta) \otimes v - \qquad (1.1.40)$$
$$(\mathbf{L}_v \alpha \wedge \beta) \otimes u + (-1)^r (d\alpha \wedge u \rfloor \beta) \otimes v + (-1)^r (v \rfloor \alpha \wedge d\beta) \otimes u,$$
$$\alpha \in \mathcal{O}^r(Z), \qquad \beta \in \mathcal{O}^s(Z), \qquad u,v \in \mathcal{T}(Z).$$

Its coordinate expression is

$$[\phi,\sigma]_{\mathrm{FN}} = \frac{1}{r!s!}(\phi^\nu_{\lambda_1\ldots\lambda_r} \partial_\nu \sigma^\mu_{\lambda_{r+1}\ldots\lambda_{r+s}} - \sigma^\nu_{\lambda_{r+1}\ldots\lambda_{r+s}} \partial_\nu \phi^\mu_{\lambda_1\ldots\lambda_r} -$$
$$r\phi^\mu_{\lambda_1\ldots\lambda_{r-1}\nu} \partial_{\lambda_r} \sigma^\nu_{\lambda_{r+1}\ldots\lambda_{r+s}} + s\sigma^\mu_{\nu\lambda_{r+2}\ldots\lambda_{r+s}} \partial_{\lambda_{r+1}} \phi^\nu_{\lambda_1\ldots\lambda_r})$$
$$dz^{\lambda_1} \wedge \cdots \wedge dz^{\lambda_{r+s}} \otimes \partial_\mu,$$
$$\phi \in \mathcal{O}^r(Z) \otimes \mathcal{T}(Z), \qquad \sigma \in \mathcal{O}^s(Z) \otimes \mathcal{T}(Z).$$

There are the relations

$$[\phi,\sigma]_{\mathrm{FN}} = (-1)^{|\phi||\psi|+1} [\sigma,\phi]_{\mathrm{FN}}, \qquad (1.1.41)$$
$$[\phi,[\sigma,\theta]_{\mathrm{FN}}]_{\mathrm{FN}} = [[\phi,\sigma]_{\mathrm{FN}},\theta]_{\mathrm{FN}} + (-1)^{|\phi||\sigma|}[\sigma,[\phi,\theta]_{\mathrm{FN}}]_{\mathrm{FN}}, \quad (1.1.42)$$
$$\phi,\sigma,\theta \in \mathcal{O}^*(Z) \otimes \mathcal{T}(Z).$$

Given a tangent-valued form θ, the *Nijenhuis differential* on $\mathcal{O}^*(Z) \otimes \mathcal{T}(Z)$ is defined as the morphism

$$d_\theta : \psi \to d_\theta \psi = [\theta,\psi]_{\mathrm{FN}}, \qquad \psi \in \mathcal{O}^*(Z) \otimes \mathcal{T}(Z).$$

By virtue of (1.1.42), it has the property

$$d_\phi [\psi,\theta]_{\mathrm{FN}} = [d_\phi \psi, \theta]_{\mathrm{FN}} + (-1)^{|\phi||\psi|} [\psi, d_\phi \theta]_{\mathrm{FN}}.$$

In particular, if $\theta = u$ is a vector field, the Nijenhuis differential is the *Lie derivative of tangent-valued forms*

$$\mathbf{L}_u \sigma = d_u \sigma = [u,\sigma]_{\mathrm{FN}} = \frac{1}{s!}(u^\nu \partial_\nu \sigma^\mu_{\lambda_1\ldots\lambda_s} - \sigma^\nu_{\lambda_1\ldots\lambda_s} \partial_\nu u^\mu +$$
$$s\sigma^\mu_{\nu\lambda_2\ldots\lambda_s} \partial_{\lambda_1} u^\nu)dx^{\lambda_1} \wedge \cdots \wedge dx^{\lambda_s} \otimes \partial_\mu, \qquad \sigma \in \mathcal{O}^s(Z) \otimes \mathcal{T}(Z).$$

Let $Y \to X$ be a fibre bundle. We consider the following subspaces of the space $\mathcal{O}^*(Y) \otimes \mathcal{T}(Y)$ of tangent-valued forms on Y:

• *horizontal tangent-valued forms*

$$\phi : Y \to \overset{r}{\wedge} T^*X \underset{Y}{\otimes} TY,$$
$$\phi = dx^{\lambda_1} \wedge \cdots \wedge dx^{\lambda_r} \otimes \frac{1}{r!}[\phi^\mu_{\lambda_1\ldots\lambda_r}(y)\partial_\mu + \phi^i_{\lambda_1\ldots\lambda_r}(y)\partial_i],$$

1.1. Geometry of fibre bundles

- projectable horizontal tangent-valued forms

$$\phi = dx^{\lambda_1} \wedge \cdots \wedge dx^{\lambda_r} \otimes \frac{1}{r!}[\phi^\mu_{\lambda_1\ldots\lambda_r}(x)\partial_\mu + \phi^i_{\lambda_1\ldots\lambda_r}(y)\partial_i],$$

- vertical-valued form

$$\phi : Y \to \overset{r}{\wedge} T^*X \underset{Y}{\otimes} VY,$$

$$\phi = \frac{1}{r!}\phi^i_{\lambda_1\ldots\lambda_r}(y)dx^{\lambda_1} \wedge \cdots \wedge dx^{\lambda_r} \otimes \partial_i,$$

- vertical-valued one-forms, called *soldering forms*,

$$\sigma = \sigma^i_\lambda(y)dx^\lambda \otimes \partial_i,$$

- basic soldering forms

$$\sigma = \sigma^i_\lambda(x)dx^\lambda \otimes \partial_i.$$

Remark 1.1.6. The tangent bundle TX is provided with the canonical soldering form θ_J (1.1.39). Due to the canonical vertical splitting

$$VTX = TX \underset{X}{\times} TX, \qquad (1.1.43)$$

the canonical soldering form (1.1.39) on TX defines the canonical tangent-valued form θ_X (1.1.37) on X. By this reason, tangent-valued one-forms on a manifold X also are called soldering forms.

Remark 1.1.7. Let $Y \to X$ be a fibre bundle, $f : X' \to X$ a manifold morphism, $f^*Y \to X'$ the pull-back of Y by f, and

$$f_Y : f^*Y \to Y$$

the corresponding bundle morphism (1.1.9). Since

$$Vf^*Y = f^*VY = f_Y^*VY, \qquad V_{y'}Y' = V_{f_Y(y')}Y,$$

one can define the *pull-back* $f^*\phi$ onto f^*Y of any vertical-valued form f on Y in accordance with the relation

$$f^*\phi(v^1,\ldots,v^r)(y') = \phi(Tf_Y(v^1),\ldots,Tf_Y(v^r))(f_Y(y')).$$

We also mention the TX-valued forms

$$\phi : Y \to \overset{r}{\wedge} T^*X \underset{Y}{\otimes} TX, \qquad (1.1.44)$$

$$\phi = \frac{1}{r!}\phi^\mu_{\lambda_1\ldots\lambda_r}dx^{\lambda_1} \wedge \cdots \wedge dx^{\lambda_r} \otimes \partial_\mu,$$

and V^*Y-valued forms

$$\phi : Y \to \overset{r}{\wedge} T^*X \underset{Y}{\otimes} V^*Y, \qquad (1.1.45)$$

$$\phi = \frac{1}{r!} \phi_{\lambda_1 \ldots \lambda_r i} dx^{\lambda_1} \wedge \cdots \wedge dx^{\lambda_r} \otimes \overline{d}y^i.$$

It should be emphasized that (1.1.44) are not tangent-valued forms, while (1.1.45) are not exterior forms. They exemplify *vector-valued forms*. Given a vector bundle $E \to X$, by a E-valued k-form on X, is meant a section of the fibre bundle

$$(\overset{k}{\wedge} T^*X) \underset{X}{\otimes} E^* \to X.$$

1.2 Jet manifolds

This Section addresses first and second order jet manifolds of sections of fibre bundles [94; 145].

Given a fibre bundle $Y \to X$ with bundle coordinates (x^λ, y^i), let us consider the equivalence classes $j_x^1 s$ of its sections s, which are identified by their values $s^i(x)$ and the values of their partial derivatives $\partial_\mu s^i(x)$ at a point $x \in X$. They are called the *first order jets* of sections at x. One can justify that the definition of jets is coordinate-independent. The key point is that the set $J^1 Y$ of first order jets $j_x^1 s$, $x \in X$, is a smooth manifold with respect to the adapted coordinates $(x^\lambda, y^i, y_\lambda^i)$ such that

$$y_\lambda^i(j_x^1 s) = \partial_\lambda s^i(x), \qquad y'^i_\lambda = \frac{\partial x^\mu}{\partial x'^\lambda}(\partial_\mu + y_\mu^j \partial_j) y'^i. \qquad (1.2.1)$$

It is called the first order *jet manifold* of a fibre bundle $Y \to X$. We call (y_λ^i) the *jet coordinate*.

Remark 1.2.1. Note that there are different notions of jets. Jets of sections are the particular jets of maps [94; 126] and the jets of submanifolds [53; 96] (see Section 8.4). Let us also mention the jets of modules over a commutative ring [96; 112] which are representative objects of differential operators on modules [71; 96]. In particular, given a smooth manifold X, the jets of a projective $C^\infty(X)$-module P of finite rank are exactly the jets of sections of the vector bundle over X whose module of sections is P in accordance with the Serre–Swan Theorem 10.9.3. The notion of jets is extended to modules over graded commutative rings [60]. However, the jets of modules over a noncommutative ring can not be defined [60].

1.2. Jet manifolds

The jet manifold J^1Y admits the natural fibrations

$$\pi^1 : J^1Y \ni j_x^1 s \to x \in X, \qquad (1.2.2)$$
$$\pi_0^1 : J^1Y \ni j_x^1 s \to s(x) \in Y. \qquad (1.2.3)$$

A glance at the transformation law (1.2.1) shows that π_0^1 is an affine bundle modelled over the vector bundle

$$T^*X \underset{Y}{\otimes} VY \to Y. \qquad (1.2.4)$$

It is convenient to call π^1 (1.2.2) the *jet bundle*, while π_0^1 (1.2.3) is said to be the *affine jet bundle*.

Let us note that, if $Y \to X$ is a vector or an affine bundle, the jet bundle π_1 (1.2.2) is so.

Jets can be expressed in terms of familiar tangent-valued forms as follows. There are the canonical imbeddings

$$\lambda_{(1)} : J^1Y \underset{Y}{\to} T^*X \underset{Y}{\otimes} TY,$$
$$\lambda_{(1)} = dx^\lambda \otimes (\partial_\lambda + y^i_\lambda \partial_i) = dx^\lambda \otimes d_\lambda, \qquad (1.2.5)$$
$$\theta_{(1)} : J^1Y \underset{Y}{\to} T^*Y \underset{Y}{\otimes} VY,$$
$$\theta_{(1)} = (dy^i - y^i_\lambda dx^\lambda) \otimes \partial_i = \theta^i \otimes \partial_i, \qquad (1.2.6)$$

where d_λ are said to be *total derivatives*, and θ^i are called *local contact forms*.

Remark 1.2.2. We further identify the jet manifold J^1Y with its images under the canonical morphisms (1.2.5) and (1.2.6), and represent the jets $j_x^1 s = (x^\lambda, y^i, y^i_\mu)$ by the tangent-valued forms $\lambda_{(1)}$ (1.2.5) and $\theta_{(1)}$ (1.2.6).

Sections and morphisms of fibre bundles admit prolongations to jet manifolds as follows.

Any section s of a fibre bundle $Y \to X$ has the *jet prolongation* to the section

$$(J^1 s)(x) = j_x^1 s, \qquad y^i_\lambda \circ J^1 s = \partial_\lambda s^i(x),$$

of the jet bundle $J^1Y \to X$. A section of the jet bundle $J^1Y \to X$ is called *integrable* if it is the jet prolongation of some section of a fibre bundle $Y \to X$.

Remark 1.2.3. By virtue of Theorem 1.1.4, the affine jet bundle $J^1Y \to Y$ admits global sections. For instance, if $Y = X \times V$ is a trivial bundle, there is the canonical zero section $\widehat{0}(Y)$ of $J^1Y \to Y$ which takes its values into centers of its affine fibres.

Any bundle morphism $\Phi : Y \to Y'$ over a diffeomorphism f admits a *jet prolongation* to a bundle morphism of affine jet bundles

$$J^1\Phi : J^1Y \xrightarrow{\Phi} J^1Y', \qquad (1.2.7)$$

$$y'^i_\lambda \circ J^1\Phi = \frac{\partial(f^{-1})^\mu}{\partial x'^\lambda} d_\mu \Phi^i.$$

Any projectable vector field u (1.1.25) on a fibre bundle $Y \to X$ has a *jet prolongation* to the projectable vector field

$$J^1 u = r_1 \circ J^1 u : J^1Y \to J^1 TY \to TJ^1Y,$$
$$J^1 u = u^\lambda \partial_\lambda + u^i \partial_i + (d_\lambda u^i - y^i_\mu \partial_\lambda u^\mu) \partial_i^\lambda, \qquad (1.2.8)$$

on the jet manifold J^1Y. In order to obtain (1.2.8), the canonical bundle morphism

$$r_1 : J^1 TY \to TJ^1Y, \qquad \dot{y}^i_\lambda \circ r_1 = (\dot{y}^i)_\lambda - y^i_\mu \dot{x}^\mu_\lambda$$

is used. In particular, there is the canonical isomorphism

$$VJ^1Y = J^1 VY, \qquad \dot{y}^i_\lambda = (\dot{y}^i)_\lambda. \qquad (1.2.9)$$

Taking the first order jet manifold of the jet bundle $J^1Y \to X$, we obtain the *repeated jet manifold* $J^1 J^1 Y$ provided with the adapted coordinates

$$(x^\lambda, y^i, y^i_\lambda, \widehat{y}^i_\mu, y^i_{\mu\lambda})$$

possessing transition functions

$$y'^i_\lambda = \frac{\partial x^\alpha}{\partial x'^\lambda} d_\alpha y'^i, \qquad \widehat{y}'^i_\lambda = \frac{\partial x^\alpha}{\partial x'^\lambda} \widehat{d}_\alpha y'^i, \qquad y'^i_{\mu\lambda} = \frac{\partial x^\alpha}{\partial x'^\mu} \widehat{d}_\alpha y'^i_\lambda,$$
$$d_\alpha = \partial_\alpha + \widehat{y}^j_\alpha \partial_j + y^j_{\nu\alpha} \partial^\nu_j, \qquad \widehat{d}_\alpha = \partial_\alpha + \widehat{y}^j_\alpha \partial_j + y^j_{\nu\alpha} \partial^\nu_j.$$

There exist two different affine fibrations of $J^1 J^1 Y$ over $J^1 Y$:

• the familiar affine jet bundle (1.2.3):

$$\pi_{11} : J^1 J^1 Y \to J^1 Y, \qquad y^i_\lambda \circ \pi_{11} = y^i_\lambda, \qquad (1.2.10)$$

• the affine bundle

$$J^1 \pi^1_0 : J^1 J^1 Y \to J^1 Y, \qquad y^i_\lambda \circ J^1 \pi^1_0 = \widehat{y}^i_\lambda. \qquad (1.2.11)$$

In general, there is no canonical identification of these fibrations. The points $q \in J^1 J^1 Y$, where

$$\pi_{11}(q) = J^1 \pi^1_0(q),$$

form an affine subbundle $\widehat{J}^2 Y \to J^1 Y$ of $J^1 J^1 Y$ called the *sesquiholonomic jet manifold*. It is given by the coordinate conditions $\widehat{y}^i_\lambda = y^i_\lambda$, and is coordinated by $(x^\lambda, y^i, y^i_\lambda, y^i_{\mu\lambda})$.

1.3. Connections on fibre bundles

The *second order jet manifold* J^2Y of a fibre bundle $Y \to X$ can be defined as the affine subbundle of the fibre bundle $\widehat{J}^2Y \to J^1Y$ given by the coordinate conditions
$$y^i_{\lambda\mu} = y^i_{\mu\lambda}.$$
It is modelled over the vector bundle
$$\overset{2}{\vee} T^*X \underset{J^1Y}{\otimes} VY \to J^1Y,$$
and is endowed with adapted coordinates $(x^\lambda, y^i, y^i_\lambda, y^i_{\lambda\mu} = y^i_{\mu\lambda})$, possessing transition functions
$$y'^i_\lambda = \frac{\partial x^\alpha}{\partial x'^\lambda} d_\alpha y'^i, \qquad y'^i_{\mu\lambda} = \frac{\partial x^\alpha}{\partial x'^\mu} d_\alpha y'^i_\lambda. \tag{1.2.12}$$

The second order jet manifold J^2Y also can be introduced as the set of the equivalence classes $j^2_x s$ of sections s of the fibre bundle $Y \to X$, which are identified by their values and the values of their first and second order partial derivatives at points $x \in X$, i.e.,
$$y^i_\lambda(j^2_x s) = \partial_\lambda s^i(x), \qquad y^i_{\lambda\mu}(j^2_x s) = \partial_\lambda \partial_\mu s^i(x).$$
The equivalence classes $j^2_x s$ are called the *second order jets* of sections.

Let s be a section of a fibre bundle $Y \to X$, and let $J^1 s$ be its jet prolongation to a section of the jet bundle $J^1Y \to X$. The latter gives rise to the section $J^1 J^1 s$ of the repeated jet bundle $J^1 J^1 Y \to X$. This section takes its values into the second order jet manifold J^2Y. It is called the *second order jet prolongation of a section* s, and is denoted by $J^2 s$.

Theorem 1.2.1. *Let $\overline{s}$ be a section of the jet bundle $J^1Y \to X$, and let $J^1\overline{s}$ be its jet prolongation to a section of the repeated jet bundle $J^1 J^1 Y \to X$. The following three facts are equivalent:*
- $\overline{s} = J^1 s$ *where* s *is a section of a fibre bundle* $Y \to X$,
- $J^1 \overline{s}$ *takes its values into* $\widehat{J}^2 Y$,
- $J^1 \overline{s}$ *takes its values into* $J^2 Y$.

1.3 Connections on fibre bundles

There are several equivalent definitions of a connection on a fibre bundle. We start with the traditional notion of a connection as a splitting of the exact sequences (1.1.19) – (1.1.20), but then follow its definition as a global section of an affine jet bundle [94; 112; 145]. In the case of vector bundles, there is an equivalent definition (10.9.10) of a linear connection on their structure modules.

1.3.1 Connections as tangent-valued forms

A *connection* on a fibre bundle $Y \to X$ is defined traditionally as a linear bundle monomorphism

$$\Gamma : Y \underset{X}{\times} TX \to TY, \qquad (1.3.1)$$

$$\Gamma : \dot{x}^\lambda \partial_\lambda \to \dot{x}^\lambda (\partial_\lambda + \Gamma^i_\lambda \partial_i),$$

over Y which splits the exact sequence (1.1.19), i.e.,

$$\pi_T \circ \Gamma = \mathrm{Id}\,(Y \underset{X}{\times} TX).$$

This also is a definition of connections on fibred manifolds (see Remark 1.3.2).

By virtue of Theorem 1.1.12, a connection always exists. The local functions $\Gamma^i_\lambda(y)$ in (1.3.1) are said to be *components* of the connection Γ with respect to the bundle coordinates (x^λ, y^i) on $Y \to X$.

The image of $Y \times TX$ by the connection Γ defines the *horizontal distribution* $HY \subset TY$ which splits the tangent bundle TY as follows:

$$TY = HY \underset{Y}{\oplus} VY, \qquad (1.3.2)$$

$$\dot{x}^\lambda \partial_\lambda + \dot{y}^i \partial_i = \dot{x}^\lambda(\partial_\lambda + \Gamma^i_\lambda \partial_i) + (\dot{y}^i - \dot{x}^\lambda \Gamma^i_\lambda)\partial_i.$$

Its annihilator is locally generated by the one-forms $dy^i - \Gamma^i_\lambda dx^\lambda$.

Given the *horizontal splitting* (1.3.2), the surjection

$$\Gamma : TY \underset{Y}{\to} VY, \qquad (1.3.3)$$

$$\dot{y}^i \circ \Gamma = \dot{y}^i - \Gamma^i_\lambda \dot{x}^\lambda,$$

defines a connection on $Y \to X$ in an equivalent way.

The linear morphism Γ over Y (1.3.1) yields uniquely the horizontal tangent-valued one-form

$$\Gamma = dx^\lambda \otimes (\partial_\lambda + \Gamma^i_\lambda \partial_i) \qquad (1.3.4)$$

on Y which projects onto the canonical tangent-valued form θ_X (1.1.37) on X. With this form called the *connection form*, the morphism (1.3.1) reads

$$\Gamma : \partial_\lambda \to \partial_\lambda \rfloor \Gamma = \partial_\lambda + \Gamma^i_\lambda \partial_i.$$

Given a connection Γ and the corresponding horizontal distribution (1.3.2), a vector field u on a fibre bundle $Y \to X$ is called *horizontal* if it lives in HY. A horizontal vector field takes the form

$$u = u^\lambda(y)(\partial_\lambda + \Gamma^i_\lambda \partial_i). \qquad (1.3.5)$$

1.3. Connections on fibre bundles

In particular, let τ be a vector field on the base X. By means of the connection form Γ (1.3.4), we obtain the projectable horizontal vector field

$$\Gamma\tau = \tau\rfloor\Gamma = \tau^\lambda(\partial_\lambda + \Gamma^i_\lambda\partial_i) \tag{1.3.6}$$

on Y, called the *horizontal lift* of τ by means of a connection Γ. Conversely, any projectable horizontal vector field u on Y is the horizontal lift $\Gamma\tau$ of its projection τ on X. Moreover, the horizontal distribution HY is generated by the horizontal lifts $\Gamma\tau$ (1.3.6) of vector fields τ on X. The horizontal lift

$$\mathcal{T}(X) \ni \tau \to \Gamma\tau \in \mathcal{T}(Y) \tag{1.3.7}$$

is a $C^\infty(X)$-linear module morphism.

Given the splitting (1.3.1), the dual splitting of the exact sequence (1.1.20) is

$$\Gamma : V^*Y \to T^*Y, \qquad \Gamma : \overline{dy}^i \to dy^i - \Gamma^i_\lambda dx^\lambda. \tag{1.3.8}$$

Hence, a connection Γ on $Y \to X$ is represented by the vertical-valued form

$$\Gamma = (dy^i - \Gamma^i_\lambda dx^\lambda) \otimes \partial_i \tag{1.3.9}$$

such that the morphism (1.3.8) reads

$$\Gamma : \overline{dy}^i \to \Gamma\rfloor\overline{dy}^i = dy^i - \Gamma^i_\lambda dx^\lambda.$$

We call Γ (1.3.9) the *vertical connection form*. The corresponding horizontal splitting of the cotangent bundle T^*Y takes the form

$$T^*Y = T^*X \oplus_Y \Gamma(V^*Y), \tag{1.3.10}$$

$$\dot x_\lambda dx^\lambda + \dot y_i dy^i = (\dot x_\lambda + \dot y_i \Gamma^i_\lambda)dx^\lambda + \dot y_i(dy^i - \Gamma^i_\lambda dx^\lambda).$$

Then we have the surjection

$$\Gamma = \mathrm{pr}_1 : T^*Y \to T^*X, \qquad \dot x_\lambda \circ \Gamma = \dot x_\lambda + \dot y_i \Gamma^i_\lambda, \tag{1.3.11}$$

which also defines a connection on a fibre bundle $Y \to X$.

Remark 1.3.1. Treating a connection as the vertical-valued form (1.3.9), we come to the following important construction. Given a fibre bundle $Y \to X$, let $f : X' \to X$ be a morphism and $f^*Y \to X'$ the pull-back of Y by f. Any connection Γ (1.3.9) on $Y \to X$ induces the *pull-back connection*

$$f^*\Gamma = \left(dy^i - (\Gamma \circ f_Y)^i_\lambda \frac{\partial f^\lambda}{\partial x'^\mu} dx'^\mu\right) \otimes \partial_i \tag{1.3.12}$$

on $f^*Y \to X'$ (see Remark 1.1.7).

Remark 1.3.2. Let $\pi : Y \to X$ be a fibred manifold. Any connection Γ on $Y \to X$ yields a horizontal lift of a vector field on X onto Y, but need not defines the similar lift of a *path* in X into Y. Let

$$\mathbb{R} \supset [] \ni t \to x(t) \in X, \qquad \mathbb{R} \ni t \to y(t) \in Y,$$

be smooth paths in X and Y, respectively. Then $t \to y(t)$ is called a *horizontal lift* of $x(t)$ if

$$\pi(y(t)) = x(t), \qquad \dot{y}(t) \in H_{y(t)}Y, \qquad t \in \mathbb{R},$$

where $HY \subset TY$ is the horizontal subbundle associated to the connection Γ. If, for each path $x(t)$ ($t_0 \leq t \leq t_1$) and for any $y_0 \in \pi^{-1}(x(t_0))$, there exists a horizontal lift $y(t)$ ($t_0 \leq t \leq t_1$) such that $y(t_0) = y_0$, then Γ is called the *Ehresmann connection*. A fibred manifold is a fibre bundle if and only if it admits an Ehresmann connection [69].

1.3.2 Connections as jet bundle sections

Throughout the book, we follow the equivalent definition of connections on a fibre bundle $Y \to X$ as sections of the affine jet bundle $J^1Y \to Y$.

Let $Y \to X$ be a fibre bundle, and J^1Y its first order jet manifold. Given the canonical morphisms (1.2.5) and (1.2.6), we have the corresponding morphisms

$$\widehat{\lambda}_{(1)} : J^1Y \underset{X}{\times} TX \ni \partial_\lambda \to d_\lambda = \partial_\lambda \rfloor \lambda_{(1)} \in J^1Y \underset{Y}{\times} TY, \qquad (1.3.13)$$

$$\widehat{\theta}_{(1)} : J^1Y \underset{Y}{\times} V^*Y \ni \overline{d}y^i \to \theta^i = \theta_{(1)} \rfloor dy^i \in J^1Y \underset{Y}{\times} T^*Y \qquad (1.3.14)$$

(see Remark 1.1.2). These morphisms yield the *canonical horizontal splittings* of the pull-back bundles

$$J^1Y \underset{Y}{\times} TY = \widehat{\lambda}_{(1)}(TX) \underset{J^1Y}{\oplus} VY, \qquad (1.3.15)$$

$$\dot{x}^\lambda \partial_\lambda + \dot{y}^i \partial_i = \dot{x}^\lambda(\partial_\lambda + y^i_\lambda \partial_i) + (\dot{y}^i - \dot{x}^\lambda y^i_\lambda)\partial_i,$$

$$J^1Y \underset{Y}{\times} T^*Y = T^*X \underset{J^1Y}{\oplus} \widehat{\theta}_{(1)}(V^*Y), \qquad (1.3.16)$$

$$\dot{x}_\lambda dx^\lambda + \dot{y}_i dy^i = (\dot{x}_\lambda + \dot{y}_i y^i_\lambda)dx^\lambda + \dot{y}_i(dy^i - y^i_\lambda dx^\lambda).$$

Let Γ be a global section of $J^1Y \to Y$. Substituting the tangent-valued form

$$\lambda_{(1)} \circ \Gamma = dx^\lambda \otimes (\partial_\lambda + \Gamma^i_\lambda \partial_i)$$

1.3. Connections on fibre bundles

in the canonical splitting (1.3.15), we obtain the familiar horizontal splitting (1.3.2) of TY by means of a connection Γ on $Y \to X$. Accordingly, substitution of the tangent-valued form

$$\theta_{(1)} \circ \Gamma = (dy^i - \Gamma^i_\lambda dx^\lambda) \otimes \partial_i$$

in the canonical splitting (1.3.16) leads to the dual splitting (1.3.10) of T^*Y by means of a connection Γ.

Theorem 1.3.1. *There is one-to-one correspondence between the connections Γ on a fibre bundle $Y \to X$ and the global sections*

$$\Gamma : Y \to J^1Y, \qquad (x^\lambda, y^i, y^i_\lambda) \circ \Gamma = (x^\lambda, y^i, \Gamma^i_\lambda), \qquad (1.3.17)$$

of the affine jet bundle $J^1Y \to Y$.

There are the following corollaries of this theorem.

• Since $J^1Y \to Y$ is affine, a connection on a fibre bundle $Y \to X$ exists in accordance with Theorem 1.1.4.

• Connections on a fibre bundle $Y \to X$ make up an affine space modelled over the vector space of soldering forms on $Y \to X$, i.e., sections of the vector bundle (1.2.4).

• Connection components possess the coordinate transformation law

$$\Gamma'^i_\lambda = \frac{\partial x^\mu}{\partial x'^\lambda}(\partial_\mu + \Gamma^j_\mu \partial_j)y'^i.$$

• Every connection Γ (1.3.17) on a fibre bundle $Y \to X$ yields the first order differential operator

$$D_\Gamma : J^1Y \underset{Y}{\to} T^*X \underset{Y}{\otimes} VY, \qquad (1.3.18)$$

$$D_\Gamma = \lambda_{(1)} - \Gamma \circ \pi^1_0 = (y^i_\lambda - \Gamma^i_\lambda)dx^\lambda \otimes \partial_i,$$

on Y called the *covariant differential* relative to the connection Γ. If $s : X \to Y$ is a section, from (1.3.18) we obtain its covariant differential

$$\nabla^\Gamma s = D_\Gamma \circ J^1 s : X \to T^*X \otimes VY, \qquad (1.3.19)$$

$$\nabla^\Gamma s = (\partial_\lambda s^i - \Gamma^i_\lambda \circ s)dx^\lambda \otimes \partial_i,$$

and the *covariant derivative*

$$\nabla^\Gamma_\tau = \tau \rfloor \nabla^\Gamma$$

along a vector field τ on X. A section s is said to be an *integral section* of a connection Γ if it belongs to the kernel of the covariant differential D_Γ (1.3.18), i.e.,

$$\nabla^\Gamma s = 0 \quad \text{or} \quad J^1 s = \Gamma \circ s. \qquad (1.3.20)$$

Theorem 1.3.2. *For any global section $s : X \to Y$, there always exists a connection Γ such that s is an integral section of Γ.*

Proof. This connection Γ is an extension of the local section
$$s(x) \to J^1 s(x)$$
of the affine jet bundle $J^1Y \to Y$ over the closed imbedded submanifold $s(X) \subset Y$ in accordance with Theorem 1.1.4. $\square$

Treating connections as jet bundle sections, one comes to the following two constructions.

(i) Let Y and Y' be fibre bundles over the same base X. Given a connection Γ on $Y \to X$ and a connection Γ' on $Y' \to X$, the fibre bundle
$$Y \underset{X}{\times} Y' \to X$$
is provided with the *product connection*
$$\Gamma \times \Gamma' : Y \underset{X}{\times} Y' \to J^1(Y \underset{X}{\times} Y') = J^1Y \underset{X}{\times} J^1Y',$$
$$\Gamma \times \Gamma' = dx^\lambda \otimes \left(\partial_\lambda + \Gamma^i_\lambda \frac{\partial}{\partial y^i} + \Gamma'^j_\lambda \frac{\partial}{\partial y'^j} \right). \tag{1.3.21}$$

(ii) Let $i_Y : Y \to Y'$ be a subbundle of a fibre bundle $Y' \to X$ and Γ' a connection on $Y' \to X$. If there exists a connection Γ on $Y \to X$ such that the diagram

$$\begin{array}{ccc} Y' & \xrightarrow{\Gamma'} & J^1Y \\ {\scriptstyle i_Y} \uparrow & & \uparrow {\scriptstyle J^1 i_Y} \\ Y & \xrightarrow{\Gamma} & J^1Y' \end{array}$$

is commutative, we say that Γ' is *reducible* to a connection Γ. The following conditions are equivalent:

• Γ' is reducible to Γ;
• $Ti_Y(HY) = HY'|_{i_Y(Y)}$, where $HY \subset TY$ and $HY' \subset TY'$ are the horizontal subbundles determined by Γ and Γ', respectively;
• for every vector field τ on X, the vector fields $\Gamma\tau$ and $\Gamma'\tau$ are related as follows:
$$Ti_Y \circ \Gamma\tau = \Gamma'\tau \circ i_Y. \tag{1.3.22}$$

1.3.3 Curvature and torsion

Let Γ be a connection on a fibre bundle $Y \to X$. Its *curvature* is defined as the Nijenhuis differential
$$R = \frac{1}{2} d_\Gamma \Gamma = \frac{1}{2}[\Gamma,\Gamma]_{\text{FN}} : Y \to \overset{2}{\wedge} T^*X \otimes VY, \tag{1.3.23}$$
$$R = \frac{1}{2} R^i_{\lambda\mu} dx^\lambda \wedge dx^\mu \otimes \partial_i, \tag{1.3.24}$$
$$R^i_{\lambda\mu} = \partial_\lambda \Gamma^i_\mu - \partial_\mu \Gamma^i_\lambda + \Gamma^j_\lambda \partial_j \Gamma^i_\mu - \Gamma^j_\mu \partial_j \Gamma^i_\lambda.$$

1.3. Connections on fibre bundles

This is a VY-valued horizontal two-form on Y. Given vector fields τ, τ' on X and their horizontal lifts $\Gamma\tau$ and $\Gamma\tau'$ (1.3.6) on Y, we have the relation

$$R(\tau, \tau') = -\Gamma[\tau, \tau'] + [\Gamma\tau, \Gamma\tau'] = \tau^\lambda \tau'^\mu R^i_{\lambda\mu} \partial_i. \tag{1.3.25}$$

The curvature (1.3.23) obeys the identities

$$[R, R]_{\mathrm{FN}} = 0, \tag{1.3.26}$$

$$d_\Gamma R = [\Gamma, R]_{\mathrm{FN}} = 0. \tag{1.3.27}$$

They result from the identity (1.1.41) and the graded Jacobi identity (1.1.42), respectively. The identity (1.3.27) is called the *second Bianchi identity*. It takes the coordinate form

$$\sum_{(\lambda\mu\nu)} (\partial_\lambda R^i_{\mu\nu} + \Gamma^j_\lambda \partial_j R^i_{\mu\nu} - \partial_j \Gamma^i_\lambda R^j_{\mu\nu}) = 0, \tag{1.3.28}$$

where the sum is cyclic over the indices λ, μ and ν.

In the same manner, given a soldering form σ, one defines the *soldered curvature*

$$\rho = \frac{1}{2} d_\sigma \sigma = \frac{1}{2}[\sigma, \sigma]_{\mathrm{FN}} : Y \to \overset{2}{\wedge} T^*X \otimes VY, \tag{1.3.29}$$

$$\rho = \frac{1}{2} \rho^i_{\lambda\mu} dx^\lambda \wedge dx^\mu \otimes \partial_i,$$

$$\rho^i_{\lambda\mu} = \sigma^j_\lambda \partial_j \sigma^i_\mu - \sigma^j_\mu \partial_j \sigma^i_\lambda.$$

It fulfills the identities

$$[\rho, \rho]_{\mathrm{FN}} = 0,$$

$$d_\sigma \rho = [\sigma, \rho]_{\mathrm{FN}} = 0,$$

similar to (1.3.26) – (1.3.27).

Given a connection Γ and a soldering form σ, the *torsion form* of Γ with respect to σ is defined as

$$T = d_\Gamma \sigma = d_\sigma \Gamma : Y \to \overset{2}{\wedge} T^*X \otimes VY,$$

$$T = (\partial_\lambda \sigma^i_\mu + \Gamma^j_\lambda \partial_j \sigma^i_\mu - \partial_j \Gamma^i_\lambda \sigma^j_\mu) dx^\lambda \wedge dx^\mu \otimes \partial_i. \tag{1.3.30}$$

It obeys the *first Bianchi identity*

$$d_\Gamma T = d_\Gamma^2 \sigma = [R, \sigma]_{\mathrm{FN}} = -d_\sigma R. \tag{1.3.31}$$

If $\Gamma' = \Gamma + \sigma$, we have the relations

$$T' = T + 2\rho, \tag{1.3.32}$$

$$R' = R + \rho + T. \tag{1.3.33}$$

1.3.4 Linear connections

A connection Γ on a vector bundle $Y \to X$ is called the *linear connection* if the section

$$\Gamma : Y \to J^1 Y, \qquad \Gamma = dx^\lambda \otimes (\partial_\lambda + \Gamma_\lambda{}^i{}_j(x) y^j \partial_i), \qquad (1.3.34)$$

is a linear bundle morphism over X. Note that linear connections are principal connections, and they always exist (see Assertion 5.4.1).

The curvature R (1.3.24) of a linear connection Γ (1.3.34) reads

$$R = \frac{1}{2} R_{\lambda\mu}{}^i{}_j(x) y^j dx^\lambda \wedge dx^\mu \otimes \partial_i,$$

$$R_{\lambda\mu}{}^i{}_j = \partial_\lambda \Gamma_\mu{}^i{}_j - \partial_\mu \Gamma_\lambda{}^i{}_j + \Gamma_\lambda{}^h{}_j \Gamma_\mu{}^i{}_h - \Gamma_\mu{}^h{}_j \Gamma_\lambda{}^i{}_h. \qquad (1.3.35)$$

Due to the vertical splitting (1.1.17), we have the linear morphism

$$R : Y \ni y^i e_i \to \frac{1}{2} R_{\lambda\mu}{}^i{}_j y^j dx^\lambda \wedge dx^\mu \otimes e_i \in \mathcal{O}^2(X) \otimes Y. \qquad (1.3.36)$$

There are the following standard constructions of new linear connections from the old ones.

• Let $Y \to X$ be a vector bundle, coordinated by (x^λ, y^i), and $Y^* \to X$ its dual, coordinated by (x^λ, y_i). Any linear connection Γ (1.3.34) on a vector bundle $Y \to X$ defines the *dual linear connection*

$$\Gamma^* = dx^\lambda \otimes (\partial_\lambda - \Gamma_\lambda{}^j{}_i(x) y_j \partial^i) \qquad (1.3.37)$$

on $Y^* \to X$.

• Let Γ and Γ' be linear connections on vector bundles $Y \to X$ and $Y' \to X$, respectively. The direct sum connection $\Gamma \oplus \Gamma'$ on the Whitney sum $Y \oplus Y'$ of these vector bundles is defined as the product connection (1.3.21).

• Let Y coordinated by (x^λ, y^i) and Y' coordinated by (x^λ, y^a) be vector bundles over the same base X. Their tensor product $Y \otimes Y'$ is endowed with the bundle coordinates (x^λ, y^{ia}). Linear connections Γ and Γ' on $Y \to X$ and $Y' \to X$ define the linear *tensor product connection*

$$\Gamma \otimes \Gamma' = dx^\lambda \otimes \left[\partial_\lambda + (\Gamma_\lambda{}^i{}_j y^{ja} + \Gamma'_\lambda{}^a{}_b y^{ib}) \frac{\partial}{\partial y^{ia}} \right] \qquad (1.3.38)$$

on

$$Y \underset{X}{\otimes} Y' \to X.$$

An important example of linear connections is a linear connection

$$\Gamma = dx^\lambda \otimes (\partial_\lambda + \Gamma_\lambda{}^\mu{}_\nu \dot{x}^\nu \dot{\partial}_\mu) \qquad (1.3.39)$$

1.3. Connections on fibre bundles

on the tangent bundle TX of a manifold X. We agree to call it a *world connection* on a manifold X. The *dual world connection* (1.3.37) on the cotangent bundle T^*X is

$$\Gamma^* = dx^\lambda \otimes (\partial_\lambda - \Gamma_\lambda{}^\mu{}_\nu \dot{x}_\mu \dot{\partial}^\nu). \tag{1.3.40}$$

Then, using the construction of the tensor product connection (1.3.38), one can introduce the corresponding linear *world connection on an arbitrary tensor bundle* T (1.1.14).

Remark 1.3.3. It should be emphasized that the expressions (1.3.39) and (1.3.40) for a world connection differ in a minus sign from those usually used in the physical literature.

The *curvature of a world connection* is defined as the curvature R (1.3.35) of the connection Γ (1.3.39) on the tangent bundle TX. It reads

$$R = \frac{1}{2} R_{\lambda\mu}{}^\alpha{}_\beta \dot{x}^\beta dx^\lambda \wedge dx^\mu \otimes \dot{\partial}_\alpha, \tag{1.3.41}$$
$$R_{\lambda\mu}{}^\alpha{}_\beta = \partial_\lambda \Gamma_\mu{}^\alpha{}_\beta - \partial_\mu \Gamma_\lambda{}^\alpha{}_\beta + \Gamma_\lambda{}^\gamma{}_\beta \Gamma_\mu{}^\alpha{}_\gamma - \Gamma_\mu{}^\gamma{}_\beta \Gamma_\lambda{}^\alpha{}_\gamma.$$

By the *torsion of a world connection* is meant the torsion (1.3.30) of the connection Γ (1.3.39) on the tangent bundle TX with respect to the canonical soldering form θ_J (1.1.39):

$$T = \frac{1}{2} T_\mu{}^\nu{}_\lambda dx^\lambda \wedge dx^\mu \otimes \dot{\partial}_\nu, \tag{1.3.42}$$
$$T_\mu{}^\nu{}_\lambda = \Gamma_\mu{}^\nu{}_\lambda - \Gamma_\lambda{}^\nu{}_\mu.$$

A world connection is said to be *symmetric* if its torsion (1.3.42) vanishes, i.e.,

$$\Gamma_\mu{}^\nu{}_\lambda = \Gamma_\lambda{}^\nu{}_\mu.$$

Remark 1.3.4. For any vector field τ on a manifold X, there exists a connection Γ on the tangent bundle $TX \to X$ such that τ is an integral section of Γ, but this connection is not necessarily linear. If a vector field τ is non-vanishing at a point $x \in X$, then there exists a local symmetric world connection Γ (1.3.39) around x for which τ is an integral section

$$\partial_\nu \tau^\alpha = \Gamma_\nu{}^\alpha{}_\beta \tau^\beta. \tag{1.3.43}$$

Then the canonical lift $\widetilde{\tau}$ (1.1.28) of τ onto TX can be seen locally as the horizontal lift $\Gamma\tau$ (1.3.6) of τ by means of this connection.

Remark 1.3.5. Every manifold X can be provided with a non-degenerate fibre metric

$$g \in \overset{2}{\vee} \mathcal{O}^1(X), \qquad g = g_{\lambda\mu} dx^\lambda \otimes dx^\mu,$$

in the tangent bundle TX, and with the corresponding metric

$$g \in \overset{2}{\vee} \mathcal{T}^1(X), \qquad g = g^{\lambda\mu} \partial_\lambda \otimes \partial_\mu,$$

in the cotangent bundle T^*X. We call it a *world metric* on X. For any world metric g, there exists a unique symmetric world connection Γ (1.3.39) with the components

$$\Gamma_\lambda{}^\nu{}_\mu = \{_\lambda{}^\nu{}_\mu\} = -\frac{1}{2} g^{\nu\rho} (\partial_\lambda g_{\rho\mu} + \partial_\mu g_{\rho\lambda} - \partial_\rho g_{\lambda\mu}), \qquad (1.3.44)$$

called the *Christoffel symbols*, such that g is an integral section of Γ, i.e.

$$\partial_\lambda g^{\alpha\beta} = g^{\alpha\gamma} \{_\lambda{}^\beta{}_\gamma\} + g^{\beta\gamma} \{_\lambda{}^\alpha{}_\gamma\}.$$

It is called the *Levi–Civita connection* associated to g.

1.3.5 Affine connections

Let $Y \to X$ be an affine bundle modelled over a vector bundle $\overline{Y} \to X$. A connection Γ on $Y \to X$ is called an *affine connection* if the section $\Gamma : Y \to J^1 Y$ (1.3.17) is an affine bundle morphism over X. Affine connections are associated to principal connections, and they always exist (see Assertion 5.4.1).

For any affine connection $\Gamma : Y \to J^1 Y$, the corresponding linear derivative $\overline{\Gamma} : \overline{Y} \to J^1 \overline{Y}$ (1.1.23) defines a unique associated linear connection on the vector bundle $\overline{Y} \to X$. Since every vector bundle has a natural structure of an affine bundle, any linear connection on a vector bundle also is an affine connection.

With respect to affine bundle coordinates (x^λ, y^i) on Y, an affine connection Γ on $Y \to X$ reads

$$\Gamma_\lambda^i = \Gamma_\lambda{}^i{}_j(x) y^j + \sigma_\lambda^i(x). \qquad (1.3.45)$$

The coordinate expression of the associated linear connection is

$$\overline{\Gamma}_\lambda^i = \Gamma_\lambda{}^i{}_j(x) \overline{y}^j, \qquad (1.3.46)$$

where $(x^\lambda, \overline{y}^i)$ are the associated linear bundle coordinates on $\overline{Y}$.

Affine connections on an affine bundle $Y \to X$ constitute an affine space modelled over the soldering forms on $Y \to X$. In view of the vertical splitting (1.1.24), these soldering forms can be seen as global sections of the vector bundle

$$T^*X \underset{X}{\otimes} \overline{Y} \to X.$$

If $Y \to X$ is a vector bundle, both the affine connection Γ (1.3.45) and the associated linear connection $\overline{\Gamma}$ are connections on the same vector bundle $Y \to X$, and their difference is a basic soldering form on Y. Thus, every affine connection on a vector bundle $Y \to X$ is the sum of a linear connection and a basic soldering form on $Y \to X$.

Given an affine connection Γ on a vector bundle $Y \to X$, let R and $\overline{R}$ be the curvatures of a connection Γ and the associated linear connection $\overline{\Gamma}$, respectively. It is readily observed that

$$R = \overline{R} + T,$$

where the VY-valued two-form

$$T = d_\Gamma \sigma = d_\sigma \Gamma : X \to \overset{2}{\wedge} T^*X \underset{X}{\otimes} VY, \tag{1.3.47}$$

$$T = \frac{1}{2} T^i_{\lambda\mu} dx^\lambda \wedge dx^\mu \otimes \partial_i,$$

$$T^i_{\lambda\mu} = \partial_\lambda \sigma^i_\mu - \partial_\mu \sigma^i_\lambda + \sigma^h_\lambda \Gamma_\mu{}^i{}_h - \sigma^h_\mu \Gamma_\lambda{}^i{}_h,$$

is the torsion (1.3.30) of the connection Γ with respect to the basic soldering form σ.

In particular, let us consider the tangent bundle TX of a manifold X. We have the canonical soldering form $\sigma = \theta_J = \theta_X$ (1.1.39) on TX. Given an arbitrary world connection Γ (1.3.39) on TX, the corresponding affine connection

$$A = \Gamma + \theta_X, \qquad A^\mu_\lambda = \Gamma_\lambda{}^\mu{}_\nu \dot{x}^\nu + \delta^\mu_\lambda, \tag{1.3.48}$$

on TX is called the *Cartan connection*. Since the soldered curvature ρ (1.3.29) of θ_J equals zero, the torsion (1.3.32) of the Cartan connection coincides with the torsion T (1.3.42) of the world connection Γ, while its curvature (1.3.33) is the sum $R + T$ of the curvature and the torsion of Γ.

1.3.6 Flat connections

By a *flat* or *curvature-free* connection is meant a connection which satisfies the following equivalent conditions.

Theorem 1.3.3. *Let Γ be a connection on a fibre bundle $Y \to X$. The following assertions are equivalent.*

(i) The curvature R of a connection Γ vanishes identically, i.e., $R \equiv 0$.

(ii) The horizontal lift (1.3.7) of vector fields on X onto Y is an $\mathbb{R}$-linear Lie algebra morphism (in accordance with the formula (1.3.25)).

(iii) The horizontal distribution is involutive.

(iv) There exists a local integral section for a connection Γ through any point $y \in Y$.

By virtue of Theorem 1.1.16 and item (iii) of Theorem 1.3.3, a flat connection Γ on a fibre bundle $Y \to X$ yields a *horizontal foliation* on Y, transversal to the fibration $Y \to X$. The leaf of this foliation through a point $y \in Y$ is defined locally by an integral section s_y for the connection Γ through y. Conversely, let a fibre bundle $Y \to X$ admit a transversal foliation such that, for each point $y \in Y$, the leaf of this foliation through y is locally defined by a section s_y of $Y \to X$ through y. Then the map

$$\Gamma : Y \to J^1 Y, \qquad \Gamma(y) = j_x^1 s_y, \qquad \pi(y) = x,$$

introduces a flat connection on $Y \to X$. Thus, there is one-to-one correspondence between the flat connections and the transversal foliations of a fibre bundle $Y \to X$.

Given a transversal foliation on a fibre bundle $Y \to X$, there exists the associated atlas of bundle coordinates (x^λ, y^i) of Y such that every leaf of this foliation is locally generated by the equations $y^i =$const., and the transition functions $y^i \to y'^i(y^j)$ are independent of the base coordinates x^λ [53]. This is called the *atlas of constant local trivializations*. Two such atlases are said to be equivalent if their union also is an atlas of constant local trivializations. They are associated to the same horizontal foliation. Thus, we come to the following assertion.

Theorem 1.3.4. *There is one-to-one correspondence between the flat connections Γ on a fibre bundle $Y \to X$ and the equivalence classes of atlases of constant local trivializations of Y such that*

$$\Gamma = dx^\lambda \otimes \partial_\lambda$$

relative to these atlases.

In particular, if $Y \to X$ is a trivial bundle, one associates to each its trivialization a flat connection represented by the global zero section $\widehat{0}(Y)$ of $J^1 Y \to Y$ with respect to this trivialization (see Remark 1.2.3).

1.3.7 Second order connections

A *second order connection* $\widetilde{\Gamma}$ on a fibre bundle $Y \to X$ is defined as a connection

$$\widetilde{\Gamma} = dx^\lambda \otimes (\partial_\lambda + \widetilde{\Gamma}^i_\lambda \partial_i + \widetilde{\Gamma}^i_{\lambda\mu} \partial^\mu_i) \qquad (1.3.49)$$

on the jet bundle $J^1Y \to X$, i.e., this is a section of the affine bundle

$$\pi_{11} : J^1 J^1 Y \to J^1 Y.$$

Every connection on a fibre bundle $Y \to X$ gives rise to the second order one by means of a world connection on X as follows. The first order jet prolongation $J^1\Gamma$ of a connection Γ on $Y \to X$ is a section of the repeated jet bundle $J^1\pi^1_0$ (1.2.11), but not of π_{11}. Given a world connection K (1.3.40) on X, one can construct the affine morphism

$$s_K : J^1 J^1 Y \to J^1 J^1 Y,$$
$$(x^\lambda, y^i, y^i_\lambda, \widehat{y}^i_\lambda, y^i_{\lambda\mu}) \circ s_K = (x^\lambda, y^i, \widehat{y}^i_\lambda, y^i_\lambda, y^i_{\mu\lambda} - K_\lambda{}^\nu{}_\mu(\widehat{y}^i_\nu - y^i_\nu)),$$

such that

$$\pi_{11} = J^1\pi^1_0 \circ s_K$$

[53]. Then Γ gives rise to the second order connection

$$\widetilde{\Gamma} = s_K \circ J^1\Gamma : J^1Y \to J^1J^1Y, \qquad (1.3.50)$$
$$\widetilde{\Gamma} = dx^\lambda \otimes (\partial_\lambda + \Gamma^i_\lambda \partial_i + [\partial_\lambda \Gamma^i_\mu + y^j_\lambda \partial_j \Gamma^i_\mu + K_\lambda{}^\nu{}_\mu(y^i_\nu - \Gamma^i_\nu)]\partial^\mu_i),$$

which is an affine morphism

$$\begin{array}{ccc} J^1Y & \xrightarrow{\widetilde{\Gamma}} & J^1J^1Y \\ \pi^1_0 \downarrow & & \downarrow \pi_{11} \\ Y & \xrightarrow{\Gamma} & J^1Y \end{array}$$

over the connection Γ. Note that the curvature R (1.3.23) of a connection Γ on a fibre bundle $Y \to X$ can be seen as a soldering form

$$R = R^i_{\lambda\mu} dx^\lambda \otimes \partial^\mu_i$$

on the jet bundle $J^1Y \to X$. Therefore, $\widetilde{\Gamma} - R$ also is a connection on $J^1Y \to X$.

A second order connection $\widetilde{\Gamma}$ (1.3.49) is said to be *holonomic* if it takes its values into the subbundle J^2Y of J^1J^1Y. There is one-to-one correspondence between the global sections of the jet bundle $J^2Y \to J^1Y$ and the holonomic second order connections on $Y \to X$. Since the jet bundle

$J^2Y \to J^1Y$ is affine, a holonomic second order connection on a fibre bundle $Y \to X$ always exists. It is characterized by the coordinate conditions

$$\widetilde{\Gamma}^i_\lambda = y^i_\lambda, \qquad \widetilde{\Gamma}^i_{\lambda\mu} = \widetilde{\Gamma}^i_{\mu\lambda},$$

and takes the form

$$\widetilde{\Gamma} = dx^\lambda \otimes (\partial_\lambda + y^i_\lambda \partial_i + \widetilde{\Gamma}^i_{\lambda\mu} \partial^\mu_i). \qquad (1.3.51)$$

By virtue of Theorem 1.2.1, every integral section

$$\overline{s} : X \to J^1Y$$

of the holonomic second order connection (1.3.51) is integrable, i.e., $\overline{s} = J^1 s$.

1.4 Composite bundles

Let us consider the composition

$$\pi : Y \to \Sigma \to X \qquad (1.4.1)$$

of fibre bundles

$$\pi_{Y\Sigma} : Y \to \Sigma, \qquad (1.4.2)$$

$$\pi_{\Sigma X} : \Sigma \to X. \qquad (1.4.3)$$

One can show that it is a fibre bundle, called the *composite bundle* [53]. It is provided with bundle coordinates $(x^\lambda, \sigma^m, y^i)$, where (x^λ, σ^m) are bundle coordinates on the fibre bundle (1.4.3), i.e., transition functions of coordinates σ^m are independent of coordinates y^i.

For instance, the tangent bundle TY of a fibre bundle $Y \to X$ is a composite bundle

$$TY \to Y \to X.$$

The following two assertions make composite bundles useful for physical applications.

Theorem 1.4.1. *Given a composite bundle (1.4.1), let h be a global section of the fibre bundle $\Sigma \to X$. Then the restriction*

$$Y^h = h^*Y \qquad (1.4.4)$$

of the fibre bundle $Y \to \Sigma$ to $h(X) \subset \Sigma$ is a subbundle of the fibre bundle $Y \to X$.

1.4. Composite bundles

Theorem 1.4.2. *Given a section h of the fibre bundle $\Sigma \to X$ and a section s_Σ of the fibre bundle $Y \to \Sigma$, their composition*

$$s = s_\Sigma \circ h$$

is a section of the composite bundle $Y \to X$ (1.4.1). Conversely, every section s of the fibre bundle $Y \to X$ is a composition of the section

$$h = \pi_{Y\Sigma} \circ s$$

of the fibre bundle $\Sigma \to X$ and some section s_Σ of the fibre bundle $Y \to \Sigma$ over the closed imbedded submanifold $h(X) \subset \Sigma$.

Let us consider the jet manifolds $J^1\Sigma$, $J^1_\Sigma Y$, and $J^1 Y$ of the fibre bundles

$$\Sigma \to X, \qquad Y \to \Sigma, \qquad Y \to X,$$

respectively. They are provided with the adapted coordinates

$$(x^\lambda, \sigma^m, \sigma^m_\lambda), \quad (x^\lambda, \sigma^m, y^i, \widetilde{y}^i_\lambda, y^i_m), \quad (x^\lambda, \sigma^m, y^i, \sigma^m_\lambda, y^i_\lambda).$$

One can show the following [145].

Theorem 1.4.3. *There is the canonical map*

$$\varrho : J^1\Sigma \underset{\Sigma}{\times} J^1_\Sigma Y \underset{Y}{\longrightarrow} J^1 Y, \tag{1.4.5}$$

$$y^i_\lambda \circ \varrho = y^i_m \sigma^m_\lambda + \widetilde{y}^i_\lambda.$$

Using the canonical map (1.4.5), we can get the relations between connections on the fibre bundles $Y \to X$, $Y \to \Sigma$ and $\Sigma \to X$. These connections are given by the corresponding connection forms

$$\gamma = dx^\lambda \otimes (\partial_\lambda + \gamma^m_\lambda \partial_m + \gamma^i_\lambda \partial_i), \tag{1.4.6}$$

$$A_\Sigma = dx^\lambda \otimes (\partial_\lambda + A^i_\lambda \partial_i) + d\sigma^m \otimes (\partial_m + A^i_m \partial_i), \tag{1.4.7}$$

$$\Gamma = dx^\lambda \otimes (\partial_\lambda + \Gamma^m_\lambda \partial_m). \tag{1.4.8}$$

A connection γ (1.4.6) on the fibre bundle $Y \to X$ is called *projectable* onto a connection Γ (1.4.8) on the fibre bundle $\Sigma \to X$ if, for any vector field τ on X, its horizontal lift $\gamma\tau$ on Y by means of the connection γ is a projectable vector field over the horizontal lift $\Gamma\tau$ of τ on Σ by means of the connection Γ. This property takes place if and only if $\gamma^m_\lambda = \Gamma^m_\lambda$, i.e., components γ^m_λ of the connection γ (1.4.6) must be independent of the fibre coordinates y^i.

A connection A_Σ (1.4.7) on the fibre bundle $Y \to \Sigma$ and a connection Γ (1.4.8) on the fibre bundle $\Sigma \to X$ define a connection on the composite bundle $Y \to X$ as the composition of bundle morphisms

$$\gamma : Y \underset{X}{\times} TX \xrightarrow{(\mathrm{Id},\Gamma)} Y \underset{\Sigma}{\times} T\Sigma \xrightarrow{A_\Sigma} TY.$$

It is called the *composite connection* [112; 145]. This composite connection reads

$$\gamma = dx^\lambda \otimes (\partial_\lambda + \Gamma_\lambda^m \partial_m + (A_\lambda^i + A_m^i \Gamma_\lambda^m)\partial_i). \quad (1.4.9)$$

It is projectable onto Γ. Moreover, this is a unique connection such that the horizontal lift $\gamma\tau$ on Y of a vector field τ on X by means of the composite connection γ (1.4.9) coincides with the composition $A_\Sigma(\Gamma\tau)$ of horizontal lifts of τ on Σ by means of the connection Γ and then on Y by means of the connection A_Σ. For the sake of brevity, let us write $\gamma = A_\Sigma \circ \Gamma$.

Given a composite bundle Y (1.4.1), there are the exact sequences of vector bundles over Y:

$$0 \to V_\Sigma Y \longrightarrow VY \to Y \underset{\Sigma}{\times} V\Sigma \to 0, \quad (1.4.10)$$

$$0 \to Y \underset{\Sigma}{\times} V^*\Sigma \longrightarrow V^*Y \to V_\Sigma^* Y \to 0, \quad (1.4.11)$$

where $V_\Sigma Y$ and $V_\Sigma^* Y$ are the vertical tangent and the vertical cotangent bundles of the fibre bundle $Y \to \Sigma$ which are coordinated by $(x^\lambda, \sigma^m, y^i, \dot y^i)$ and $(x^\lambda, \sigma^m, y^i, \dot y_i)$, respectively. Let us consider a splitting of these exact sequences

$$B : VY \ni \dot y^i \partial_i + \dot\sigma^m \partial_m \to (\dot y^i \partial_i + \dot\sigma^m \partial_m)\rfloor B = \quad (1.4.12)$$
$$(\dot y^i - \dot\sigma^m B_m^i)\partial_i \in V_\Sigma Y,$$

$$B : V_\Sigma^* Y \ni \overline{d}y^i \to B\rfloor \overline{d}y^i = \overline{d}y^i - B_m^i \overline{d}\sigma^m \in V^*Y, \quad (1.4.13)$$

given by the form

$$B = (\overline{d}y^i - B_m^i \overline{d}\sigma^m) \otimes \partial_i. \quad (1.4.14)$$

Then the connection γ (1.4.6) on $Y \to X$ and the splitting B (1.4.12) define the connection

$$A_\Sigma = B \circ \gamma : TY \to VY \to V_\Sigma Y, \quad (1.4.15)$$
$$A_\Sigma = dx^\lambda \otimes (\partial_\lambda + (\gamma_\lambda^i - B_m^i \gamma_\lambda^m)\partial_i) + d\sigma^m \otimes (\partial_m + B_m^i \partial_i),$$

on the fibre bundle $Y \to \Sigma$.

Conversely, every connection A_Σ (1.4.7) on the fibre bundle $Y \to \Sigma$ yields the splitting

$$A_\Sigma : TY \supset VY \ni \dot y^i \partial_i + \dot\sigma^m \partial_m \to (\dot y^i - A_m^i \dot\sigma^m)\partial_i \quad (1.4.16)$$

of the exact sequence (1.4.10). Using this splitting, one can construct a first order differential operator

$$\widetilde D : J^1 Y \to T^* X \underset{Y}{\otimes} V_\Sigma Y, \quad (1.4.17)$$

$$\widetilde D = dx^\lambda \otimes (y_\lambda^i - A_\lambda^i - A_m^i \sigma_\lambda^m)\partial_i,$$

1.4. Composite bundles

on the composite bundle $Y \to X$. It is called the *vertical covariant differential*. This operator also can be defined as the composition

$$\widetilde{D} = \mathrm{pr}_1 \circ D^\gamma : J^1Y \to T^*X \underset{Y}{\otimes} VY \to T^*X \underset{Y}{\otimes} VY_\Sigma,$$

where D^γ is the covariant differential (1.3.18) relative to some composite connection $A_\Sigma \circ \Gamma$ (1.4.9), but $\widetilde{D}$ does not depend on the choice of a connection Γ on the fibre bundle $\Sigma \to X$.

The vertical covariant differential (1.4.17) possesses the following important property. Let h be a section of the fibre bundle $\Sigma \to X$, and let $Y^h \to X$ be the restriction (1.4.4) of the fibre bundle $Y \to \Sigma$ to $h(X) \subset \Sigma$. This is a subbundle

$$i_h : Y^h \to Y$$

of the fibre bundle $Y \to X$. Every connection A_Σ (1.4.7) induces the pullback connection

$$A_h = i_h^* A_\Sigma = dx^\lambda \otimes [\partial_\lambda + ((A_m^i \circ h)\partial_\lambda h^m + (A \circ h)_\lambda^i)\partial_i] \qquad (1.4.18)$$

on $Y^h \to X$. Then the restriction of the vertical covariant differential $\widetilde{D}$ (1.4.17) to

$$J^1 i_h(J^1 Y^h) \subset J^1 Y$$

coincides with the familiar covariant differential D^{A_h} (1.3.18) on Y^h relative to the pull-back connection A_h (1.4.18).

Remark 1.4.1. Let $\Gamma : Y \to J^1Y$ be a connection on a fibre bundle $Y \to X$. In accordance with the canonical isomorphism $VJ^1Y = J^1VY$ (1.2.9), the vertical tangent map

$$V\Gamma : VY \to VJ^1Y$$

to Γ defines the connection

$$V\Gamma : VY \to J^1VY,$$
$$V\Gamma = dx^\lambda \otimes (\partial_\lambda + \Gamma_\lambda^i \partial_i + \partial_j \Gamma_\lambda^i \dot{y}^j \dot{\partial}_i), \qquad (1.4.19)$$

on the composite vertical tangent bundle

$$VY \to Y \to X.$$

This is called the *vertical connection* to Γ. Of course, the connection $V\Gamma$ projects onto Γ. Moreover, $V\Gamma$ is linear over Γ. Then the dual connection of $V\Gamma$ on the composite vertical cotangent bundle

$$V^*Y \to Y \to X$$

reads
$$V^*\Gamma : V^*Y \to J^1 V^*Y,$$
$$V^*\Gamma = dx^\lambda \otimes (\partial_\lambda + \Gamma^i_\lambda \partial_i - \partial_j \Gamma^i_\lambda \dot{y}_i \dot\partial^j). \quad (1.4.20)$$

It is called the *covertical connection* to Γ. If $Y \to X$ is an affine bundle, the connection $V\Gamma$ (1.4.19) can be seen as the composite connection generated by the connection Γ on $Y \to X$ and the linear connection
$$\widetilde{\Gamma} = dx^\lambda \otimes (\partial_\lambda + \partial_j \Gamma^i_\lambda \dot{y}^j \dot\partial_i) + dy^i \otimes \partial_i \quad (1.4.21)$$
on the vertical tangent bundle $VY \to Y$.

1.5 Higher order jet manifolds

The notion of first and second order jets of sections of a fibre bundle is naturally extended to higher order jets [53; 94; 145].

Let $Y \to X$ be a fibre bundle. Given its bundle coordinates (x^λ, y^i), a *multi-index* Λ of the length $|\Lambda| = k$ throughout denotes a collection of indices $(\lambda_1 ... \lambda_k)$ modulo permutations. By $\Lambda + \Sigma$ is meant a multi-index $(\lambda_1 ... \lambda_k \sigma_1 ... \sigma_r)$. For instance $\lambda + \Lambda = (\lambda\lambda_1...\lambda_r)$. By $\Lambda\Sigma$ is denoted the union of collections $(\lambda_1 ... \lambda_k; \sigma_1 ... \sigma_r)$ where the indices λ_i and σ_j are not permitted. Summation over a multi-index Λ means separate summation over each its index λ_i. We use the compact notation
$$\partial_\Lambda = \partial_{\lambda_k} \circ \cdots \circ \partial_{\lambda_1}, \qquad \Lambda = (\lambda_1...\lambda_k).$$

The *r-order jet manifold* $J^r Y$ of sections of a fibre bundle $Y \to X$ is defined as the disjoint union of the equivalence classes $j^r_x s$ of sections s of $Y \to X$ such that sections s and s' belong to the same equivalence class $j^r_x s$ if and only if
$$s^i(x) = s'^i(x), \qquad \partial_\Lambda s^i(x) = \partial_\Lambda s'^i(x), \qquad 0 < |\Lambda| \leq r.$$

In brief, one can say that sections of $Y \to X$ are identified by the $r+1$ terms of their Taylor series at points of X. The particular choice of coordinates does not matter for this definition. The equivalence classes $j^r_x s$ are called the *r-order jets* of sections. Their set $J^r Y$ is endowed with an atlas of the adapted coordinates
$$(x^\lambda, y^i_\Lambda), \qquad y^i_\Lambda \circ s = \partial_\Lambda s^i(x), \qquad 0 \leq |\Lambda| \leq r, \quad (1.5.1)$$
possessing transition functions
$$y'^i_{\lambda+\Lambda} = \frac{\partial x^\mu}{\partial' x^\lambda} d_\mu y'^i_\Lambda, \quad (1.5.2)$$

1.5. Higher order jet manifolds

where the symbol d_λ stands for the *higher order total derivative*

$$d_\lambda = \partial_\lambda + \sum_{0 \leq |\Lambda| \leq r-1} y^i_{\Lambda+\lambda} \partial_i^\Lambda, \qquad d'_\lambda = \frac{\partial x^\mu}{\partial x'^\lambda} d_\mu. \qquad (1.5.3)$$

These derivatives act on exterior forms on $J^r Y$ and obey the relations

$$[d_\lambda, d_\mu] = 0, \qquad d_\lambda \circ d = d \circ d_\lambda,$$
$$d_\lambda(\phi \wedge \sigma) = d_\lambda(\phi) \wedge \sigma + \phi \wedge d_\lambda(\sigma),$$
$$d_\lambda(d\phi) = d(d_\lambda(\phi)).$$

For instance,

$$d_\lambda(dx^\mu) = 0, \qquad d_\lambda(dy^i_\Lambda) = dy^i_{\lambda+\Lambda}.$$

We use the compact notation

$$d_\Lambda = d_{\lambda_r} \circ \cdots \circ d_{\lambda_1}, \qquad \Lambda = (\lambda_r ... \lambda_1).$$

The coordinates (1.5.1) bring the set $J^r Y$ into a smooth manifold of finite dimension

$$\dim J^r Y = n + m \sum_{i=0}^{r} \frac{(n+i-1)!}{i!(n-1)!}.$$

The coordinates (1.5.1) are compatible with the natural surjections

$$\pi^r_k : J^r Y \to J^k Y, \quad r > k,$$

which form the composite bundle

$$\pi^r : J^r Y \xrightarrow{\pi^r_{r-1}} J^{r-1} Y \xrightarrow{\pi^{r-1}_{r-2}} \cdots \xrightarrow{\pi^1_0} Y \xrightarrow{\pi} X$$

with the properties

$$\pi^k_s \circ \pi^r_k = \pi^r_s, \qquad \pi^s \circ \pi^r_s = \pi^r.$$

A glance at the transition functions (1.5.2), when $|\Lambda| = r$, shows that the fibration

$$\pi^r_{r-1} : J^r Y \to J^{r-1} Y$$

is an affine bundle modelled over the vector bundle

$$\overset{r}{\vee} T^* X \underset{J^{r-1}Y}{\otimes} VY \to J^{r-1} Y. \qquad (1.5.4)$$

Remark 1.5.1. Let us recall that a base of any affine bundle is a strong deformation retract of its total space. Consequently, Y is a strong deformation retract of $J^1 Y$, which in turn is a strong deformation retract of $J^2 Y$, and so on. It follows that a fibre bundle Y is a strong deformation retract of any finite order jet manifold $J^r Y$. Therefore, by virtue of the Vietoris–Begle theorem [22], there is an isomorphism

$$H^*(J^r Y; \mathbb{R}) = H^*(Y; \mathbb{R}) \qquad (1.5.5)$$

of cohomology groups of $J^r Y$, $1 \leq r$, and Y with coefficients in the constant sheaf $\mathbb{R}$.

Remark 1.5.2. To introduce higher order jet manifolds, one can use the construction of repeated jet manifolds. Let us consider the r-order jet manifold $J^r J^k Y$ of a jet bundle $J^k Y \to X$. It is coordinated by $(x^\mu, y^i_{\Sigma\Lambda})$, $|\Lambda| \leq k$, $|\Sigma| \leq r$. There is a canonical monomorphism

$$\sigma_{rk} : J^{r+k} Y \to J^r J^k Y, \qquad y^i_{\Sigma\Lambda} \circ \sigma_{rk} = y^i_{\Sigma+\Lambda}.$$

In the calculus in higher order jets, we have the r-order *jet prolongation functor* such that, given fibre bundles Y and Y' over X, every bundle morphism $\Phi : Y \to Y'$ over a diffeomorphism f of X admits the r-order jet prolongation to a morphism of r-order jet manifolds

$$J^r \Phi : J^r Y \ni j^r_x s \to j^r_{f(x)} (\Phi \circ s \circ f^{-1}) \in J^r Y'. \qquad (1.5.6)$$

The jet prolongation functor is exact. If Φ is an injection or a surjection, so is $J^r \Phi$. It also preserves an algebraic structure. In particular, if $Y \to X$ is a vector bundle, $J^r Y \to X$ is well. If $Y \to X$ is an affine bundle modelled over the vector bundle $\overline{Y} \to X$, then $J^r Y \to X$ is an affine bundle modelled over the vector bundle $J^r \overline{Y} \to X$.

Every section s of a fibre bundle $Y \to X$ admits the r-order *jet prolongation* to the *integrable section*

$$(J^r s)(x) = j^r_x s$$

of the jet bundle $J^r Y \to X$.

Let $\mathcal{O}^*_k = \mathcal{O}^*(J^k Y)$ be the differential graded algebra of exterior forms on a jet manifold $J^k Y$. Every exterior form ϕ on a jet manifold $J^k Y$ gives rise to the pull-back form $\pi^{k+i*}_k \phi$ on a jet manifold $J^{k+i} Y$. We have the direct sequence of $C^\infty(X)$-algebras

$$\mathcal{O}^*(X) \xrightarrow{\pi^*} \mathcal{O}^*(Y) \xrightarrow{\pi^{1*}_0} \mathcal{O}^*_1 \xrightarrow{\pi^{2*}_1} \cdots \xrightarrow{\pi^{r*}_{r-1}} \mathcal{O}^*_r.$$

Remark 1.5.3. By virtue of de Rham Theorem 10.9.4, the cohomology of the de Rham complex of $\mathcal{O}^*_k$ equals the cohomology $H^*(J^k Y; \mathbb{R})$ of $J^k Y$ with coefficients in the constant sheaf $\mathbb{R}$. The latter in turn coincides with the sheaf cohomology $H^*(Y; \mathbb{R})$ of Y (see Remark 1.5.1) and, thus, it equals the de Rham cohomology $H^*_{\mathrm{DR}}(Y)$ of Y.

Given a k-order jet manifold $J^k Y$ of $Y \to X$, there exists the canonical bundle morphism

$$r_{(k)} : J^k TY \to TJ^k Y$$

over a surjection

$$J^k Y \underset{X}{\times} J^k TX \to J^k Y \underset{X}{\times} TX$$

1.5. Higher order jet manifolds

whose coordinate expression is

$$\dot{y}^i_\Lambda \circ r_{(k)} = (\dot{y}^i)_\Lambda - \sum (\dot{y}^i)_{\mu+\Sigma}(\dot{x}^\mu)_\Xi, \qquad 0 \leq |\Lambda| \leq k,$$

where the sum is taken over all partitions $\Sigma + \Xi = \Lambda$ and $0 < |\Xi|$. In particular, we have the canonical isomorphism over $J^k Y$

$$r_{(k)} : J^k VY \to V J^k Y, \qquad (\dot{y}^i)_\Lambda = \dot{y}^i_\Lambda \circ r_{(k)}. \tag{1.5.7}$$

As a consequence, every projectable vector field u (1.1.25) on a fibre bundle $Y \to X$ has the following k-order *jet prolongation* to a vector field on $J^k Y$:

$$J^k u = r_{(k)} \circ J^k u : J^k Y \to T J^k Y,$$

$$J^k u = u^\lambda \partial_\lambda + u^i \partial_i + \sum_{0 < |\Lambda| \leq k} (d_\Lambda (u^i - y^i_\mu u^\mu) + y^i_{\mu+\Lambda} u^\mu) \partial_i^\Lambda, \tag{1.5.8}$$

(cf. (1.2.8) for $k = 1$). In particular, the k-order jet prolongation (1.5.8) of a vertical vector field $u = u^i \partial_i$ on $Y \to X$ is a vertical vector field

$$J^k u = u^i \partial_i + \sum_{0 < |\Lambda| \leq k} d_\Lambda u^i \partial_i^\Lambda \tag{1.5.9}$$

on $J^k Y \to X$ due to the isomorphism (1.5.7).

A vector field u_r on an r-order jet manifold $J^r Y$ is called *projectable* if, for any $k < r$, there exists a projectable vector field u_k on $J^k Y$ such that

$$u_k \circ \pi^r_k = T\pi^r_k \circ u_r.$$

A projectable vector field u_k on $J^k Y$ has the coordinate expression

$$u_k = u^\lambda \partial_\lambda + \sum_{0 \leq |\Lambda| \leq k} u^i_\Lambda \partial_i^\Lambda$$

such that u_λ depends only on coordinates x^μ and every component u^i_Λ is independent of coordinates y^i_Ξ, $|\Xi| > |\Lambda|$. In particular, the k-order jet prolongation $J^k u$ (1.5.8) of a projectable vector field on Y is a projectable vector field on $J^k Y$. It is called an *integrable vector field*.

Let $\mathcal{P}^k$ denote a vector space of projectable vector fields on a jet manifold $J^k Y$. It is easily seen that $\mathcal{P}^r$ is a real Lie algebra and that the morphisms $T\pi^r_k$, $k < r$, constitute the inverse system

$$\mathcal{P}^0 \xleftarrow{T\pi^1_0} \mathcal{P}^1 \xleftarrow{T\pi^2_1} \cdots \xleftarrow{T\pi^{r-1}_{r-2}} \mathcal{P}^{r-1} \xleftarrow{T\pi^r_{r-1}} \mathcal{P}^r \tag{1.5.10}$$

of these Lie algebras. One can show the following [149].

Theorem 1.5.1. *The k-order jet prolongation (1.5.8) is a Lie algebra monomorphism of the Lie algebra $\mathcal{P}^0$ of projectable vector fields on $Y \to X$ to the Lie algebra $\mathcal{P}^k$ of projectable vector fields on $J^k Y$ such that*

$$T\pi^r_k(J^r u) = J^k u \circ \pi^r_k. \tag{1.5.11}$$

Every projectable vector field u_k on $J^k Y$ is decomposed into the sum
$$u_k = J^k(T\pi_0^k(u_k)) + v_k \tag{1.5.12}$$
of the integrable vector field $J^k(T\pi_0^k(u_k))$ and a projectable vector field v_k which is vertical with respect to a fibration $J^k Y \to Y$.

Similarly to the canonical monomorphisms (1.2.5) – (1.2.6), there are the canonical bundle monomorphisms over $J^k Y$:
$$\lambda_{(k)} : J^{k+1}Y \longrightarrow T^*X \underset{J^k Y}{\otimes} TJ^k Y,$$
$$\lambda_{(k)} = dx^\lambda \otimes d_\lambda, \tag{1.5.13}$$
$$\theta_{(k)} : J^{k+1}Y \longrightarrow T^*J^k Y \underset{J^k Y}{\otimes} VJ^k Y,$$
$$\theta_{(k)} = \sum_{|\Lambda| \le k} (dy_\Lambda^i - y_{\lambda+\Lambda}^i dx^\lambda) \otimes \partial_i^\Lambda. \tag{1.5.14}$$

The one-forms
$$\theta_\Lambda^i = dy_\Lambda^i - y_{\lambda+\Lambda}^i dx^\lambda \tag{1.5.15}$$
are called the *local contact forms*. The monomorphisms (1.5.13) – (1.5.14) yield the bundle monomorphisms over $J^{k+1}Y$:
$$\widehat{\lambda}_{(k)} : TX \underset{X}{\times} J^{k+1}Y \longrightarrow TJ^k Y \underset{J^k Y}{\times} J^{k+1}Y,$$
$$\widehat{\theta}_{(k)} : V^* J^k Y \underset{J^k Y}{\times} \longrightarrow T^* J^k Y \underset{J^k Y}{\times} J^{k+1}Y$$
(cf. (1.3.13) – (1.3.14) for $k = 1$). These monomorphisms in turn define the *canonical horizontal splittings* of the pull-back bundles
$$\pi_k^{k+1*} TJ^k Y = \widehat{\lambda}_{(k)}(TX \underset{X}{\times} J^{k+1}Y) \underset{J^{k+1}Y}{\oplus} VJ^k Y, \tag{1.5.16}$$
$$\dot{x}^\lambda \partial_\lambda + \sum_{|\Lambda| \le k} \dot{y}_\Lambda^i \partial_i^\Lambda = \dot{x}^\lambda d_\lambda + \sum_{|\Lambda| \le k} (\dot{y}_\Lambda^i - \dot{x}^\lambda y_{\lambda+\Lambda}^i) \partial_i^\Lambda,$$
$$\pi_k^{k+1*} T^* J^k Y = T^*X \underset{J^{k+1}Y}{\oplus} \widehat{\theta}_{(k)}(V^* J^k Y \underset{J^k Y}{\times} J^{k+1}Y), \tag{1.5.17}$$
$$\dot{x}_\lambda dx^\lambda + \sum_{|\Lambda| \le k} \dot{y}_i^\Lambda dy_\Lambda^i = (\dot{x}_\lambda + \sum_{|\Lambda| \le k} \dot{y}_i^\Lambda y_{\lambda+\Lambda}^i) dx^\lambda + \sum \dot{y}_i^\Lambda \theta_\Lambda^i.$$

For instance, it follows from the canonical horizontal splitting (1.5.16) that any vector field u_k on $J^k Y$ admits the canonical decomposition
$$u_k = u_H + u_V = (u^\lambda \partial_\lambda + \sum_{|\Lambda| \le k} y_{\lambda+\Lambda}^i \partial_i^\Lambda) + \tag{1.5.18}$$
$$\sum_{|\Lambda| \le k} (u_\Lambda^i - u^\lambda y_{\lambda+\Lambda}^i) \partial_i^\Lambda$$

over $J^{k+1}Y$ into the horizontal and vertical parts.

By virtue of the canonical horizontal splitting (1.5.17), every exterior one-form ϕ on J^kY admits the canonical splitting of its pull-back onto $J^{k+1}Y$ into the horizontal and vertical parts:

$$\pi_k^{k+1*}\phi = \phi_H + \phi_V = h_0\phi + (\phi - h_0(\phi)), \tag{1.5.19}$$

where h_0 is the *horizontal projection*

$$h_0(dx^\lambda) = dx^\lambda, \qquad h_0(dy^i_{\lambda_1\ldots\lambda_k}) = y^i_{\mu\lambda_1\ldots\lambda_k}dx^\mu.$$

The vertical part of the splitting is called a *contact one-form* on $J^{k+1}Y$.

Let us consider an ideal of the algebra $\mathcal{O}_k^*$ of exterior forms on J^kY which is generated by the contact one-forms on J^kY. This ideal, called the *ideal of contact forms*, is locally generated by the contact forms θ^i_Λ (1.5.15). One can show that an exterior form ϕ on the a manifold J^kY is a contact form if and only if its pull-back $\overline{s}^*\phi$ onto a base X by means of any integrable section $\overline{s}$ of $J^kY \to X$ vanishes.

1.6 Differential operators and equations

Jet manifolds provides the conventional language of theory of differential equations and differential operators if they need not be linear [24; 53; 96].

Definition 1.6.1. A system of k-order partial differential equations on a fibre bundle $Y \to X$ is defined as a closed subbundle $\mathfrak{E}$ of a jet bundle $J^kY \to X$. For the sake of brevity, we agree to call $\mathfrak{E}$ a *differential equation*.

Let J^kY be provided with the adapted coordinates (x^λ, y^i_Λ). There exists a local coordinate system (z^A), $A = 1, \ldots, \text{codim}\mathfrak{E}$, on J^kY such that $\mathfrak{E}$ is locally given (in the sense of item (i) of Theorem 1.1.1) by equations

$$\mathcal{E}^A(x^\lambda, y^i_\Lambda) = 0, \qquad A = 1, \ldots, \text{codim}\mathfrak{E}. \tag{1.6.1}$$

Given a k-order differential equation $\mathfrak{E}$, one can always construct its r-order jet prolongation as follows. Let us consider a repeated jet manifold

$$\sigma_k^r : J^rJ^kY \to J^kY. \tag{1.6.2}$$

The s-order jet prolongation of the differential equation $\mathfrak{E}$ is defined as a subset

$$\mathfrak{E}^{(r)} = (\sigma_k^r)^{-1}(\mathfrak{E}) \bigcap J^{k+r}Y.$$

In particular, if $\mathfrak{E}^{(r)}$ is a smooth submanifold of $J^{k+r}Y$, then the s-order jet prolongation $(\mathfrak{E}^{(r)})^{(s)}$ of $\mathfrak{E}^{(r)}$ coincides with the $(r+s)$-order jet prolongation $\mathfrak{E}^{(s+r)}$ of $\mathfrak{E}$.

A differential equation $\mathfrak{E}$ is called *regular* if all finite order jet prolongations $\mathfrak{E}^{(r)}$ of $\mathfrak{E}$ also are differential equations. Let $\mathfrak{E} \subset J^k Y$ be a regular k-order differential equation. If it is locally described by the system of equations (1.6.1), its r-order jet prolongation $\mathfrak{E}^{(r)}$ is given by the system of equations

$$\mathcal{E}^A = 0, \qquad d_{\alpha_1} \mathcal{E}^A = 0, \qquad \cdots, \qquad d_{\alpha_1} \cdots d_{\alpha_r} \mathcal{E}^A = 0.$$

By a *classical solution* of a differential equation $\mathfrak{E}$ on $Y \to X$ is meant a section s of $Y \to X$ such that its k-order jet prolongation $J^k s$ lives in $\mathfrak{E}$. If a differential equation $\mathfrak{E}$ has a classical solution through a point $q \in \mathfrak{E}$, this point gives rise to an element of every finite order jet prolongation $\mathfrak{E}^{(r)}$ of a differential equation $\mathfrak{E}$. It follows that a necessary condition for a differential equation $\mathfrak{E}$ to admit a solution through everyone of its point is that the mappings

$$\rho_k^{k+r} = \sigma_k^r|_{\mathfrak{E}^{(r)}} : \mathfrak{E}^{(r)} \to \mathfrak{E} \tag{1.6.3}$$

are surjections. In this case, if $\mathfrak{E}$ is a regular differential equation, there is one-to-one correspondence between classical solutions of $\mathfrak{E}$ and those of its k-order jet prolongation $\mathfrak{E}^{(k)}$. If additionally every tangent vector to a differential equation $\mathfrak{E}$ is tangent to some classical solution of $\mathfrak{E}$, then the mapping (1.6.3) is a submersion. A regular k-order differential equation $\mathfrak{E}$ is called *formally integrable* if the morphisms

$$\rho_{k+r}^{k+r+1} : \mathfrak{E}^{(r+1)} \to \mathfrak{E}^{(r)}, \qquad r \in \mathbb{N},$$

are fibred manifolds. One can show that if a differential equation $\mathfrak{E}$ is formally integrable and analytic, it admits an analytic classical solution through any its point [53; 109].

In classical field theory, differential equations are mostly associated to differential operators. There are several equivalent definitions of (non-linear) differential operators. We start with the following.

Definition 1.6.2. Let $Y \to X$ and $E \to X$ be fibre bundles, which are assumed to have global sections. A k-order E-valued *differential operator* on a fibre bundle $Y \to X$ is defined as a section $\mathcal{E}$ of the pull-back bundle

$$\mathrm{pr}_1 : E_Y^k = J^k Y \underset{X}{\times} E \to J^k Y. \tag{1.6.4}$$

Given bundle coordinates (x^λ, y^i) on Y and (x^λ, χ^a) on E, the pull-back (1.6.4) is provided with coordinates $(x^\lambda, y^j_\Sigma, \chi^a)$, $0 \leq |\Sigma| \leq k$. With respect to these coordinates, a differential operator $\mathcal{E}$ seen as a closed imbedded submanifold $\mathcal{E} \subset E^k_Y$ is given by the equalities

$$\chi^a = \mathcal{E}^a(x^\lambda, y^j_\Sigma). \tag{1.6.5}$$

There is obvious one-to-one correspondence between the sections $\mathcal{E}$ (1.6.5) of the fibre bundle (1.6.4) and the bundle morphisms

$$\Phi : J^k Y \underset{X}{\longrightarrow} E, \tag{1.6.6}$$

$$\Phi = \mathrm{pr}_2 \circ \mathcal{E} \iff \mathcal{E} = (\mathrm{Id}\, J^k Y, \Phi).$$

Therefore, we come to the following equivalent definition of differential operators on $Y \to X$.

Definition 1.6.3. Let $Y \to X$ and $E \to X$ be fibre bundles. A bundle morphism $J^k Y \to E$ over X is called a E-valued k-order *differential operator* on $Y \to X$.

It is readily observed that the differential operator Φ (1.6.6) sends each section s of $Y \to X$ onto the section $\Phi \circ J^k s$ of $E \to X$. The mapping

$$\Delta_\Phi : \mathcal{S}(Y) \to \mathcal{S}(E),$$
$$\Delta_\Phi : s \to \Phi \circ J^k s, \qquad \chi^a(x) = \mathcal{E}^a(x^\lambda, \partial_\Sigma s^j(x)),$$

is called the *standard form of a differential operator*.

Let e be a global section of a fibre bundle $E \to X$, the *kernel* of a E-valued differential operator Φ is defined as the kernel

$$\mathrm{Ker}_e \Phi = \Phi^{-1}(e(X)) \tag{1.6.7}$$

of the bundle morphism Φ (1.6.6). If it is a closed subbundle of the jet bundle $J^k Y \to X$, one says that $\mathrm{Ker}_e \Phi$ (1.6.7) is a *differential equation* associated to the differential operator Φ. By virtue of Theorem 1.1.10, this condition holds if Φ is a bundle morphism of constant rank.

If $E \to X$ is a vector bundle, by the kernel of a E-valued differential operator is usually meant its kernel with respect to the canonical zero-valued section $\widehat{0}$ of $E \to X$.

In the framework of Lagrangian formalism, we deal with differential operators of the following type. Let

$$F \to Y \to X, \qquad E \to Y \to X$$

be composite bundles where $E \to Y$ is a vector bundle. By a k-order differential operator on $F \to X$ taking its values into $E \to X$ is meant a bundle morphism

$$\Phi : J^k F \underset{Y}{\longrightarrow} E, \qquad (1.6.8)$$

which certainly is a bundle morphism over X in accordance with Definition 1.6.3. Its kernel $\operatorname{Ker} \Phi$ is defined as the inverse image of the canonical zero-valued section of $E \to Y$. In an equivalent way, the differential operator (1.6.8) is represented by a section $\mathcal{E}_\Phi$ of the vector bundle

$$J^k F \underset{Y}{\times} E \to J^k F.$$

Given bundle coordinates (x^λ, y^i, w^r) on F and (x^λ, y^i, c^A) on E with respect to the fibre basis $\{e_A\}$ for $E \to Y$, this section reads

$$\mathcal{E}_\Phi = \mathcal{E}^A(x^\lambda, y^i_\Lambda, w^r_\Lambda) e_A, \qquad 0 \leq |\Lambda| \leq k. \qquad (1.6.9)$$

Then the differential operator (1.6.8) also is represented by a function

$$\mathcal{E}_\Phi = \mathcal{E}^A(x^\lambda, y^i_\Lambda, w^r_\Lambda) c_A \in C^\infty(F \underset{Y}{\times} E^*) \qquad (1.6.10)$$

on the product $F \times_Y E^*$, where $E^* \to Y$ is the dual of $E \to Y$ coordinated by (x^λ, y^i, c_A).

If $F \to Y$ is a vector bundle, a differential operator Φ (1.6.8) on the composite bundle

$$F \to Y \to X$$

is called linear if it is linear on the fibres of the vector bundle $J^k F \to J^k Y$. In this case, its representations (1.6.9) and (1.6.10) take the form

$$\mathcal{E}_\Phi = \sum_{0 \leq |\Xi| \leq k} \mathcal{E}^{A,\Xi}_r(x^\lambda, y^i_\Lambda) w^r_\Xi e_A, \qquad 0 \leq |\Lambda| \leq k, \qquad (1.6.11)$$

$$\mathcal{E}_\Phi = \sum_{0 \leq |\Xi| \leq k} \mathcal{E}^{A,\Xi}_r(x^\lambda, y^i_\Lambda) w^r_\Xi c_A, \qquad 0 \leq |\Lambda| \leq k. \qquad (1.6.12)$$

1.7 Infinite order jet formalism

The finite order jet manifolds $J^k Y$ of a fibre bundle $Y \to X$ form the inverse sequence

$$Y \xleftarrow{\pi} J^1 Y \longleftarrow \cdots J^{r-1} Y \xleftarrow{\pi^r_{r-1}} J^r Y \longleftarrow \cdots , \qquad (1.7.1)$$

1.7. Infinite order jet formalism

where π_{r-1}^r are affine bundles modelled over the vector bundles (1.5.4). Its inductive limit $J^\infty Y$ is defined as a minimal set such that there exist surjections

$$\pi^\infty : J^\infty Y \to X, \quad \pi_0^\infty : J^\infty Y \to Y, \quad \pi_k^\infty : J^\infty Y \to J^k Y, \qquad (1.7.2)$$

obeying the relations

$$\pi_r^\infty = \pi_r^k \circ \pi_k^\infty$$

for all admissible k and $r < k$. A projective limit of the inverse system (1.7.1) always exists. It consists of those elements

$$(\ldots, z_r, \ldots, z_k, \ldots), \qquad z_r \in J^r Y, \quad z_k \in J^k Y,$$

of the Cartesian product $\prod_k J^k Y$ which satisfy the relations $z_r = \pi_r^k(z_k)$ for all $k > r$. One can think of elements of $J^\infty Y$ as being *infinite order jets* of sections of $Y \to X$ identified by their Taylor series at points of X.

The set $J^\infty Y$ is provided with the projective limit topology. This is the coarsest topology such that the surjections π_r^∞ (1.7.2) are continuous. Its base consists of inverse images of open subsets of $J^r Y$, $r = 0, \ldots$, under the maps π_r^∞. With this topology, $J^\infty Y$ is a paracompact Fréchet (complete metrizable, but not Banach) manifold modelled over a locally convex vector space of formal number series $\{a^\lambda, a^i, a_\lambda^i, \cdots\}$ [150]. It is called the *infinite order jet manifold*. One can show that the surjections π_r^∞ are open maps admitting local sections, i.e., $J^\infty Y \to J^r Y$ are continuous bundles. A bundle coordinate atlas $\{U_Y, (x^\lambda, y^i)\}$ of $Y \to X$ provides $J^\infty Y$ with the manifold coordinate atlas

$$\{(\pi_0^\infty)^{-1}(U_Y), (x^\lambda, y_\Lambda^i)\}_{0 \leq |\Lambda|}, \qquad y'^i_{\lambda+\Lambda} = \frac{\partial x^\mu}{\partial x'^\lambda} d_\mu y'^i_\Lambda. \qquad (1.7.3)$$

Theorem 1.7.1. *A fibre bundle Y is a strong deformation retract of the infinite order jet manifold $J^\infty Y$ [4; 56].*

Proof. To show that Y is a strong deformation retract of $J^\infty Y$, let us construct a homotopy from $J^\infty Y$ to Y in an explicit form. Let $\gamma_{(k)}$, $k \leq 1$, be global sections of the affine jet bundles $J^k Y \to J^{k-1} Y$. Then we have a global section

$$\gamma : Y \ni (x^\lambda, y^i) \to (x^\lambda, y^i, y_\Lambda^i = \gamma_{(|\Lambda|)}{}_\Lambda^i \circ \gamma_{(|\Lambda|-1)} \circ \cdots \circ \gamma_{(1)}) \in J^\infty Y. \quad (1.7.4)$$

of the open surjection $\pi_0^\infty : J^\infty Y \to Y$. Let us consider the map

$$[0,1] \times J^\infty Y \ni (t; x^\lambda, y^i, y_\Lambda^i) \to (x^\lambda, y^i, y'^i_\Lambda) \in J^\infty Y, \qquad (1.7.5)$$

$$y'^i_\Lambda = f_k(t) y_\Lambda^i + (1 - f_k(t)) \gamma_{(k)}{}_\Lambda^i (x^\lambda, y^i, y_\Sigma^i), \qquad |\Sigma| < k = |\Lambda|,$$

where $f_k(t)$ is a continuous monotone real function on $[0,1]$ such that

$$f_k(t) = \begin{cases} 0, & t \leq 1 - 2^{-k}, \\ 1, & t \geq 1 - 2^{-(k+1)}. \end{cases} \qquad (1.7.6)$$

A glance at the transition functions (1.7.3) shows that, although written in a coordinate form, this map is globally defined. It is continuous because, given an open subset $U_k \subset J^k Y$, the inverse image of the open set

$$(\pi_k^\infty)^{-1}(U_k) \subset J^\infty Y$$

is an open subset

$$(t_k, 1] \times (\pi_k^\infty)^{-1}(U_k) \cup (t_{k-1}, 1] \times (\pi_{k-1}^\infty)^{-1}(\pi_{k-1}^k [U_k \cap \gamma_{(k)}(J^{k-1}Y)]) \cup$$
$$\cdots \cup [0,1] \times (\pi_0^\infty)^{-1}(\pi_0^k [U_k \cap \gamma_{(k)} \circ \cdots \circ \gamma_{(1)}(Y)])$$

of $[0,1] \times J^\infty Y$, where $[t_r, 1] = \operatorname{supp} f_r$. Then, the map (1.7.5) is a desired homotopy from $J^\infty Y$ to Y which is identified with its image under the global section (1.7.4). $\square$

Corollary 1.7.1. *By virtue of the Vietoris–Begle theorem [22], there is an isomorphism*

$$H^*(J^\infty Y; \mathbb{R}) = H^*(Y; \mathbb{R}) \qquad (1.7.7)$$

between the cohomology of $J^\infty Y$ with coefficients in the constant sheaf $\mathbb{R}$ and that of Y.

The inverse sequence (1.7.1) of jet manifolds yields the direct sequence of graded differential algebras $\mathcal{O}_r^*$ of exterior forms on finite order jet manifolds

$$\mathcal{O}^*(X) \xrightarrow{\pi^*} \mathcal{O}^*(Y) \xrightarrow{\pi_0^{1*}} \mathcal{O}_1^* \longrightarrow \cdots \mathcal{O}_{r-1}^* \xrightarrow{\pi_{r-1}^{r\ *}} \mathcal{O}_r^* \longrightarrow \cdots, \qquad (1.7.8)$$

where $\pi_{r-1}^{r\ *}$ are the pull-back monomorphisms. Its direct limit

$$\mathcal{O}_\infty^* Y = \varinjlim \mathcal{O}_r^* \qquad (1.7.9)$$

exists and consists of all exterior forms on finite order jet manifolds modulo the pull-back identification. In accordance with Theorem 10.1.5, $\mathcal{O}_\infty^* Y$ is a differential graded algebra which inherits the operations of the exterior differential d and exterior product $\wedge$ of exterior algebras $\mathcal{O}_r^*$. If there is no danger of confusion, we denote $\mathcal{O}_\infty^* = \mathcal{O}_\infty^* Y$.

Theorem 1.7.2. *The cohomology $H^*(\mathcal{O}_\infty^*)$ of the de Rham complex*

$$0 \longrightarrow \mathbb{R} \longrightarrow \mathcal{O}_\infty^0 \xrightarrow{d} \mathcal{O}_\infty^1 \xrightarrow{d} \cdots \qquad (1.7.10)$$

of the differential graded algebra $\mathcal{O}_\infty^*$ equals the de Rham cohomology $H_{\mathrm{DR}}^*(Y)$ of a fibre bundle Y [3; 17].

Proof. By virtue of Theorem 10.3.2, the operation of taking homology groups of cochain complexes commutes with the passage to a direct limit. Since the differential graded algebra $\mathcal{O}_\infty^*$ is a direct limit of differential graded algebras $\mathcal{O}_r^*$, its cohomology $H^*(\mathcal{O}_\infty^*)$ is isomorphic to the direct limit of the direct sequence

$$H^*_{\mathrm{DR}}(Y) \longrightarrow H^*_{\mathrm{DR}}(J^1 Y) \longrightarrow \cdots \qquad (1.7.11)$$
$$H^*_{\mathrm{DR}}(J^{r-1} Y) \longrightarrow H^*_{\mathrm{DR}}(J^r Y) \longrightarrow \cdots$$

of the de Rham cohomology groups $H^*_{\mathrm{DR}}(J^r Y)$ of finite order jet manifolds $J^r Y$. In accordance with Remark 1.5.3, all these groups equal the de Rham cohomology $H^*_{\mathrm{DR}}(Y)$ of Y, and so is its direct limit $H^*(\mathcal{O}_\infty^*)$. $\square$

Corollary 1.7.2. *Any closed form $\phi \in \mathcal{O}_\infty^*$ is decomposed into the sum $\phi = \sigma + d\xi$, where σ is a closed form on Y.*

One can think of elements of $\mathcal{O}_\infty^*$ as being differential forms on the infinite order jet manifold $J^\infty Y$ as follows. Let $\mathfrak{D}_r^*$ be a sheaf of germs of exterior forms on $J^r Y$ and $\overline{\mathfrak{D}}_r^*$ the canonical presheaf of local sections of $\mathfrak{D}_r^*$. Since π_{r-1}^r are open maps, there is the direct sequence of presheaves

$$\overline{\mathfrak{D}}_0^* \xrightarrow{\pi_0^{0*}} \overline{\mathfrak{D}}_1^* \cdots \xrightarrow{\pi_{r-1}^{r\,*}} \overline{\mathfrak{D}}_r^* \longrightarrow \cdots.$$

Its direct limit $\overline{\mathfrak{D}}_\infty^*$ is a presheaf of differential graded algebras on $J^\infty Y$. Let $\mathfrak{Q}_\infty^*$ be the sheaf of differential graded algebras of germs of $\overline{\mathfrak{D}}_\infty^*$ on $J^\infty Y$. The structure module

$$\mathcal{Q}_\infty^* = \Gamma(\mathfrak{Q}_\infty^*) \qquad (1.7.12)$$

of global sections of $\mathfrak{Q}_\infty^*$ is a differential graded algebra such that, given an element $\phi \in \mathcal{Q}_\infty^*$ and a point $z \in J^\infty Y$, there exist an open neighbourhood U of z and an exterior form $\phi^{(k)}$ on some finite order jet manifold $J^k Y$ so that

$$\phi|_U = \pi_k^{\infty *} \phi^{(k)}|_U.$$

Therefore, one can think of $\mathcal{Q}_\infty^*$ as being an algebra of locally exterior forms on finite order jet manifolds. In particular, there is a monomorphism $\mathcal{O}_\infty^* \to \mathcal{Q}_\infty^*$.

Theorem 1.7.3. *The paracompact space $J^\infty Y$ admits a partition of unity by elements of the ring $\mathcal{Q}_\infty^0$* [150].

Since elements of the differential graded algebra $\mathcal{Q}_\infty^*$ are locally exterior forms on finite order jet manifolds, the following Poincaré lemma holds.

Lemma 1.7.1. *Given a closed element $\phi \in \mathcal{Q}_\infty^*$, there exists a neighbourhood U of each point $z \in J^\infty Y$ such that $\phi|_U$ is exact.*

Theorem 1.7.4. *The cohomology $H^*(\mathcal{Q}_\infty^*)$ of the de Rham complex*

$$0 \longrightarrow \mathbb{R} \longrightarrow \mathcal{Q}_\infty^0 \xrightarrow{d} \mathcal{Q}_\infty^1 \xrightarrow{d} \cdots . \qquad (1.7.13)$$

of the differential graded algebra $\mathcal{Q}_\infty^*$ equals the de Rham cohomology of a fibre bundle Y [4; 150].

Proof. Let us consider the de Rham complex of sheaves

$$0 \longrightarrow \mathbb{R} \longrightarrow \mathfrak{Q}_\infty^0 \xrightarrow{d} \mathfrak{Q}_\infty^1 \xrightarrow{d} \cdots \qquad (1.7.14)$$

on $J^\infty Y$. By virtue of Lemma 1.7.1, it is exact at all terms, except $\mathbb{R}$. Being sheaves of $\mathcal{Q}_\infty^0$-modules, the sheaves $\mathfrak{Q}_\infty^r$ are fine and, consequently acyclic because the paracompact space $J^\infty Y$ admits the partition of unity by elements of the ring $\mathcal{Q}_\infty^0$. Thus, the complex (1.7.14) is a resolution of the constant sheaf $\mathbb{R}$ on $J^\infty Y$. In accordance with abstract de Rham Theorem 10.7.5, cohomology $H^*(\mathcal{Q}_\infty^*)$ of the complex (1.7.13) equals the cohomology $H^*(J^\infty Y; \mathbb{R})$ of $J^\infty Y$ with coefficients in the constant sheaf $\mathbb{R}$. Since Y is a strong deformation retract of $J^\infty Y$, there is the isomorphism (1.5.5) and, consequently, a desired isomorphism

$$H^*(\mathcal{Q}_\infty^*) = H^*_{DR}(Y). \qquad \square$$

Due to a monomorphism $\mathcal{O}_\infty^* \to \mathcal{Q}_\infty^*$, one can restrict $\mathcal{O}_\infty^*$ to the coordinate chart (1.7.3) where horizontal forms dx^λ and contact one-forms

$$\theta_\Lambda^i = dy_\Lambda^i - y_{\lambda+\Lambda}^i dx^\lambda$$

make up a local basis for the $\mathcal{O}_\infty^0$-algebra $\mathcal{O}_\infty^*$. Though $J^\infty Y$ is not a smooth manifold, elements of $\mathcal{O}_\infty^*$ are exterior forms on finite order jet manifolds and, therefore, their coordinate transformations are smooth. Moreover, there is the canonical decomposition

$$\mathcal{O}_\infty^* = \oplus \mathcal{O}_\infty^{k,m}$$

of $\mathcal{O}_\infty^*$ into $\mathcal{O}_\infty^0$-modules $\mathcal{O}_\infty^{k,m}$ of k-contact and m-horizontal forms together with the corresponding projectors

$$h_k : \mathcal{O}_\infty^* \to \mathcal{O}_\infty^{k,*}, \qquad h^m : \mathcal{O}_\infty^* \to \mathcal{O}_\infty^{*,m}.$$

1.7. Infinite order jet formalism

Accordingly, the exterior differential on $\mathcal{O}_\infty^*$ is decomposed into the sum

$$d = d_V + d_H$$

of the *vertical differential*

$$d_V \circ h^m = h^m \circ d \circ h^m, \qquad d_V(\phi) = \theta_\Lambda^i \wedge \partial_i^\Lambda \phi, \qquad \phi \in \mathcal{O}_\infty^*,$$

and the *total differential*

$$d_H \circ h_k = h_k \circ d \circ h_k, \qquad d_H \circ h_0 = h_0 \circ d, \qquad d_H(\phi) = dx^\lambda \wedge d_\lambda(\phi),$$

where

$$d_\lambda = \partial_\lambda + y_\lambda^i \partial_i + \sum_{0<|\Lambda|} y_{\lambda+\Lambda}^i \partial_i^\Lambda \qquad (1.7.15)$$

are the *infinite order total derivatives*. These differentials obey the nilpotent conditions

$$d_H \circ d_H = 0, \qquad d_V \circ d_V = 0, \qquad d_H \circ d_V + d_V \circ d_H = 0, \qquad (1.7.16)$$

and make $\mathcal{O}_\infty^{*,*}$ into a bicomplex.

Let us consider the $\mathcal{O}_\infty^0$-module $\mathfrak{d}\mathcal{O}_\infty^0$ of derivations of the real ring $\mathcal{O}_\infty^0$.

Theorem 1.7.5. *The derivation module $\mathfrak{d}\mathcal{O}_\infty^0$ is isomorphic to the $\mathcal{O}_\infty^0$-dual $(\mathcal{O}_\infty^1)^*$ of the module of one-forms $\mathcal{O}_\infty^1$ [59].*

Proof. At first, let us show that $\mathcal{O}_\infty^*$ is generated by elements df, $f \in \mathcal{O}_\infty^0$. It suffices to justify that any element of $\mathcal{O}_\infty^1$ is a finite $\mathcal{O}_\infty^0$-linear combination of elements df, $f \in \mathcal{O}_\infty^0$. Indeed, every $\psi \in \mathcal{O}_\infty^1$ is an exterior form on some finite order jet manifold $J^r Y$. By virtue of Serre–Swan Theorem 10.9.3, the $C^\infty(J^r Y)$-module $\mathcal{O}_r^1$ of one-forms on $J^r Y$ is a projective module of finite rank, i.e., ψ is represented by a finite $C^\infty(J^r Y)$-linear combination of elements df, $f \in C^\infty(J^r Y) \subset \mathcal{O}_\infty^0$. Any element $\Phi \in (\mathcal{O}_\infty^1)^*$ yields a derivation $\vartheta_\Phi(f) = \Phi(df)$ of the real ring $\mathcal{O}_\infty^0$. Since the module $\mathcal{O}_\infty^1$ is generated by elements df, $f \in \mathcal{O}_\infty^0$, different elements of $(\mathcal{O}_\infty^1)^*$ provide different derivations of $\mathcal{O}_\infty^0$, i.e., there is a monomorphism $(\mathcal{O}_\infty^1)^* \to \mathfrak{d}\mathcal{O}_\infty^0$. By the same formula, any derivation $\vartheta \in \mathfrak{d}\mathcal{O}_\infty^0$ sends $df \to \vartheta(f)$ and, since $\mathcal{O}_\infty^0$ is generated by elements df, it defines a morphism $\Phi_\vartheta : \mathcal{O}_\infty^1 \to \mathcal{O}_\infty^0$. Moreover, different derivations ϑ provide different morphisms Φ_ϑ. Thus, we have a monomorphism $\mathfrak{d}\mathcal{O}_\infty^0 \to (\mathcal{O}_\infty^1)^*$ and, consequently, isomorphism $\mathfrak{d}\mathcal{O}_\infty^0 = (\mathcal{O}_\infty^1)^*$. □

The proof of Theorem 1.7.5 gives something more. The differential graded algebra $\mathcal{O}_\infty^*$ is a minimal Chevalley–Eilenberg differential calculus $\mathcal{O}^*\mathcal{A}$ over the real ring $\mathcal{A} = \mathcal{O}_\infty^0$ of smooth real functions on finite order jet manifolds of $Y \to X$. Let $\vartheta \rfloor \phi$, $\vartheta \in \mathfrak{d}\mathcal{O}_\infty^0$, $\phi \in \mathcal{O}_\infty^1$, denote the interior product. Extended to the differential graded algebra $\mathcal{O}_\infty^*$, the interior product $\rfloor$ obeys the rule

$$\vartheta \rfloor (\phi \wedge \sigma) = (\vartheta \rfloor \phi) \wedge \sigma + (-1)^{|\phi|} \phi \wedge (\vartheta \rfloor \sigma).$$

Restricted to a coordinate chart (1.7.3), $\mathcal{O}_\infty^1$ is a free $\mathcal{O}_\infty^0$-module generated by one-forms dx^λ, θ_Λ^i. Since $\mathfrak{d}\mathcal{O}_\infty^0 = (\mathcal{O}_\infty^1)^*$, any derivation of the real ring $\mathcal{O}_\infty^0$ takes the coordinate form

$$\vartheta = \vartheta^\lambda \partial_\lambda + \vartheta^i \partial_i + \sum_{0<|\Lambda|} \vartheta_\Lambda^i \partial_i^\Lambda, \tag{1.7.17}$$

where

$$\partial_i^\Lambda(y_\Sigma^j) = \partial_i^\Lambda \rfloor dy_\Sigma^j = \delta_i^j \delta_\Sigma^\Lambda$$

up to permutations of multi-indices Λ and Σ. Its coefficients ϑ^λ, ϑ^i, ϑ_Λ^i are local smooth functions of finite jet order possessing the transformation law

$$\vartheta'^\lambda = \frac{\partial x'^\lambda}{\partial x^\mu} \vartheta^\mu, \qquad \vartheta'^i = \frac{\partial y'^i}{\partial y^j} \vartheta^j + \frac{\partial y'^i}{\partial x^\mu} \vartheta^\mu,$$

$$\vartheta_\Lambda'^i = \sum_{|\Sigma| \le |\Lambda|} \frac{\partial y_\Lambda'^i}{\partial y_\Sigma^j} \vartheta_\Sigma^j + \frac{\partial y_\Lambda'^i}{\partial x^\mu} \vartheta^\mu. \tag{1.7.18}$$

Any derivation ϑ (1.7.17) of the ring $\mathcal{O}_\infty^0$ yields a derivation (called the Lie derivative) $\mathbf{L}_\vartheta$ of the differential graded algebra $\mathcal{O}_\infty^*$ given by the relations

$$\mathbf{L}_\vartheta \phi = \vartheta \rfloor d\phi + d(\vartheta \rfloor \phi),$$
$$\mathbf{L}_\vartheta(\phi \wedge \phi') = \mathbf{L}_\vartheta(\phi) \wedge \phi' + \phi \wedge \mathbf{L}_\vartheta(\phi').$$

Remark 1.7.1. In particular, the total derivatives (1.7.15) are defined as the local derivations of $\mathcal{O}_\infty^0$ and the corresponding Lie derivatives

$$d_\lambda \phi = \mathbf{L}_{d_\lambda} \phi$$

of $\mathcal{O}_\infty^*$. Moreover, the $C^\infty(X)$-ring $\mathcal{O}_\infty^0$ possesses the canonical connection

$$\nabla = dx^\lambda \otimes d_\lambda \tag{1.7.19}$$

in the sense of Definition 10.2.3 [112].

Chapter 2

Lagrangian field theory on fibre bundles

This Chapter addresses general formulation of Lagrangian theory of even fields on an arbitrary smooth manifold X, except the second Noether theorems which involve odd antifields and ghosts. For the sake of convenience, we call X a *world manifold*. Hereafter, it is assumed that $\dim X > 1$ because $\dim X = 1$ is the case of non-relativistic time-dependent mechanics [111; 137]. We consider Lagrangian field theory of finite order, but it is conventionally formulated in the framework of infinite order jet formalism. Section 2.4 is especially devoted to first order Lagrangian field theory because the basic classical field models are of this type.

2.1 Variational bicomplex

Let $Y \to X$ be a fibre bundle. The graded differential algebra $\mathcal{O}^*_\infty$ (1.7.9), decomposed into the variational bicomplex, describes finite order Lagrangian theories on $Y \to X$ [3; 17; 56; 59; 123; 157]. One also considers the variational bicomplex of the graded differential algebra $\mathcal{Q}^*_\infty$ (1.7.12) [4; 150] and different variants of the variational sequence of finite jet order [3; 97; 162; 163].

In order to transform the bicomplex $\mathcal{O}^{*,*}_\infty$ into the variational one, let us consider the following two operators acting on $\mathcal{O}^{*,n}_\infty$ [53; 157].

(i) There exists an $\mathbb{R}$-module endomorphism

$$\varrho = \sum_{k>0} \frac{1}{k} \overline{\varrho} \circ h_k \circ h^n : \mathcal{O}^{*>0,n}_\infty \to \mathcal{O}^{*>0,n}_\infty, \qquad (2.1.1)$$

$$\overline{\varrho}(\phi) = \sum_{0 \leq |\Lambda|} (-1)^{|\Lambda|} \theta^i \wedge [d_\Lambda(\partial^\Lambda_i \rfloor \phi)], \qquad \phi \in \mathcal{O}^{>0,n}_\infty,$$

possessing the following properties.

Lemma 2.1.1. *For any $\phi \in \mathcal{O}_\infty^{>0,n}$, the form $\phi - \varrho(\phi)$ is locally d_H-exact on each coordinate chart (1.7.3).*

Lemma 2.1.2. *The operator ϱ obeys the relation*

$$(\varrho \circ d_H)(\psi) = 0, \qquad \psi \in \mathcal{O}_\infty^{>0,n-1}. \tag{2.1.2}$$

It follows from Lemmas 2.1.1 and 2.1.2 that ϱ (2.1.1) is a projector, i.e., $\varrho \circ \varrho = \varrho$.

(ii) One defines the *variational operator*

$$\delta = \varrho \circ d : \mathcal{O}_\infty^{*,n} \to \mathcal{O}_\infty^{*+1,n}. \tag{2.1.3}$$

Lemma 2.1.3. *The variational operator δ (2.1.3) is nilpotent, i.e., $\delta \circ \delta = 0$, and it obeys the relation $\delta \circ \varrho = \delta$.*

Let us denote $\mathbf{E}_k = \varrho(\mathcal{O}_\infty^{k,n})$. Provided with the operators d_H, d_V, ϱ and δ, the differential graded algebra $\mathcal{O}_\infty^*$ is decomposed into the *variational bicomplex*

$$
\begin{array}{ccccccccccc}
& & \vdots & & \vdots & & \vdots & & & & \vdots \\
& & d_V \uparrow & & d_V \uparrow & & d_V \uparrow & & & & -\delta \uparrow \\
0 & \to & \mathcal{O}_\infty^{1,0} & \xrightarrow{d_H} & \mathcal{O}_\infty^{1,1} & \xrightarrow{d_H} & \cdots & \mathcal{O}_\infty^{1,n} & \xrightarrow{\varrho} & \mathbf{E}_1 & \to 0 \\
& & d_V \uparrow & & d_V \uparrow & & & d_V \uparrow & & & -\delta \uparrow \\
0 \to \mathbb{R} \to & & \mathcal{O}_\infty^0 & \xrightarrow{d_H} & \mathcal{O}_\infty^{0,1} & \xrightarrow{d_H} & \cdots & \mathcal{O}_\infty^{0,n} & \equiv & \mathcal{O}_\infty^{0,n} & \\
& & \uparrow & & \uparrow & & & \uparrow & & & \\
0 \to \mathbb{R} \to & & \mathcal{O}^0(X) & \xrightarrow{d} & \mathcal{O}^1(X) & \xrightarrow{d} & \cdots & \mathcal{O}^n(X) & \xrightarrow{d} & 0 & \\
& & \uparrow & & \uparrow & & & \uparrow & & & \\
& & 0 & & 0 & & & 0 & & &
\end{array}
$$
$$\tag{2.1.4}$$

It possesses the following cohomology [56; 139] (see Section 2.5 for the proof).

Theorem 2.1.1. *The second row from the bottom and the last column of the variational bicomplex (2.1.4) make up the variational complex*

$$0 \to \mathbb{R} \to \mathcal{O}_\infty^0 \xrightarrow{d_H} \mathcal{O}_\infty^{0,1} \cdots \xrightarrow{d_H} \mathcal{O}_\infty^{0,n} \xrightarrow{\delta} \mathbf{E}_1 \xrightarrow{\delta} \mathbf{E}_2 \longrightarrow \cdots. \tag{2.1.5}$$

Its cohomology is isomorphic to the de Rham cohomology of a fibre bundle Y, namely,

$$H^{k<n}(d_H; \mathcal{O}_\infty^*) = H_{\mathrm{DR}}^{k<n}(Y), \qquad H^{k\geq n}(\delta; \mathcal{O}_\infty^*) = H_{\mathrm{DR}}^{k\geq n}(Y). \tag{2.1.6}$$

2.1. Variational bicomplex

Theorem 2.1.2. *The rows of contact forms of the variational bicomplex (2.1.4) are exact sequences.*

Note that the cohomology isomorphism (2.1.6) gives something more. Due to the relations $d_H \circ h_0 = h_0 \circ d$ and $\delta \circ \varrho = \delta$, we have the cochain morphism

$$
\begin{array}{ccccccccc}
\cdots & \to & \mathcal{O}_\infty^{n-1} & \xrightarrow{d} & \mathcal{O}_\infty^n & \xrightarrow{d} & \mathcal{O}_\infty^{n+1} & \xrightarrow{d} & \mathcal{O}_\infty^{n+2} & \to \cdots \\
& & \downarrow h_0 & & \downarrow h_0 & & \downarrow \varrho & & \downarrow \varrho & \\
\cdots & \to & \mathcal{O}_\infty^{0,n-1} & \xrightarrow{d_H} & \mathcal{O}_\infty^{0,n} & \xrightarrow{\delta} & \mathbf{E}_1 & \xrightarrow{\delta} & \mathbf{E}_2 & \to \cdots
\end{array}
$$

of the de Rham complex (1.7.10) of the differential graded algebra $\mathcal{O}_\infty^*$ to its variational complex (2.1.5). By virtue of Theorems 1.7.2 and 2.1.1, the corresponding homomorphism of their cohomology groups is an isomorphism. Then the splitting of a closed form $\phi \in \mathcal{O}_\infty^*$ in Corollary 1.7.2 leads to the following decompositions.

Theorem 2.1.3. *Any d_H-closed form $\phi \in \mathcal{O}^{0,m}$, $m < n$, is represented by a sum*

$$\phi = h_0\sigma + d_H\xi, \qquad \xi \in \mathcal{O}_\infty^{m-1}, \tag{2.1.7}$$

where σ is a closed m-form on Y. Any δ-closed form $\phi \in \mathcal{O}^{k,n}$ is split into

$$\phi = h_0\sigma + d_H\xi, \qquad k = 0, \qquad \xi \in \mathcal{O}_\infty^{0,n-1}, \tag{2.1.8}$$
$$\phi = \varrho(\sigma) + \delta(\xi), \qquad k = 1, \qquad \xi \in \mathcal{O}_\infty^{0,n}, \tag{2.1.9}$$
$$\phi = \varrho(\sigma) + \delta(\xi), \qquad k > 1, \qquad \xi \in \mathbf{E}_{k-1}, \tag{2.1.10}$$

where σ is a closed $(n+k)$-form on Y.

In Lagrangian formalism on fibre bundles, a finite order *Lagrangian* and its *Euler–Lagrange operator* are defined as elements

$$L = \mathcal{L}\omega \in \mathcal{O}_\infty^{0,n}, \tag{2.1.11}$$
$$\delta L = \mathcal{E}_L = \mathcal{E}_i \theta^i \wedge \omega \in \mathbf{E}_1, \tag{2.1.12}$$
$$\mathcal{E}_i = \sum_{0 \leq |\Lambda|} (-1)^{|\Lambda|} d_\Lambda(\partial_i^\Lambda \mathcal{L}), \tag{2.1.13}$$

of the variational complex (2.1.5) (see the notation (1.1.33)). Components $\mathcal{E}_i$ (2.1.13) of the Euler–Lagrange operator (2.1.12) are called the *variational derivatives*. Elements of $\mathbf{E}_1$ are called the *Euler–Lagrange-type operators*.

Hereafter, we call a pair $(\mathcal{O}_\infty^*, L)$ the *Lagrangian system*. The following are corollaries of Theorem 2.1.3.

Corollary 2.1.1. *A finite order Lagrangian L (2.1.11) is variationally trivial, i.e., $\delta(L) = 0$ if and only if*

$$L = h_0\sigma + d_H\xi, \qquad \xi \in \mathcal{O}_\infty^{0,n-1}, \qquad (2.1.14)$$

where σ is a closed n-form on Y.

Corollary 2.1.2. *A finite order Euler–Lagrange-type operator $\mathcal{E} \in \mathbf{E}_1$ satisfies the Helmholtz condition $\delta(\mathcal{E}) = 0$ if and only if*

$$\mathcal{E} = \delta L + \varrho(\sigma), \qquad L \in \mathcal{O}_\infty^{0,n},$$

where σ is a closed $(n+1)$-form on Y.

Remark 2.1.1. Corollaries 2.1.1 and 2.1.2 provide a solution of the so called global inverse problem of the calculus of variations. This solution agrees with that of [3] obtained by computing cohomology of a variational sequence of bounded jet order, but without minimizing an order of a Lagrangian (see also particular results of [98; 161]). A solution of the global inverse problem of the calculus of variations in the case of a graded differential algebra $\mathcal{Q}_\infty^*$ (1.7.12) has been found in [4; 150] (see Theorem 2.5.1).

A glance at the expression (2.1.12) shows that, if a Lagrangian L (2.1.11) is of r-order, its Euler–Lagrange operator $\mathcal{E}_L$ is of $2r$-order. Its kernel Ker $\mathcal{E}_L \subset J^{2r}Y$ is called the *Euler–Lagrange equation*. It is locally given by the equalities

$$\mathcal{E}_i = \sum_{0 \leq |\Lambda|} (-1)^{|\Lambda|} d_\Lambda (\partial_i^\Lambda \mathcal{L}) = 0. \qquad (2.1.15)$$

However, it may happen that the Euler–Lagrange equation is not a differential equation in the strict sense of Definition 1.6.1 because Ker $\mathcal{E}_L$ need not be a closed subbundle of $J^{2r}Y \to X$.

Euler–Lagrange equations (2.1.15) traditionally came from the *variational formula*

$$dL = \delta L - d_H \Xi_L \qquad (2.1.16)$$

of the calculus of variations. In formalism of the variational bicomplex, this formula is a corollary of Theorem 2.1.2.

Corollary 2.1.3. *The exactness of the row of one-contact forms of the variational bicomplex (2.1.4) at the term $\mathcal{O}_\infty^{1,n}$ relative to the projector ϱ provides the $\mathbb{R}$-module decomposition*

$$\mathcal{O}_\infty^{1,n} = \mathbf{E}_1 \oplus d_H(\mathcal{O}_\infty^{1,n-1}).$$

In particular, any Lagrangian L admits the decomposition (2.1.16).

2.1. Variational bicomplex

Defined up to a d_H-closed term, a form $\Xi_L \in \mathcal{O}_\infty^n$ in the variational formula (2.1.16) reads

$$\Xi_L = L + [(\partial_i^\lambda \mathcal{L} - d_\mu F_i^{\mu\lambda})\theta^i + \sum_{s=1} F_i^{\lambda\nu_s\ldots\nu_1} \theta^i_{\nu_s\ldots\nu_1}] \wedge \omega_\lambda, \quad (2.1.17)$$

$$F_i^{\nu_k\ldots\nu_1} = \partial_i^{\nu_k\ldots\nu_1}\mathcal{L} - d_\mu F_i^{\mu\nu_k\ldots\nu_1} + \sigma_i^{\nu_k\ldots\nu_1}, \qquad k = 2, 3, \ldots,$$

where $\sigma_i^{\nu_k\ldots\nu_1}$ are local functions such that

$$\sigma_i^{(\nu_k\nu_{k-1})\ldots\nu_1} = 0.$$

It is readily observed that the form Ξ_L (2.1.17) possesses the following properties:
- $h_0(\Xi_L) = L$,
- $h_0(\vartheta \rfloor d\Xi_L) = \vartheta^i \mathcal{E}_i \omega$ for any derivation ϑ (1.7.17).

Consequently, Ξ_L is a Lepage equivalent of a Lagrangian L.

Remark 2.1.2. Following the terminology of finite order jet formalism [53; 65; 99], we call an exterior n-form $\rho \in \mathcal{O}_\infty^n$ the *Lepage form* if, for any derivation ϑ (1.7.17), the density $h_0(\vartheta \rfloor d\rho)$ depends only on the restriction of ϑ to a derivation $\vartheta^\lambda \partial_\lambda + \vartheta^i \partial_i$ of the subring $C^\infty(Y) \subset \mathcal{O}_\infty^0$. The Lepage forms constitute a real vector space. In particular, closed n-forms and $(2 \leq k)$-contact n-forms are Lepage forms. Given a Lagrangian L, a Lepage form ρ is called the *Lepage equivalent* of L if $h_0(\rho) = L$. Any Lepage form ρ is a Lepage equivalent of the Lagrangian $h_0(\rho)$. Conversely, any r-order Lagrangian possesses a Lepage equivalent of $(2r-1)$-order [65]. The Lepage equivalents of a Lagrangian L constitute an affine space modelled over a vector space of contact Lepage forms. In particular, one can locally put $\sigma_i^{\nu_k\ldots\nu_1} = 0$ in the formula (2.1.17).

Our special interest is concerned with Lagrangian theories on an affine bundle $Y \to X$. Since X is a strong deformation retract of an affine bundle Y, the de Rham cohomology of Y equals that of X. In this case, the cohomology (2.1.6) of the variational complex (2.1.5) equals the de Rham cohomology of X, namely,

$$H^{<n}(d_H; \mathcal{O}_\infty^*) = H_{DR}^{<n}(X),$$
$$H^n(\delta; \mathcal{O}_\infty^*) = H_{DR}^n(X), \qquad (2.1.18)$$
$$H^{>n}(\delta; \mathcal{O}_\infty^*) = 0.$$

It follows that every d_H-closed form $\phi \in \mathcal{O}_\infty^{0,m<n}$ is represented by the sum

$$\phi = \sigma + d_H \xi, \qquad \xi \in \mathcal{O}_\infty^{0,m-1}, \qquad (2.1.19)$$

where σ is a closed m-form on X. Similarly, any variationally trivial Lagrangian takes the form

$$L = \sigma + d_H \xi, \qquad \xi \in \mathcal{O}_\infty^{0,n-1}, \qquad (2.1.20)$$

where σ is an n-form on X.

In view of the cohomology isomorphism (2.1.18), if $Y \to X$ is an affine bundle, let us restrict our consideration to the *short variational complex*

$$0 \to \mathbb{R} \to \mathcal{O}_\infty^0 \xrightarrow{d_H} \mathcal{O}_\infty^{0,1} \cdots \xrightarrow{d_H} \mathcal{O}_\infty^{0,n} \xrightarrow{\delta} \mathbf{E}_1, \qquad (2.1.21)$$

whose non-trivial cohomology equals that of the variational complex (2.1.5). Let us consider a differential graded subalgebra $\mathcal{P}_\infty^* \subset \mathcal{O}_\infty^*$ of exterior forms whose coefficients are polynomials in jet coordinates y_Λ^i, $0 \leq |\Lambda|$, of the continuous bundle $J^\infty Y \to X$. This property is coordinate-independent due to the transition functions (1.7.3).

Theorem 2.1.4. *The cohomology of the short variational complex*

$$0 \to \mathbb{R} \to \mathcal{P}_\infty^0 \xrightarrow{d_H} \mathcal{P}_\infty^{0,1} \cdots \xrightarrow{d_H} \mathcal{P}_\infty^{0,n} \xrightarrow{\delta} 0 \qquad (2.1.22)$$

of the polynomial algebra $\mathcal{P}_\infty^*$ equals that of the complex (2.1.21), i.e., the de Rham cohomology of X [56] (see Section 2.5 for the proofs).

2.2 Lagrangian symmetries

Given a Lagrangian system $(\mathcal{O}_\infty^*, L)$, its *infinitesimal transformations* are defined to be contact derivations of the ring $\mathcal{O}_\infty^0$.

A derivation $\vartheta \in \mathfrak{d}\mathcal{O}_\infty^0$ (1.7.17) is called *contact* if the Lie derivative $\mathbf{L}_\vartheta$ preserves the ideal of contact forms of the differential graded algebra $\mathcal{O}_\infty^*$, i.e., the Lie derivative $\mathbf{L}_\vartheta$ of a contact form is a contact form.

Lemma 2.2.1. *A derivation ϑ (1.7.17) is contact if and only if it takes the form*

$$\vartheta = v^\lambda \partial_\lambda + v^i \partial_i + \sum_{0 < |\Lambda|} [d_\Lambda(v^i - y_\mu^i v^\mu) + y_{\mu+\Lambda}^i v^\mu] \partial_i^\Lambda. \qquad (2.2.1)$$

Proof. The expression (2.2.1) results from a direct computation similar to that of the first part of Bäcklund's theorem [81]. One can then justify that local functions (2.2.1) satisfy the transformation law (1.7.18). □

2.2. Lagrangian symmetries

A glance at the expression (1.5.8) enables one to regard a contact derivation ϑ (2.2.1) as an infinite order jet prolongation $\vartheta = J^\infty v$ of its restriction

$$v = v^\lambda \partial_\lambda + v^i \partial_i \qquad (2.2.2)$$

to the ring $C^\infty(Y)$. Since coefficients v^λ and v^i of v (2.2.2) depend generally on jet coordinates y^i_Λ, $0 < |\Lambda|$, one calls v (2.2.2) a *generalized vector field*. It can be represented as a section of some pull-back bundle

$$J^rY \underset{Y}{\times} TY \to J^rY.$$

A contact derivation ϑ (2.2.1) is called *projectable*, if the generalized vector field v (2.2.2) projects onto a vector field $v^\lambda \partial_\lambda$ on X, i.e., its components ϑ^λ depend only on coordinates on X.

Any contact derivation ϑ (2.2.1) admits the horizontal splitting

$$\vartheta = \vartheta_H + \vartheta_V = v^\lambda d_\lambda + [v^i_V \partial_i + \sum_{0<|\Lambda|} d_\Lambda v^i_V \partial^\Lambda_i], \qquad (2.2.3)$$

$$v = v_H + v_V = v^\lambda d_\lambda + (v^i - y^i_\mu v^\mu) \partial_i, \qquad (2.2.4)$$

relative to the canonical connection ∇ (1.7.19) on the $C^\infty(X)$-ring $\mathcal{O}^0_\infty$.

Lemma 2.2.2. *Any vertical contact derivation*

$$\vartheta = v^i \partial_i + \sum_{0<|\Lambda|} d_\Lambda v^i \partial^\Lambda_i \qquad (2.2.5)$$

obeys the relations

$$\vartheta \rfloor d_H \phi = -d_H(\vartheta \rfloor \phi), \qquad (2.2.6)$$

$$\mathbf{L}_\vartheta(d_H \phi) = d_H(\mathbf{L}_\vartheta \phi), \qquad \phi \in \mathcal{O}^*_\infty. \qquad (2.2.7)$$

Proof. It is easily justified that, if ϕ and ϕ' satisfy the relation (2.2.6), then $\phi \wedge \phi'$ does well. Then it suffices to prove the relation (2.2.6) when ϕ is a function and $\phi = \theta^i_\Lambda$. The result follows from the equalities

$$\vartheta \rfloor \theta^i_\Lambda = v^i_\Lambda, \quad d_H(v^i_\Lambda) = v^i_{\lambda+\Lambda} dx^\lambda, \quad d_H \theta^i_\Lambda = dx^\lambda \wedge \theta^i_{\lambda+\Lambda}, \qquad (2.2.8)$$

$$d_\lambda \circ v^i_\Lambda \partial^\Lambda_i = v^i_\Lambda \partial^\Lambda_i \circ d_\lambda. \qquad (2.2.9)$$

The relation (2.2.7) is a corollary of the equality (2.2.6). □

The global decomposition (2.1.16) leads to the following first variational formula (Theorem 2.2.1) and the first Noether theorem (Theorem 2.2.2).

Theorem 2.2.1. *Given a Lagrangian $L \in \mathcal{O}^{0,n}_\infty$, its Lie derivative $\mathbf{L}_v L$ along a contact derivation v (2.2.3) fulfils the first variational formula*

$$\mathbf{L}_\vartheta L = v_V \rfloor \delta L + d_H(h_0(\vartheta \rfloor \Xi_L)) + \mathcal{L} d_V(v_H \rfloor \omega), \qquad (2.2.10)$$

where Ξ_L is the Lepage equivalent (2.1.17) of L.

Proof. The formula (2.2.10) comes from the splitting (2.1.16) and the relations (2.2.6) as follows:

$$\mathbf{L}_\vartheta L = \vartheta\rfloor dL + d(\vartheta\rfloor L) = [\vartheta_V\rfloor dL - d_V\mathcal{L} \wedge v_H\rfloor\omega] + [d_H(v_H\rfloor L) +$$
$$d_V(\mathcal{L}v_H\rfloor\omega)] = \vartheta_V\rfloor dL + d_H(v_H\rfloor L) + \mathcal{L}d_V(v_H\rfloor\omega) =$$
$$v_V\rfloor\delta L - \vartheta_V\rfloor d_H\Xi_L + d_H(v_H\rfloor L) + \mathcal{L}d_V(v_H\rfloor\omega) =$$
$$v_V\rfloor\delta L + d_H(\vartheta_V\rfloor\Xi_L + v_H\rfloor L) + \mathcal{L}d_V(v_H\rfloor\omega),$$

where

$$v_V\rfloor\Xi_L = h_0(v_V\rfloor\Xi_L)$$

since $\Xi_L - L$ is a one-contact form and

$$v_H\rfloor L = h_0(v_H\rfloor\Xi_L). \qquad \square$$

A contact derivation ϑ (2.2.1) is called a *variational symmetry* of a Lagrangian L if the Lie derivative $\mathbf{L}_\vartheta L$ is d_H-exact, i.e.,

$$\mathbf{L}_\vartheta L = d_H\sigma. \qquad (2.2.11)$$

Lemma 2.2.3. *A glance at the expression (2.2.10) shows the following.*

(i) A contact derivation ϑ is a variational symmetry only if it is projectable.

(ii) Any projectable contact derivation is a variational symmetry of a variationally trivial Lagrangian. It follows that, if ϑ is a variational symmetry of a Lagrangian L, it also is a variational symmetry of a Lagrangian $L + L_0$, where L_0 is a variationally trivial Lagrangian.

(iii) A projectable contact derivations ϑ is a variational symmetry if and only if its vertical part v_V (2.2.3) is well.

(iv) A projectable contact derivations ϑ is a variational symmetry if and only if the density $v_V\rfloor\delta L$ is d_H-exact.

It is readily observed that variational symmetries of a Lagrangian L constitute a real vector subspace $\mathcal{G}_L$ of the derivation module $\mathfrak{d}\mathcal{O}^0_\infty$. By virtue of item (ii) of Lemma 2.2.3, the Lie bracket

$$\mathbf{L}_{[\vartheta,\vartheta']} = [\mathbf{L}_\vartheta, \mathbf{L}_{\vartheta'}]$$

of variational symmetries is a variational symmetry and, therefore, their vector space $\mathcal{G}_L$ is a real Lie algebra.

The following is the first Noether theorem.

Theorem 2.2.2. *If a contact derivation ϑ (2.2.1) is a variational symmetry (2.2.11) of a Lagrangian L, the first variational formula (2.2.10)*

2.2. Lagrangian symmetries

restricted to the kernel of the Euler–Lagrange operator $\operatorname{Ker} \mathcal{E}_L$ leads to the weak conservation law

$$0 \approx d_H(h_0(\vartheta \rfloor \Xi_L) - \sigma) \qquad (2.2.12)$$

on the shell $\delta L = 0$.

A variational symmetry ϑ of a Lagrangian L is called its *exact symmetry* or, simply, a *symmetry* if

$$\mathbf{L}_\vartheta L = 0. \qquad (2.2.13)$$

Symmetries of a Lagrangian L constitute a real vector space, which is a real Lie algebra. However, its vertical symmetries v (2.2.5) obey the relation

$$\mathbf{L}_v L = v \rfloor dL$$

and, therefore, make up a $\mathcal{O}_\infty^0$-module which is a Lie $C^\infty(X)$-algebra.

If ϑ is an exact symmetry of a Lagrangian L, the weak conservation law (2.2.12) takes the form

$$0 \approx d_H(h_0(\vartheta \rfloor \Xi_L)) = -d_H \mathcal{J}_v, \qquad (2.2.14)$$

where

$$\mathcal{J}_v = \mathcal{J}_v^\mu \omega_\mu = -h_0(\vartheta \rfloor \Xi_L) \qquad (2.2.15)$$

is called the *symmetry current*. Of course, the symmetry current (2.2.15) is defined with the accuracy of a d_H-closed term. Therefore, a Lagrangian symmetry ϑ fails to define a unique conserved current, but $\mathcal{J}_v$ (2.2.15) is surely conserved.

Let ϑ be an exact symmetry of a Lagrangian L. Whenever L_0 is a variationally trivial Lagrangian, ϑ is a variational symmetry of the Lagrangian $L + L_0$ such that the weak conservation law (2.2.12) for this Lagrangian is reduced to the weak conservation law (2.2.14) for a Lagrangian L as follows:

$$\mathbf{L}_\vartheta(L + L_0) = d_H \sigma \approx d_H \sigma - d_H \mathcal{J}_v.$$

Remark 2.2.1. In accordance with the standard terminology, variational and exact symmetries generated by generalized vector fields (2.2.2) are called *generalized symmetries* because they depend on derivatives of variables. Generalized symmetries of differential equations and Lagrangian systems have been intensively investigated [24; 40; 59; 81; 96; 123]. Accordingly, by variational symmetries and symmetries one means only those generated by vector fields u on Y. We agree to call them *classical symmetries*.

Let ϑ be a classical variational symmetry of a Lagrangian L, i.e., ϑ (2.2.1) is the jet prolongation of a vector field u on Y. Then the relation

$$\mathbf{L}_\vartheta \mathcal{E}_L = \delta(\mathbf{L}_\vartheta L) \qquad (2.2.16)$$

holds [53; 123]. It follows that ϑ also is a symmetry of the Euler–Lagrange operator $\mathcal{E}_L$ of L, i.e.,

$$\mathbf{L}_\vartheta \mathcal{E}_L = 0.$$

However, the equality (2.2.16) fails to be true in the case of generalized symmetries.

2.3 Gauge symmetries

Treating gauge symmetries of Lagrangian field theory, one is traditionally based on an example of the Yang–Mills gauge theory of principal connections on a principal bundle (see Section 5.8). This notion of gauge symmetries is generalized to Lagrangian field theory on an arbitrary fibre bundle $Y \to X$ as follows [12; 13].

Definition 2.3.1. Let $E \to X$ be a vector bundle and $E(X)$ the $C^\infty(X)$ module $E(X)$ of sections of $E \to X$. Let ζ be a linear differential operator on $E(X)$ taking values into the vector space $\mathcal{G}_L$ of variational symmetries of a Lagrangian L (see Definition 10.2.1). Elements

$$u_\xi = \zeta(\xi) \qquad (2.3.1)$$

of $\mathrm{Im}\,\zeta$ are called the *gauge symmetry* of a Lagrangian L parameterized by sections ξ of $E \to X$. They are called the *gauge parameters*.

Remark 2.3.1. The differential operator ζ in Definition 2.3.1 takes its values into the vector space $\mathcal{G}_L$ as a subspace of the $C^\infty(X)$-module $\mathfrak{dO}^0_\infty$, but it sends the $C^\infty(X)$-module $E(X)$ into the real vector space $\mathcal{G}_L \subset \mathfrak{dO}^0_\infty$. The differential operator ζ is assumed to be at least of first order (see Remark 2.3.2).

Equivalently, the gauge symmetry (2.3.1) is given by a section $\widetilde{\zeta}$ of the fibre bundle

$$(J^r Y \underset{Y}{\times} J^m E) \underset{Y}{\times} TY \to J^r Y \underset{Y}{\times} J^m E$$

(see Definition 1.6.2) such that

$$u_\xi = \zeta(\xi) = \widetilde{\zeta} \circ \xi$$

2.3. Gauge symmetries

for any section ξ of $E \to X$. Hence, it is a generalized vector field u_ζ on the product $Y \times E$ represented by a section of the pull-back bundle

$$J^k(Y \underset{X}{\times} E) \underset{Y}{\times} T(Y \underset{X}{\times} E) \to J^k(Y \underset{X}{\times} E), \qquad k = \max(r, m),$$

which lives in

$$TY \subset T(Y \times E).$$

This generalized vector field yields a contact derivation $J^\infty u_\zeta$ (2.2.1) of the real ring $\mathcal{O}^0_\infty[Y \times E]$ which obeys the following condition.

Condition 2.3.1. Given a Lagrangian

$$L \in \mathcal{O}^{0,n}_\infty E \subset \mathcal{O}^{0,n}_\infty[Y \times E],$$

let us consider its Lie derivative

$$\mathbf{L}_{J^\infty u_\zeta} L = J^\infty u_\zeta \rfloor dL + d(J^\infty u_\zeta \rfloor L) \qquad (2.3.2)$$

where d is the exterior differential of $\mathcal{O}^0_\infty[Y \times E]$. Then, for any section ξ of $E \to X$, the pull-back $\xi^* \mathbf{L}_\vartheta$ is d_H-exact.

It follows at once from the first variational formula (2.2.10) for the Lie derivative (2.3.2) that Condition 2.3.1 holds only if u_ζ projects onto a generalized vector field on E and, in this case, if and only if the density $(u_\zeta)_V \rfloor \mathcal{E}$ is d_H-exact. Thus, we come to the following equivalent definition of gauge symmetries.

Definition 2.3.2. Let $E \to X$ be a vector bundle. A gauge symmetry of a Lagrangian L parameterized by sections ξ of $E \to X$ is defined as a contact derivation $\vartheta = J^\infty u$ of the real ring $\mathcal{O}^0_\infty[Y \times E]$ such that:

(i) it vanishes on the subring $\mathcal{O}^0_\infty E$,

(ii) the generalized vector field u is linear in coordinates χ^a_Λ on $J^\infty E$, and it projects onto a generalized vector field on E, i.e., it takes the form

$$u = \left(\sum_{0 \leq |\Lambda| \leq m} u_a^{\lambda \Lambda}(x^\mu) \chi^a_\Lambda \right) \partial_\lambda + \left(\sum_{0 \leq |\Lambda| \leq m} u_a^{i\Lambda}(x^\mu, y^j_\Sigma) \chi^a_\Lambda \right) \partial_i, \qquad (2.3.3)$$

(iii) the vertical part of u (2.3.3) obeys the equality

$$u_V \rfloor \mathcal{E} = d_H \sigma. \qquad (2.3.4)$$

For the sake of convenience, we also call a generalized vector field (2.3.3) the gauge symmetry. By virtue of item (iii) of Definition 2.3.2, u (2.3.3) is a gauge symmetry if and only if its vertical part is so.

Gauge symmetries possess the following particular properties.

(i) Let $E' \to X$ be another vector bundle and ζ' a linear $E(X)$-valued differential operator on the $C^\infty(X)$-module $E'(X)$ of sections of $E' \to X$. Then

$$u_{\zeta'(\xi')} = (\zeta \circ \zeta')(\xi')$$

also is a gauge symmetry of L parameterized by sections ξ' of $E' \to X$. It factorizes through the gauge symmetries u_ϕ (2.3.1).

(ii) If a gauge symmetry is an exact Lagrangian symmetry, the corresponding conserved symmetry current $\mathcal{J}_u$ (2.2.15) is reduced to a superpotential (see Theorems 2.4.2 and 4.2.3).

(iii) The *direct second Noether theorem* associates to a gauge symmetry of a Lagrangian L the Noether identities of its Euler–Lagrange operator δL.

Theorem 2.3.1. *Let u (2.3.3) be a gauge symmetry of a Lagrangian L, then its Euler–Lagrange operator δL obeys the Noether identities (2.3.5).*

Proof. The density (2.3.4) is variationally trivial and, therefore, its variational derivatives with respect to variables χ^a vanish, i.e.,

$$\mathcal{E}_a = \sum_{0 \leq |\Lambda|} (-1)^{|\Lambda|} d_\Lambda [(u_a^{i\Lambda} - y_\lambda^i u_a^{\lambda\Lambda})\mathcal{E}_i] = \qquad (2.3.5)$$

$$\sum_{0 \leq |\Lambda|} \eta(u_a^i - y_\lambda^i u_a^\lambda)^\Lambda d_\Lambda \mathcal{E}_i = 0$$

(see Notation 4.2.2). In accordance with Definition 4.5.1, the equalities (2.3.5) are the Noether identities for the Euler–Lagrange operator δL. □

For instance, if the gauge symmetry u (2.3.3) is of second jet order in gauge parameters, i.e.,

$$u_V = (u_a^i \chi^a + u_a^{i\mu} \chi_\mu^a + u_a^{i\nu\mu} \chi_{\nu\mu}^a)\partial_i, \qquad (2.3.6)$$

the corresponding Noether identities (2.3.5) take the form

$$u_a^i \mathcal{E}_i - d_\mu(u_a^{i\mu} \mathcal{E}_i) + d_{\nu\mu}(u_a^{i\nu\mu} \mathcal{E}_i) = 0, \qquad (2.3.7)$$

and *vice versa*.

Remark 2.3.2. A glance at the expression (2.3.7) shows that, if a gauge symmetry is independent of derivatives of gauge parameters (i.e., the differential operator ζ in Definition 2.3.1 is of zero order), then all variational derivatives of a Lagrangian equals zero, i.e., this Lagrangian is variationally trivial. Therefore, such gauge symmetries usually are not considered (see Example 4.3.1).

Remark 2.3.3. The notion of gauge symmetries can be generalized as follows. Let a differential operator ζ in Definition 2.3.1 need not be linear. Then elements of $\operatorname{Im}\zeta$ are called a *generalized gauge symmetry*. However, direct second Noether Theorem 2.3.1 is not relevant to generalized gauge symmetries because, in this case, an Euler–Lagrange operator satisfies the identities depending on gauge parameters.

It follows from direct second Noether Theorem 2.3.1 that gauge symmetries of Lagrangian field theory characterize its degeneracy. A problem is that any Lagrangian possesses gauge symmetries and, therefore, one must separate them into the trivial and non-trivial ones. Moreover, gauge symmetries can be *reducible*, i.e., $\operatorname{Ker}\zeta \neq 0$. Another problem is that gauge symmetries need not form an algebra [48; 61; 63]. The Lie bracket $[u_\phi, u_{\phi'}]$ of gauge symmetries $u_\phi, u_{\phi'} \in \operatorname{Im}\zeta$ is a variational symmetry, but it need not belong to $\operatorname{Im}\zeta$.

To solve these problems, we follow a different definition of gauge symmetries as those associated to non-trivial Noether identities by means of inverse second Noether Theorem 4.2.1.

2.4 First order Lagrangian field theory

In first order Lagrangian field theory on a fibre bundle $Y \to X$, a *first order Lagrangian*

$$L = \mathcal{L}\omega : J^1 Y \to \overset{n}{\wedge} T^* X \tag{2.4.1}$$

is defined on the first order jet manifold $J^1 Y$, called the *configuration space*. The corresponding *second-order Euler–Lagrange operator* (2.1.12) reads

$$\mathcal{E}_L : J^2 Y \to T^* Y \wedge (\overset{n}{\wedge} T^* X),$$
$$\mathcal{E}_L = (\partial_i \mathcal{L} - d_\lambda \pi_i^\lambda)\theta^i \wedge \omega, \qquad \pi_i^\lambda = \partial_i^\lambda \mathcal{L}. \tag{2.4.2}$$

Its kernel defines the *second order Euler–Lagrange equation*

$$(\partial_i - d_\lambda \partial_i^\lambda)\mathcal{L} = 0. \tag{2.4.3}$$

Remark 2.4.1. Given a Lagrangian L, a holonomic second order connection $\widetilde{\Gamma}$ (1.3.51) on a fibre bundle $Y \to X$ is called a *Lagrangian connection* if it takes its values into the kernel of the Euler–Lagrange operator $\mathcal{E}_L$, i.e., if it satisfies the equation

$$\partial_i \mathcal{L} - \partial_\lambda \pi_i^\lambda - y_\lambda^j \partial_j \pi_i^\lambda - \widetilde{\Gamma}_{\lambda\mu}^j \partial_j^\mu \pi_i^\lambda = 0. \tag{2.4.4}$$

If a Lagrangian connection $\widehat{\Gamma}$ exists, it defines the second order dynamic equation

$$y_{\lambda\mu}^i = \widetilde{\Gamma}_{\lambda\mu}^i(x^\nu, y^j, y_\nu^j) \tag{2.4.5}$$

on $Y \to X$, whose solutions also are solutions of the Euler–Lagrange equation (2.4.3). Conversely, since the jet bundle $J^2Y \to J^1Y$ is affine, every solution s of the Euler–Lagrange equation also is an integral section of a holonomic second order connection $\widetilde{\Gamma}$ which is the global extension of the local section $J^1s(X) \to J^2s(X)$ of this jet bundle over the closed imbedded submanifold $J^1s(X) \subset J^1Y$. Hence, every solution of the Euler–Lagrange equations also is a solution of some second order dynamic equation, but it is not necessarily a Lagrangian connection.

Given a Lagrangian L, let us consider the vertical tangent map VL (1.1.15) to L (2.4.1). Since $J^1Y \to Y$ is an affine bundle, VL yields the linear morphism

$$J^1Y \underset{Y}{\times} (T^*X \otimes VY) \longrightarrow J^1Y \underset{Y}{\times} (\overset{n}{\wedge} T^*X)$$

over J^1Y and the corresponding morphism

$$\widehat{L} : J^1Y \to V^*Y \underset{Y}{\otimes} (\overset{n}{\wedge} T^*X) \underset{Y}{\otimes} TX \tag{2.4.6}$$

over Y. It is called the *Legendre map* associated to a Lagrangian L. The fibre bundle

$$\Pi = V^*Y \underset{Y}{\otimes} (\overset{n}{\wedge} T^*X) \underset{Y}{\otimes} TX = V^*Y \underset{Y}{\wedge} (\overset{n-1}{\wedge} T^*X) \tag{2.4.7}$$

over Y is called the *Legendre bundle*. It is provided with the holonomic coordinates $(x^\lambda, y^i, p_i^\lambda)$, where the fibre coordinates p_i^λ have the transition functions

$$p'^\lambda_i = \det\left(\frac{\partial x^\varepsilon}{\partial x'^\nu}\right) \frac{\partial y^j}{\partial y'^i} \frac{\partial x'^\lambda}{\partial x^\mu} p_j^\mu. \tag{2.4.8}$$

With respect to these coordinates, the Legendre map (2.4.6) reads

$$p_i^\lambda \circ \widehat{L} = \pi_i^\lambda. \tag{2.4.9}$$

Its range $N_L = \widehat{L}(J^1Y)$ is called the *Lagrangian constraint space*.

2.4. First order Lagrangian field theory

Definition 2.4.1. A Lagrangian L is said to be:
- *hyperregular* if the Legendre map $\widehat{L}$ is a diffeomorphism;
- *regular* if $\widehat{L}$ is a local diffeomorphism, i.e., $\det(\partial_i^\mu \partial_j^\nu \mathcal{L}) \neq 0$;
- *semiregular* if the inverse image $\widehat{L}^{-1}(q)$ of any point $q \in N_L$ is a connected submanifold of J^1Y;
- *almost regular* if the Lagrangian constraint space N_L is a closed imbedded subbundle

$$i_N : N_L \to \Pi$$

of the Legendre bundle $\Pi \to Y$ and the Legendre map

$$\widehat{L} : J^1 Y \to N_L \qquad (2.4.10)$$

is a fibred manifold with connected fibres (i.e., a Lagrangian is semiregular).

Remark 2.4.2. A glance at the equation (2.4.4) shows that a regular Lagrangian L admits a unique Lagrangian connection. In this case, the Euler–Lagrange equation for L is equivalent to the second order dynamic equation associated to this Lagrangian connection.

2.4.1 Cartan and Hamilton–De Donder equations

Given a first order Lagrangian L, its Lepage equivalents Ξ_L (2.1.17) in the variational formula (2.1.16) read

$$\Xi_L = L + (\pi_i^\lambda - d_\mu \sigma_i^{\mu\lambda})\theta^i \wedge \omega_\lambda + \sigma_i^{\lambda\mu}\theta_\mu^i \wedge \omega_\lambda, \qquad (2.4.11)$$

where $\sigma_i^{\mu\lambda} = -\sigma_i^{\lambda\mu}$ are skew-symmetric local functions on Y. These Lepage equivalents constitute an affine space modelled over a vector space of d_H-exact one-contact Lepage forms

$$\rho = -d_\mu \sigma_i^{\mu\lambda} \theta^i \wedge \omega_\lambda + \sigma_i^{\lambda\mu}\theta_\mu^i \wedge \omega_\lambda.$$

Let us choose the *Poincaré–Cartan form*

$$H_L = \mathcal{L}\omega + \pi_i^\lambda \theta^i \wedge \omega_\lambda \qquad (2.4.12)$$

as the origin of this affine space because it is defined on J^1Y. In a general setting, one also considers other Lepage equivalents of L [100; 101].

The Poincaré–Cartan form (2.4.12) takes its values into the subbundle

$$J^1 Y \underset{Y}{\times} (T^*Y \underset{Y}{\wedge} (\overset{n-1}{\wedge} T^*X))$$

of $\overset{n}{\wedge} T^* J^1 Y$. Hence, it defines a morphism

$$\widehat{H}_L : J^1 Y \to Z_Y = T^*Y \wedge (\overset{n-1}{\wedge} T^*X), \qquad (2.4.13)$$

whose range
$$Z_L = \widehat{H}_L(J^1Y) \qquad (2.4.14)$$
is an imbedded subbundle $i_L : Z_L \to Z_Y$ of the fibre bundle $Z_Y \to Y$. This morphism is called the *homogeneous Legendre map*. Accordingly, the fibre bundle $Z_Y \to Y$ (2.4.13) is said to be the *homogeneous Legendre bundle*. It is equipped with holonomic coordinates $(x^\lambda, y^i, p_i^\lambda, p)$ possessing the transition functions

$$p'^\lambda_i = \det\left(\frac{\partial x^\varepsilon}{\partial x'^\nu}\right) \frac{\partial y^j}{\partial y'^i} \frac{\partial x'^\lambda}{\partial x^\mu} p_j^\mu, \qquad (2.4.15)$$

$$p' = \det\left(\frac{\partial x^\varepsilon}{\partial x'^\nu}\right) \left(p - \frac{\partial y^j}{\partial y'^i} \frac{\partial y'^i}{\partial x^\mu} p_j^\mu\right).$$

With respect to these coordinates, the morphism $\widehat{H}_L$ (2.4.13) reads

$$(p_i^\mu, p) \circ \widehat{H}_L = (\pi_i^\mu, \mathcal{L} - y_\mu^i \pi_i^\mu).$$

A glance at the transition functions (2.4.15) shows that Z_Y (2.4.13) is a one-dimensional affine bundle

$$\pi_{Z\Pi} : Z_Y \to \Pi \qquad (2.4.16)$$

over the Legendre bundle Π (2.4.7). Moreover, the Legendre map $\widehat{L}$ (2.4.6) is exactly the composition of morphisms

$$\widehat{L} = \pi_{Z\Pi} \circ H_L : J^1Y \underset{Y}{\to} \Pi. \qquad (2.4.17)$$

Being a Lepage equivalent of L, the Poincaré–Cartan form H_L (2.4.12) also is a Lepage equivalent of the first order Lagrangian

$$\overline{L} = \widehat{h}_0(H_L) = (\mathcal{L} + (\widehat{y}_\lambda^i - y_\lambda^i)\pi_i^\lambda)\omega, \qquad (2.4.18)$$
$$\widehat{h}_0(dy^i) = \widehat{y}_\lambda^i dx^\lambda,$$

on the repeated jet manifold J^1J^1Y, coordinated by

$$(x^\lambda, y^i, y_\lambda^i, \widehat{y}_\mu^i, y_{\mu\lambda}^i).$$

The Euler–Lagrange operator for $\overline{L}$ (called the *Euler–Lagrange–Cartan operator*) reads

$$\mathcal{E}_{\overline{L}} : J^1J^1Y \to T^*J^1Y \wedge (\overset{n}{\wedge} T^*X),$$
$$\mathcal{E}_{\overline{L}} = [(\partial_i\mathcal{L} - \widehat{d}_\lambda\pi_i^\lambda + \partial_i\pi_j^\lambda(\widehat{y}_\lambda^j - y_\lambda^j))dy^i + \qquad (2.4.19)$$
$$\partial_i^\lambda \pi_j^\mu(\widehat{y}_\mu^j - y_\mu^j)dy_\lambda^i] \wedge \omega,$$
$$\widehat{d}_\lambda = \partial_\lambda + \widehat{y}_\lambda^i\partial_i + y_{\lambda\mu}^i\partial_i^\mu.$$

2.4. First order Lagrangian field theory

Its kernel $\operatorname{Ker} \mathcal{E}_{\overline{L}} \subset J^1 J^1 Y$ defines the *Cartan equation*

$$\partial_i^\lambda \pi_j^\mu (\widehat{y}_\mu^j - y_\mu^j) = 0, \qquad (2.4.20)$$

$$\partial_i \mathcal{L} - \widehat{d}_\lambda \pi_i^\lambda + (\widehat{y}_\lambda^j - y_\lambda^j) \partial_i \pi_j^\lambda = 0 \qquad (2.4.21)$$

on $J^1 Y$. Since

$$\mathcal{E}_{\overline{L}}|_{J^2 Y} = \mathcal{E}_L,$$

the Cartan equation (2.4.20) – (2.4.21) is equivalent to the Euler–Lagrange equation (2.4.3) on integrable sections of $J^1 Y \to X$. These equations are equivalent if a Lagrangian is regular. The Cartan equation (2.4.20) – (2.4.21) on sections

$$\overline{s} : X \to J^1 Y$$

is equivalent to the relation

$$\overline{s}^*(u \rfloor dH_L) = 0, \qquad (2.4.22)$$

which is assumed to hold for all vertical vector fields u on $J^1 Y \to X$.

The homogeneous Legendre bundle Z_Y (2.4.13) admits the canonical *multisymplectic Liouville form*

$$\Xi_Y = p\omega + p_i^\lambda dy^i \wedge \omega_\lambda. \qquad (2.4.23)$$

Accordingly, its imbedded subbundle Z_L (2.4.14) is provided with the pullback *De Donder form* $\Xi_L = i_L^* \Xi_Y$. There is the equality

$$H_L = \widehat{H}_L^* \Xi_L = \widehat{H}_L^*(i_L^* \Xi_Y). \qquad (2.4.24)$$

By analogy with the Cartan equation (2.4.22), the *Hamilton–De Donder equation* for sections $\overline{r}$ of $Z_L \to X$ is written as

$$\overline{r}^*(u \rfloor d\Xi_L) = 0, \qquad (2.4.25)$$

where u is an arbitrary vertical vector field on $Z_L \to X$.

Theorem 2.4.1. *Let the homogeneous Legendre map $\widehat{H}_L$ be a submersion. Then a section $\overline{s}$ of $J^1 Y \to X$ is a solution of the Cartan equation (2.4.22) if and only if $\widehat{H}_L \circ \overline{s}$ is a solution of the Hamilton–De Donder equation (2.4.25), i.e., the Cartan and Hamilton–De Donder equations are quasi-equivalent [65].*

Remark 2.4.3. The Legendre bundle Π (2.4.7) and the homogeneous Legendre bundle Z_Y (2.4.13) play the role of a momentum phase space and homogeneous momentum phase space in polysymplectic and multisymplectic Hamiltonian field theory, respectively (see Chapter 9).

2.4.2 Lagrangian conservation laws

We restrict our study of symmetries of first order Lagrangian field theory to classical symmetries, generated by projectable vector fields on a fibre bundle $Y \to X$. This is the case of all basic classical field models.

Let
$$u = u^\lambda \partial_\lambda + u^i \partial_i$$
be a projectable vector field on a fibre bundle $Y \to X$. Its canonical decomposition (1.5.18) into the horizontal and vertical parts over $J^1 Y$ reads
$$u = u_H + u_V = (u^\lambda \partial_\lambda + y^i_\lambda \partial_i^\lambda) + (u^i \partial_i - y^i_\lambda \partial_i^\lambda). \tag{2.4.26}$$

Its first order jet prolongation (1.5.8) onto $J^1 Y$ is
$$J^1 u = u^\lambda \partial_\lambda + u^i \partial_i + (d_\lambda u^i - y^i_\mu d_\lambda u^\mu) \partial_i^\lambda. \tag{2.4.27}$$

Given a first order Lagrangian L, the first variational formula (2.2.10) takes the form
$$\mathbf{L}_{J^1 u} L = u_V \rfloor \mathcal{E}_L + d_H (h_0(u \rfloor H_L)), \tag{2.4.28}$$
where $\Xi_L = H_L$ is the Poincaré–Cartan form (2.4.12). Its coordinate expression reads
$$\partial_\lambda u^\lambda \mathcal{L} + [u^\lambda \partial_\lambda + u^i \partial_i + (d_\lambda u^i - y^i_\mu \partial_\lambda u^\mu) \partial_i^\lambda] \mathcal{L} = \tag{2.4.29}$$
$$(u^i - y^i_\lambda u^\lambda) \mathcal{E}_i - d_\lambda [\pi_i^\lambda (u^\mu y^i_\mu - u^i) - u^\lambda \mathcal{L}].$$

If u is an exact symmetry of L, we obtain the weak conservation law
$$0 \approx -d_\lambda [\pi_i^\lambda (u^\mu y^i_\mu - u^i) - u^\lambda \mathcal{L}] \tag{2.4.30}$$
(2.2.14) of the symmetry current
$$\mathcal{J}_u = [\pi_i^\lambda (u^\mu y^i_\mu - u^i) - u^\lambda \mathcal{L}] \omega_\lambda \tag{2.4.31}$$
(2.2.15) along a vector field u.

The weak conservation law (2.4.30) leads to the *differential conservation law*
$$\partial_\lambda (\mathcal{J}^\lambda \circ s) = 0$$
on solutions s of the Euler–Lagrange equation (2.4.3). This differential conservation law yields the *integral conservation law*
$$\int_{\partial N} s^* \mathcal{J}_u = 0, \tag{2.4.32}$$

2.4. First order Lagrangian field theory

where N is an n-dimensional oriented compact submanifold of X with a boundary ∂N.

Remark 2.4.4. If we choose a different Lepage equivalent Ξ_L (2.4.11) in the first variational formula, the corresponding symmetry current differs from $\mathcal{J}_u$ (2.4.31) in the d_H-exact term

$$d_\mu(\sigma_i^{\mu\lambda}(u^i - y_\nu^i u^\nu))\omega_\lambda.$$

This term is independent of a Lagrangian, and it does not contribute to the integral conservation law (2.4.32).

It is readily observed that the symmetry current $\mathcal{J}_u$ (2.4.31) is linear in a vector field u. Therefore, one can consider a superposition of symmetry currents

$$\mathcal{J}_u + \mathcal{J}_{u'} = \mathcal{J}_{u+u'}, \qquad \mathcal{J}_{cu} = c\mathcal{J}_u, \qquad c \in \mathbb{R},$$

and a superposition of weak conservation laws (2.4.30) associated to different symmetries u.

For instance, let $u = u^i \partial_i$ be a vertical vector field on $Y \to X$. If it is a symmetry of L, the weak conservation law (2.4.30) takes the form

$$0 \approx -d_\lambda(\pi_i^\lambda u^i). \tag{2.4.33}$$

It is called the *Noether conservation law* of of the *Noether current*

$$\mathcal{J}^\lambda = -\pi_i^\lambda u^i. \tag{2.4.34}$$

Given a connection Γ (1.3.4) on a fibre bundle $Y \to X$, a vector field τ on X gives rise to the projectable vector field $\Gamma\tau$ (1.3.6) on Y. The corresponding symmetry current (2.4.31) along $\Gamma\tau$ reads

$$\mathcal{J}_\Gamma^\lambda = \tau^\mu \mathcal{J}_\Gamma{}^\lambda{}_\mu = \tau^\mu(\pi_i^\lambda(y_\mu^i - \Gamma_\mu^i) - \delta_\mu^\lambda \mathcal{L}). \tag{2.4.35}$$

Its coefficients $\mathcal{J}_\Gamma{}^\lambda{}_\mu$ are components of the tensor field

$$\mathcal{J}_\Gamma = \mathcal{J}_\Gamma{}^\lambda{}_\mu dx^\mu \otimes \omega_\lambda, \tag{2.4.36}$$
$$\mathcal{J}_\Gamma{}^\lambda{}_\mu = \pi_i^\lambda(y_\mu^i - \Gamma_\mu^i) - \delta_\mu^\lambda \mathcal{L},$$

called the energy-momentum tensor relative to a connection Γ [43; 112; 135]. If $\Gamma\tau$ (1.3.6) is a symmetry of a Lagrangian L, we have the energy-momentum conservation law

$$0 \approx -d_\lambda[\pi_i^\lambda \tau^\mu(y_\mu^i - \Gamma_\mu^i) - \delta_\mu^\lambda \tau^\mu \mathcal{L}]. \tag{2.4.37}$$

For instance, let a fibre bundle $Y \to X$ admit a flat connection Γ. By virtue of Theorem 1.3.4, there exist bundle coordinates such that $\Gamma^i_\lambda = 0$, and the corresponding energy-momentum tensor (2.4.36) takes the form

$$\mathcal{J}_0{}^\lambda{}_\mu = \pi^\lambda_i y^i_\mu - \delta^\lambda_\mu \mathcal{L}$$

of the familiar *canonical energy–momentum tensor*. It obeys the first variational formula (2.4.29) which leads to the weak transformation law

$$\partial_\mu \mathcal{L} \approx -d_\lambda \mathcal{J}_0{}^\lambda{}_\mu. \qquad (2.4.38)$$

In a general setting, let

$$\gamma : \mathcal{T}(X) \to \mathcal{T}(Y)$$

be an $\mathbb{R}$-linear module morphism which sends a vector field τ on X onto a vector field

$$\gamma\tau = \tau^\lambda \partial_\lambda + (\gamma\tau)^i \partial_i \qquad (2.4.39)$$

on Y projected onto τ. Then we agree to call the symmetry current

$$\mathcal{J}_{\gamma\tau} = \pi^\lambda_i(\tau^\mu y^i_\mu - (\gamma\tau)^i) - \tau^\lambda \mathcal{L} \qquad (2.4.40)$$

along $\gamma\tau$ (2.4.39) the *energy-momentum current*. For instance, this is the case both of the above mentioned horizontal lift by means of a connection and a functorial lift on natural bundles (see Section 6.6), e.g., the canonical lift (1.1.26) on a tensor bundle.

2.4.3 Gauge conservation laws. Superpotential

If a Lagrangian L admits a gauge symmetry u (2.3.3) and if this is an exact symmetry of L, i.e., $\mathbf{L}_{J^1 u} L = 0$, the weak conservation law (2.4.30) of the corresponding symmetry current $\mathcal{J}_u$ (2.4.31) holds. We call it the *gauge conservation law*. Because gauge symmetries depend on derivatives of gauge parameters, all gauge conservation laws in first order Lagrangian field theory possess the following peculiarity.

Theorem 2.4.2. *If u (2.3.3) is an exact gauge symmetry of a first order Lagrangian L, the corresponding conserved symmetry current $\mathcal{J}_u$ (2.4.31) takes the form*

$$\mathcal{J}_u = W + d_H U = (W^\mu + d_\nu U^{\nu\mu})\omega_\mu, \qquad (2.4.41)$$

where the term W vanishes on-shell, i.e., $W \approx 0$, and $U = U^{\nu\mu}\omega_{\nu\mu}$ is a horizontal $(n-2)$-form.

2.4. First order Lagrangian field theory

Proof. Let a gauge symmetry be at most of jet order N in gauge parameters. Then the symmetry current $\mathcal{J}_u$ is decomposed into the sum

$$\mathcal{J}_u^\mu = J_a^{\mu\mu_1\ldots\mu_N}\chi^a_{\mu_1\ldots\mu_N} + \sum_{1<k<N} J_a^{\mu\mu_k\ldots\mu_N}\chi^a_{\mu_k\ldots\mu_N} + \quad (2.4.42)$$
$$J_a^{\mu\mu_N}\chi^a_{\mu_N} + J_a^\mu \chi^a.$$

The first variational formula (2.4.29) takes the form

$$0 = \left[\sum_{k=1}^N u_V{}^i{}_a^{\mu_k\ldots\mu_N}\chi^a_{\mu_k\ldots\mu_N} + u_V{}^i{}_a\chi^a\right]\mathcal{E}_i +$$
$$d_\mu\left(\sum_{k=1}^N J_a^{\mu\mu_k\ldots\mu_N}\chi^a_{\mu_k\ldots\mu_N} + J_a^\mu\chi^a\right).$$

It falls into the set of equalities for each $\chi^a_{\mu\mu_k\ldots\mu_N}$, $\chi^a_{\mu_k\ldots\mu_N}$, $k=1,\ldots,N$, and χ^a as follows:

$$0 = J_a^{(\mu\mu_1)\ldots\mu_N}, \qquad (2.4.43)$$
$$0 = u_V{}^i{}_a^{\mu_k\ldots\mu_N}\mathcal{E}_i - J_a^{(\mu_k\mu_{k+1})\ldots\mu_N} - d_\nu J_a^{\nu\mu_k\ldots\mu_N}, \qquad (2.4.44)$$
$$0 = u_V{}^i{}_a^\mu\mathcal{E}_i - J_a^\mu - d_\nu J_a^{\nu\mu}, \qquad (2.4.45)$$
$$0 = u_V{}^i{}_a\mathcal{E}_i - d_\mu J_a^\mu, \qquad (2.4.46)$$

where $(\mu\nu)$ means symmetrization of indices in accordance with the splitting

$$J_a^{\mu_k\mu_{k+1}\ldots\mu_N} = J_a^{(\mu_k\mu_{k+1})\ldots\mu_N} + J_a^{[\mu_k\mu_{k+1}]\ldots\mu_N}.$$

With the equalities (2.4.43) – (2.4.45), the decomposition (2.4.42) takes the form

$$\mathcal{J}_u^\mu = J_a^{[\mu\mu_1]\ldots\mu_N}\chi^a_{\mu_1\ldots\mu_N} +$$
$$\sum_{1<k<N}[(u_V{}^i{}_a^{\mu\mu_k\ldots\mu_N}\mathcal{E}_i - d_\nu J_a^{\nu\mu\mu_k\ldots\mu_N} + J_a^{[\mu\mu_k]\ldots\mu_N})\chi^a_{\mu_k\ldots\mu_N}] +$$
$$(u_V{}^i{}_a^{\mu\mu_N}\mathcal{E}_i - d_\nu J_a^{\nu\mu\mu_N} + J_a^{[\mu\mu_N]})\chi^a_{\mu_N} - (u_V{}^i{}_a^\mu\mathcal{E}_i + d_\nu J_a^{\nu\mu})\chi^a.$$

A direct computation

$$\begin{aligned}\mathcal{J}_u^\mu &= d_\nu(J_a^{[\mu\nu]\mu_2\ldots\mu_N}\chi^a_{\mu_2\ldots\mu_N}) - d_\nu J_a^{[\mu\nu]\mu_2\ldots\mu_N}\chi^a_{\mu_2\ldots\mu_N} + \\ &\quad \sum_{1<k<N}[(u_V^i{}_a^{\mu\mu_k\ldots\mu_N}\mathcal{E}_i - d_\nu J_a^{\nu\mu\mu_k\ldots\mu_N})\chi^a_{\mu_k\ldots\mu_N} + \\ &\quad d_\nu(J_a^{[\mu\nu]\mu_{k+1}\ldots\mu_N}\chi^a_{\mu_{k+1}\ldots\mu_N}) - d_\nu J_a^{[\mu\nu]\mu_{k+1}\ldots\mu_N}\chi^a_{\mu_{k+1}\ldots\mu_N}] + \\ &\quad [(u_V^i{}_a^{\mu\mu_N}\mathcal{E}_i - d_\nu J_a^{\nu\mu\mu_N})\chi^a_{\mu_N} + d_\nu(J_a^{[\mu\nu]}\chi^a) - d_\nu J_a^{[\mu\nu]}\chi^a] + \\ &\quad (u_V^i{}_a^{\mu}\mathcal{E}_i - d_\nu J_a^{\nu\mu})\chi^a \\ &= d_\nu(J_a^{[\mu\nu]\mu_2\ldots\mu_N}\chi^a_{\mu_2\ldots\mu_N}) + \\ &\quad \sum_{1<k<N}[(u_V^i{}_a^{\mu\mu_k\ldots\mu_N}\mathcal{E}_i - d_\nu J_a^{(\nu\mu)\mu_k\ldots\mu_N})\chi^a_{\mu_k\ldots\mu_N} + \\ &\quad d_\nu(J_a^{[\mu\nu]\mu_{k+1}\ldots\mu_N}\chi^a_{\mu_{k+1}\ldots\mu_N})] + \\ &\quad [(u_V^i{}_a^{\mu\mu_N}\mathcal{E}_i - d_\nu J_a^{(\nu\mu)\mu_N})\chi^a_{\mu_N} + d_\nu(J_a^{[\mu\nu]}\chi^a)] + (u_V^i{}_a^{\mu}\mathcal{E}_i - d_\nu J_a^{(\nu\mu)})\chi^a \end{aligned}$$

leads to the expression

$$\mathcal{J}_u^\mu = \left(\sum_{1<k\leq N} u_V^i{}_a^{\mu\mu_k\ldots\mu_N}\chi^a_{\mu_k\ldots\mu_N} + u_V^i{}_a^{\mu}\chi^a\right)\mathcal{E}_i - \quad (2.4.47)$$

$$\left(\sum_{1<k\leq N} d_\nu J_a^{(\nu\mu)\mu_k\ldots\mu_N}\chi^a_{\mu_k\ldots\mu_N} + d_\nu J_a^{(\nu\mu)}\chi^a\right) -$$

$$d_\nu\left(\sum_{1<k\leq N} J_a^{[\nu\mu]\mu_k\ldots\mu_N}\chi^a_{\mu_k\ldots\mu_N} + J_a^{[\nu\mu]}\chi^a\right).$$

The first summand of this expression vanishes on-shell. Its second one contains the terms $d_\nu J^{(\nu\mu_k)\mu_{k+1}\ldots\mu_N}$, $k=1,\ldots,N$. By virtue of the equalities (2.4.44), every $d_\nu J^{(\nu\mu_k)\mu_{k+1}\ldots\mu_N}$ is expressed into the terms vanishing on-shell and the term $d_\nu J^{(\nu\mu_{k-1})\mu_k\ldots\mu_N}$. Iterating the procedure and bearing in mind the equality (2.4.43), one can easily show that the second summand of the expression (2.4.47) also vanishes on-shell. Thus, the symmetry current $\mathcal{J}_u$ takes the form (2.4.41), where

$$U^{\nu\mu} = -\sum_{1<k\leq N} J_a^{[\nu\mu]\mu_k\ldots\mu_N}\chi^a_{\mu_k\ldots\mu_N} - J_a^{[\nu\mu]}\chi^a. \quad (2.4.48)$$

□

The term U in the expression (2.4.41) is called the *superpotential*. If a symmetry current admits the decomposition (2.4.41), one says that it is reduced to a superpotential [39; 53; 135].

2.4. First order Lagrangian field theory

For instance, if a gauge symmetry

$$u = (u_a^\lambda \chi^a + u_a^{\lambda\mu}\chi_\mu^a)\partial_\lambda + (u_a^i \chi^a + u_a^{i\mu}\chi_\mu^a)\partial_i \qquad (2.4.49)$$

of a Lagrangian L depends at most on the first jets of gauge parameters, then the decomposition (2.4.41) takes the form

$$\mathcal{J}_u^\mu = u_V{}_a^{i\mu}\mathcal{E}_i\chi^a - d_\nu(J_a^{[\nu\mu]}\chi^a) = \qquad (2.4.50)$$
$$(u_a^{i\mu} - y_\lambda^i u_a^{\lambda\mu})\chi^a \mathcal{E}_i + d_\nu[(u_a^{i[\mu} - y_\lambda^i u_a^{\lambda[\mu})\chi^a \pi_i^{\nu]} + u_a^{[\nu\mu]}\chi^a \mathcal{L}].$$

Remark 2.4.5. Theorem 2.4.2 generalizes the result in [67] for gauge symmetries u whose gauge parameters $\chi^\lambda = u^\lambda$ are components of the projection $u^\lambda \partial_\lambda$ of u onto X. In Section 4.3, Theorem 2.4.2 is extended to Grassmann-graded Lagrangian theories of any order (see Theorem 4.2.3).

Remark 2.4.6. The proof of Theorem 2.4.2 gives something more. Let us substitute the equality (2.4.46) into the equality (2.4.45), then the latter into the equality (2.4.44) for $k = N - 1$, and so on. Then we obtain the Noether identities (2.3.5). However, it should be emphasized that the conditions (2.4.43) – (2.4.46) and the Noether identities (2.3.5) are not equivalent. The Noether identities characterize a variational gauge symmetry of a Lagrangian, while (2.4.43) – (2.4.46) are the conditions of a gauge symmetry to be exact.

If a symmetry current $\mathcal{J}$ reduces to a superpotential, the integral conservation law (2.4.32) becomes tautological. At the same time, the superpotential form (2.4.41) of $\mathcal{J}_u$ implies the following integral relation

$$\int_{N^{n-1}} s^* \mathcal{J}_u = \int_{\partial N^{n-1}} s^* U, \qquad (2.4.51)$$

where N^{n-1} is an $(n-1)$-dimensional oriented compact submanifold of X with the boundary ∂N^{n-1}.

2.4.4 Non-regular quadratic Lagrangians

This Section is devoted to the physically relevant case of almost regular quadratic Lagrangians [112].

Given a fibre bundle $Y \to X$, let us consider a quadratic Lagrangian L given by the coordinate expression

$$\mathcal{L} = \frac{1}{2} a_{ij}^{\lambda\mu}(x^\nu, y^k) y_\lambda^i y_\mu^j + b_i^\lambda(x^\nu, y^k) y_\lambda^i + c(x^\nu, y^k), \qquad (2.4.52)$$

where a, b and c are local functions on Y. This property is coordinate-independent due to the affine transformation law (1.2.1) of the jet coordinates y^i_λ. The associated Legendre map $\widehat{L}$ (2.4.6) is given by the coordinate expression

$$p^\lambda_i \circ \widehat{L} = a^{\lambda\mu}_{ij} y^j_\mu + b^\lambda_i, \qquad (2.4.53)$$

and is an affine morphism over Y. It defines the corresponding linear morphism

$$\widehat{a} : T^*X \underset{Y}{\otimes} VY \to \Pi, \qquad (2.4.54)$$

$$p^\lambda_i \circ \widehat{a} = a^{\lambda\mu}_{ij} \overline{y}^j_\mu,$$

where $\overline{y}^j_\mu$ are fibred coordinates on the vector bundle

$$T^*X \underset{Y}{\otimes} VY \to Y.$$

Let the Lagrangian L (2.4.52) be almost regular, i.e., the morphism $\widehat{a}$ (2.4.54) is of constant rank. Then the Lagrangian constraint space N_L (2.4.53) is an affine subbundle of the Legendre bundle $\Pi \to Y$, modelled over the vector subbundle $\overline{N}_L$ (2.4.54) of $\Pi \to Y$. Hence, $N_L \to Y$ has a global section s. For the sake of simplicity, let us assume that $s = \widehat{0}$ is the canonical zero section of $\Pi \to Y$. Then $\overline{N}_L = N_L$. Accordingly, the kernel of the Legendre map (2.4.53) is an affine subbundle of the affine jet bundle $J^1Y \to Y$, modelled over the kernel of the linear morphism $\widehat{a}$ (2.4.54). Then there exists a connection

$$\Gamma : Y \to \operatorname{Ker}\widehat{L} \subset J^1Y, \qquad (2.4.55)$$

$$a^{\lambda\mu}_{ij} \Gamma^j_\mu + b^\lambda_i = 0, \qquad (2.4.56)$$

on $Y \to X$. Connections (2.4.55) constitute an affine space modelled over the linear space of soldering forms

$$\phi = \phi^i_\lambda dx^\lambda \otimes \partial_i$$

on $Y \to X$, satisfying the conditions

$$a^{\lambda\mu}_{ij} \phi^j_\mu = 0 \qquad (2.4.57)$$

and, as a consequence, the conditions

$$\phi^i_\lambda b^\lambda_i = 0.$$

If the Lagrangian (2.4.52) is regular, the connection (2.4.55) is unique.

Remark 2.4.7. If $s \neq \widehat{0}$, one can consider connections Γ taking their values into $\operatorname{Ker}{}_s\widehat{L}$.

The matrix a in the Lagrangian L (2.4.52) can be seen as a global section of constant rank of the tensor bundle

$$\overset{n}{\wedge} T^*X \underset{Y}{\otimes} [\overset{2}{\vee}(TX \underset{Y}{\otimes} V^*Y)] \to Y.$$

Then it satisfies the following corollary of Theorem 1.1.12.

Corollary 2.4.1. *Given a k-dimensional vector bundle $E \to Z$, let a be a fibre metric of rank r in E. There is a splitting*

$$E = \operatorname{Ker} a \underset{Z}{\oplus} E' \qquad (2.4.58)$$

where $E' = E/\operatorname{Ker} a$ is the quotient bundle, and a is a non-degenerate fibre metric in E'.

Theorem 2.4.3. *There exists a linear bundle map*

$$\sigma : \Pi \underset{Y}{\to} T^*X \underset{Y}{\otimes} VY, \qquad \overline{y}^i_\lambda \circ \sigma = \sigma^{ij}_{\lambda\mu} p^\mu_j, \qquad (2.4.59)$$

such that

$$\widehat{a} \circ \sigma \circ i_N = i_N.$$

Proof. The map (2.4.59) is a solution of the algebraic equations

$$a^{\lambda\mu}_{ij} \sigma^{jk}_{\mu\alpha} a^{\alpha\nu}_{kb} = a^{\lambda\nu}_{ib}. \qquad (2.4.60)$$

By virtue of Corollary 2.4.1, there exists the bundle splitting

$$TX^* \underset{Y}{\otimes} VY = \operatorname{Ker} a \underset{Y}{\oplus} E' \qquad (2.4.61)$$

and an atlas of this bundle such that transition functions of $\operatorname{Ker} a$ and E' are independent. Since a is a non-degenerate section of

$$\overset{n}{\wedge} T^*X \underset{Y}{\otimes} (\overset{2}{\vee} E'^*) \to Y,$$

there exist fibre coordinates $(\overline{y}^A)$ on E' such that a is brought into a diagonal matrix with non-vanishing components a_{AA}. Due to the splitting (2.4.61), we have the corresponding bundle splitting

$$TX \underset{Y}{\otimes} V^*Y = (\operatorname{Ker} a)^* \underset{Y}{\oplus} E'^*.$$

Then a desired map σ is represented by a direct sum $\sigma_1 \oplus \sigma_0$ of an arbitrary section σ_1 of the fibre bundle

$$\overset{n}{\wedge} TX \underset{Y}{\otimes} (\overset{2}{\vee} \operatorname{Ker} a) \to Y$$

and the section σ_0 of the fibre bundle

$$\overset{n}{\wedge} TX \underset{Y}{\otimes} (\overset{2}{\vee} E') \to Y$$

which has non-vanishing components

$$\sigma^{AA} = (a_{AA})^{-1}$$

with respect to the fibre coordinates $(\overline{y}^A)$ on E'. The relations

$$\sigma_0 = \sigma_0 \circ a \circ \sigma_0, \qquad a \circ \sigma_1 = 0, \qquad \sigma_1 \circ a = 0 \qquad (2.4.62)$$

hold. $\square$

Remark 2.4.8. Using the relations (2.4.62), one can write the above assumption, that the Lagrangian constraint space $N_L \to Y$ admits a global zero section, in the form

$$b_i^\mu = a_{ij}^{\mu\lambda} \sigma_{\lambda\nu}^{jk} b_k^\nu. \qquad (2.4.63)$$

With the relations (2.4.56), (2.4.60) and (2.4.62), we obtain the splitting

$$J^1 Y = \mathcal{S}(J^1 Y) \underset{Y}{\oplus} \mathcal{F}(J^1 Y) = \operatorname{Ker} \widehat{L} \underset{Y}{\oplus} \operatorname{Im}(\sigma \circ \widehat{L}), \qquad (2.4.64)$$

$$y_\lambda^i = \mathcal{S}_\lambda^i + \mathcal{F}_\lambda^i = [y_\lambda^i - \sigma_{\lambda\alpha}^{ik}(a_{kj}^{\alpha\mu} y_\mu^j + b_k^\alpha)] + \qquad (2.4.65)$$
$$[\sigma_{\lambda\alpha}^{ik}(a_{kj}^{\alpha\mu} y_\mu^j + b_k^\alpha)],$$

where, in fact, $\sigma = \sigma_0$ owing to the relations (2.4.62) and (2.4.63). Then with respect to the coordinates $\mathcal{S}_\lambda^i$ and $\mathcal{F}_\lambda^i$ (2.4.65), the Lagrangian (2.4.52) reads

$$\mathcal{L} = \frac{1}{2} a_{ij}^{\lambda\mu} \mathcal{F}_\lambda^i \mathcal{F}_\mu^j + c', \qquad (2.4.66)$$

where

$$\mathcal{F}_\lambda^i = \sigma_{0\lambda\alpha}^{ik} a_{kj}^{\alpha\mu} (y_\mu^j - \Gamma_\mu^j) \qquad (2.4.67)$$

for some $(\operatorname{Ker} \widehat{L})$-valued connection Γ (2.4.55) on $Y \to X$. Thus, the Lagrangian (2.4.52), written in the form (2.4.66), factorizes through the covariant differential relative to any such connection.

Note that, in gauge theory of principal connections (see Section 5.5), we have the canonical (independent of a Lagrangian) variant (5.5.11) of the splitting (2.4.64) where $\mathcal{F}$ is the strength form (5.5.8). The Yang–Mills Lagrangian (9.5.4) of gauge theory is exactly of the form (2.4.66) where $c' = 0$.

2.4.5 Reduced second order Lagrangians

Let us consider second order Lagrangians on the second order jet manifold J^2Y of Y which, however, lead to second order Euler-Lagrange equations.

Given a second order Lagrangian L, its four-order Euler–Lagrange operator (2.1.12) reads

$$\mathcal{E}_L = (\partial_i - d_\lambda \partial_i^\lambda + d_\mu d_\lambda \partial_i^{\mu\lambda})\mathcal{L}\theta^i \wedge \omega.$$

This operator is reduced to the second order one if a Lagrangian L obeys the conditions

$$\partial_j^{\alpha\beta}\partial_i^{\mu\nu}\mathcal{L} = 0, \qquad (2.4.68)$$

$$(\partial_j^\nu \partial_i^{\mu\lambda} - \partial_i^\mu \partial_j^{\nu\lambda})\mathcal{L} = 0. \qquad (2.4.69)$$

The relation (2.4.68) means that a Lagrangian L is linear in the jet coordinates $y^i_{\lambda\mu}$, i.e., it is given by the coordinate expression

$$L = (\mathcal{L}' + \sigma_i^{\mu\lambda} y^i_{\mu\lambda})\omega, \qquad (2.4.70)$$

where $\mathcal{L}'$ and $\sigma_i^{\mu\lambda}$ are local functions on J^1Y. Since this expression is maintained under coordinate transformations (1.2.12), the functions $\sigma_i^{\mu\lambda}$ satisfy the transformation law

$$\sigma_i^{\prime\mu\lambda} = \frac{\partial x'^\mu}{\partial x^\gamma}\frac{\partial y^j_\nu}{\partial y'^i_\lambda}\sigma_j^{\nu\gamma}.$$

Therefore, one can define the fibrewise form

$$\sigma = \sigma_i^{\mu\lambda}\overline{d}y^i_\mu \wedge \omega_\lambda$$

on the affine bundle $J^1Y \to Y$ (see Section 10.10). Then the condition (2.4.69) means that this form is $\overline{d}$-closed, i.e.,

$$\overline{d}\sigma = \partial_j^\nu \sigma_i^{\mu\lambda}\overline{d}y^j_\nu \wedge \overline{d}y^i_\mu \wedge \omega_\lambda = 0.$$

In accordance with Theorem 10.10.2, any $\overline{d}$-closed fibrewise form on an affine bundle is $\overline{d}$-exact. Consequently, there exists a form $\phi = \phi^\lambda \omega_\lambda$ on J^1Y such that

$$\sigma = \overline{d}\phi = \partial_i^\mu \phi^\lambda \overline{d}y^i_\mu \wedge \omega_\lambda.$$

Let us consider the variationally trivial second order Lagrangian $d_H\phi$. It is readily observed that the Lagrangian $L - d_H\phi$ is of first order, but it possesses the same second order Euler–Lagrange operator $\mathcal{E}_L$ as the second order Lagrangian L (2.4.70). Thus, the following has been proved.

Theorem 2.4.4. *If an Euler–Lagrange operator of a second order Lagrangian also is of second order, it is an Euler–Lagrange operator of some first order Lagrangian.*

In particular, if the functions $\sigma_i^{\mu\lambda}$ are independent of the jet coordinates y_μ^i, one can take

$$\phi = \sigma_i^{\lambda\mu}(y_\mu^i - \Gamma_\mu^i)\omega_\lambda,$$

where Γ is a connection on $Y \to X$. Then a desired first order Lagrangian reads

$$L - d_\lambda[\pi_i^{\lambda\mu}(y_\mu^i - \Gamma_\mu^i)]\omega.$$

One can think of a connection Γ in this Lagrangian as a background field.

For instance, this is the case of the Einstein–Hilbert gravitation Lagrangian of General Relativity (see Remark 6.5.2).

2.4.6 Background fields

In Lagrangian field theory on a fibre bundle $Y \to X$, by *background fields* are meant classical fields which do not obey Euler–Lagrange equations. Let these fields be represented by sections of a fibre bundle $\Sigma \to X$ endowed with bundle coordinates (x^λ, σ^m). In order to formulate Lagrangian field theory in the presence of background fields, let us consider the bundle product

$$Y_{\text{tot}} = \Sigma \underset{X}{\times} Y \to X \qquad (2.4.71)$$

coordinated by $(x^\lambda, \sigma^m, y^i)$ and its jet manifold

$$J^1 Y_{\text{tot}} = J^1\Sigma \underset{X}{\times} J^1 Y.$$

Let L be a first order Lagrangian on the configuration space $J^1 Y_{\text{tot}}$. It can be regarded as a total Lagrangian of field theory where background fields are treated as the dynamic ones. Given a section h of $\Sigma \to X$, we obtain the pull-back Lagrangian

$$L_h = (J^1 h)^* L$$

on $J^1 Y$. It can be regarded as a Lagrangian of first order field theory on the configuration space $J^1 Y$ in the presence of a background field h.

Let us consider the variational formula (2.1.16):

$$dL - \mathcal{E}_L - d_H\Xi = 0,$$

for a total Lagrangian L. Its pull-back

$$(J^2 h)^*(dL - \delta L - d_H\Xi) = dL_h - \delta L_h - d_H\Xi_h \qquad (2.4.72)$$

2.4. First order Lagrangian field theory

is exactly the variational formula for the Lagrangian L_h in the presence of a background field h. The corresponding Euler–Lagrange operator in the presence of a background field h reads

$$\delta L_h = (J^2 h)^*(\delta L) = (J^2 h)^*(\mathcal{E}_i \theta^i + \mathcal{E}_m \theta^m) \wedge \omega = (J^1 h)^*(\mathcal{E}_i) \theta^i \omega.$$

The variational formula (2.4.72) enables us to obtain conservation laws in the presence of a background field. Let

$$u = u^\lambda(x^\mu)\partial_\lambda + u^m(x^\mu, \sigma^n)\partial_m + u^i(x^\mu, \sigma^n, y^j)\partial_i \qquad (2.4.73)$$

be a vector field on Y projected onto Σ and X. Its restriction

$$u_h : Y \underset{X}{\times} h(X) \xrightarrow{u} TY \underset{X}{\times} T\Sigma \longrightarrow TY,$$
$$u_h = u = u^\lambda \partial_\lambda + u_h^i \partial_i, \qquad (2.4.74)$$
$$u_h^i(x^\mu, y^j) = u^i(x^\mu, h^n(x), y^j),$$

is a vector field on Y. Let us suppose that u (2.4.73) is an exact symmetry of a total Lagrangian L. The corresponding first variational formula (2.4.28) leads to the equality

$$0 = (u^m - y_\lambda^m u^\lambda)\partial_m \mathcal{L} + \pi_m^\lambda d_\lambda(u^m - y_\mu^m u^\mu) + \\ (u^i - y_\lambda^i u^\lambda)\delta_i \mathcal{L} - d_\lambda[\partial_i^\lambda \mathcal{L}(u^\mu y_\mu^i - u^i) - u^\lambda \mathcal{L}].$$

Putting $\sigma^m = h^m(x)$, we obtain the equality

$$0 = (J^1 h)^*[(u^m - y_\lambda^m u^\lambda)\partial_m \mathcal{L} + \pi_m^\lambda d_\lambda(u^m - y_\mu^m u^\mu)] + \\ (u_h^i - y_\lambda^i u^\lambda)\delta_i \mathcal{L}_h - d_\lambda[\partial_i^\lambda \mathcal{L}_h(u^\mu y_\mu^i - u_h^i) - u^\lambda \mathcal{L}_h].$$

On the shell $\delta_i L_h = 0$, this equality is brought into the weak transformation law

$$0 \approx (J^1 h)^*[(u^m - y_\lambda^m u^\lambda)\partial_m \mathcal{L} + \pi_m^\lambda d_\lambda(u^m - y_\mu^m u^\mu)] - \\ d_\lambda[\partial_i^\lambda \mathcal{L}_h(u^\mu y_\mu^i - u_h^i) - u^\lambda \mathcal{L}_h] \qquad (2.4.75)$$

of the symmetry current

$$\mathcal{J}^\lambda = \partial_i^\lambda \mathcal{L}_h(u^\mu y_\mu^i - u_h^i) - u^\lambda \mathcal{L}_h \qquad (2.4.76)$$

of dynamic fields y^i along the vector field (2.4.74) in the presence of a background field h.

2.4.7 Variation Euler–Lagrange equation. Jacobi fields

The vertical extension of Lagrangian field theory on a fibre bundle $Y \to X$ onto the vertical tangent bundle VY of $Y \to X$ describes the linear deviations of solutions of the Euler–Lagrange equation which are Jacobi fields.

The configuration space of field theory on $VY \to X$ is the jet manifold J^1VY. Due to the canonical isomorphism $J^1VY = VJ^1Y$ (1.2.9), this configuration space is provided with the coordinates $(x^\lambda, y^i, y^i_\lambda, \dot{y}^i, \dot{y}^i_\lambda)$. It follows that Lagrangian theory on J^1VY can be developed as the vertical extension of Lagrangian theory on J^1Y.

Lemma 2.4.1. *Similar to the canonical isomorphism between fibre bundles TT^*Z and T^*TX [90], the isomorphism*

$$VV^*Y \underset{VY}{=} V^*VY, \qquad (2.4.77)$$

$$p_i \longleftrightarrow \dot{v}_i, \quad \dot{p}_i \longleftrightarrow \dot{y}_i,$$

can be established by inspection of the transformation laws of the holonomic coordinates $(x^\lambda, y^i, p_i = \dot{y}_i)$ on V^*Y and $(x^\lambda, y^i, v^i = \dot{y}^i)$ on VY.

It follows that any exterior form ϕ on a fibre bundle Y gives rise to the exterior form

$$\phi_V = \partial_V(\phi) = \dot{y}^i \partial_i(\phi) \qquad (2.4.78)$$

on VY so that $d\phi_V = (d\phi)_V$. For instance,

$$\partial_V f = \dot{y}^i \partial_i f, \qquad f \in C^\infty(Y),$$

$$\partial_V(dy^i) = d\dot{y}^i.$$

The form ϕ_V (2.4.78) is called the *vertical extension* of ϕ on Y.

Let L be a Lagrangian on J^1Y. Its vertical extension (2.4.78) onto the vertical configuration space VJ^1Y reads

$$L_V = \partial_V L = (\dot{y}^i \partial_i + \dot{y}^i_\lambda \partial^\lambda_i)\mathcal{L}\omega. \qquad (2.4.79)$$

The corresponding Euler–Lagrange equation (2.4.3) takes the form

$$\dot{\delta}_i \mathcal{L}_V = \delta_i \mathcal{L} = 0, \qquad (2.4.80)$$

$$\delta_i \mathcal{L}_V = \partial_V \delta_i \mathcal{L} = 0, \qquad (2.4.81)$$

$$\partial_V = \dot{y}^i \partial_i + \dot{y}^i_\lambda \partial^\lambda_i + \dot{y}^i_{\mu\lambda} \partial^{\mu\lambda}_i$$

(see the compact notation (1.1.27)). The equation (2.4.80) is exactly the Euler–Larange equation (2.4.3) for the original Lagrangian L. In order to clarify the meaning of the equation (2.4.81), let us suppose that $Y \to X$ is a vector bundle. Given a solution s of the Euler–Lagrange equation (2.4.80), let δs be a *Jacobi field*, i.e., $s + \varepsilon \delta s$ also is a solution of the Euler–Lagrange equation (2.4.80) modulo the terms of order > 1 in the small parameter ε. Then it is readily observed that the Jacobi field δs satisfies the Euler–Lagrange equation (2.4.81), which therefore is called the *variation equation* of the equation (2.4.80) [33; 112].

The Lagrangian L_V (2.4.79) yields the Legendre map

$$\widehat{L}_V : VJ^1Y \xrightarrow[VY]{} \Pi_{VY} = V^*VY \underset{VY}{\wedge} (\overset{n-1}{\wedge} T^*X), \qquad (2.4.82)$$

where Π_{VY} is called the *vertical Legendre bundle*.

Lemma 2.4.2. *Due to the isomorphism (2.4.77) there exists the bundle isomorphism*

$$\Pi_{VY} \underset{VY}{=} V\Pi, \qquad (2.4.83)$$

$$p_i^\lambda \longleftrightarrow \dot{p}_i^\lambda, \qquad q_i^\lambda \longleftrightarrow p_i^\lambda,$$

written with respect to the holonomic coordinates $(x^\lambda, y^i, \dot{y}^i, p_i^\lambda, q_i^\lambda)$ *on* Π_{VY} *and* $(x^\lambda, y^i, p_i^\lambda, \dot{y}^i, \dot{p}_i^\lambda)$ *on* $V\Pi$.

In view of the isomorphism (2.4.83), the Legendre map (2.4.82) takes the form

$$\widehat{L}_V = V\widehat{L} : VJ^1Y \xrightarrow[VY]{} \Pi_{VY} = V\Pi, \qquad (2.4.84)$$

$$p_i^\lambda = \partial_i^\lambda \mathcal{L}_V = \pi_i^\lambda, \qquad \dot{p}_i^\lambda = \partial_\lambda^i \mathcal{L} = \partial_V \pi_i^\lambda.$$

It is called the *vertical Legendre map*.

Let Z_{VY} be the homogeneous Legendre bundle (2.4.13) over VY endowed with the corresponding coordinates

$$(x^\lambda, y^i, \dot{y}^i, p_i^\lambda, q_i^\lambda, p).$$

There is the fibre bundle

$$\zeta : VZ_Y \to Z_{VY}, \qquad (2.4.85)$$

$$(x^\lambda, y^i, \dot{y}^i, p_i^\lambda, q_i^\lambda, p) \circ \zeta = (x^\lambda, y^i, \dot{y}^i, \dot{p}_i^\lambda, p_i^\lambda, \dot{p}).$$

Then the vertical tangent morphism $V\pi_{Z\Pi}$ to $\pi_{Z\Pi}$ (2.4.16) factorizes through the composition of fibre bundles

$$V\pi_{Z\Pi} : VZ_Y \to Z_{VY} \to \Pi_{VY} = V\Pi. \qquad (2.4.86)$$

Owing to this fact, one can develop Hamiltonian field theory on a momentum phase space Π_{VY} as the vertical extension of polysymplectic Hamiltonian field theory on a momentum phase space Π (see Section 9.6).

2.5 Appendix. Cohomology of the variational bicomplex

This Section is devoted to the proof of Theorems 2.1.1, 2.1.2 and 2.1.4 on the relevant cohomology of the variational bicomplex (2.1.4) of the differential graded algebra $\mathcal{O}_\infty^*$. At first, we obtain the corresponding cohomology of the differential graded algebra $\mathcal{Q}_\infty^*$ (1.7.12). For this purpose, one can use abstract de Rham Theorem 10.7.4 because, as was mentioned above, the paracompact infinite jet order manifold $J^\infty Y$ admits the partition of unity by elements of $\mathcal{Q}_\infty^0$, but not $\mathcal{O}_\infty^0$ [4; 150]. After that, we show that cohomology of $\mathcal{O}_\infty^* \subset \mathcal{Q}_\infty^*$ equals that of $\mathcal{Q}_\infty^*$ [55; 56; 139].

Let us start with the so called *algebraic Poincaré lemma* [123; 157].

Lemma 2.5.1. *If Y is a contractible bundle $\mathbb{R}^{n+p} \to \mathbb{R}^n$, the variational bicomplex (2.1.4) is exact at all terms, except $\mathbb{R}$.*

Proof. The homotopy operators for d_V, d_H, δ and ϱ are given by the formulas (5.72), (5.109), (5.84) in [123] and (4.5) in [157], respectively. □

Let $\mathfrak{Q}_\infty^*$ be the sheaf of germs of differential forms $\phi \in \mathcal{O}_\infty^*$ on $J^\infty Y$. It is decomposed into the variational bicomplex $\mathfrak{Q}_\infty^{*,*}$. The differential graded algebra $\mathcal{Q}_\infty^*$ of global sections of $\mathfrak{Q}_\infty^*$ also is decomposed into the variational bicomplex $\mathcal{Q}_\infty^{*,*}$ similar to the bicomplex (2.1.4). Let us consider the variational subcomplex

$$0 \to \mathbb{R} \to \mathfrak{Q}_\infty^0 \xrightarrow{d_H} \mathfrak{Q}_\infty^{0,1} \cdots \xrightarrow{d_H} \mathfrak{Q}_\infty^{0,n} \xrightarrow{\delta} \mathbf{E}_1 \xrightarrow{\delta} \mathbf{E}_2 \longrightarrow \cdots \quad (2.5.1)$$

of $\mathfrak{Q}_\infty^{*,*}$ and the subcomplexes of sheaves of contact forms

$$0 \to \mathfrak{Q}_\infty^{k,0} \xrightarrow{d_H} \mathfrak{Q}_\infty^{k,1} \cdots \xrightarrow{d_H} \mathfrak{Q}_\infty^{k,n} \xrightarrow{\varrho} \mathbf{E}_k \to 0, \quad k = 1, \ldots, \quad (2.5.2)$$

where

$$\mathbf{E}_k = \varrho(\mathfrak{Q}_\infty^{k,n}).$$

By virtue of Lemma 2.5.1, these complexes are exact at all terms, except $\mathbb{R}$.

Since the paracompact space $J^\infty Y$ admits a partition of unity by elements of the ring $\mathcal{Q}_\infty^0$, the sheaves $\mathfrak{Q}_\infty^{m,k}$ of $\mathcal{Q}_\infty^0$-modules are fine (see Theorem 10.7.7) and, consequently, acyclic (see Theorem 10.7.6). Let us show that the sheaves $\mathbf{E}_k$ also are fine [56]. Though the $\mathbb{R}$-modules $\Gamma(\mathbf{E}_{k>1})$ fail to be $\mathcal{Q}_\infty^0$-modules [157], one can use the fact that the sheaves $\mathbf{E}_{k>0}$ are projections $\varrho(\mathfrak{Q}_\infty^{k,n})$ of sheaves of $\mathcal{Q}_\infty^0$-modules. Let $\{U_i\}_{i \in I}$ be a locally finite open cover of $J^\infty Y$ and $\{f_i \in \mathcal{Q}_\infty^0\}$ the associated partition of unity.

2.5. Appendix. Cohomology of the variational bicomplex

For any open subset $U \subset J^\infty Y$ and any section φ of the sheaf $\mathfrak{Q}_\infty^{k,n}$ over U, let us put $g_i(\varphi) = f_i \varphi$. The endomorphisms g_i of $\mathfrak{Q}_\infty^{k,n}$ yield the $\mathbb{R}$-module endomorphisms

$$\overline{g}_i = \varrho \circ g_i : \mathbf{E}_k \xrightarrow{\text{in}} \mathfrak{Q}_\infty^{k,n} \xrightarrow{g_i} \mathfrak{Q}_\infty^{k,n} \xrightarrow{\varrho} \mathbf{E}_k$$

of the sheaves $\mathbf{E}_k$. They possess the properties required for $\mathbf{E}_k$ to be a fine sheaf. Indeed, for each $i \in I$, $\operatorname{supp} f_i \subset U_i$ provides a closed set such that $\overline{g}_i$ is zero outside this set, while the sum $\sum_{i \in I} \overline{g}_i$ is the identity morphism.

Consequently, all sheaves, except $\mathbb{R}$, in the complexes (2.5.1) – (2.5.2) are acyclic. Therefore, these complexes are resolutions of the constant sheaf $\mathbb{R}$ and the zero sheaf on $J^\infty Y$, respectively. Let us consider the corresponding subcomplexes

$$0 \to \mathbb{R} \to \mathcal{Q}_\infty^0 \xrightarrow{d_H} \mathcal{Q}_\infty^{0,1} \xrightarrow{d_H} \cdots \xrightarrow{d_H} \mathcal{Q}_\infty^{0,n} \xrightarrow{\delta} \Gamma(\mathbf{E}_1) \xrightarrow{\delta} \Gamma(\mathbf{E}_2) \to \cdots, \quad (2.5.3)$$

$$0 \to \mathcal{Q}_\infty^{k,0} \xrightarrow{d_H} \mathcal{Q}_\infty^{k,1} \xrightarrow{d_H} \cdots \xrightarrow{d_H} \mathcal{Q}_\infty^{k,n} \xrightarrow{\varrho} \Gamma(\mathbf{E}_k) \to 0, \quad k = 1, \ldots, \quad (2.5.4)$$

of the differential graded algebra $\mathcal{Q}_\infty^*$. In accordance with abstract de Rham Theorem 10.7.5, cohomology of the complex (2.5.3) equals the cohomology of $J^\infty Y$ with coefficients in the constant sheaf $\mathbb{R}$, while the complex (2.5.4) is exact. Since Y is a strong deformation retract of $J^\infty Y$, cohomology of the complex (2.5.3) equals the de Rham cohomology of Y (see Remark 1.5.3).

Thus, the following has been proved.

Theorem 2.5.1. *The cohomology of the variational complex (2.5.3) equals the de Rham cohomology of a fibre bundle Y. All the complexes (2.5.4) are exact.*

Now, let us show the following.

Theorem 2.5.2. *The subalgebra $\mathcal{O}_\infty^* \subset \mathcal{Q}_\infty^*$ has the same d_H- and δ-cohomology as $\mathcal{Q}_\infty^*$.*

Let the common symbol D stand for d_H and δ. Bearing in mind the decompositions (2.1.7) – (2.1.10), it suffices to show that, if an element $\phi \in \mathcal{O}_\infty^*$ is D-exact in the algebra $\mathcal{Q}_\infty^*$, then it is so in the algebra $\mathcal{O}_\infty^*$.

Lemma 2.5.1 states that, if Y is a contractible bundle and a D-exact form ϕ on $J^\infty Y$ is of finite jet order $[\phi]$ (i.e., $\phi \in \mathcal{O}_\infty^*$), there exists a differential form $\varphi \in \mathcal{O}_\infty^*$ on $J^\infty Y$ such that $\phi = D\varphi$. Moreover, a glance at the homotopy operators for d_H and δ shows that the jet order $[\varphi]$ of φ is bounded by an integer $N([\phi])$, depending only on the jet order of ϕ. Let us call this fact the *finite exactness* of the operator D. Lemma 2.5.1 shows that the finite exactness takes place on $J^\infty Y|_U$ over any domain $U \subset Y$. Let us prove the following.

Lemma 2.5.2. *Given a family $\{U_\alpha\}$ of disjoint open subsets of Y, let us suppose that the finite exactness takes place on $J^\infty Y|_{U_\alpha}$ over every subset U_α from this family. Then, it is true on $J^\infty Y$ over the union $\cup_\alpha U_\alpha$ of these subsets.*

Proof. Let $\phi \in \mathcal{O}_\infty^*$ be a D-exact form on $J^\infty Y$. The finite exactness on $(\pi_0^\infty)^{-1}(\cup U_\alpha)$ holds since $\phi = D\varphi_\alpha$ on every $(\pi_0^\infty)^{-1}(U_\alpha)$ and $[\varphi_\alpha] < N([\phi])$. □

Lemma 2.5.3. *Suppose that the finite exactness of the operator D takes place on $J^\infty Y$ over open subsets U, V of Y and their non-empty overlap $U \cap V$. Then, it also is true on $J^\infty Y|_{U \cup V}$.*

Proof. Let $\phi = D\varphi \in \mathcal{O}_\infty^*$ be a D-exact form on $J^\infty Y$. By assumption, it can be brought into the form $D\varphi_U$ on $(\pi_0^\infty)^{-1}(U)$ and $D\varphi_V$ on $(\pi_0^\infty)^{-1}(V)$, where φ_U and φ_V are differential forms of bounded jet order. Let us consider their difference $\varphi_U - \varphi_V$ on $(\pi_0^\infty)^{-1}(U \cap V)$. It is a D-exact form of bounded jet order

$$[\varphi_U - \varphi_V] < N([\phi])$$

which, by assumption, can be written as

$$\varphi_U - \varphi_V = D\sigma$$

where σ also is of bounded jet order

$$[\sigma] < N(N([\phi])).$$

Lemma 2.5.4 below shows that $\sigma = \sigma_U + \sigma_V$ where σ_U and σ_V are differential forms of bounded jet order on $(\pi_0^\infty)^{-1}(U)$ and $(\pi_0^\infty)^{-1}(V)$, respectively. Then, putting

$$\varphi'|_U = \varphi_U - D\sigma_U, \qquad \varphi'|_V = \varphi_V + D\sigma_V,$$

we have the form ϕ, equal to $D\varphi'_U$ on $(\pi_0^\infty)^{-1}(U)$ and $D\varphi'_V$ on $(\pi_0^\infty)^{-1}(V)$, respectively. Since the difference $\varphi'_U - \varphi'_V$ on $(\pi_0^\infty)^{-1}(U \cap V)$ vanishes, we obtain $\phi = D\varphi'$ on $(\pi_0^\infty)^{-1}(U \cup V)$ where

$$\varphi' = \begin{cases} \varphi'|_U = \varphi'_U, \\ \varphi'|_V = \varphi'_V \end{cases}$$

is of bounded jet order $[\varphi'] < N(N([\phi]))$. □

Lemma 2.5.4. *Let U and V be open subsets of a bundle Y and $\sigma \in \mathfrak{D}_\infty^*$ a differential form of bounded jet order on*

$$(\pi_0^\infty)^{-1}(U \cap V) \subset J^\infty Y.$$

Then, σ is decomposed into a sum $\sigma_U + \sigma_V$ of differential forms σ_U and σ_V of bounded jet order on $(\pi_0^\infty)^{-1}(U)$ and $(\pi_0^\infty)^{-1}(V)$, respectively.

Proof. By taking a smooth partition of unity on $U \cup V$ subordinate to the cover $\{U, V\}$ and passing to the function with support in V, one gets a smooth real function f on $U \cup V$ which equals 0 on a neighborhood of $U \setminus V$ and 1 on a neighborhood of $V \setminus U$ in $U \cup V$. Let $(\pi_0^\infty)^* f$ be the pullback of f onto $(\pi_0^\infty)^{-1}(U \cup V)$. The differential form $((\pi_0^\infty)^* f)\sigma$ equals 0 on a neighborhood of $(\pi_0^\infty)^{-1}(U)$ and, therefore, can be extended by 0 to $(\pi_0^\infty)^{-1}(U)$. Let us denote it σ_U. Accordingly, the differential form $(1 - (\pi_0^\infty)^* f)\sigma$ has an extension σ_V by 0 to $(\pi_0^\infty)^{-1}(V)$. Then, $\sigma = \sigma_U + \sigma_V$ is a desired decomposition because σ_U and σ_V are of the jet order which does not exceed that of σ. □

To prove the finite exactness of D on $J^\infty Y$, it remains to choose an appropriate cover of Y. A smooth manifold Y admits a countable cover $\{U_\xi\}$ by domains U_ξ, $\xi \in \mathbf{N}$, and its refinement $\{U_{ij}\}$, where $j \in \mathbf{N}$ and i runs through a finite set, such that $U_{ij} \cap U_{ik} = \emptyset$, $j \neq k$ [69]. Then Y has a finite cover $\{U_i = \cup_j U_{ij}\}$. Since the finite exactness of the operator D takes place over any domain U_ξ, it also holds over any member U_{ij} of the refinement $\{U_{ij}\}$ of $\{U_\xi\}$ and, in accordance with Lemma 2.5.2, over any member of the finite cover $\{U_i\}$ of Y. Then by virtue of Lemma 2.5.3, the finite exactness of D takes place on $J^\infty Y$ over Y.

Similarly, one can show that:

Theorem 2.5.3. *Restricted to $\mathcal{O}_\infty^{k,n}$, the operator ϱ remains exact.*

Theorems 2.5.1 – 2.5.3 result in Theorems 2.1.1 – 2.1.2.

Turn now to the proof of Theorem 2.1.4. Given the short variational complex (2.1.21), let us consider the corresponding complex of sheaves

$$0 \to \mathbb{R} \to \mathfrak{Q}_\infty^0 \xrightarrow{d_H} \mathfrak{Q}_\infty^{0,1} \cdots \xrightarrow{d_H} \mathfrak{Q}_\infty^{0,n} \xrightarrow{\delta} 0. \quad (2.5.5)$$

In the case of an affine bundle $Y \to X$, we can lower this complex onto the base X as follows.

Let us consider the open surjection

$$\pi^\infty : J^\infty Y \to X$$

and the direct image $\mathfrak{X}^*_\infty = \pi^\infty_* \mathfrak{Q}^*_\infty$ on X of the sheaf $\mathfrak{Q}^*_\infty$. Its stalk over a point $x \in X$ consists of the equivalence classes of sections of the sheaf $\mathfrak{Q}^*_\infty$ which coincide on the inverse images $(\pi^\infty)^{-1}(U_x)$ of neighbourhoods U_x of x. Since $\pi^\infty_* \mathbb{R} = \mathbb{R}$, we have the following complex of sheaves on X:

$$0 \to \mathbb{R} \to \mathfrak{X}^0_\infty \xrightarrow{d_H} \mathfrak{X}^{0,1}_\infty \xrightarrow{d_H} \cdots \xrightarrow{d_H} \mathfrak{X}^{0,n}_\infty \xrightarrow{\delta} 0. \qquad (2.5.6)$$

Every point $x \in X$ has a base of open contractible neighbourhoods $\{U_x\}$ such that the sheaves $\mathfrak{Q}^{0,*}_\infty$ of $\mathcal{Q}^*_\infty$-modules are acyclic on the inverse images $(\pi^\infty)^{-1}(U_x)$ of these neighbourhoods. Then, in accordance with the Leray theorem [62], cohomology of $J^\infty Y$ with coefficients in the sheaves $\mathfrak{Q}^{0,*}_\infty$ are isomorphic to that of X with coefficients in their direct images $\mathfrak{X}^{0,*}_\infty$, i.e., the sheaves $\mathfrak{X}^{0,*}_\infty$ on X are acyclic. Furthermore, Lemma 2.5.1 also shows that the complexes of sections of sheaves $\mathfrak{Q}^{0,*}_\infty$ over $(\pi^\infty_0)^{-1}(U_x)$ are exact. It follows that the complex (2.5.6) on X is exact at all terms, except $\mathbb{R}$, and it is a resolution of the constant sheaf $\mathbb{R}$ on X. Due to the $\mathbb{R}$-algebra isomorphism $\mathcal{Q}^*_\infty = \Gamma(\mathfrak{X}^*_\infty)$, one can think of the short variational subcomplex of the complex (2.5.1) as being the complex of the structure algebras of the sheaves (2.5.6) on X.

Given the sheaf $\mathfrak{X}^*_\infty$ on X, let us consider its subsheaf $\mathfrak{P}^*_\infty$ of germs of exterior forms which are polynomials in the fibre coordinates y^i_Λ, $|\Lambda| \geq 0$, of the continuous bundle $J^\infty Y \to X$. The sheaf $\mathfrak{P}^*_\infty$ is a sheaf of $C^\infty(X)$-modules. The differential graded algebra P^*_∞ of its global sections is a $C^\infty(X)$-subalgebra of $\mathcal{Q}^*_\infty$. We have the subcomplex

$$0 \to \mathbb{R} \longrightarrow \mathfrak{P}^0_\infty \xrightarrow{d_H} \mathfrak{P}^{0,1}_\infty \xrightarrow{d_H} \cdots \xrightarrow{d_H} \mathfrak{P}^{0,n}_\infty \xrightarrow{\delta} 0 \qquad (2.5.7)$$

of the complex (2.5.6) on X. As a particular variant of the algebraic Poincaré lemma, the exactness of the complex (2.5.7) at all terms, except $\mathbb{R}$, follows from the form of the homotopy operator for d_H or can be proved in a straightforward way [9]. Since the sheaves $\mathfrak{P}^{0,*}_\infty$ of $C^\infty(X)$-modules on X are acyclic, the complex (2.5.7) is a resolution of the constant sheaf $\mathbb{R}$ on X. Hence, cohomology of the complex

$$0 \to \mathbb{R} \longrightarrow P^0_\infty \xrightarrow{d_H} P^{0,1}_\infty \xrightarrow{d_H} \cdots \xrightarrow{d_H} P^{0,n}_\infty \xrightarrow{\delta} 0 \qquad (2.5.8)$$

of the differential graded algebras $P^{0,<n}_\infty$ equals the de Rham cohomology of X. It follows that every d_H-closed polynomial form $\phi \in P^{0,m<n}_\infty$ is decomposed into the sum

$$\phi = \sigma + d_H \xi, \qquad \xi \in \mathcal{P}^{0,m-1}_\infty, \qquad (2.5.9)$$

where σ is a closed form on X.

2.5. Appendix. Cohomology of the variational bicomplex

Let $\mathcal{P}_\infty^*$ be $C^\infty(X)$-subalgebra of the polynomial algebra P_∞^* which consists of exterior forms which are polynomials in the fibre coordinates y_Λ^i. Obviously, $\mathcal{P}_\infty^*$ is a subalgebra of $\mathcal{O}_\infty^*$. One can show that $\mathcal{P}_\infty^*$ have the same cohomology as P_∞^*, i.e., if ϕ in the decomposition (2.5.9) is an element of $\mathcal{P}_\infty^{0,*}$ then ξ is so. The proof of this fact follows the proof of Theorem 2.5.2, but differential forms on X (not $J^\infty Y$) are considered.

Chapter 3

Grassmann-graded Lagrangian field theory

In classical field theory, there are different descriptions of odd fields on graded manifolds [27; 118] and supermanifolds [29; 45]. Both graded manifolds and supermanifolds are phrased in terms of sheaves of graded commutative algebras [10]. However, graded manifolds are characterized by sheaves on smooth manifolds, while supermanifolds are constructed by gluing of sheaves on supervector spaces. Treating odd fields on a smooth manifold X, we follow the Serre–Swan theorem generalized to graded manifolds (Theorem 3.3.2). It states that, if a Grassmann algebra is an exterior algebra of some projective $C^\infty(X)$-module of finite rank, it is isomorphic to the algebra of graded functions on a graded manifold whose body is X. By virtue of this theorem, we describe odd fields and their jets on an arbitrary smooth manifold X as generating elements of the structure ring of a graded manifold whose body is X [13; 14; 59]. This definition differs from that of jets of a graded fibre bundle [78; 118], but reproduces the heuristic notion of jets of ghosts in the field-antifield BRST theory [9; 21].

3.1 Grassmann-graded algebraic calculus

Throughout the book, by the Grassmann gradation is meant $\mathbb{Z}_2$-gradation. Hereafter, the symbol [.] stands for the Grassmann parity. In the literature, a $\mathbb{Z}_2$-graded structure is simply called the graded structure if there is no danger of confusion. Let us summarize the relevant notions of the Grassmann-graded algebraic calculus [10; 28].

An algebra $\mathcal{A}$ is called *graded* if it is endowed with a *grading automorphism* γ such that $\gamma^2 = \mathrm{Id}$. A graded algebra falls into the direct sum

$\mathcal{A} = \mathcal{A}_0 \oplus \mathcal{A}_1$ of $\mathbb{Z}$-modules $\mathcal{A}_0$ and $\mathcal{A}_1$ of *even* and *odd* elements such that

$$\gamma(a) = (-1)^i a, \qquad a \in \mathcal{A}_i, \qquad i = 0, 1,$$
$$[aa'] = ([a] + [a']) \bmod 2, \qquad a \in \mathcal{A}_{[a]}, \qquad a' \in \mathcal{A}_{[a']}.$$

One calls $\mathcal{A}_0$ and $\mathcal{A}_1$ the even and odd parts of $\mathcal{A}$, respectively. The even part $\mathcal{A}_0$ is a subalgebra of $\mathcal{A}$ and the odd one $\mathcal{A}_1$ is an $\mathcal{A}_0$-module. If $\mathcal{A}$ is a *graded ring*, then $[\mathbf{1}] = 0$.

A graded algebra $\mathcal{A}$ is called *graded commutative* if

$$aa' = (-1)^{[a][a']} a'a,$$

where a and a' are *graded-homogeneous elements* of $\mathcal{A}$, i.e., they are either even or odd.

Given a graded algebra $\mathcal{A}$, a left *graded $\mathcal{A}$-module* Q is defined as a left $\mathcal{A}$-module provided with the grading automorphism γ such that

$$\gamma(aq) = \gamma(a)\gamma(q), \qquad a \in \mathcal{A}, \qquad q \in Q,$$
$$[aq] = ([a] + [q]) \bmod 2.$$

A graded module Q is split into the direct sum $Q = Q_0 \oplus Q_1$ of two $\mathcal{A}_0$-modules Q_0 and Q_1 of even and odd elements. Similarly, right graded modules are defined.

If $\mathcal{K}$ is a graded commutative ring, a graded $\mathcal{K}$-module can be provided with a graded *$\mathcal{K}$-bimodule* structure by letting

$$qa = (-1)^{[a][q]} aq, \qquad a \in \mathcal{K}, \qquad q \in Q.$$

A graded $\mathcal{K}$-module is called *free* if it has a basis generated by graded-homogeneous elements. This basis is said to be of type (n, m) if it contains n even and m odd elements.

In particular, by a real *graded vector space* $B = B_0 \oplus B_1$ is meant a graded $\mathbb{R}$-module. A real graded vector space is said to be (n,m)-dimensional if $B_0 = \mathbb{R}^n$ and $B_1 = \mathbb{R}^m$.

Given a graded commutative ring $\mathcal{K}$, the following are standard constructions of new graded modules from old ones.

• The direct sum of graded modules and a graded factor module are defined just as those of modules over a commutative ring.

• The *tensor product* $P \otimes Q$ of graded $\mathcal{K}$-modules P and Q is an additive group generated by elements $p \otimes q$, $p \in P$, $q \in Q$, obeying the relations

$$(p + p') \otimes q = p \otimes q + p' \otimes q,$$
$$p \otimes (q + q') = p \otimes q + p \otimes q',$$
$$ap \otimes q = (-1)^{[p][a]} pa \otimes q = (-1)^{[p][a]} p \otimes aq, \qquad a \in \mathcal{K}.$$

3.1. Grassmann-graded algebraic calculus

In particular, the tensor algebra $\otimes P$ of a graded $\mathcal{K}$-module P is defined as that (10.1.5) of a module over a commutative ring. Its quotient $\wedge P$ with respect to the ideal generated by elements

$$p \otimes p' + (-1)^{[p][p']} p' \otimes p, \qquad p, p' \in P,$$

is the *bigraded exterior algebra* of a graded module P with respect to the *graded exterior product*

$$p \wedge p' = -(-1)^{[p][p']} p' \wedge p.$$

- A morphism $\Phi : P \to Q$ of graded $\mathcal{K}$-modules seen as additive groups is said to be *even graded morphism* (resp. *odd graded morphism*) if Φ preserves (resp. change) the Grassmann parity of all graded-homogeneous elements of P and obeys the relations

$$\Phi(ap) = (-1)^{[\Phi][a]} a\Phi(p), \qquad p \in P, \qquad a \in \mathcal{K}.$$

A morphism $\Phi : P \to Q$ of graded $\mathcal{K}$-modules as additive groups is called a *graded $\mathcal{K}$-module morphism* if it is represented by a sum of even and odd graded morphisms. The set $\mathrm{Hom}\,_{\mathcal{K}}(P, Q)$ of graded morphisms of a graded $\mathcal{K}$-module P to a graded $\mathcal{K}$-module Q is naturally a graded $\mathcal{K}$-module. The graded $\mathcal{K}$-module

$$P^* = \mathrm{Hom}\,_{\mathcal{K}}(P, \mathcal{K})$$

is called the *dual* of a graded $\mathcal{K}$-module P.

A *graded commutative $\mathcal{K}$-ring* $\mathcal{A}$ is a graded commutative ring which also is a graded $\mathcal{K}$-module. A *real graded commutative ring* is said to be of rank N if it is a free algebra generated by the unit element $\mathbf{1}$ and N odd elements. A *graded commutative Banach ring* $\mathcal{A}$ is a real graded commutative ring which is a real Banach algebra whose norm obeys the condition

$$\|a_0 + a_1\| = \|a_0\| + \|a_1\|, \qquad a_0 \in \mathcal{A}_0, \quad a_1 \in \mathcal{A}_1.$$

Let V be a real vector space, and let $\Lambda = \wedge V$ be its exterior algebra endowed with the Grassmann gradation

$$\Lambda = \Lambda_0 \oplus \Lambda_1, \qquad \Lambda_0 = \mathbb{R} \bigoplus_{k=1}^{2k} \wedge V, \qquad \Lambda_1 = \bigoplus_{k=1}^{2k-1} \wedge V. \qquad (3.1.1)$$

It is a real graded commutative ring, called the *Grassmann algebra*. A Grassmann algebra, seen as an additive group, admits the decomposition

$$\Lambda = \mathbb{R} \oplus R = \mathbb{R} \oplus R_0 \oplus R_1 = \mathbb{R} \oplus (\Lambda_1)^2 \oplus \Lambda_1, \qquad (3.1.2)$$

where R is the *ideal of nilpotents* of Λ. The corresponding projections $\sigma : \Lambda \to \mathbb{R}$ and $s : \Lambda \to R$ are called the *body* and *soul* maps, respectively.

Remark 3.1.1. There is a different definition of a Grassmann algebra [85] which is equivalent to the above mentioned one only in the case of an infinite-dimensional vector space V [28].

Hereafter, we restrict our consideration to Grassmann algebras of finite rank. Given a basis $\{c^i\}$ for the vector space V, the elements of the Grassmann algebra Λ (3.1.1) take the form

$$a = \sum_{k=0,1,\ldots} \sum_{(i_1 \cdots i_k)} a_{i_1 \cdots i_k} c^{i_1} \cdots c^{i_k}, \qquad (3.1.3)$$

where the second sum runs through all the tuples $(i_1 \cdots i_k)$ such that no two of them are permutations of each other. The Grassmann algebra Λ becomes a graded commutative Banach ring if its elements (3.1.3) are endowed with the norm

$$\|a\| = \sum_{k=0} \sum_{(i_1 \cdots i_k)} |a_{i_1 \cdots i_k}|.$$

Let B be a graded vector space. Given a Grassmann algebra Λ, it can be brought into a graded Λ-module

$$\Lambda B = (\Lambda B)_0 \oplus (\Lambda B)_1 = (\Lambda_0 \otimes B_0 \oplus \Lambda_1 \otimes B_1) \oplus (\Lambda_1 \otimes B_0 \oplus \Lambda_0 \otimes B_1),$$

called a *superspace*. The superspace

$$B^{n|m} = [(\overset{n}{\oplus}\Lambda_0) \oplus (\overset{m}{\oplus}\Lambda_1)] \oplus [(\overset{n}{\oplus}\Lambda_1) \oplus (\overset{m}{\oplus}\Lambda_0)] \qquad (3.1.4)$$

is said to be (n,m)-dimensional. The graded Λ_0-module

$$B^{n,m} = (\overset{n}{\oplus}\Lambda_0) \oplus (\overset{m}{\oplus}\Lambda_1)$$

is called an (n,m)-dimensional *supervector space*. Whenever referring to a topology on a supervector space $B^{n,m}$, we will mean the Euclidean topology on a $2^{N-1}[n+m]$-dimensional real vector space.

Given a superspace $B^{n|m}$ over a Grassmann algebra Λ, any Λ-module endomorphism of $B^{n|m}$ is represented by an $(n+m) \times (n+m)$ matrix

$$L = \begin{pmatrix} L_1 & L_2 \\ L_3 & L_4 \end{pmatrix} \qquad (3.1.5)$$

with entries in Λ. It is called a *supermatrix*. A supermatrix L (3.1.5) is
- *even* if L_1 and L_4 have even entries, while L_2 and L_3 have the odd ones;

- *odd* if L_1 and L_4 have odd entries, while L_2 and L_3 have the even ones.

Endowed with this gradation, the set of supermatrices (3.1.5) is a graded Λ-ring.

The notion of a trace is extended to supermatrices (3.1.5) as the *supertrace*
$$\operatorname{Str} L = \operatorname{Tr} L_1 - (-1)^{[L]} \operatorname{Tr} L_4.$$
For instance, $\operatorname{Str}(\mathbf{1}) = n - m$.

A *supertransposition* L^{st} of a supermatrix L is defined as the supermatrix
$$L^{st} = \begin{pmatrix} L_1^t & (-1)^{[L]} L_3^t \\ -(-1)^{[L]} L_2^t & L_4^t \end{pmatrix},$$
where L^t denotes the ordinary matrix transposition. There are the relations
$$\operatorname{Str}(L^{st}) = \operatorname{Str} L,$$
$$(LL')^{st} = (-1)^{[L][L']} L'^{st} L^{st},$$
$$\operatorname{Str}([L, L']) = 0.$$

Let us consider invertible supermatrices L (3.1.5). One can show that a supermatrix L is invertible only if it is even and if and only if either the matrices L_1 and L_4 are invertible or the real matrix $\sigma(L)$ is invertible, where σ is the body map. A *superdeterminant* of $L \in GL(n|m; \Lambda)$ is defined as
$$\operatorname{Sdet} L = \det(L_1 - L_2 L_4^{-1} L_3)(\det L_4^{-1}).$$
It satisfies the relations
$$\operatorname{Sdet}(LL') = (\operatorname{Sdet} L)(\operatorname{Sdet} L'),$$
$$\operatorname{Sdet}(L^{st}) = \operatorname{Sdet} L.$$
Invertible supermatrices constitute a group $GL(n|m; \Lambda)$, called the *general linear supergroup*.

Let $\mathcal{K}$ be a graded commutative ring. A graded commutative (non-associative) $\mathcal{K}$-algebra $\mathfrak{g}$ is called a *Lie $\mathcal{K}$-superalgebra* if its product $[.,.]$, called the *Lie superbracket*, obeys the relations
$$[\varepsilon, \varepsilon'] = -(-1)^{[\varepsilon][\varepsilon']} [\varepsilon', \varepsilon],$$
$$(-1)^{[\varepsilon][\varepsilon'']} [\varepsilon, [\varepsilon', \varepsilon'']] + (-1)^{[\varepsilon'][\varepsilon]} [\varepsilon', [\varepsilon'', \varepsilon]] + (-1)^{[\varepsilon''][\varepsilon']} [\varepsilon'', [\varepsilon, \varepsilon']] = 0.$$
Obviously, the even part $\mathfrak{g}_0$ of a Lie $\mathcal{K}$-superalgebra $\mathfrak{g}$ is a Lie $\mathcal{K}_0$-algebra. A graded $\mathcal{K}$-module P is called a $\mathfrak{g}$-*module* if it is provided with a $\mathcal{K}$-bilinear map
$$\mathfrak{g} \times P \ni (\varepsilon, p) \to \varepsilon p \in P, \qquad [\varepsilon p] = ([\varepsilon] + [p]) \operatorname{mod} 2,$$
$$[\varepsilon, \varepsilon'] p = (\varepsilon \circ \varepsilon' - (-1)^{[\varepsilon][\varepsilon']} \varepsilon' \circ \varepsilon) p.$$

3.2 Grassmann-graded differential calculus

Linear differential operators on graded modules over a graded commutative ring are defined similarly to those in commutative geometry (see Section 10.2).

Let $\mathcal{K}$ be a graded commutative ring and $\mathcal{A}$ a graded commutative $\mathcal{K}$-ring. Let P and Q be graded $\mathcal{A}$-modules. The graded $\mathcal{K}$-module $\text{Hom}_{\mathcal{K}}(P,Q)$ of graded $\mathcal{K}$-module homomorphisms $\Phi : P \to Q$ can be endowed with the two graded $\mathcal{A}$-module structures

$$(a\Phi)(p) = a\Phi(p), \qquad (\Phi \bullet a)(p) = \Phi(ap), \qquad a \in \mathcal{A}, \quad p \in P, \qquad (3.2.1)$$

called $\mathcal{A}$- and $\mathcal{A}^{\bullet}$-module structures, respectively. Let us put

$$\delta_a \Phi = a\Phi - (-1)^{[a][\Phi]}\Phi \bullet a, \qquad a \in \mathcal{A}. \qquad (3.2.2)$$

An element $\Delta \in \text{Hom}_{\mathcal{K}}(P,Q)$ is said to be a Q-valued *graded differential operator* of order s on P if

$$\delta_{a_0} \circ \cdots \circ \delta_{a_s} \Delta = 0$$

for any tuple of $s+1$ elements $a_0, \ldots, a_s$ of $\mathcal{A}$. The set $\text{Diff}_s(P,Q)$ of these operators inherits the graded module structures (3.2.1).

In particular, zero order graded differential operators obey the condition

$$\delta_a \Delta(p) = a\Delta(p) - (-1)^{[a][\Delta]}\Delta(ap) = 0, \qquad a \in \mathcal{A}, \quad p \in P,$$

i.e., they coincide with graded $\mathcal{A}$-module morphisms $P \to Q$. A first order graded differential operator Δ satisfies the relation

$$\delta_a \circ \delta_b \Delta(p) = ab\Delta(p) - (-1)^{([b]+[\Delta])[a]}b\Delta(ap) - (-1)^{[b][\Delta]}a\Delta(bp) +$$
$$(-1)^{[b][\Delta]+([\Delta]+[b])[a]} = 0, \qquad a,b \in \mathcal{A}, \quad p \in P.$$

For instance, let $P = \mathcal{A}$. Any zero order Q-valued graded differential operator Δ on $\mathcal{A}$ is defined by its value $\Delta(1)$. Then there is a graded $\mathcal{A}$-module isomorphism

$$\text{Diff}_0(\mathcal{A},Q) = Q, \qquad Q \ni q \to \Delta_q \in \text{Diff}_0(\mathcal{A},Q),$$

where Δ_q is given by the equality $\Delta_q(1) = q$. A first order Q-valued graded differential operator Δ on $\mathcal{A}$ fulfils the condition

$$\Delta(ab) = \Delta(a)b + (-1)^{[a][\Delta]}a\Delta(b) - (-1)^{([b]+[a])[\Delta]}ab\Delta(1), \qquad a,b \in \mathcal{A}.$$

It is called a Q-valued *graded derivation* of $\mathcal{A}$ if $\Delta(1) = 0$, i.e., the Grassmann-graded Leibniz rule

$$\Delta(ab) = \Delta(a)b + (-1)^{[a][\Delta]}a\Delta(b), \qquad a,b \in \mathcal{A}, \qquad (3.2.3)$$

holds. One obtains at once that any first order graded differential operator on $\mathcal{A}$ falls into the sum
$$\Delta(a) = \Delta(1)a + [\Delta(a) - \Delta(1)a]$$
of a zero order graded differential operator $\Delta(1)a$ and a graded derivation $\Delta(a) - \Delta(1)a$. If ∂ is a graded derivation of $\mathcal{A}$, then $a\partial$ is so for any $a \in \mathcal{A}$. Hence, graded derivations of $\mathcal{A}$ constitute a graded $\mathcal{A}$-module $\mathfrak{d}(\mathcal{A}, Q)$, called the *graded derivation module*.

If $Q = \mathcal{A}$, the graded derivation module $\mathfrak{d}\mathcal{A}$ also is a Lie superalgebra over the graded commutative ring $\mathcal{K}$ with respect to the superbracket
$$[u, u'] = u \circ u' - (-1)^{[u][u']} u' \circ u, \qquad u, u' \in \mathcal{A}. \tag{3.2.4}$$
We have the graded $\mathcal{A}$-module decomposition
$$\mathrm{Diff}\,_1(\mathcal{A}) = \mathcal{A} \oplus \mathfrak{d}\mathcal{A}. \tag{3.2.5}$$

Since $\mathfrak{d}\mathcal{A}$ is a Lie $\mathcal{K}$-superalgebra, let us consider the Chevalley–Eilenberg complex $C^*[\mathfrak{d}\mathcal{A}; \mathcal{A}]$ where the graded commutative ring $\mathcal{A}$ is a regarded as a $\mathfrak{d}\mathcal{A}$-module [46; 60]. It is the complex
$$0 \to \mathcal{A} \xrightarrow{d} C^1[\mathfrak{d}\mathcal{A}; \mathcal{A}] \xrightarrow{d} \cdots C^k[\mathfrak{d}\mathcal{A}; \mathcal{A}] \xrightarrow{d} \cdots \tag{3.2.6}$$
where
$$C^k[\mathfrak{d}\mathcal{A}; \mathcal{A}] = \mathrm{Hom}_{\mathcal{K}}(\overset{k}{\wedge}\mathfrak{d}\mathcal{A}, \mathcal{A})$$
are $\mathfrak{d}\mathcal{A}$-modules of $\mathcal{K}$-linear graded morphisms of the graded exterior products $\overset{k}{\wedge}\mathfrak{d}\mathcal{A}$ of the $\mathcal{K}$-module $\mathfrak{d}\mathcal{A}$ to $\mathcal{A}$. Let us bring homogeneous elements of $\overset{k}{\wedge}\mathfrak{d}\mathcal{A}$ into the form
$$\varepsilon_1 \wedge \cdots \wedge \varepsilon_r \wedge \epsilon_{r+1} \wedge \cdots \wedge \epsilon_k, \qquad \varepsilon_i \in \mathfrak{d}\mathcal{A}_0, \quad \epsilon_j \in \mathfrak{d}\mathcal{A}_1.$$
Then the even coboundary operator d of the complex (3.2.6) is given by the expression
$$dc(\varepsilon_1 \wedge \cdots \wedge \varepsilon_r \wedge \epsilon_1 \wedge \cdots \wedge \epsilon_s) = \tag{3.2.7}$$
$$\sum_{i=1}^{r} (-1)^{i-1} \varepsilon_i c(\varepsilon_1 \wedge \cdots \widehat{\varepsilon}_i \cdots \wedge \varepsilon_r \wedge \epsilon_1 \wedge \cdots \epsilon_s) +$$
$$\sum_{j=1}^{s} (-1)^r \varepsilon_i c(\varepsilon_1 \wedge \cdots \wedge \varepsilon_r \wedge \epsilon_1 \wedge \cdots \widehat{\epsilon}_j \cdots \wedge \epsilon_s) +$$
$$\sum_{1 \leq i < j \leq r} (-1)^{i+j} c([\varepsilon_i, \varepsilon_j] \wedge \varepsilon_1 \wedge \cdots \widehat{\varepsilon}_i \cdots \widehat{\varepsilon}_j \cdots \wedge \varepsilon_r \wedge \epsilon_1 \wedge \cdots \wedge \epsilon_s) +$$
$$\sum_{1 \leq i < j \leq s} c([\epsilon_i, \epsilon_j] \wedge \varepsilon_1 \wedge \cdots \wedge \varepsilon_r \wedge \epsilon_1 \wedge \cdots \widehat{\epsilon}_i \cdots \widehat{\epsilon}_j \cdots \wedge \epsilon_s) +$$
$$\sum_{1 \leq i < r, 1 \leq j \leq s} (-1)^{i+r+1} c([\varepsilon_i, \epsilon_j] \wedge \varepsilon_1 \wedge \cdots \widehat{\varepsilon}_i \cdots \wedge \varepsilon_r \wedge \epsilon_1 \wedge \cdots \widehat{\epsilon}_j \cdots \wedge \epsilon_s),$$

where the caret $\widehat{}$ denotes omission. This operator is called the *graded Chevalley–Eilenberg coboundary operator*.

Let us consider the extended Chevalley–Eilenberg complex

$$0 \to \mathcal{K} \xrightarrow{\text{in}} C^*[\mathfrak{d}\mathcal{A}; \mathcal{A}].$$

It is easily justified that this complex contains a subcomplex $\mathcal{O}^*[\mathfrak{d}\mathcal{A}]$ of $\mathcal{A}$-linear graded morphisms. The $\mathbb{N}$-graded module $\mathcal{O}^*[\mathfrak{d}\mathcal{A}]$ is provided with the structure of a bigraded $\mathcal{A}$-algebra with respect to the graded exterior product

$$\phi \wedge \phi'(u_1, ..., u_{r+s}) = \tag{3.2.8}$$
$$\sum_{i_1 < \cdots < i_r; j_1 < \cdots < j_s} \mathrm{Sgn}^{i_1 \cdots i_r j_1 \cdots j_s}_{1 \cdots r+s} \phi(u_{i_1}, \ldots, u_{i_r}) \phi'(u_{j_1}, \ldots, u_{j_s}),$$

$$\phi \in \mathcal{O}^r[\mathfrak{d}\mathcal{A}], \qquad \phi' \in \mathcal{O}^s[\mathfrak{d}\mathcal{A}], \qquad u_k \in \mathfrak{d}\mathcal{A},$$

where $u_1, \ldots, u_{r+s}$ are graded-homogeneous elements of $\mathfrak{d}\mathcal{A}$ and

$$u_1 \wedge \cdots \wedge u_{r+s} = \mathrm{Sgn}^{i_1 \cdots i_r j_1 \cdots j_s}_{1 \cdots r+s} u_{i_1} \wedge \cdots \wedge u_{i_r} \wedge u_{j_1} \wedge \cdots \wedge u_{j_s}.$$

The graded Chevalley–Eilenberg coboundary operator d (3.2.7) and the graded exterior product $\wedge$ (3.2.8) bring $\mathcal{O}^*[\mathfrak{d}\mathcal{A}]$ into a *differential bigraded algebra* whose elements obey the relations

$$\phi \wedge \phi' = (-1)^{|\phi||\phi'|+[\phi][\phi']} \phi' \wedge \phi, \tag{3.2.9}$$
$$d(\phi \wedge \phi') = d\phi \wedge \phi' + (-1)^{|\phi||\phi'|} \phi \wedge d\phi'. \tag{3.2.10}$$

It is called the *graded Chevalley–Eilenberg differential calculus* over a graded commutative $\mathcal{K}$-ring $\mathcal{A}$. In particular, we have

$$\mathcal{O}^1[\mathfrak{d}\mathcal{A}] = \mathrm{Hom}_{\mathcal{A}}(\mathfrak{d}\mathcal{A}, \mathcal{A}) = \mathfrak{d}\mathcal{A}^*. \tag{3.2.11}$$

One can extend this duality relation to the *graded interior product* of $u \in \mathfrak{d}\mathcal{A}$ with any element $\phi \in \mathcal{O}^*[\mathfrak{d}\mathcal{A}]$ by the rules

$$u \rfloor (bda) = (-1)^{[u][b]} u(a), \qquad a, b \in \mathcal{A},$$
$$u \rfloor (\phi \wedge \phi') = (u \rfloor \phi) \wedge \phi' + (-1)^{|\phi|+[\phi][u]} \phi \wedge (u \rfloor \phi'). \tag{3.2.12}$$

As a consequence, any graded derivation $u \in \mathfrak{d}\mathcal{A}$ of $\mathcal{A}$ yields a derivation

$$\mathbf{L}_u \phi = u \rfloor d\phi + d(u \rfloor \phi), \qquad \phi \in \mathcal{O}^*, \qquad u \in \mathfrak{d}\mathcal{A}, \tag{3.2.13}$$
$$\mathbf{L}_u(\phi \wedge \phi') = \mathbf{L}_u(\phi) \wedge \phi' + (-1)^{[u][\phi]} \phi \wedge \mathbf{L}_u(\phi'),$$

called the *graded Lie derivative* of the differential bigraded algebra $\mathcal{O}^*[\mathfrak{d}\mathcal{A}]$.

The minimal graded Chevalley–Eilenberg differential calculus $\mathcal{O}^*\mathcal{A} \subset \mathcal{O}^*[\mathfrak{d}\mathcal{A}]$ over a graded commutative ring $\mathcal{A}$ consists of the monomials

$$a_0 da_1 \wedge \cdots \wedge da_k, \qquad a_i \in \mathcal{A}.$$

The corresponding complex

$$0 \to \mathcal{K} \longrightarrow \mathcal{A} \xrightarrow{d} \mathcal{O}^1\mathcal{A} \xrightarrow{d} \cdots \mathcal{O}^k\mathcal{A} \xrightarrow{d} \cdots \qquad (3.2.14)$$

is called the *bigraded de Rham complex* of a graded commutative $\mathcal{K}$-ring $\mathcal{A}$.

Following the construction of a connection in commutative geometry (see Section 10.2), one comes to the notion of a connection on modules over a real graded commutative ring $\mathcal{A}$. The following are the straightforward counterparts of Definitions 10.2.2 and 10.2.3.

Definition 3.2.1. A *connection* on a graded $\mathcal{A}$-module P is a graded $\mathcal{A}$-module morphism

$$\mathfrak{d}\mathcal{A} \ni u \to \nabla_u \in \mathrm{Diff}_1(P, P) \qquad (3.2.15)$$

such that the first order differential operators ∇_u obey the *Grassmann-graded Leibniz rule*

$$\nabla_u(ap) = u(a)p + (-1)^{[a][u]} a \nabla_u(p), \quad a \in \mathcal{A}, \quad p \in P. \qquad (3.2.16)$$

Definition 3.2.2. Let P in Definition 3.2.1 be a graded commutative $\mathcal{A}$-ring and $\mathfrak{d}P$ the derivation module of P as a graded commutative $\mathcal{K}$-ring. A *connection* on a graded commutative $\mathcal{A}$-ring P is a graded $\mathcal{A}$-module morphism

$$\mathfrak{d}\mathcal{A} \ni u \to \nabla_u \in \mathfrak{d}P, \qquad (3.2.17)$$

which is a connection on P as an $\mathcal{A}$-module, i.e., it obeys the graded Leibniz rule (3.2.16).

3.3 Geometry of graded manifolds

In accordance with Serre–Swan Theorem 3.3.2 below, if a real graded commutative algebra $\mathcal{A}$ is generated by a projective module of finite rank over the ring $C^\infty(Z)$ of smooth functions on some manifold Z, then $\mathcal{A}$ is isomorphic to the algebra of graded functions on a graded manifold with a body Z, and *vice versa*. Then the minimal graded Chevalley–Eilenberg differential calculus $\mathcal{O}^*\mathcal{A}$ over $\mathcal{A}$ is the differential bigraded algebra of graded exterior forms on this graded manifold.

A *graded manifold* of dimension (n, m) is defined as a local-ringed space $(Z, \mathfrak{A})$ where Z is an n-dimensional smooth manifold Z and $\mathfrak{A} = \mathfrak{A}_0 \oplus \mathfrak{A}_1$ is a sheaf of graded commutative algebras of rank m such that [10]:

- there is the exact sequence of sheaves

$$0 \to \mathcal{R} \to \mathfrak{A} \xrightarrow{\sigma} C_Z^\infty \to 0, \qquad \mathcal{R} = \mathfrak{A}_1 + (\mathfrak{A}_1)^2, \qquad (3.3.1)$$

where C_Z^∞ is the sheaf of smooth real functions on Z;
- $\mathcal{R}/\mathcal{R}^2$ is a locally free sheaf of C_Z^∞-modules of finite rank (with respect to pointwise operations), and the sheaf $\mathfrak{A}$ is locally isomorphic to the exterior product $\wedge_{C_Z^\infty}(\mathcal{R}/\mathcal{R}^2)$.

The sheaf $\mathfrak{A}$ is called a *structure sheaf* of a graded manifold $(Z,\mathfrak{A})$, and a manifold Z is said to be the *body* of $(Z,\mathfrak{A})$. Sections of the sheaf $\mathfrak{A}$ are called *graded functions* on a graded manifold $(Z,\mathfrak{A})$. They make up a graded commutative $C^\infty(Z)$-ring $\mathfrak{A}(Z)$ called the *structure ring* of $(Z,\mathfrak{A})$.

A graded manifold $(Z,\mathfrak{A})$ possesses the following local structure. Given a point $z \in Z$, there exists its open neighborhood U, called a *splitting domain*, such that

$$\mathfrak{A}(U) = C^\infty(U) \otimes \wedge \mathbb{R}^m. \qquad (3.3.2)$$

This means that the restriction $\mathfrak{A}|_U$ of the structure sheaf $\mathfrak{A}$ to U is isomorphic to the sheaf $C_U^\infty \otimes \wedge \mathbb{R}^m$ of sections of some exterior bundle

$$\wedge E_U^* = U \times \wedge \mathbb{R}^m \to U.$$

The well-known *Batchelor theorem* [10; 16] states that such a structure of a graded manifold is global as follows.

Theorem 3.3.1. *Let $(Z,\mathfrak{A})$ be a graded manifold. There exists a vector bundle $E \to Z$ with an m-dimensional typical fibre V such that the structure sheaf $\mathfrak{A}$ of $(Z,\mathfrak{A})$ is isomorphic to the structure sheaf $\mathfrak{A}_E = S_{\wedge E^*}$ of germs of sections of the exterior bundle $\wedge E^*$ (1.1.11), whose typical fibre is the Grassmann algebra $\wedge V^*$.*

Proof. The local sheaves $C_U^\infty \otimes \wedge \mathbb{R}^m$ are glued into the global structure sheaf $\mathfrak{A}$ of the graded manifold $(Z,\mathfrak{A})$ by means of transition functions in Theorem 10.7.1, which are assembled into a cocycle of the sheaf $\mathrm{Aut}\,(\wedge \mathbb{R}^m)^\infty$ of smooth mappings from Z to $\mathrm{Aut}\,(\wedge \mathbb{R}^m)$. The proof is based on the bijection between the cohomology sets $H^1(Z;\mathrm{Aut}\,(\wedge \mathbb{R}^m)^\infty)$ and $H^1(Z; GL(m,\mathbb{R}^m)^\infty)$. □

It should be emphasized that Batchelor's isomorphism in Theorem 3.3.1 fails to be canonical. In field models, it however is fixed from the beginning. Therefore, we restrict our consideration to graded manifolds $(Z,\mathfrak{A}_E)$ whose structure sheaf is the sheaf of germs of sections of some exterior bundle $\wedge E^*$. We agree to call $(Z,\mathfrak{A}_E)$ a *simple graded manifold* modelled over a vector bundle $E \to Z$, called its *characteristic vector bundle*. Accordingly, the structure ring $\mathcal{A}_E$ of a simple graded manifold $(Z, \mathfrak{A}_E)$ is the structure module

$$\mathcal{A}_E = \mathfrak{A}_E(Z) = \wedge E^*(Z) \qquad (3.3.3)$$

of sections of the exterior bundle $\wedge E^*$. Automorphisms of a simple graded manifold $(Z, \mathfrak{A}_E)$ are restricted to those induced by automorphisms of its characteristic vector bundles $E \to Z$ (see Remark 3.3.2).

Combining Batchelor Theorem 3.3.1 and classical Serre–Swan Theorem 10.9.3, we come to the following *Serre–Swan theorem for graded manifolds* [14].

Theorem 3.3.2. *Let Z be a smooth manifold. A graded commutative $C^\infty(Z)$-algebra $\mathcal{A}$ is isomorphic to the structure ring of a graded manifold with a body Z if and only if it is the exterior algebra of some projective $C^\infty(Z)$-module of finite rank.*

Proof. By virtue of the Batchelor theorem, any graded manifold is isomorphic to a simple graded manifold $(Z, \mathfrak{A}_E)$ modelled over some vector bundle $E \to Z$. Its structure ring $\mathcal{A}_E$ (3.3.3) of graded functions consists of sections of the exterior bundle $\wedge E^*$ (1.1.11). The classical Serre–Swan theorem states that a $C^\infty(Z)$-module is isomorphic to the module of sections of a smooth vector bundle over Z if and only if it is a projective module of finite rank. $\square$

Given a graded manifold $(Z, \mathfrak{A}_E)$, every trivialization chart $(U; z^A, y^a)$ of the vector bundle $E \to Z$ yields a splitting domain $(U; z^A, c^a)$ of $(Z, \mathfrak{A}_E)$. Graded functions on such a chart are Λ-valued functions

$$f = \sum_{k=0}^m \frac{1}{k!} f_{a_1 \ldots a_k}(z) c^{a_1} \cdots c^{a_k}, \qquad (3.3.4)$$

where $f_{a_1 \ldots a_k}(z)$ are smooth functions on U and $\{c^a\}$ is the fibre basis for E^*. In particular, the sheaf epimorphism σ in (3.3.1) is induced by the body map of Λ. One calls $\{z^A, c^a\}$ the *local basis for the graded manifold* $(Z, \mathfrak{A}_E)$ [10]. Transition functions $y'^a = \rho^a_b(z^A) y^b$ of bundle coordinates on $E \to Z$ induce the corresponding transformation

$$c'^a = \rho^a_b(z^A) c^b \qquad (3.3.5)$$

of the associated local basis for the graded manifold $(Z, \mathfrak{A}_E)$ and the according coordinate transformation law of graded functions (3.3.4).

Remark 3.3.1. Strictly speaking, elements c^a of the local basis for a graded manifold are locally constant sections c^a of $E^* \to X$ such that $y_b \circ c^a = \delta_b^a$. Therefore, graded functions are locally represented by Λ-valued functions (3.3.4), but they are not Λ-valued functions on a manifold Z because of the transformation law (3.3.5).

Remark 3.3.2. In general, automorphisms of a graded manifold take the form

$$c'^a = \rho^a(z^A, c^b), \qquad (3.3.6)$$

where $\rho^a(z^A, c^b)$ are local graded functions. Considering a simple graded manifold $(Z, \mathfrak{A}_E)$, we restrict the class of graded manifold transformations (3.3.6) to the linear ones (3.3.5), compatible with given Batchelor's isomorphism.

Let $E \to Z$ and $E' \to Z$ be vector bundles and $\Phi : E \to E'$ their bundle morphism over a morphism $\varphi : Z \to Z'$. Then every section s^* of the dual bundle $E'^* \to Z'$ defines the pull-back section $\Phi^* s^*$ of the dual bundle $E^* \to Z$ by the law

$$v_z \rfloor \Phi^* s^*(z) = \Phi(v_z) \rfloor s^*(\varphi(z)), \qquad v_z \in E_z.$$

It follows that the bundle morphism (Φ, φ) yields a *morphism of simple graded manifolds*

$$\widehat{\Phi} : (Z, \mathfrak{A}_E) \to (Z', \mathfrak{A}_{E'}) \qquad (3.3.7)$$

treated as local-ringed spaces (see Section 10.8). This is a pair $(\varphi, \varphi_* \circ \Phi^*)$ of a morphism φ of body manifolds and the composition $\varphi_* \circ \Phi^*$ of the pull-back

$$\mathcal{A}_{E'} \ni f \to \Phi^* f \in \mathcal{A}_E$$

of graded functions and the direct image φ_* of the sheaf $\mathfrak{A}_E$ onto Z'. Relative to local bases (z^A, c^a) and (z'^A, c'^a) for $(Z, \mathfrak{A}_E)$ and $(Z', \mathfrak{A}_{E'})$, the morphism (3.3.7) of graded manifolds reads

$$\widehat{\Phi}(z) = \varphi(z), \qquad \widehat{\Phi}(c'^a) = \Phi_b^a(z) c^b.$$

Given a graded manifold $(Z, \mathfrak{A})$, by the *sheaf $\mathfrak{d}\mathfrak{A}$ of graded derivations* of $\mathfrak{A}$ is meant a subsheaf of endomorphisms of the structure sheaf $\mathfrak{A}$ such that any section $u \in \mathfrak{d}\mathfrak{A}(U)$ of $\mathfrak{d}\mathfrak{A}$ over an open subset $U \subset Z$ is a graded derivation of the real graded commutative algebra $\mathfrak{A}(U)$, i.e., $u \in \mathfrak{d}(\mathfrak{A}(U))$.

3.3. Geometry of graded manifolds

Conversely, one can show that, given open sets $U' \subset U$, there is a surjection of the graded derivation modules

$$\mathfrak{d}(\mathfrak{A}(U)) \to \mathfrak{d}(\mathfrak{A}(U'))$$

[10]. It follows that any graded derivation of the local graded algebra $\mathfrak{A}(U)$ also is a local section over U of the sheaf $\mathfrak{d}\mathfrak{A}$. Global sections of $\mathfrak{d}\mathfrak{A}$ are called *graded vector fields* on the graded manifold $(Z, \mathfrak{A})$. They make up the graded derivation module $\mathfrak{d}\mathfrak{A}(Z)$ of the real graded commutative ring $\mathfrak{A}(Z)$. This module is a real Lie superalgebra with respect to the superbracket (3.2.4).

A key point is that graded vector fields $u \in \mathfrak{d}\mathcal{A}_E$ on a simple graded manifold $(Z, \mathfrak{A}_E)$ can be represented by sections of some vector bundle as follows [60].

Due to the canonical splitting $VE = E \times E$, the vertical tangent bundle VE of $E \to Z$ can be provided with the fibre bases $\{\partial/\partial c^a\}$, which are the duals of the bases $\{c^a\}$. Then graded vector fields on a splitting domain $(U; z^A, c^a)$ of $(Z, \mathfrak{A}_E)$ read

$$u = u^A \partial_A + u^a \frac{\partial}{\partial c^a}, \qquad (3.3.8)$$

where u^λ, u^a are local graded functions on U. In particular,

$$\frac{\partial}{\partial c^a} \circ \frac{\partial}{\partial c^b} = -\frac{\partial}{\partial c^b} \circ \frac{\partial}{\partial c^a}, \qquad \partial_A \circ \frac{\partial}{\partial c^a} = \frac{\partial}{\partial c^a} \circ \partial_A.$$

The graded derivations (3.3.8) act on graded functions $f \in \mathfrak{A}_E(U)$ (3.3.4) by the rule

$$u(f_{a...b} c^a \cdots c^b) = u^A \partial_A(f_{a...b}) c^a \cdots c^b + u^k f_{a...b} \frac{\partial}{\partial c^k} \rfloor (c^a \cdots c^b). \qquad (3.3.9)$$

This rule implies the corresponding coordinate transformation law

$$u'^A = u^A, \qquad u'^a = \rho^a_j u^j + u^A \partial_A(\rho^a_j) c^j$$

of graded vector fields. It follows that graded vector fields (3.3.8) can be represented by sections of the following vector bundle $\mathcal{V}_E \to Z$. This vector bundle is locally isomorphic to the vector bundle

$$\mathcal{V}_E|_U \approx \wedge E^* \underset{Z}{\otimes} (E \underset{Z}{\oplus} TZ)|_U, \qquad (3.3.10)$$

and is characterized by an atlas of bundle coordinates

$$(z^A, z^A_{a_1...a_k}, v^i_{b_1...b_k}), \qquad k = 0, \ldots, m,$$

possessing the transition functions
$$z'^A_{i_1\ldots i_k} = \rho^{-1}{}^{a_1}_{i_1} \cdots \rho^{-1}{}^{a_k}_{i_k} z^A_{a_1\ldots a_k},$$
$$v'^i_{j_1\ldots j_k} = \rho^{-1}{}^{b_1}_{j_1} \cdots \rho^{-1}{}^{b_k}_{j_k} \left[\rho^i_j v^j_{b_1\ldots b_k} + \frac{k!}{(k-1)!} z^A_{b_1\ldots b_{k-1}} \partial_A \rho^i_{b_k}\right],$$
which fulfil the cocycle condition (1.1.4). Thus, the graded derivation module $\mathfrak{d}\mathcal{A}_E$ is isomorphic to the structure module $\mathcal{V}_E(Z)$ of global sections of the vector bundle $\mathcal{V}_E \to Z$.

There is the exact sequence
$$0 \to \wedge E^* \underset{Z}{\otimes} E \to \mathcal{V}_E \to \wedge E^* \underset{Z}{\otimes} TZ \to 0 \qquad (3.3.11)$$
of vector bundles over Z. Its splitting
$$\widetilde{\gamma}: \dot{z}^A \partial_A \to \dot{z}^A \left(\partial_A + \widetilde{\gamma}^a_A \frac{\partial}{\partial c^a}\right) \qquad (3.3.12)$$
transforms every vector field τ on Z into the graded vector field
$$\tau = \tau^A \partial_A \to \nabla_\tau = \tau^A \left(\partial_A + \widetilde{\gamma}^a_A \frac{\partial}{\partial c^a}\right), \qquad (3.3.13)$$
which is a graded derivation of the real graded commutative ring $\mathcal{A}_E$ (3.3.3) satisfying the Leibniz rule
$$\nabla_\tau(sf) = (\tau\rfloor ds)f + s\nabla_\tau(f), \quad f \in \mathcal{A}_E, \quad s \in C^\infty(Z).$$
It follows that the splitting (3.3.12) of the exact sequence (3.3.11) yields a connection on the graded commutative $C^\infty(Z)$-ring $\mathcal{A}_E$ in accordance with Definition 3.2.2 [60]. It is called a *graded connection* on the simple graded manifold $(Z, \mathfrak{A}_E)$. In particular, this connection provides the corresponding horizontal splitting
$$u = u^A \partial_A + u^a \frac{\partial}{\partial c^a} = u^A \left(\partial_A + \widetilde{\gamma}^a_A \frac{\partial}{\partial c^a}\right) + (u^a - u^A \widetilde{\gamma}^a_A)\frac{\partial}{\partial c^a}$$
of graded vector fields. In accordance with Theorem 1.1.12, a graded connection (3.3.12) always exists.

Remark 3.3.3. By virtue of the isomorphism (3.3.2), any connection $\widetilde{\gamma}$ on a graded manifold $(Z, \mathfrak{A})$, restricted to a splitting domain U, takes the form (3.3.12). Given two splitting domains U and U' of $(Z, \mathfrak{A})$ with the transition functions (3.3.6), the connection components $\widetilde{\gamma}^a_A$ obey the transformation law
$$\widetilde{\gamma}'^a_A = \widetilde{\gamma}^b_A \frac{\partial}{\partial c^b}\rho^a + \partial_A \rho^a. \qquad (3.3.14)$$
If U and U' are the trivialization charts of the same vector bundle E in Theorem 3.3.1 together with the transition functions (3.3.5), the transformation law (3.3.14) takes the form
$$\widetilde{\gamma}'^a_A = \rho^a_b(z)\widetilde{\gamma}^b_A + \partial_A \rho^a_b(z)c^b. \qquad (3.3.15)$$

3.3. Geometry of graded manifolds

Remark 3.3.4. It should be emphasized that the above notion of a graded connection is a connection on the graded commutative ring $\mathcal{A}_E$ seen as a $C^\infty(Z)$-module. It differs from that of a connection on a graded fibre bundle $(Z, \mathfrak{A}) \to (X, \mathcal{B})$ [2]. The latter is a connection on a graded $\mathcal{B}(X)$-module.

Remark 3.3.5. Every linear connection
$$\gamma = dz^A \otimes (\partial_A + \gamma_A{}^a{}_b y^b \partial_a)$$
on a vector bundle $E \to Z$ yields the graded connection
$$\gamma_S = dz^A \otimes \left(\partial_A + \gamma_A{}^a{}_b c^b \frac{\partial}{\partial c^a}\right) \quad (3.3.16)$$
on the simple graded manifold $(Z, \mathfrak{A}_E)$ modelled over E. In view of Remark 3.3.3, γ_S also is a graded connection on the graded manifold $(Z, \mathfrak{A}) \cong (Z, \mathfrak{A}_E)$, but its linear form (3.3.16) is not maintained under the transformation law (3.3.14).

Given the structure ring $\mathcal{A}_E$ of graded functions on a simple graded manifold $(Z, \mathfrak{A}_E)$ and the real Lie superalgebra $\mathfrak{d}\mathcal{A}_E$ of its graded derivations, let us consider the graded Chevalley–Eilenberg differential calculus
$$\mathcal{S}^*[E; Z] = \mathcal{O}^*[\mathfrak{d}\mathcal{A}_E] \quad (3.3.17)$$
over $\mathcal{A}_E$. Since the graded derivation module $\mathfrak{d}\mathcal{A}_E$ is isomorphic to the structure module of sections of the vector bundle $\mathcal{V}_E \to Z$, elements of $\mathcal{S}^*[E; Z]$ are represented by sections of the exterior bundle $\wedge \overline{\mathcal{V}}_E$ of the $\mathcal{A}_E$-dual $\overline{\mathcal{V}}_E \to Z$ of $\mathcal{V}_E$. The bundle $\overline{\mathcal{V}}_E$ is locally isomorphic to the vector bundle
$$\overline{\mathcal{V}}_E|_U \approx (E^* \underset{Z}{\oplus} T^*Z)|_U. \quad (3.3.18)$$
With respect to the dual fibre bases $\{dz^A\}$ for T^*Z and $\{dc^b\}$ for E^*, sections of $\overline{\mathcal{V}}_E$ take the coordinate form
$$\phi = \phi_A dz^A + \phi_a dc^a,$$
together with transition functions
$$\phi'_a = \rho^{-1}{}^b{}_a \phi_b, \qquad \phi'_A = \phi_A + \rho^{-1}{}^b{}_a \partial_A(\rho^a_j)\phi_b c^j.$$
The duality isomorphism (3.2.11):
$$\mathcal{S}^1[E; Z] = \mathfrak{d}\mathcal{A}_E^*$$
is given by the graded interior product
$$u \rfloor \phi = u^A \phi_A + (-1)^{[\phi_a]} u^a \phi_a. \quad (3.3.19)$$

Elements of $\mathcal{S}^*[E;Z]$ are called *graded exterior forms* on on the graded manifold $(Z,\mathfrak{A}_E)$.

Seen as an $\mathcal{A}_E$-algebra, the differential bigraded algebra $\mathcal{S}^*[E;Z]$ (3.3.17) on a splitting domain $(U; z^A, c^a)$ is locally generated by the graded one-forms dz^A, dc^i such that

$$dz^A \wedge dc^i = -dc^i \wedge dz^A, \qquad dc^i \wedge dc^j = dc^j \wedge dc^i. \qquad (3.3.20)$$

Accordingly, the graded Chevalley–Eilenberg coboundary operator d (3.2.7), called the *graded!exterior differential*, reads

$$d\phi = dz^A \wedge \partial_A \phi + dc^a \wedge \frac{\partial}{\partial c^a}\phi,$$

where the derivatives ∂_λ, $\partial/\partial c^a$ act on coefficients of graded exterior forms by the formula (3.3.9), and they are graded commutative with the graded forms dz^A and dc^a. The formulas (3.2.9) – (3.2.13) hold.

Theorem 3.3.3. *The differential bigraded algebra $\mathcal{S}^*[E;Z]$ (3.3.17) is a minimal differential calculus over $\mathcal{A}_E$, i.e., it is generated by elements df, $f \in \mathcal{A}_E$.*

Proof. The proof follows that of Theorem 1.7.5. Since

$$\eth \mathcal{A}_E = \mathcal{V}_E(Z),$$

it is a projective $C^\infty(Z)$- and $\mathcal{A}_E$-module of finite rank, and so is its $\mathcal{A}_E$-dual $\mathcal{S}^1[E;Z]$. Hence, $\eth \mathcal{A}_E$ is the $\mathcal{A}_E$-dual of $\mathcal{S}^1[E;Z]$ and, consequently, $\mathcal{S}^1[E;Z]$ is generated by elements df, $f \in \mathcal{A}_E$. □

The bigraded de Rham complex (3.2.14) of the minimal graded Chevalley–Eilenberg differential calculus $\mathcal{S}^*[E;Z]$ reads

$$0 \to \mathbb{R} \to \mathcal{A}_E \xrightarrow{d} \mathcal{S}^1[E;Z] \xrightarrow{d} \cdots \mathcal{S}^k[E;Z] \xrightarrow{d} \cdots . \qquad (3.3.21)$$

Its cohomology $H^*(\mathcal{A}_E)$ is called the *de Rham cohomology of a simple graded manifold* $(Z,\mathfrak{A}_E)$.

In particular, given the differential graded algebra $\mathcal{O}^*(Z)$ of exterior forms on Z, there exist the canonical monomorphism

$$\mathcal{O}^*(Z) \to \mathcal{S}^*[E;Z] \qquad (3.3.22)$$

and the body epimorphism

$$\mathcal{S}^*[E;Z] \to \mathcal{O}^*(Z)$$

which are cochain morphisms of the de Rham complexes (3.3.21) and (10.9.12).

Theorem 3.3.4. *The de Rham cohomology of a simple graded manifold $(Z,\mathfrak{A}_E)$ equals the de Rham cohomology of its body Z.*

Proof. Let $\mathfrak{A}_E^k$ denote the sheaf of germs of graded k-forms on $(Z, \mathfrak{A}_E)$. Its structure module is $\mathcal{S}^k[E; Z]$. These sheaves constitute the complex

$$0 \to \mathbb{R} \longrightarrow \mathfrak{A}_E \xrightarrow{d} \mathfrak{A}_E^1 \xrightarrow{d} \cdots \mathfrak{A}_E^k \xrightarrow{d} \cdots . \qquad (3.3.23)$$

Its members $\mathfrak{A}_E^k$ are sheaves of C_Z^∞-modules on Z and, consequently, are fine and acyclic. Furthermore, the Poincaré lemma for graded exterior forms holds [10]. It follows that the complex (3.3.23) is a fine resolution of the constant sheaf $\mathbb{R}$ on a manifold Z. Then, by virtue of Theorem 10.7.4, there is an isomorphism

$$H^*(\mathcal{A}_E) = H^*(Z; \mathbb{R}) = H^*_{\mathrm{DR}}(Z) \qquad (3.3.24)$$

of the cohomology $H^*(\mathcal{A}_E)$ to the de Rham cohomology $H^*_{\mathrm{DR}}(Z)$ of the smooth manifold Z. □

Corollary 3.3.1. *The cohomology isomorphism (3.3.24) accompanies the cochain monomorphism (3.3.22). Hence, any closed graded exterior form is decomposed into a sum $\phi = \sigma + d\xi$ where σ is a closed exterior form on Z.*

3.4 Grassmann-graded variational bicomplex

As was mentioned above, extending jet formalism to odd variables, we consider graded manifolds of jets of smooth fibre bundles, but not jets of fibred graded manifolds.

Remark 3.4.1. To motivate this construction, let us return to the case of even variables in Section 2.1 when $Y \to X$ is a vector bundle. The jet bundles $J^k Y \to X$ also are vector bundles. Let $\mathcal{P}_\infty^* \subset \mathcal{O}_\infty^*$ be a subalgebra of exterior forms on these bundles whose coefficients are polynomial in their fibre coordinates. In particular, $\mathcal{P}_\infty^0$ is the ring of polynomials of these coordinates with coefficients in the ring $C^\infty(X)$. One can associate to such a polynomial of degree m a section of the symmetric product $\overset{m}{\vee}(J^k Y)^*$ of the dual to some jet bundle $J^k Y \to X$, and vice versa. Moreover, any element of $\mathcal{P}_\infty^*$ is an element of the Chevalley–Eilenberg differential calculus over $\mathcal{P}_\infty^0$.

Following this example, let us consider a vector bundle $F \to X$ and the simple graded manifolds $(X, \mathcal{A}_{J^r F})$ modelled over the vector bundles

$J^rF \to X$. There is the direct system of the corresponding differential bigraded algebras

$$\mathcal{S}^*[F;X] \longrightarrow \mathcal{S}^*[J^1F;X] \longrightarrow \cdots \mathcal{S}^*[J^rF;X] \longrightarrow \cdots$$

of graded exterior forms on graded manifolds $(X, \mathcal{A}_{J^rF})$. Its direct limit $\mathcal{S}^*_\infty[F;X]$ is the Grassmann-graded counterpart of the above mentioned differential graded algebra $\mathcal{P}^*_\infty$.

In order to describe Lagrangian theories both of even and odd fields, let us consider a composite bundle

$$F \to Y \to X \qquad (3.4.1)$$

where $F \to Y$ is a vector bundle provided with bundle coordinates (x^λ, y^i, q^a). We call the simple graded manifold $(Y, \mathfrak{A}_F)$ modelled over $F \to Y$ the *composite graded manifold*. Let us associate to this graded manifold the following differential bigraded algebra $\mathcal{S}^*_\infty[F;Y]$.

It is readily observed that the jet manifold J^rF of $F \to X$ is a vector bundle $J^rF \to J^rY$ coordinated by $(x^\lambda, y^i_\Lambda, q^a_\Lambda)$, $0 \le |\Lambda| \le r$. Let $(J^rY, \mathfrak{A}_r)$ be a simple graded manifold modelled over this vector bundle. Its local basis is $(x^\lambda, y^i_\Lambda, c^a_\Lambda)$, $0 \le |\Lambda| \le r$. Let

$$\mathcal{S}^*_r[F;Y] = \mathcal{S}^*_r[J^rF; J^rY] \qquad (3.4.2)$$

denote the bigraded differential algebra of graded exterior forms on the simple graded manifold $(J^rY, \mathfrak{A}_r)$. In particular, there is a cochain monomorphism

$$\mathcal{O}^*_r = \mathcal{O}^*(J^rY) \to \mathcal{S}^*_r[F;Y]. \qquad (3.4.3)$$

The surjection

$$\pi^{r+1}_r : J^{r+1}Y \to J^rY$$

yields an epimorphism of graded manifolds

$$(\pi^{r+1}_r, \widehat{\pi}^{r+1}_r) : (J^{r+1}Y, \mathfrak{A}_{r+1}) \to (J^rY, \mathfrak{A}_r),$$

including the sheaf monomorphism

$$\widehat{\pi}^{r+1}_r : \pi^{r+1*}_r \mathfrak{A}_r \to \mathfrak{A}_{r+1},$$

where $\pi^{r+1*}_r \mathfrak{A}_r$ is the pull-back onto $J^{r+1}Y$ of the continuous fibre bundle $\mathfrak{A}_r \to J^rY$. This sheaf monomorphism induces the monomorphism of the canonical presheaves $\overline{\mathfrak{A}}_r \to \overline{\mathfrak{A}}_{r+1}$, which associates to each open subset $U \subset J^{r+1}Y$ the ring of sections of $\mathfrak{A}_r$ over $\pi^{r+1}_r(U)$. Accordingly, there is a monomorphism of the structure rings

$$\pi^{r+1*}_r : \mathcal{S}^0_r[F;Y] \to \mathcal{S}^0_{r+1}[F;Y] \qquad (3.4.4)$$

3.4. Grassmann-graded variational bicomplex

of graded functions on graded manifolds $(J^rY, \mathfrak{A}_r)$ and $(J^{r+1}Y, \mathfrak{A}_{r+1})$. By virtue of Lemma 3.3.3, the differential calculus $\mathcal{S}_r^*[F;Y]$ and $\mathcal{S}_{r+1}^*[F;Y]$ are minimal. Therefore, the monomorphism (3.4.4) yields a monomorphism of differential bigraded algebras

$$\pi_r^{r+1*} : \mathcal{S}_r^*[F;Y] \to \mathcal{S}_{r+1}^*[F;Y]. \tag{3.4.5}$$

As a consequence, we have the direct system of differential bigraded algebras

$$\mathcal{S}^*[F;Y] \xrightarrow{\pi^*} \mathcal{S}_1^*[F;Y] \longrightarrow \cdots \mathcal{S}_{r-1}^*[F;Y] \xrightarrow{\pi_{r-1}^{r*}} \mathcal{S}_r^*[F;Y] \longrightarrow \cdots. \tag{3.4.6}$$

The differential bigraded algebra $\mathcal{S}_\infty^*[F;Y]$ that we associate to the composite graded manifold $(Y, \mathfrak{A}_F)$ is defined as the direct limit

$$\mathcal{S}_\infty^*[F;Y] = \varinjlim \mathcal{S}_r^*[F;Y] \tag{3.4.7}$$

of the direct system (3.4.6). It consists of all graded exterior forms $\phi \in \mathcal{S}^*[F_r; J^rY]$ on graded manifolds $(J^rY, \mathfrak{A}_r)$ modulo the monomorphisms (3.4.5). Its elements obey the relations (3.2.9) – (3.2.10).

The cochain monomorphisms $\mathcal{O}_r^* \to \mathcal{S}_r^*[F;Y]$ (3.4.3) provide a monomorphism of the direct system (1.7.8) to the direct system (3.4.6) and, consequently, the monomorphism

$$\mathcal{O}_\infty^* \to \mathcal{S}_\infty^*[F;Y] \tag{3.4.8}$$

of their direct limits. In particular, $\mathcal{S}_\infty^*[F;Y]$ is an $\mathcal{O}_\infty^0$-algebra. Accordingly, the body epimorphisms

$$\mathcal{S}_r^*[F;Y] \to \mathcal{O}_r^*$$

yield the epimorphism of $\mathcal{O}_\infty^0$-algebras

$$\mathcal{S}_\infty^*[F;Y] \to \mathcal{O}_\infty^*. \tag{3.4.9}$$

It is readily observed that the morphisms (3.4.8) and (3.4.9) are cochain morphisms between the de Rham complex (1.7.10) of the differential graded algebra $\mathcal{O}_\infty^*$ (1.7.9) and the de Rham complex

$$0 \to \mathbb{R} \longrightarrow \mathcal{S}_\infty^0[F;Y] \xrightarrow{d} \mathcal{S}_\infty^1[F;Y] \cdots \xrightarrow{d} \mathcal{S}_\infty^k[F;Y] \longrightarrow \cdots \tag{3.4.10}$$

of the differential bigraded algebra $\mathcal{S}_\infty^0[F;Y]$. Moreover, the corresponding homomorphisms of cohomology groups of these complexes are isomorphisms as follows.

Theorem 3.4.1. *There is an isomorphism*

$$H^*(\mathcal{S}_\infty^*[F;Y]) = H_{DR}^*(Y) \tag{3.4.11}$$

of the cohomology $H^*(\mathcal{S}_\infty^*[F;Y])$ of the de Rham complex (3.4.10) to the de Rham cohomology $H_{DR}^*(Y)$ of Y.

Proof. The complex (3.4.10) is the direct limit of the de Rham complexes of the differential graded algebras $\mathcal{S}_r^*[F;Y]$. Therefore, the direct limit of cohomology groups of these complexes is the cohomology of the de Rham complex (3.4.10) in accordance with Theorem 10.3.2. By virtue of Theorem 3.3.4, cohomology of the de Rham complex of $\mathcal{S}_r^*[F;Y]$ equals the de Rham cohomology of J^rY and, consequently, that of Y, which is the strong deformation retract of any jet manifold J^rY (see Remark 1.5.1). Hence, the isomorphism (3.4.11) holds. □

Corollary 3.4.1. *Any closed graded form $\phi \in \mathcal{S}_\infty^*[F;Y]$ is decomposed into the sum $\phi = \sigma + d\xi$ where σ is a closed exterior form on Y.*

One can think of elements of $\mathcal{S}_\infty^*[F;Y]$ as being *graded differential forms* on the infinite order jet manifold $J^\infty Y$. Indeed, let $\mathfrak{S}_r^*[F;Y]$ be the sheaf of differential bigraded algebras on J^rY and $\overline{\mathfrak{S}}_r^*[F;Y]$ its canonical presheaf. Then the above mentioned presheaf monomorphisms $\overline{\mathfrak{A}}_r \to \overline{\mathfrak{A}}_{r+1}$ yield the direct system of presheaves

$$\overline{\mathfrak{S}}^*[F;Y] \longrightarrow \overline{\mathfrak{S}}_1^*[F;Y] \longrightarrow \cdots \overline{\mathfrak{S}}_r^*[F;Y] \longrightarrow \cdots, \qquad (3.4.12)$$

whose direct limit $\overline{\mathfrak{S}}_\infty^*[F;Y]$ is a presheaf of differential bigraded algebras on the infinite order jet manifold $J^\infty Y$. Let $\mathfrak{Q}_\infty^*[F;Y]$ be the sheaf of differential bigraded algebras of germs of the presheaf $\overline{\mathfrak{S}}_\infty^*[F;Y]$. One can think of the pair $(J^\infty Y, \mathfrak{Q}_\infty^0[F;Y])$ as being a graded Fréchet manifold, whose body is the infinite order jet manifold $J^\infty Y$ and the structure sheaf $\mathfrak{Q}_\infty^0[F;Y]$ is the sheaf of germs of graded functions on graded manifolds $(J^rY, \mathfrak{A}_r)$. We agree to call $(J^\infty Y, \mathfrak{Q}_\infty^0[F;Y])$ the *graded infinite order jet manifold*. The structure module $\mathcal{Q}_\infty^*[F;Y]$ of sections of $\mathfrak{Q}_\infty^*[F;Y]$ is a differential bigraded algebra such that, given an element $\phi \in \mathcal{Q}_\infty^*[F;Y]$ and a point $z \in J^\infty Y$, there exist an open neighbourhood U of z and a graded exterior form $\phi^{(k)}$ on some finite order jet manifold J^kY so that $\phi|_U = \pi_k^{\infty*}\phi^{(k)}|_U$. In particular, there is the monomorphism

$$\mathcal{S}_\infty^*[F;Y] \to \mathcal{Q}_\infty^*[F;Y]. \qquad (3.4.13)$$

Due to this monomorphism, one can restrict $\mathcal{S}_\infty^*[F;Y]$ to the coordinate chart (1.7.3) of $J^\infty Y$ and say that $\mathcal{S}_\infty^*[F;Y]$ as an $\mathcal{O}_\infty^0$-algebra is locally generated by the elements

$$(c_\Lambda^a, dx^\lambda, \theta_\Lambda^a = dc_\Lambda^a - c_{\lambda+\Lambda}^a dx^\lambda, \theta_\Lambda^i = dy_\Lambda^i - y_{\lambda+\Lambda}^i dx^\lambda), \qquad 0 \leq |\Lambda|,$$

where c_Λ^a, θ_Λ^a are odd and dx^λ, θ_Λ^i are even. We agree to call (y^i, c^a) the local *generating basis* for $\mathcal{S}_\infty^*[F;Y]$. Let the collective symbol s^A stand for its elements. Accordingly, the notation s_Λ^A and

$$\theta_\Lambda^A = ds_\Lambda^A - s_{\lambda+\Lambda}^A dx^\lambda$$

is introduced. For the sake of simplicity, we further denote $[A] = [s^A]$.

Remark 3.4.2. Strictly speaking, elements y^i and c^a of the generating basis are of different mathematical origin. However, they can be treated on the same level if $Y \to X$ is an affine bundle and polynomial functions in y^i are only considered. In this case, the both of them can be seen as locally constant functions of vector bundles $Y \to X$ and $F \to X$, respectively (see Remarks 3.3.1 and 3.4.1).

The differential bigraded algebra $\mathcal{S}^*_\infty[F;Y]$ is decomposed into $\mathcal{S}^0_\infty[F;Y]$-modules $\mathcal{S}^{k,r}_\infty[F;Y]$ of k-contact and r-horizontal graded forms together with the corresponding projections

$$h_k : \mathcal{S}^*_\infty[F;Y] \to \mathcal{S}^{k,*}_\infty[F;Y], \qquad h^m : \mathcal{S}^*_\infty[F;Y] \to \mathcal{S}^{*,m}_\infty[F;Y].$$

Accordingly, the graded exterior differential d on $\mathcal{S}^*_\infty[F;Y]$ falls into the sum $d = d_V + d_H$ of the vertical graded differential

$$d_V \circ h^m = h^m \circ d \circ h^m, \qquad d_V(\phi) = \theta^A_\Lambda \wedge \partial^\Lambda_A \phi, \qquad \phi \in \mathcal{S}^*_\infty[F;Y],$$

and the total graded differential

$$d_H \circ h_k = h_k \circ d \circ h_k, \qquad d_H \circ h_0 = h_0 \circ d, \qquad d_H(\phi) = dx^\lambda \wedge d_\lambda(\phi),$$

where

$$d_\lambda = \partial_\lambda + \sum_{0 \le |\Lambda|} s^A_{\lambda+\Lambda} \partial^\Lambda_A$$

are the *graded total derivatives*. These differentials obey the nilpotent relations (1.7.16).

Similarly to the differential graded algebra $\mathcal{O}^*_\infty$, the differential bigraded algebra $\mathcal{S}^*_\infty[F;Y]$ is provided with the graded projection endomorphism

$$\varrho = \sum_{k>0} \frac{1}{k} \overline{\varrho} \circ h_k \circ h^n : \mathcal{S}^{*>0,n}_\infty[F;Y] \to \mathcal{S}^{*>0,n}_\infty[F;Y],$$

$$\overline{\varrho}(\phi) = \sum_{0 \le |\Lambda|} (-1)^{|\Lambda|} \theta^A \wedge [d_\Lambda(\partial^\Lambda_A \rfloor \phi)], \qquad \phi \in \mathcal{S}^{>0,n}_\infty[F;Y],$$

such that $\varrho \circ d_H = 0$, and with the nilpotent graded variational operator

$$\delta = \varrho \circ d\, \mathcal{S}^{*,n}_\infty[F;Y] \to \mathcal{S}^{*+1,n}_\infty[F;Y]. \tag{3.4.14}$$

With these operators the differential bigraded algebra $\mathcal{S}^*_\infty[F;Y]$ is decomposed into the Grassmann-graded variational bicomplex. We restrict our consideration to its short *variational subcomplex*

$$0 \to \mathbb{R} \to \mathcal{S}^0_\infty[F;Y] \xrightarrow{d_H} \mathcal{S}^{0,1}_\infty[F;Y] \cdots \xrightarrow{d_H} \tag{3.4.15}$$

$$\mathcal{S}^{0,n}_\infty[F;Y] \xrightarrow{\delta} \mathbf{E}_1, \qquad \mathbf{E}_1 = \varrho(\mathcal{S}^{1,n}_\infty[F;Y]),$$

and the subcomplex of one-contact graded forms

$$0 \to \mathcal{S}^{1,0}_\infty[F;Y] \xrightarrow{d_H} \mathcal{S}^{1,1}_\infty[F;Y] \cdots \xrightarrow{d_H} \mathcal{S}^{1,n}_\infty[F;Y] \xrightarrow{\varrho} \mathbf{E}_1 \to 0. \qquad (3.4.16)$$

They possess the following cohomology [59; 144].

Theorem 3.4.2. *Cohomology of the complex (3.4.15) equals the de Rham cohomology $H^*_{DR}(Y)$ of Y.*

Theorem 3.4.3. *The complex (3.4.16) is exact.*

The proof of these theorems follow that of Theorems 2.1.1 – 2.1.2 (see Section 3.6).

Remark 3.4.3. If $Y \to X$ is an affine bundle, one can consider the subalgebra $\mathcal{P}^*_\infty[F;Y] \subset \mathcal{S}^*_\infty[F;Y]$ of graded differential forms whose coefficients are polynomials in fibre coordinates of $Y \to X$ and their jets. This subalgebra also is decomposed into the Grassmann-graded variational bicomplex. Following the proof of Theorem 2.1.4, one can show that the cohomology of its short variational subcomplex as like as that of the complex (2.1.22 equals the de Rham cohomology of X.

3.5 Lagrangian theory of even and odd fields

Decomposed into the variational bicomplex, the differential bigraded algebra $\mathcal{S}^*_\infty[F;Y]$ describes Grassmann-graded field theory on the composite graded manifold $(Y, \mathfrak{A}_F)$. Its *graded Lagrangian* is defined as an element

$$L = \mathcal{L}\omega \in \mathcal{S}^{0,n}_\infty[F;Y] \qquad (3.5.1)$$

of the graded variational complex (3.4.15). Accordingly, the graded exterior form

$$\delta L = \theta^A \wedge \mathcal{E}_A \omega = \sum_{0 \leq |\Lambda|} (-1)^{|\Lambda|} \theta^A \wedge d_\Lambda (\partial^\Lambda_A L) \omega \in \mathbf{E}_1 \qquad (3.5.2)$$

is said to be its *graded Euler–Lagrange operator*. We agree to call a pair $(\mathcal{S}^{0,n}_\infty[F;Y], L)$ the *Grassmann-graded Lagrangian system* and $\mathcal{S}^*_\infty[F;Y]$ the *field system algebra*.

The following is a corollary of Theorem 3.4.2.

Theorem 3.5.1. *Every d_H-closed graded form $\phi \in \mathcal{S}^{0,m<n}_\infty[F;Y]$ falls into the sum*

$$\phi = h_0 \sigma + d_H \xi, \qquad \xi \in \mathcal{S}^{0,m-1}_\infty[F;Y], \qquad (3.5.3)$$

where σ is a closed m-form on Y. Any δ-closed (i.e., variationally trivial) Grassmann-graded Lagrangian $L \in \mathcal{S}_\infty^{0,n}[F;Y]$ is the sum

$$L = h_0\sigma + d_H\xi, \qquad \xi \in \mathcal{S}_\infty^{0,n-1}[F;Y], \qquad (3.5.4)$$

where σ is a closed n-form on Y.

Proof. The complex (3.4.15) possesses the same cohomology as the short variational complex

$$0 \to \mathbb{R} \to \mathcal{O}_\infty^0 \xrightarrow{d_H} \mathcal{O}_\infty^{0,1} \cdots \xrightarrow{d_H} \mathcal{O}_\infty^{0,n} \xrightarrow{\delta} \mathbf{E}_1 \qquad (3.5.5)$$

of the differential graded algebra $\mathcal{O}_\infty^*$. The monomorphism (3.4.8) and the body epimorphism (3.4.9) yield the corresponding cochain morphisms of the complexes (3.4.15) and (3.5.5). Therefore, cohomology of the complex (3.4.15) is the image of the cohomology of $\mathcal{O}_\infty^*$. $\square$

Corollary 3.5.1. *Any variationally trivial odd Lagrangian is d_H-exact.*

The exactness of the complex (3.4.16) at the term $\mathcal{S}_\infty^{1,n}[F;Y]$ results in the following [59].

Theorem 3.5.2. *Given a graded Lagrangian L, there is the decomposition*

$$dL = \delta L - d_H \Xi_L, \qquad \Xi \in \mathcal{S}_\infty^{n-1}[F;Y], \qquad (3.5.6)$$

$$\Xi_L = L + \sum_{s=0} \theta_{\nu_s\ldots\nu_1}^A \wedge F_A^{\lambda\nu_s\ldots\nu_1}\omega_\lambda, \qquad (3.5.7)$$

$$F_A^{\nu_k\ldots\nu_1} = \partial_A^{\nu_k\ldots\nu_1}\mathcal{L} - d_\lambda F_A^{\lambda\nu_k\ldots\nu_1} + \sigma_A^{\nu_k\ldots\nu_1}, \qquad k=1,2,\ldots,$$

where local graded functions σ obey the relations

$$\sigma_A^\nu = 0, \qquad \sigma_A^{(\nu_k\nu_{k-1})\ldots\nu_1} = 0.$$

The form Ξ_L (3.5.7) provides a global Lepage equivalent of a graded Lagrangian L. In particular, one can locally choose Ξ_L (3.5.7) where all functions σ vanish.

Given a Grassmann-graded Lagrangian system $(\mathcal{S}_\infty^*[F;Y], L)$, by its infinitesimal transformations are meant contact graded derivations of the real graded commutative ring $\mathcal{S}_\infty^0[F;Y]$. They constitute a $\mathcal{S}_\infty^0[F;Y]$-module $\partial\mathcal{S}_\infty^0[F;Y]$ which is a real Lie superalgebra with respect to the Lie superbracket (3.2.4).

Theorem 3.5.3. *The derivation module $\partial\mathcal{S}_\infty^0[F;Y]$ is isomorphic to the $\mathcal{S}_\infty^0[F;Y]$-dual $(\mathcal{S}_\infty^1[F;Y])^*$ of the module of graded one-forms $\mathcal{S}_\infty^1[F;Y]$. It follows that the differential bigraded algebra $\mathcal{S}_\infty^*[F;Y]$ is minimal differential calculus over the real graded commutative ring $\mathcal{S}_\infty^0[F;Y]$.*

The proof of this theorem is a repetition of that of Theorem 1.7.5.

Let $\vartheta \rfloor \phi$, $\vartheta \in \mathfrak{d}\mathcal{S}^0_\infty[F;Y]$, $\phi \in \mathcal{S}^1_\infty[F;Y]$, denote the corresponding interior product. Extended to the differential bigraded algebra $\mathcal{S}^*_\infty[F;Y]$, it obeys the rule

$$\vartheta\rfloor(\phi\wedge\sigma) = (\vartheta\rfloor\phi)\wedge\sigma + (-1)^{|\phi|+[\phi][\vartheta]}\phi\wedge(\vartheta\rfloor\sigma), \qquad \phi,\sigma \in \mathcal{S}^*_\infty[F;Y].$$

Restricted to a coordinate chart (1.7.3) of $J^\infty Y$, the algebra $\mathcal{S}^*_\infty[F;Y]$ is a free $\mathcal{S}^0_\infty[F;Y]$-module generated by one-forms dx^λ, θ^A_Λ. Due to the isomorphism stated in Theorem 3.5.3, any graded derivation $\vartheta \in \mathfrak{d}\mathcal{S}^0_\infty[F;Y]$ takes the local form

$$\vartheta = \vartheta^\lambda \partial_\lambda + \vartheta^A \partial_A + \sum_{0<|\Lambda|} \vartheta^A_\Lambda \partial^\Lambda_A, \qquad (3.5.8)$$

where

$$\partial^\Lambda_A \rfloor dy^B_\Sigma = \delta^B_A \delta^\Lambda_\Sigma$$

up to permutations of multi-indices Λ and Σ.

Every graded derivation ϑ (3.5.8) yields the graded Lie derivative

$$\mathbf{L}_\vartheta \phi = \vartheta\rfloor d\phi + d(\vartheta\rfloor\phi), \qquad \phi \in \mathcal{S}^*_\infty[F;Y],$$
$$\mathbf{L}_\vartheta(\phi\wedge\sigma) = \mathbf{L}_\vartheta(\phi)\wedge\sigma + (-1)^{[\vartheta][\phi]}\phi\wedge\mathbf{L}_\vartheta(\sigma),$$

of the differential bigraded algebra $\mathcal{S}^*_\infty[F;Y]$. A graded derivation ϑ (3.5.8) is called *contact* if the Lie derivative $\mathbf{L}_\vartheta$ preserves the ideal of contact graded forms of the differential bigraded algebra $\mathcal{S}^*_\infty[F;Y]$.

Lemma 3.5.1. *With respect to the local generating basis (s^A) for the differential bigraded algebra $\mathcal{S}^*_\infty[F;Y]$, any its contact graded derivation takes the form*

$$\vartheta = \upsilon_H + \upsilon_V = \upsilon^\lambda d_\lambda + [\upsilon^A \partial_A + \sum_{|\Lambda|>0} d_\Lambda(\upsilon^A - s^A_\mu \upsilon^\mu)\partial^\Lambda_A], \qquad (3.5.9)$$

where υ_H and υ_V denotes the horizontal and vertical parts of ϑ.

The proof is similar to that of the formula (2.2.1). A glance at the expression (3.5.9) shows that a contact graded derivation ϑ as an infinite order jet prolongation of its restriction

$$\upsilon = \upsilon^\lambda \partial_\lambda + \upsilon^A \partial_A \qquad (3.5.10)$$

to the graded commutative ring $\mathcal{S}^0[F;Y]$. We call υ (3.5.10) the *generalized graded vector field*. It is readily justified the following (see Lemma 2.2.5).

Lemma 3.5.2. *Any vertical contact graded derivation*

$$\vartheta = v^A \partial_A + \sum_{|\Lambda|>0} d_\Lambda v^A \partial_A^\Lambda \qquad (3.5.11)$$

satisfies the relations

$$\vartheta \rfloor d_H \phi = -d_H(\vartheta \rfloor \phi), \qquad (3.5.12)$$

$$\mathbf{L}_\vartheta(d_H \phi) = d_H(\mathbf{L}_\vartheta \phi) \qquad (3.5.13)$$

for all $\phi \in \mathcal{S}_\infty^*[F; Y]$.

Then the forthcoming assertions are the straightforward generalizations of Theorem 2.2.1, Lemma 2.2.3 and Theorem 2.2.2.

A corollary of the decomposition (3.5.6) is the first variational formula for a graded Lagrangian [13; 59].

Theorem 3.5.4. *The Lie derivative of a graded Lagrangian along any contact graded derivation (3.5.9) obeys the first variational formula*

$$\mathbf{L}_\vartheta L = v_V \rfloor \delta L + d_H(h_0(\vartheta \rfloor \Xi_L)) + d_V(v_H \rfloor \omega) \mathcal{L}, \qquad (3.5.14)$$

where Ξ_L *is the Lepage equivalent (3.5.7) of* L.

A contact graded derivation ϑ (3.5.9) is called a *variational symmetry* (strictly speaking, a variational *supersymmetry*) of a graded Lagrangian L if the Lie derivative $\mathbf{L}_\vartheta L$ is d_H-exact, i.e.,

$$\mathbf{L}_\vartheta L = d_H \sigma. \qquad (3.5.15)$$

Lemma 3.5.3. *A glance at the expression (3.5.14) shows the following.*

(i) A contact graded derivation ϑ *is a variational symmetry only if it is projected onto* X.

(ii) Any projectable contact graded derivation is a variational symmetry of a variationally trivial graded Lagrangian. It follows that, if ϑ *is a variational symmetry of a graded Lagrangian* L, *it also is a variational symmetry of a Lagrangian* $L + L_0$, *where* L_0 *is a variationally trivial graded Lagrangian.*

(iii) A contact graded derivations ϑ *is a variational symmetry if and only if its vertical part* v_V *(3.5.9) is well.*

(iv) It is a variational symmetry if and only if the graded density $v_V \rfloor \delta L$ *is* d_H-*exact.*

Variational symmetries of a graded Lagrangian L constitute a real vector subspace $\mathcal{G}_L$ of the graded derivation module $\mathfrak{d}\mathcal{S}_\infty^0[F;Y]$. By virtue of item (ii) of Lemma 3.5.3, the Lie superbracket

$$\mathbf{L}_{[\vartheta,\vartheta']} = [\mathbf{L}_\vartheta, \mathbf{L}_{\vartheta'}]$$

of variational symmetries is a variational symmetry and, therefore, their vector space $\mathcal{G}_L$ is a real Lie superalgebra.

A corollary of the first variational formula (3.5.14) is the *first Noether theorem* for graded Lagrangians.

Theorem 3.5.5. *If a contact graded derivation ϑ (3.5.9) is a variational symmetry (3.5.15) of a graded Lagrangian L, the first variational formula (3.5.14) restricted to $\mathrm{Ker}\,\delta L$ leads to the weak conservation law*

$$0 \approx d_H(h_0(\vartheta\rfloor\Xi_L) - \sigma). \tag{3.5.16}$$

A vertical contact graded derivation ϑ (3.5.11) is said to be *nilpotent* if

$$\mathbf{L}_\vartheta(\mathbf{L}_\vartheta\phi) = \sum_{0\leq|\Sigma|,0\leq|\Lambda|} (v_\Sigma^B \partial_B^\Sigma(v_\Lambda^A)\partial_A^\Lambda + \tag{3.5.17}$$

$$(-1)^{[s^B][v^A]} v_\Sigma^B v_\Lambda^A \partial_B^\Sigma \partial_A^\Lambda)\phi = 0$$

for any horizontal graded form $\phi \in \mathcal{S}_\infty^{0,*}$.

Lemma 3.5.4. *A vertical contact graded derivation (3.5.11) is nilpotent only if it is odd and if and only if the equality*

$$\mathbf{L}_\vartheta(v^A) = \sum_{0\leq|\Sigma|} v_\Sigma^B \partial_B^\Sigma(v^A) = 0$$

holds for all v^A.

Proof. There is the relation

$$d_\lambda \circ v_\Lambda^A \partial_A^\lambda = v_\Lambda^A \partial_A^\lambda \circ d_\lambda, \tag{3.5.18}$$

similar to that (2.2.9). Then the result follows from the equality (3.5.17) where one puts $\phi = s^A$ and $\phi = s_\Lambda^A s_\Sigma^B$. □

For the sake of brevity, the common symbol v further stands for a generalized graded vector field v, the contact graded derivation ϑ determined by v, and the Lie derivative $\mathbf{L}_\vartheta$. We agree to call all these operators, simply, a *graded derivation of a field system algebra*.

Remark 3.5.1. For the sake of convenience, *right derivations*

$$\overleftarrow{v} = \overleftarrow{\partial}_A v^A \tag{3.5.19}$$

also are considered. They act on graded functions and differential forms ϕ on the right by the rules

$$\overleftarrow{v}(\phi) = d\phi\lfloor \overleftarrow{v} + d(\phi\lfloor \overleftarrow{v}),$$
$$\overleftarrow{v}(\phi \wedge \phi') = (-1)^{[\phi']}\overleftarrow{v}(\phi) \wedge \phi' + \phi \wedge \overleftarrow{v}(\phi'),$$
$$\theta_{\Lambda A}\lfloor \overleftarrow{\partial}{}^{\Sigma B} = \delta^A_B \delta^\Sigma_\Lambda.$$

One associates to any graded right derivation $\overleftarrow{v}$ (3.5.19) the left one

$$v^l = (-1)^{[v][A]} v^A \partial_A, \qquad (3.5.20)$$
$$v^l(f) = (-1)^{[v][f]} v(f), \qquad f \in \mathcal{S}^0_\infty[F;Y].$$

3.6 Appendix. Cohomology of the Grassmann-graded variational bicomplex

The proof of Theorem 3.4.2 follows the scheme of the proof of Theorem 2.1.1 [59; 144]. It falls into the three steps.

(I) We start with showing that the complexes (3.4.15) – (3.4.16) are locally exact.

Lemma 3.6.1. *If*

$$Y = \mathbb{R}^{n+k} \to \mathbb{R}^n,$$

the complex (3.4.15) is acyclic.

Proof. Referring to [9] for the proof, we summarize a few formulas. Any horizontal graded form $\phi \in \mathcal{S}^{0,*}_\infty$ admits the decomposition

$$\phi = \phi_0 + \widetilde{\phi}, \qquad \widetilde{\phi} = \int_0^1 \frac{d\lambda}{\lambda} \sum_{0 \leq |\Lambda|} s^A_\Lambda \partial^\Lambda_A \phi, \qquad (3.6.1)$$

where ϕ_0 is an exterior form on $\mathbb{R}^{n+k}$. Let $\phi \in \mathcal{S}^{0,m<n}_\infty$ be d_H-closed. Then its component ϕ_0 (3.6.1) is an exact exterior form on $\mathbb{R}^{n+k}$ and $\widetilde{\phi} = d_H \xi$, where ξ is given by the following expressions. Let us introduce the operator

$$D^{+\nu}\widetilde{\phi} = \int_0^1 \frac{d\lambda}{\lambda} \sum_{0 \leq k} k \delta^\nu_{(\mu_1} \delta^{\alpha_1}_{\mu_2} \cdots \delta^{\alpha_{k-1}}_{\mu_k)} \lambda s^A_{(\alpha_1 \ldots \alpha_{k-1})} \qquad (3.6.2)$$

$$\partial^{\mu_1 \ldots \mu_k}_A \widetilde{\phi}(x^\mu, \lambda s^A_\Lambda, dx^\mu).$$

The relation

$$[D^{+\nu}, d_\mu]\widetilde{\phi} = \delta^\nu_\mu \widetilde{\phi}$$

holds, and it leads to the desired expression

$$\xi = \sum_{k=0} \frac{(n-m-1)!}{(n-m+k)!} D^{+\nu} P_k \partial_\nu \rfloor \widetilde{\phi}, \qquad (3.6.3)$$

$$P_0 = 1, \qquad P_k = d_{\nu_1} \cdots d_{\nu_k} D^{+\nu_1} \cdots D^{+\nu_k}.$$

Now let $\phi \in \mathcal{S}_\infty^{0,n}$ be a graded density such that $\delta \phi = 0$. Then its component ϕ_0 (3.6.1) is an exact n-form on $\mathbb{R}^{n+k}$ and $\widetilde{\phi} = d_H \xi$, where ξ is given by the expression

$$\xi = \sum_{|\Lambda| \geq 0} \sum_{\Sigma + \Xi = \Lambda} (-1)^{|\Sigma|} s_\Xi^A d_\Sigma \partial_A^{\mu+\Lambda} \widetilde{\phi} \omega_\mu. \qquad (3.6.4)$$

Since elements of $\mathcal{S}_\infty^*$ are polynomials in s_Λ^A, the sum in the expression (3.6.3) is finite. However, this expression contains a d_H-exact summand which prevents its extension to $\mathcal{O}_\infty^*$. In this respect, we also quote the homotopy operator (5.107) in [123] which leads to the expression

$$\xi = \int_0^1 I(\phi)(x^\mu, \lambda s_\Lambda^A, dx^\mu) \frac{d\lambda}{\lambda}, \qquad (3.6.5)$$

$$I(\phi) = \sum_{0 \leq |\Lambda|} \sum_\mu \frac{\Lambda_\mu + 1}{n - m + |\Lambda| + 1}$$

$$d_\Lambda \left[\sum_{0 \leq |\Xi|} (-1)^\Xi \frac{(\mu + \Lambda + \Xi)!}{(\mu + \Lambda)! \Xi!} s^A d_\Xi \partial_A^{\mu+\Lambda+\Xi} (\partial_\mu \rfloor \phi) \right],$$

where $\Lambda! = \Lambda_{\mu_1}! \cdots \Lambda_{\mu_n}!$, and Λ_μ denotes the number of occurrences of the index μ in Λ [123]. The graded forms (3.6.4) and (3.6.5) differ in a d_H-exact graded form. □

Lemma 3.6.2. *If $Y = \mathbb{R}^{n+k} \to \mathbb{R}^n$, the complex (3.4.16) is exact.*

Proof. The fact that a d_H-closed graded $(1,m)$-form $\phi \in \mathcal{S}_\infty^{1,m<n}$ is d_H-exact is derived from Lemma 3.6.1 as follows. We write

$$\phi = \sum \phi_A^\Lambda \wedge \theta_\Lambda^A, \qquad (3.6.6)$$

where $\phi_A^\Lambda \in \mathcal{S}_\infty^{0,m}$ are horizontal graded m-forms. Let us introduce additional variables $\overline{s}_\Lambda^A$ of the same Grassmann parity as s_Λ^A. Then one can associate to each graded $(1,m)$-form ϕ (3.6.6) a unique horizontal graded m-form

$$\overline{\phi} = \sum \phi_A^\Lambda \overline{s}_\Lambda^A, \qquad (3.6.7)$$

3.6. Appendix. Cohomology of the Grassmann-graded variational bicomplex

whose coefficients are linear in the variables $\overline{s}_\Lambda^A$, and *vice versa*. Let us consider the modified total differential

$$\overline{d}_H = d_H + dx^\lambda \wedge \sum_{0<|\Lambda|} \overline{s}_{\lambda+\Lambda}^A \overline{\partial}_A^\Lambda,$$

acting on graded forms (3.6.7), where $\overline{\partial}_A^\Lambda$ is the dual of $d\overline{s}_\Lambda^A$. Comparing the equality

$$\overline{d}_H \overline{s}_\Lambda^A = dx^\lambda s_{\lambda+\Lambda}^A$$

and the last equality (2.2.8), one can easily justify that

$$\overline{d_H \phi} = \overline{d}_H \overline{\phi}.$$

Let a graded $(1,m)$-form ϕ (3.6.6) be d_H-closed. Then the associated horizontal graded m-form $\overline{\phi}$ (3.6.7) is $\overline{d}_H$-closed and, by virtue of Lemma 3.6.1, it is $\overline{d}_H$-exact, i.e., $\overline{\phi} = \overline{d}_H \overline{\xi}$, where $\overline{\xi}$ is a horizontal graded $(m-1)$-form given by the expression (3.6.3) depending on additional variables $\overline{s}_\Lambda^A$. A glance at this expression shows that, since $\overline{\phi}$ is linear in the variables $\overline{s}_\Lambda^A$, so is

$$\overline{\xi} = \sum \xi_A^\Lambda \overline{s}_\Lambda^A.$$

It follows that $\phi = d_H \xi$ where

$$\xi = \sum \xi_A^\Lambda \wedge \theta_\Lambda^A.$$

It remains to prove the exactness of the complex (3.4.16) at the last term $\mathbf{E}_1$. If

$$\varrho(\sigma) = \sum_{0\leq|\Lambda|} (-1)^{|\Lambda|} \theta^A \wedge [d_\Lambda(\partial_A^\Lambda \rfloor \sigma)] =$$

$$\sum_{0\leq|\Lambda|} (-1)^{|\Lambda|} \theta^A \wedge [d_\Lambda \sigma_A^\Lambda]\omega = 0, \qquad \sigma \in \mathcal{S}_\infty^{1,n},$$

a direct computation gives

$$\sigma = d_H \xi, \qquad \xi = -\sum_{0\leq|\Lambda|} \sum_{\Sigma+\Xi=\Lambda} (-1)^{|\Sigma|} \theta_\Xi^A \wedge d_\Sigma \sigma_A^{\mu+\Lambda} \omega_\mu. \qquad (3.6.8)$$

$\square$

Remark 3.6.1. The proof of Lemma 3.6.2 fails to be extended to complexes of higher contact forms because the products $\theta_\Lambda^A \wedge \theta_\Xi^B$ and $s_\Lambda^A s_\Xi^B$ obey different commutation rules.

(II) Let us now prove Theorem 3.4.2 for the differential bigraded algebra $\mathcal{Q}^*_\infty[F;Y]$. Similarly to $\mathcal{S}^*_\infty[F;Y]$, the sheaf $\mathfrak{Q}^*_\infty[F;Y]$ and the differential bigraded algebra $\mathcal{Q}^*_\infty[F;Y]$ are decomposed into the Grassmann-graded variational bicomplexes. We consider their subcomplexes

$$0 \to \mathbb{R} \to \mathfrak{Q}^0_\infty[F;Y] \xrightarrow{d_H} \mathfrak{Q}^{0,1}_\infty[F;Y] \cdots \qquad (3.6.9)$$

$$\xrightarrow{d_H} \mathfrak{Q}^{0,n}_\infty[F;Y] \xrightarrow{\delta} \mathbf{E}_1, \qquad \mathbf{E}_1 = \varrho(\mathfrak{Q}^{1,n}_\infty[F;Y]),$$

$$0 \to \mathfrak{Q}^{1,0}_\infty[F;Y] \xrightarrow{d_H} \mathfrak{Q}^{1,1}_\infty[F;Y] \cdots \qquad (3.6.10)$$

$$\xrightarrow{d_H} \mathfrak{Q}^{1,n}_\infty[F;Y] \xrightarrow{\varrho} \mathbf{E}_1 \to 0,$$

$$0 \to \mathbb{R} \to \mathcal{Q}^0_\infty[F;Y] \xrightarrow{d_H} \mathcal{Q}^{0,1}_\infty[F;Y] \cdots \qquad (3.6.11)$$

$$\xrightarrow{d_H} \mathcal{Q}^{0,n}_\infty[F;Y] \xrightarrow{\delta} \Gamma(\mathbf{E}_1),$$

$$0 \to \mathcal{Q}^{1,0}_\infty[F;Y] \xrightarrow{d_H} \mathcal{Q}^{1,1}_\infty[F;Y] \cdots \qquad (3.6.12)$$

$$\xrightarrow{d_H} \mathcal{Q}^{1,n}_\infty[F;Y] \xrightarrow{\varrho} \Gamma(\mathbf{E}_1) \to 0.$$

By virtue of Lemmas 3.6.1 and 3.6.2, the complexes (3.6.9) – (3.6.10) are acyclic. The terms $\mathfrak{Q}^{*,*}_\infty[F;Y]$ of the complexes (3.6.9) – (3.6.10) are sheaves of $\mathcal{Q}^0_\infty$-modules. Since $J^\infty Y$ admits the partition of unity just by elements of $\mathcal{Q}^0_\infty$, these sheaves are fine and, consequently, acyclic. By virtue of abstract de Rham Theorem 10.7.4, cohomology of the complex (3.6.11) equals the cohomology of $J^\infty Y$ with coefficients in the constant sheaf $\mathbb{R}$ and, consequently, the de Rham cohomology of Y in accordance with the isomorphisms (1.5.5). Similarly, the complex (3.6.12) is proved to be exact.

(III) It remains to prove that cohomology of the complexes (3.4.15) – (3.4.16) equals that of the complexes (3.6.11) – (3.6.12). The proof of this fact straightforwardly follows the proof of Theorem 2.1.1, and it is a slight modification of the proof of [59], Theorem 4.1, where graded exterior forms on the infinite order jet manifold $J^\infty Y$ of an affine bundle are treated as those on X.

Chapter 4
Lagrangian BRST theory

Quantization of Lagrangian field theory essentially depends on its degeneracy characterized by a set of non-trivial reducible Noether identities [9; 15; 63]. However, any Euler–Lagrange operator satisfies Noether identities which therefore must be separated into the trivial and non-trivial ones. These Noether identities can obey first-stage Noether identities, which in turn are subject to the second-stage ones, and so on. Thus, there is a hierarchy of Noether and higher-stage Noether identities which also are separated into the trivial and non-trivial ones. In accordance with general analysis of Noether identities of differential operators (see Section 4.5), if certain conditions hold, one can associate to a Grassmann-graded Lagrangian system the exact antifield Koszul–Tate complex possessing the boundary operator whose nilpotentness is equivalent to all non-trivial Noether and higher-stage Noether identities [14; 15].

It should be noted that the notion of higher-stage Noether identities has come from that of reducible constraints. The Koszul–Tate complex of Noether identities has been invented similarly to that of constraints under the condition that Noether identities are locally separated into independent and dependent ones [9; 44]. This condition is relevant for constraints, defined by a finite set of functions which the inverse mapping theorem is applied to. However, Noether identities unlike constraints are differential equations. They are given by an infinite set of functions on a Fréchet manifold of infinite order jets where the inverse mapping theorem fails to be valid. Therefore, the regularity condition for the Koszul–Tate complex of constraints is replaced with homology regularity Condition 4.1.1) in order to construct the Koszul–Tate complex of Noether identities.

The inverse second Noether theorem formulated in homology terms associates to this Koszul–Tate complex the cochain sequence of ghosts with the ascent operator, called the gauge operator, whose components are nontrivial gauge and higher-stage gauge symmetries of Lagrangian field theory (see Section 4.2).

The gauge operator unlike the Koszul–Tate one is not nilpotent, unless gauge symmetries are Abelian. This is the cause why an intrinsic definition of non-trivial gauge and higher-stage gauge symmetries meets difficulties. Another problem is that gauge symmetries need not form an algebra [48; 61; 63]. Therefore, we replace the notion of the algebra of gauge symmetries with some conditions on the gauge operator. Gauge symmetries are said to be algebraically closed if the gauge operator admits a nilpotent extension, called the BRST operator (see Section 4.3).

If the BRST operator exists, the cochain sequence of ghosts is brought into the BRST complex. The Koszul–Tate and BRST complexes provide the BRST extension of original Lagrangian field theory by ghosts and antifields (see Section 4.4). This BRST extension is a first step towards quantization of classical field theory in terms of functional integrals [9; 63].

4.1 Noether identities. The Koszul–Tate complex

Let $(\mathcal{S}^*_\infty[F;Y], L)$ be a Grassmann-graded Lagrangian system. Describing its Noether identities, we follow the general analysis of Noether identities of differential operators on fibre bundles in Section 4.5.

Without a lose of generality, let a Lagrangian L be even. Its Euler–Lagrange operator δL (3.5.2) is assumed to be at least of order 1 in order to guarantee that transition functions of Y do not vanish on-shell. This Euler–Lagrange operator $\delta L \in \mathbf{E}_1 \subset \mathcal{S}^{1,n}_\infty[F;Y]$ takes its values into the graded vector bundle

$$\overline{VF} = V^*F \underset{F}{\otimes} \overset{n}{\wedge} T^*X \to F, \qquad (4.1.1)$$

where V^*F is the vertical cotangent bundle of $F \to X$. It however is not a vector bundle over Y. Therefore, we restrict our consideration to the case of a pull-back composite bundle F (3.4.1) that is

$$F = Y \underset{X}{\times} F^1 \to Y \to X, \qquad (4.1.2)$$

where $F^1 \to X$ is a vector bundle. Let us introduce the following notation.

4.1. Noether identities. The Koszul–Tate complex

Notation 4.1.1. Given the vertical tangent bundle VE of a fibre bundle $E \to X$, by its *density-dual bundle* is meant the fibre bundle

$$\overline{VE} = V^*E \underset{E}{\otimes} \overset{n}{\wedge} T^*X. \tag{4.1.3}$$

If $E \to X$ is a vector bundle, we have

$$\overline{VE} = \overline{E} \underset{X}{\times} E, \qquad \overline{E} = E^* \underset{X}{\otimes} \overset{n}{\wedge} T^*X, \tag{4.1.4}$$

where $\overline{E}$ is called the *density-dual* of E. Let

$$E = E^0 \underset{X}{\oplus} E^1$$

be a *graded vector bundle* over X. Its *graded density-dual* is defined to be

$$\overline{E} = \overline{E}^1 \underset{X}{\oplus} \overline{E}^0.$$

In these terms, we treat the composite bundle F (3.4.1) as a graded vector bundle over Y possessing only odd part. The density-dual $\overline{VF}$ (4.1.3) of the vertical tangent bundle VF of $F \to X$ is $\overline{VF}$ (4.1.1). If F (3.4.1) is the pull-back bundle (4.1.2), then

$$\overline{VF} = ((\overline{F}^1 \underset{Y}{\oplus} V^*Y) \underset{Y}{\otimes} \overset{n}{\wedge} T^*X) \underset{Y}{\oplus} F^1 \tag{4.1.5}$$

is a graded vector bundle over Y. Given a graded vector bundle

$$E = E^0 \underset{Y}{\oplus} E^1$$

over Y, we consider the composite bundle

$$E \to E^0 \to X$$

and the differential bigraded algebra (3.4.7):

$$\mathcal{P}^*_\infty[E; Y] = \mathcal{S}^*_\infty[E; E^0]. \tag{4.1.6}$$

Lemma 4.1.1. *One can associate to any Grassmann-graded Lagrangian system $(\mathcal{S}^*_\infty[F; Y], L)$ the chain complex (4.1.8) whose one-boundaries vanish on-shell.*

Proof. Let us consider the density-dual $\overline{VF}$ (4.1.5) of the vertical tangent bundle $VF \to F$, and let us enlarge the original field system algebra $\mathcal{S}^*_\infty[F; Y]$ to the differential bigraded algebra $\mathcal{P}^*_\infty[\overline{VF}; Y]$ (4.1.6) with the local generating basis

$$(s^A, \overline{s}_A), \qquad [\overline{s}_A] = ([A] + 1) \bmod 2.$$

Following the terminology of Lagrangian BRST theory [9; 63], we agree to call its elements $\overline{s}_A$ the antifields of antifield number $\text{Ant}[\overline{s}_A] = 1$. The differential bigraded algebra $\mathcal{P}^*_\infty[\overline{VF}; Y]$ is endowed with the nilpotent right graded derivation

$$\overline{\delta} = \overleftarrow{\partial}{}^A \mathcal{E}_A, \qquad (4.1.7)$$

where $\mathcal{E}_A$ are the variational derivatives (3.5.2). Then we have the chain complex

$$0 \leftarrow \text{Im}\,\overline{\delta} \xleftarrow{\overline{\delta}} \mathcal{P}^{0,n}_\infty[\overline{VF}; Y]_1 \xleftarrow{\overline{\delta}} \mathcal{P}^{0,n}_\infty[\overline{VF}; Y]_2 \qquad (4.1.8)$$

of graded densities of antifield number ≤ 2. Its one-boundaries $\overline{\delta}\Phi$, $\Phi \in \mathcal{P}^{0,n}_\infty[\overline{VF}; Y]_2$, by very definition, vanish on-shell. □

Any one-cycle

$$\Phi = \sum_{0 \leq |\Lambda|} \Phi^{A,\Lambda} \overline{s}_{\Lambda A} \omega \in \mathcal{P}^{0,n}_\infty[\overline{VF}; Y]_1 \qquad (4.1.9)$$

of the complex (4.1.8) is a differential operator on the bundle $\overline{VF}$ such that it is linear on fibres of $\overline{VF} \to F$ and its kernel contains the graded Euler–Lagrange operator δL (3.5.2), i.e.,

$$\overline{\delta}\Phi = 0, \qquad \sum_{0 \leq |\Lambda|} \Phi^{A,\Lambda} d_\Lambda \mathcal{E}_A \omega = 0. \qquad (4.1.10)$$

In accordance with Definition 4.5.1, the one-cycles (4.1.9) define the Noether identities (4.1.10) of the Euler–Lagrange operator δL, which we agree to call *Noether identities* of the Grassmann-graded Lagrangian system $(\mathcal{S}^*_\infty[F; Y], L)$.

In particular, one-chains Φ (4.1.9) are necessarily Noether identities if they are boundaries. Therefore, these Noether identities are called *trivial*. They are of the form

$$\Phi = \sum_{0 \leq |\Lambda|, |\Sigma|} T^{(A\Lambda)(B\Sigma)} d_\Sigma \mathcal{E}_B \overline{s}_{\Lambda A} \omega,$$

$$T^{(A\Lambda)(B\Sigma)} = -(-1)^{[A][B]} T^{(B\Sigma)(A\Lambda)}.$$

Accordingly, *non-trivial Noether identities* modulo the trivial ones are associated to elements of the first homology $H_1(\overline{\delta})$ of the complex (4.1.8). A Lagrangian L is called *degenerate* if there are non-trivial Noether identities.

Non-trivial Noether identities can obey first-stage Noether identities. In order to describe them, let us assume that the module $H_1(\overline{\delta})$ is finitely

4.1. Noether identities. The Koszul–Tate complex

generated. Namely, there exists a graded projective $C^\infty(X)$-module $\mathcal{C}_{(0)} \subset H_1(\overline{\delta})$ of finite rank possessing a local basis $\{\Delta_r \omega\}$:

$$\Delta_r \omega = \sum_{0 \leq |\Lambda|} \Delta_r^{A,\Lambda} \overline{s}_{\Lambda A} \omega, \qquad \Delta_r^{A,\Lambda} \in \mathcal{S}_\infty^0[F;Y], \qquad (4.1.11)$$

such that any element $\Phi \in H_1(\overline{\delta})$ factorizes as

$$\Phi = \sum_{0 \leq |\Xi|} \Phi^{r,\Xi} d_\Xi \Delta_r \omega, \qquad \Phi^{r,\Xi} \in \mathcal{S}_\infty^0[F;Y], \qquad (4.1.12)$$

through elements (4.1.11) of $\mathcal{C}_{(0)}$. Thus, all non-trivial Noether identities (4.1.10) result from the Noether identities

$$\overline{\delta}\Delta_r = \sum_{0 \leq |\Lambda|} \Delta_r^{A,\Lambda} d_\Lambda \mathcal{E}_A = 0, \qquad (4.1.13)$$

called the *complete Noether identities*. Clearly, the factorization (4.1.12) is independent of specification of a local basis $\{\Delta_r \omega\}$. Note that, being representatives of $H_1(\overline{\delta})$, the graded densities $\Delta_r \omega$ (4.1.11) are not $\overline{\delta}$-exact.

A Lagrangian system whose non-trivial Noether identities are finitely generated is called *finitely degenerate*. Hereafter, degenerate Lagrangian systems only of this type are considered.

Lemma 4.1.2. *If the homology $H_1(\overline{\delta})$ of the complex (4.1.8) is finitely generated in the above mentioned sense, this complex can be extended to the one-exact chain complex (4.1.16) with a boundary operator whose nilpotency conditions are equivalent to the complete Noether identities (4.1.13).*

Proof. By virtue of Serre–Swan Theorem 3.3.2, the graded module $\mathcal{C}_{(0)}$ is isomorphic to a module of sections of the density-dual $\overline{E}_0$ of some graded vector bundle $E_0 \to X$. Let us enlarge $\mathcal{P}_\infty^*[\overline{VF};Y]$ to the differential bigraded algebra

$$\overline{\mathcal{P}}_\infty^*\{0\} = \mathcal{P}_\infty^*[\overline{VF} \oplus_Y \overline{E}_0; Y] \qquad (4.1.14)$$

possessing the local generating basis $(s^A, \overline{s}_A, \overline{c}_r)$ where $\overline{c}_r$ are Noether antifields of Grassmann parity

$$[\overline{c}_r] = (|\Delta_r| + 1) \bmod 2$$

and antifield number $\mathrm{Ant}[\overline{c}_r] = 2$. The differential bigraded algebra (4.1.14) is provided with the odd right graded derivation

$$\delta_0 = \overline{\delta} + \overleftarrow{\partial}{}^r \Delta_r \qquad (4.1.15)$$

which is nilpotent if and only if the complete Noether identities (4.1.13) hold. Then δ_0 (4.1.15) is a boundary operator of the chain complex

$$0 \leftarrow \operatorname{Im} \overline{\delta} \xleftarrow{\overline{\delta}} \mathcal{P}_\infty^{0,n}[\overline{VF};Y]_1 \xleftarrow{\delta_0} \overline{\mathcal{P}}_\infty^{0,n}\{0\}_2 \xleftarrow{\delta_0} \overline{\mathcal{P}}_\infty^{0,n}\{0\}_3 \qquad (4.1.16)$$

of graded densities of antifield number ≤ 3. Let $H_*(\delta_0)$ denote its homology. We have

$$H_0(\delta_0) = H_0(\overline{\delta}) = 0.$$

Furthermore, any one-cycle Φ up to a boundary takes the form (4.1.12) and, therefore, it is a δ_0-boundary

$$\Phi = \sum_{0 \leq |\Sigma|} \Phi^{r,\Xi} d_\Xi \Delta_r \omega = \delta_0 \left(\sum_{0 \leq |\Sigma|} \Phi^{r,\Xi} \overline{c}_{\Xi r} \omega \right).$$

Hence, $H_1(\delta_0) = 0$, i.e., the complex (4.1.16) is one-exact. □

Let us consider the second homology $H_2(\delta_0)$ of the complex (4.1.16). Its two-chains read

$$\Phi = G + H = \sum_{0 \leq |\Lambda|} G^{r,\Lambda} \overline{c}_{\Lambda r} \omega + \sum_{0 \leq |\Lambda|,|\Sigma|} H^{(A,\Lambda)(B,\Sigma)} \overline{s}_{\Lambda A} \overline{s}_{\Sigma B} \omega. \qquad (4.1.17)$$

Its two-cycles define the *first-stage Noether identities*

$$\delta_0 \Phi = 0, \qquad \sum_{0 \leq |\Lambda|} G^{r,\Lambda} d_\Lambda \Delta_r \omega = -\overline{\delta} H. \qquad (4.1.18)$$

Conversely, let the equality (4.1.18) hold. Then it is a cycle condition of the two-chain (4.1.17).

Remark 4.1.1. Note that this definition of first-stage Noether identities is independent on specification of a generating module $\mathcal{C}_{(0)}$. Given a different one, there exists a chain isomorphism between the corresponding complexes (4.1.16).

The first-stage Noether identities (4.1.18) are *trivial* either if a two-cycle Φ (4.1.17) is a δ_0-boundary or its summand G vanishes on-shell. Therefore, non-trivial first-stage Noether identities fails to exhaust the second homology $H_2(\delta_0)$ the complex (4.1.16) in general.

Lemma 4.1.3. *Non-trivial first-stage Noether identities modulo the trivial ones are identified with elements of the homology $H_2(\delta_0)$ if and only if any $\overline{\delta}$-cycle $\phi \in \overline{\mathcal{P}}_\infty^{0,n}\{0\}_2$ is a δ_0-boundary.*

4.1. Noether identities. The Koszul–Tate complex

Proof. It suffices to show that, if the summand G of a two-cycle Φ (4.1.17) is $\overline{\delta}$-exact, then Φ is a boundary. If $G = \overline{\delta}\Psi$, let us write

$$\Phi = \delta_0 \Psi + (\overline{\delta} - \delta_0)\Psi + H. \tag{4.1.19}$$

Hence, the cycle condition (4.1.18) reads

$$\delta_0 \Phi = \overline{\delta}((\overline{\delta} - \delta_0)\Psi + H) = 0.$$

Since any $\overline{\delta}$-cycle $\phi \in \overline{\mathcal{P}}_\infty^{0,n}\{0\}_2$, by assumption, is δ_0-exact, then

$$(\overline{\delta} - \delta_0)\Psi + H$$

is a δ_0-boundary. Consequently, Φ (4.1.19) is δ_0-exact. Conversely, let $\Phi \in \overline{\mathcal{P}}_\infty^{0,n}\{0\}_2$ be a $\overline{\delta}$-cycle, i.e.,

$$\overline{\delta}\Phi = 2\Phi^{(A,\Lambda)(B,\Sigma)}\overline{s}_{\Lambda A}\overline{\delta s}_{\Sigma B}\omega = 2\Phi^{(A,\Lambda)(B,\Sigma)}\overline{s}_{\Lambda A}d_\Sigma \mathcal{E}_B \omega = 0.$$

It follows that

$$\Phi^{(A,\Lambda)(B,\Sigma)}\overline{\delta s}_{\Sigma B} = 0$$

for all indices (A, Λ). Omitting a $\overline{\delta}$-boundary term, we obtain

$$\Phi^{(A,\Lambda)(B,\Sigma)}\overline{s}_{\Sigma B} = G^{(A,\Lambda)(r,\Xi)}d_\Xi \Delta_r.$$

Hence, Φ takes the form

$$\Phi = G'^{(A,\Lambda)(r,\Xi)}d_\Xi \Delta_r \overline{s}_{\Lambda A}\omega.$$

Then there exists a three-chain

$$\Psi = G'^{(A,\Lambda)(r,\Xi)}\overline{c}_{\Xi r}\overline{s}_{\Lambda A}\omega$$

such that

$$\delta_0 \Psi = \Phi + \sigma = \Phi + G'''^{(A,\Lambda)(r,\Xi)}d_\Lambda \mathcal{E}_A \overline{c}_{\Xi r}\omega. \tag{4.1.20}$$

Owing to the equality $\overline{\delta}\Phi = 0$, we have $\delta_0 \sigma = 0$. Thus, σ in the expression (4.1.20) is $\overline{\delta}$-exact δ_0-cycle. By assumption, it is δ_0-exact, i.e., $\sigma = \delta_0 \psi$. Consequently, a $\overline{\delta}$-cycle Φ is a δ_0-boundary

$$\Phi = \delta_0 \Psi - \delta_0 \psi.$$

□

Remark 4.1.2. It is easily justified that a two-cycle Φ (4.1.17) is δ_0-exact if and only if Φ up to a $\overline{\delta}$-boundary takes the form

$$\Phi = \sum_{0 \leq |\Lambda|, |\Sigma|} G'^{(r,\Sigma)(r',\Lambda)} d_\Sigma \Delta_r d_\Lambda \Delta_{r'} \omega.$$

A degenerate Lagrangian system is called *reducible* (resp. *irreducible*) if it admits (resp. does not admit) non-trivial first stage Noether identities.

If the condition of Lemma 4.1.3 is satisfied, let us assume that non-trivial first-stage Noether identities are finitely generated as follows. There exists a graded projective $C^\infty(X)$-module $\mathcal{C}_{(1)} \subset H_2(\delta_0)$ of finite rank possessing a local basis $\{\Delta_{r_1}\omega\}$:

$$\Delta_{r_1}\omega = \sum_{0\leq|\Lambda|} \Delta_{r_1}^{r,\Lambda}\overline{c}_{\Lambda r}\omega + h_{r_1}\omega, \qquad (4.1.21)$$

such that any element $\Phi \in H_2(\delta_0)$ factorizes as

$$\Phi = \sum_{0\leq|\Xi|} \Phi^{r_1,\Xi} d_\Xi \Delta_{r_1}\omega, \qquad \Phi^{r_1,\Xi} \in \mathcal{S}^0_\infty[F;Y], \qquad (4.1.22)$$

through elements (4.1.21) of $\mathcal{C}_{(1)}$. Thus, all non-trivial first-stage Noether identities (4.1.18) result from the equalities

$$\sum_{0\leq|\Lambda|} \Delta_{r_1}^{r,\Lambda} d_\Lambda \Delta_r + \overline{\delta}h_{r_1} = 0, \qquad (4.1.23)$$

called the *complete first-stage Noether identities*. Note that, by virtue of the condition of Lemma 4.1.3, the first summands of the graded densities $\Delta_{r_1}\omega$ (4.1.21) are not $\overline{\delta}$-exact.

A degenerate Lagrangian system is called *finitely reducible* if admits finitely generated non-trivial first-stage Noether identities.

Lemma 4.1.4. *The one-exact complex (4.1.16) of a finitely reducible Lagrangian system is extended to the two-exact one (4.1.25) with a boundary operator whose nilpotency conditions are equivalent to the complete Noether identities (4.1.13) and the complete first-stage Noether identities (4.1.23).*

Proof. By virtue of Serre–Swan Theorem 3.3.2, the graded module $\mathcal{C}_{(1)}$ is isomorphic to a module of sections of the density-dual $\overline{E}_1$ of some graded vector bundle $E_1 \to X$. Let us enlarge the differential bigraded algebra $\overline{\mathcal{P}}^*_\infty\{0\}$ (4.1.14) to the differential bigraded algebra

$$\overline{\mathcal{P}}^*_\infty\{1\} = \mathcal{P}^*_\infty[\overline{VF} \underset{Y}{\oplus} \overline{E}_0 \underset{Y}{\oplus} \overline{E}_1; Y]$$

possessing the local generating basis $\{s^A, \overline{s}_A, \overline{c}_r, \overline{c}_{r_1}\}$ where $\overline{c}_{r_1}$ are first stage Noether antifields of Grassmann parity

$$[\overline{c}_{r_1}] = ([\Delta_{r_1}] + 1) \bmod 2$$

and antifield number $\mathrm{Ant}[\overline{c}_{r_1}] = 3$. This differential bigraded algebra is provided with the odd right graded derivation

$$\delta_1 = \delta_0 + \overleftarrow{\partial}{}^{r_1}\Delta_{r_1} \qquad (4.1.24)$$

4.1. Noether identities. The Koszul–Tate complex

which is nilpotent if and only if the complete Noether identities (4.1.13) and the complete first-stage Noether identities (4.5.16) hold. Then δ_1 (4.1.24) is a boundary operator of the chain complex

$$0 \leftarrow \text{Im}\,\overline{\delta} \xleftarrow{\overline{\delta}} \mathcal{P}_\infty^{0,n}[\overline{VF};Y]_1 \xleftarrow{\delta_0} \overline{\mathcal{P}}_\infty^{0,n}\{0\}_2 \xleftarrow{\delta_1} \overline{\mathcal{P}}_\infty^{0,n}\{1\}_3 \xleftarrow{\delta_1} \overline{\mathcal{P}}_\infty^{0,n}\{1\}_4 \quad (4.1.25)$$

of graded densities of antifield number ≤ 4. Let $H_*(\delta_1)$ denote its homology. It is readily observed that

$$H_0(\delta_1) = H_0(\overline{\delta}), \qquad H_1(\delta_1) = H_1(\delta_0) = 0.$$

By virtue of the expression (4.1.22), any two-cycle of the complex (4.1.25) is a boundary

$$\Phi = \sum_{0 \leq |\Xi|} \Phi^{r_1,\Xi} d_\Xi \Delta_{r_1} \omega = \delta_1 \left(\sum_{0 \leq |\Xi|} \Phi^{r_1,\Xi} \overline{c}_{\Xi r_1} \omega \right).$$

It follows that $H_2(\delta_1) = 0$, i.e., the complex (4.1.25) is two-exact. □

If the third homology $H_3(\delta_1)$ of the complex (4.1.25) is not trivial, its elements correspond to second-stage Noether identities which the complete first-stage ones satisfy, and so on. Iterating the arguments, one comes to the following.

A degenerate Grassmann-graded Lagrangian system $(\mathcal{S}_\infty^*[F;Y], L)$ is called *N-stage reducible* if it admits finitely generated non-trivial N-stage Noether identities, but no non-trivial $(N+1)$-stage ones. It is characterized as follows [14].

- There are graded vector bundles $E_0, \ldots, E_N$ over X, and the differential bigraded algebra $\mathcal{P}_\infty^*[\overline{VF};Y]$ is enlarged to the differential bigraded algebra

$$\overline{\mathcal{P}}_\infty^*\{N\} = \mathcal{P}_\infty^*[\overline{VF} \oplus_Y \overline{E}_0 \oplus_Y \cdots \oplus_Y \overline{E}_N; Y] \quad (4.1.26)$$

with the local generating basis

$$(s^A, \overline{s}_A, \overline{c}_r, \overline{c}_{r_1}, \ldots, \overline{c}_{r_N})$$

where $\overline{c}_{r_k}$ are *Noether k-stage antifields* of antifield number $\text{Ant}[\overline{c}_{r_k}] = k+2$.

- The differential bigraded algebra (4.1.26) is provided with the nilpotent right graded derivation

$$\delta_{\text{KT}} = \delta_N = \overline{\delta} + \sum_{0 \leq |\Lambda|} \overleftarrow{\partial}^r \Delta_r^{A,\Lambda} \overline{s}_{\Lambda A} + \sum_{1 \leq k \leq N} \overleftarrow{\partial}^{r_k} \Delta_{r_k}, \quad (4.1.27)$$

$$\Delta_{r_k}\omega = \sum_{0 \leq |\Lambda|} \Delta_{r_k}^{r_{k-1},\Lambda} \overline{c}_{\Lambda r_{k-1}} \omega + \quad (4.1.28)$$

$$\sum_{0 \leq |\Sigma|, |\Xi|} (h_{r_k}^{(r_{k-2},\Sigma)(A,\Xi)} \overline{c}_{\Sigma r_{k-2}} \overline{s}_{\Xi A} + \ldots)\omega \in \overline{\mathcal{P}}_\infty^{0,n}\{k-1\}_{k+1},$$

of antifield number -1. The index $k = -1$ here stands for $\overline{s}_A$. The nilpotent derivation δ_{KT} (4.1.27) is called the *Koszul–Tate operator*.

- With this graded derivation, the module $\overline{\mathcal{P}}_\infty^{0,n}\{N\}_{\leq N+3}$ of densities of antifield number $\leq (N+3)$ is decomposed into the exact *Koszul–Tate chain complex*

$$0 \leftarrow \operatorname{Im}\overline{\delta} \xleftarrow{\overline{\delta}} \mathcal{P}_\infty^{0,n}[\overline{VF};Y]_1 \xleftarrow{\delta_0} \overline{\mathcal{P}}_\infty^{0,n}\{0\}_2 \xleftarrow{\delta_1} \overline{\mathcal{P}}_\infty^{0,n}\{1\}_3 \cdots \quad (4.1.29)$$
$$\xleftarrow{\delta_{N-1}} \overline{\mathcal{P}}_\infty^{0,n}\{N-1\}_{N+1} \xleftarrow{\delta_{\mathrm{KT}}} \overline{\mathcal{P}}_\infty^{0,n}\{N\}_{N+2} \xleftarrow{\delta_{\mathrm{KT}}} \overline{\mathcal{P}}_\infty^{0,n}\{N\}_{N+3}$$

which satisfies the following *homology regularity condition*.

Condition 4.1.1. Any $\delta_{k<N}$-cycle

$$\phi \in \overline{\mathcal{P}}_\infty^{0,n}\{k\}_{k+3} \subset \overline{\mathcal{P}}_\infty^{0,n}\{k+1\}_{k+3}$$

is a δ_{k+1}-boundary.

Remark 4.1.3. The exactness of the complex (4.1.29) means that any $\delta_{k<N}$-cycle $\phi \in \mathcal{P}_\infty^{0,n}\{k\}_{k+3}$, is a δ_{k+2}-boundary, but not necessary a δ_{k+1}-one.

- The nilpotentness $\delta_{\mathrm{KT}}^2 = 0$ of the Koszul–Tate operator (4.1.27) is equivalent to the complete non-trivial Noether identities (4.1.13) and the complete non-trivial $(k \leq N)$-stage Noether identities

$$\sum_{0\leq|\Lambda|} \Delta_{r_k}^{r_{k-1},\Lambda} d_\Lambda \left(\sum_{0\leq|\Sigma|} \Delta_{r_{k-1}}^{r_{k-2},\Sigma} \overline{c}_{\Sigma r_{k-2}} \right) = \quad (4.1.30)$$
$$-\overline{\delta}\left(\sum_{0\leq|\Sigma|,|\Xi|} h_{r_k}^{(r_{k-2},\Sigma)(A,\Xi)} \overline{c}_{\Sigma r_{k-2}} \overline{s}_{\Xi A} \right).$$

This item means the following.

Assertion 4.1.1. Any δ_k-cocycle $\Phi \in \mathcal{P}_\infty^{0,n}\{k\}_{k+2}$ is a k-stage Noether identity, and *vice versa*.

Proof. Any $(k+2)$-chain $\Phi \in \mathcal{P}_\infty^{0,n}\{k\}_{k+2}$ takes the form

$$\Phi = G + H = \sum_{0\leq|\Lambda|} G^{r_k,\Lambda} \overline{c}_{\Lambda r_k} \omega + \quad (4.1.31)$$
$$\sum_{0\leq\Sigma,0\leq\Xi} (H^{(A,\Xi)(r_{k-1},\Sigma)} \overline{s}_{\Xi A} \overline{c}_{\Sigma r_{k-1}} + ...)\omega.$$

4.1. Noether identities. The Koszul–Tate complex

If it is a δ_k-cycle, then

$$\sum_{0\leq|\Lambda|} G^{r_k,\Lambda} d_\Lambda \left(\sum_{0\leq|\Sigma|} \Delta_{r_k}^{r_{k-1},\Sigma} \overline{c}_{\Sigma r_{k-1}} \right) + \qquad (4.1.32)$$

$$\overline{\delta}\left(\sum_{0\leq\Sigma, 0\leq\Xi} H^{(A,\Xi)(r_{k-1},\Sigma)} \overline{s}_{\Xi A} \overline{c}_{\Sigma r_{k-1}} \right) = 0$$

are the k-stage Noether identities. Conversely, let the condition (4.1.32) hold. Then it can be extended to a cycle condition as follows. It is brought into the form

$$\delta_k \left(\sum_{0\leq|\Lambda|} G^{r_k,\Lambda} \overline{c}_{\Lambda r_k} + \sum_{0\leq\Sigma, 0\leq\Xi} H^{(A,\Xi)(r_{k-1},\Sigma)} \overline{s}_{\Xi A} \overline{c}_{\Sigma r_{k-1}} \right) =$$
$$- \sum_{0\leq|\Lambda|} G^{r_k,\Lambda} d_\Lambda h_{r_k} + \sum_{0\leq\Sigma, 0\leq\Xi} H^{(A,\Xi)(r_{k-1},\Sigma)} \overline{s}_{\Xi A} d_\Sigma \Delta_{r_{k-1}}.$$

A glance at the expression (4.1.28) shows that the term in the right-hand side of this equality belongs to $\mathcal{P}^{0,n}_\infty\{k-2\}_{k+1}$. It is a δ_{k-2}-cycle and, consequently, a δ_{k-1}-boundary $\delta_{k-1}\Psi$ in accordance with Condition 4.1.1. Then the equality (4.1.32) is a $\overline{c}_{\Sigma r_{k-1}}$-dependent part of the cycle condition

$$\delta_k \left(\sum_{0\leq|\Lambda|} G^{r_k,\Lambda} \overline{c}_{\Lambda r_k} + \sum_{0\leq\Sigma, 0\leq\Xi} H^{(A,\Xi)(r_{k-1},\Sigma)} \overline{s}_{\Xi A} \overline{c}_{\Sigma r_{k-1}} - \Psi \right) = 0,$$

but $\delta_k \Psi$ does not make a contribution to this condition. $\square$

Assertion 4.1.2. Any trivial k-stage Noether identity is a δ_k-boundary $\Phi \in \mathcal{P}^{0,n}_\infty\{k\}_{k+2}$.

Proof. The k-stage Noether identities (4.1.32) are trivial either if a δ_k-cycle Φ (4.1.31) is a δ_k-boundary or its summand G vanishes on-shell. Let us show that, if the summand G of Φ (4.1.31) is $\overline{\delta}$-exact, then Φ is a δ_k-boundary. If $G = \overline{\delta}\Psi$, one can write

$$\Phi = \delta_k \Psi + (\overline{\delta} - \delta_k)\Psi + H. \qquad (4.1.33)$$

Hence, the δ_k-cycle condition reads

$$\delta_k \Phi = \delta_{k-1}((\overline{\delta} - \delta_k)\Psi + H) = 0.$$

By virtue of Condition 4.1.1, any δ_{k-1}-cycle $\phi \in \overline{\mathcal{P}}^{0,n}_\infty\{k-1\}_{k+2}$ is δ_k-exact. Then

$$(\overline{\delta} - \delta_k)\Psi + H$$

is a δ_k-boundary. Consequently, Φ (4.1.31) is δ_k-exact. $\square$

Assertion 4.1.3. All non-trivial k-stage Noether identity (4.1.32), by assumption, factorize as

$$\Phi = \sum_{0\leq|\Xi|} \Phi^{r_k,\Xi} d_\Xi \Delta_{r_k}\omega, \qquad \Phi^{r_1,\Xi} \in \mathcal{S}^0_\infty[F;Y],$$

through the complete ones (4.1.30).

It may happen that a Grassmann-graded Lagrangian field system possesses non-trivial Noether identities of any stage. However, we restrict our consideration to N-reducible Lagrangian systems for a finite integer N. In this case, the Koszul–Tate operator (4.1.27) and the gauge operator (4.2.8) below contain finite terms.

4.2 Second Noether theorems in a general setting

Different variants of the second Noether theorem have been suggested in order to relate reducible Noether identities and gauge symmetries [9; 12; 13; 49]. The *inverse second Noether theorem* (Theorem 4.2.1), that we formulate in homology terms, associates to the Koszul–Tate complex (4.1.29) of non-trivial Noether identities the cochain sequence (4.2.7) with the ascent operator **u** (4.2.8) whose components are non-trivial gauge and higher-stage gauge symmetries of Lagrangian system. Let us start with the following notation.

Notation 4.2.1. Given the differential bigraded algebra $\overline{\mathcal{P}}^*_\infty\{N\}$ (4.1.26), we consider the differential bigraded algebra

$$\mathcal{P}^*_\infty\{N\} = \mathcal{P}^*_\infty[F \underset{Y}{\oplus} E_0 \underset{Y}{\oplus} \cdots \underset{Y}{\oplus} E_N; Y], \qquad (4.2.1)$$

possessing the local generating basis

$$(s^A, c^r, c^{r_1}, \ldots, c^{r_N}), \qquad [c^{r_k}] = ([\overline{c}_{r_k}] + 1) \bmod 2,$$

and the differential bigraded algebra

$$P^*_\infty\{N\} = \mathcal{P}^*_\infty[\overline{VF} \underset{Y}{\oplus} E_0 \underset{Y}{\oplus} \cdots \underset{Y}{\oplus} E_N \underset{Y}{\oplus} \overline{E}_0 \underset{Y}{\oplus} \cdots \underset{Y}{\oplus} \overline{E}_N; Y] \qquad (4.2.2)$$

with the local generating basis

$$(s^A, \overline{s}_A, c^r, c^{r_1}, \ldots, c^{r_N}, \overline{c}_r, \overline{c}_{r_1}, \ldots, \overline{c}_{r_N}).$$

Their elements c^{r_k} are called k-stage *ghosts* of ghost number $\mathrm{gh}[c^{r_k}] = k+1$ and antifield number

$$\mathrm{Ant}[c^{r_k}] = -(k+1).$$

The $C^\infty(X)$-module $\mathcal{C}^{(k)}$ of k-stage ghosts is the density-dual of the module $\mathcal{C}_{(k+1)}$ of $(k+1)$-stage antifields. The differential bigraded algebras $\overline{\mathcal{P}}^*_\infty\{N\}$ (4.1.26) and $\mathcal{P}^*_\infty\{N\}$ (4.2.1) are subalgebras of $P^*_\infty\{N\}$ (4.2.2). The Koszul–Tate operator δ_{KT} (4.1.27) is naturally extended to a graded derivation of the differential bigraded algebra $P^*_\infty\{N\}$.

Notation 4.2.2. Any graded differential form $\phi \in \mathcal{S}^*_\infty[F;Y]$ and any finite tuple (f^Λ), $0 \leq |\Lambda| \leq k$, of local graded functions $f^\Lambda \in \mathcal{S}^0_\infty[F;Y]$ obey the following relations [12]:

$$\sum_{0 \leq |\Lambda| \leq k} f^\Lambda d_\Lambda \phi \wedge \omega = \sum_{0 \leq |\Lambda|} (-1)^{|\Lambda|} d_\Lambda (f^\Lambda) \phi \wedge \omega + d_H \sigma, \qquad (4.2.3)$$

$$\sum_{0 \leq |\Lambda| \leq k} (-1)^{|\Lambda|} d_\Lambda (f^\Lambda \phi) = \sum_{0 \leq |\Lambda| \leq k} \eta(f)^\Lambda d_\Lambda \phi, \qquad (4.2.4)$$

$$\eta(f)^\Lambda = \sum_{0 \leq |\Sigma| \leq k - |\Lambda|} (-1)^{|\Sigma + \Lambda|} \frac{(|\Sigma + \Lambda|)!}{|\Sigma|! |\Lambda|!} d_\Sigma f^{\Sigma + \Lambda}, \qquad (4.2.5)$$

$$\eta(\eta(f))^\Lambda = f^\Lambda. \qquad (4.2.6)$$

Theorem 4.2.1. *Given the Koszul–Tate complex (4.1.29), the module of graded densities $\mathcal{P}^{0,n}_\infty\{N\}$ is decomposed into the cochain sequence*

$$0 \to \mathcal{S}^{0,n}_\infty[F;Y] \xrightarrow{\mathbf{u}} \mathcal{P}^{0,n}_\infty\{N\}^1 \xrightarrow{\mathbf{u}} \mathcal{P}^{0,n}_\infty\{N\}^2 \xrightarrow{\mathbf{u}} \cdots, \qquad (4.2.7)$$

$$\mathbf{u} = u + u^{(1)} + \cdots + u^{(N)} = \qquad (4.2.8)$$

$$u^A \frac{\partial}{\partial s^A} + u^r \frac{\partial}{\partial c^r} + \cdots + u^{r_{N-1}} \frac{\partial}{\partial c^{r_{N-1}}},$$

graded in ghost number. Its ascent operator $\mathbf{u}$ (4.2.8) is an odd graded derivation of ghost number 1 where u (4.2.13) is a variational symmetry of a graded Lagrangian L and the graded derivations $u_{(k)}$ (4.2.16), $k = 1, \ldots, N$, obey the relations (4.2.15).

Proof. Given the Koszul–Tate operator (4.1.27), let us extend an original Lagrangian L to the Lagrangian

$$L_e = L + L_1 = L + \sum_{0 \leq k \leq N} c^{r_k} \Delta_{r_k} \omega = L + \delta_{\mathrm{KT}} \big(\sum_{0 \leq k \leq N} c^{r_k} \overline{c}_{r_k} \omega \big) \qquad (4.2.9)$$

of zero antifield number. It is readily observed that the Koszul–Tate operator δ_{KT} is an exact symmetry of the extended Lagrangian $L_e \in P^{0,n}_\infty\{N\}$ (4.2.9). Since the graded derivation δ_{KT} is vertical, it follows from the first variational formula (3.5.14) that

$$\left[\frac{\overleftarrow{\delta} \mathcal{L}_e}{\delta \overline{s}_A}\mathcal{E}_A + \sum_{0\leq k\leq N}\frac{\overleftarrow{\delta} \mathcal{L}_e}{\delta \overline{c}_{r_k}}\Delta_{r_k}\right]\omega = \qquad (4.2.10)$$

$$\left[v^A \mathcal{E}_A + \sum_{0\leq k\leq N} v^{r_k}\frac{\delta\mathcal{L}_e}{\delta c^{r_k}}\right]\omega = d_H\sigma,$$

$$v^A = \frac{\overleftarrow{\delta}\mathcal{L}_e}{\delta\overline{s}_A} = u^A + w^A = \sum_{0\leq|\Lambda|}c_\Lambda^r \eta(\Delta_r^A)^\Lambda + \sum_{1\leq i\leq N}\sum_{0\leq|\Lambda|}c_\Lambda^{r_i}\eta(\overleftarrow{\partial}^A(h_{r_i}))^\Lambda,$$

$$v^{r_k} = \frac{\overleftarrow{\delta}\mathcal{L}_e}{\delta\overline{c}_{r_k}} = u^{r_k} + w^{r_k} = \sum_{0\leq|\Lambda|}c_\Lambda^{r_{k+1}}\eta(\Delta_{r_{k+1}}^{r_k})^\Lambda +$$

$$\sum_{k+1<i\leq N}\sum_{0\leq|\Lambda|}c_\Lambda^{r_i}\eta(\overleftarrow{\partial}^{r_k}(h_{r_i}))^\Lambda.$$

The equality (4.2.10) is split into the set of equalities

$$\frac{\overleftarrow{\delta}(c^r\Delta_r)}{\delta\overline{s}_A}\mathcal{E}_A\omega = u^A\mathcal{E}_A\omega = d_H\sigma_0, \qquad (4.2.11)$$

$$\left[\frac{\overleftarrow{\delta}(c^{r_k}\Delta_{r_k})}{\delta\overline{s}_A}\mathcal{E}_A + \sum_{0\leq i<k}\frac{\overleftarrow{\delta}(c^{r_k}\Delta_{r_k})}{\delta\overline{c}_{r_i}}\Delta_{r_i}\right]\omega = d_H\sigma_k, \qquad (4.2.12)$$

where $k = 1,\ldots,N$. A glance at the equality (4.2.11) shows that, by virtue of the first variational formula (3.5.14), the odd graded derivation

$$u = u^A\frac{\partial}{\partial s^A}, \qquad u^A = \sum_{0\leq|\Lambda|}c_\Lambda^r\eta(\Delta_r^A)^\Lambda, \qquad (4.2.13)$$

of $\mathcal{P}^0\{0\}$ is a variational symmetry of a graded Lagrangian L. Every equality (4.2.12) falls into a set of equalities graded by the polynomial degree in antifields. Let us consider that of them linear in antifields $\overline{c}_{r_{k-2}}$. We have

$$\frac{\overleftarrow{\delta}}{\delta\overline{s}_A}\left(c^{r_k}\sum_{0\leq|\Sigma|,|\Xi|}h_{r_k}^{(r_{k-2},\Sigma)(A,\Xi)}\overline{c}_{\Sigma r_{k-2}}\overline{s}_{\Xi A}\right)\mathcal{E}_A\omega +$$

$$\frac{\overleftarrow{\delta}}{\delta\overline{c}_{r_{k-1}}}\left(c^{r_k}\sum_{0\leq|\Sigma|}\Delta_{r_k}^{r'_{k-1},\Sigma}\overline{c}_{\Sigma r'_{k-1}}\right)\sum_{0\leq|\Xi|}\Delta_{r_{k-1}}^{r_{k-2},\Xi}\overline{c}_{\Xi r_{k-2}}\omega = d_H\sigma_k.$$

4.2. Second Noether theorems in a general setting

This equality is brought into the form

$$\sum_{0\leq|\Xi|}(-1)^{|\Xi|}d_\Xi\left(c^{r_k}\sum_{0\leq|\Sigma|}h^{(r_k-2,\Sigma)(A,\Xi)}_{r_k}\overline{c}_{\Sigma r_k-2}\right)\mathcal{E}_A\omega +$$

$$u^{r_k-1}\sum_{0\leq|\Xi|}\Delta^{r_k-2,\Xi}_{r_k-1}\overline{c}_{\Xi r_k-2}\omega = d_H\sigma_k.$$

Using the relation (4.2.3), we obtain the equality

$$\sum_{0\leq|\Xi|}c^{r_k}\sum_{0\leq|\Sigma|}h^{(r_k-2,\Sigma)(A,\Xi)}_{r_k}\overline{c}_{\Sigma r_k-2}d_\Xi\mathcal{E}_A\omega + \quad (4.2.14)$$

$$u^{r_k-1}\sum_{0\leq|\Xi|}\Delta^{r_k-2,\Xi}_{r_k-1}\overline{c}_{\Xi r_k-2}\omega = d_H\sigma'_k.$$

The variational derivative of both its sides with respect to $\overline{c}_{r_k-2}$ leads to the relation

$$\sum_{0\leq|\Sigma|}d_\Sigma u^{r_k-1}\frac{\partial}{\partial c^{r_k-1}_\Sigma}u^{r_k-2} = \overline{\delta}(\alpha^{r_k-2}), \quad (4.2.15)$$

$$\alpha^{r_k-2} = -\sum_{0\leq|\Sigma|}\eta(h^{(r_k-2)(A,\Xi)}_{r_k})^\Sigma d_\Sigma(c^{r_k}\overline{s}_{\Xi A}),$$

which the odd graded derivation

$$u^{(k)} = u^{r_k-1}\frac{\partial}{\partial c^{r_k-1}} = \sum_{0\leq|\Lambda|}c^{r_k}_\Lambda\eta(\Delta^{r_k-1}_{r_k})^\Lambda\frac{\partial}{\partial c^{r_k-1}}, \quad k=1,\ldots,N, \quad (4.2.16)$$

satisfies. Graded derivations u (4.2.13) and $u^{(k)}$ (4.2.16) are assembled into the ascent operator **u** (4.2.8) of the cochain sequence (4.2.7). □

A glance at the expression (4.2.13) shows that the variational symmetry u is a linear differential operator on the $C^\infty(X)$-module $\mathcal{C}^{(0)}$ of ghosts with values into the real space $\mathfrak{g}_L$ of variational symmetries. Following Definition 2.3.1 extended to Lagrangian theories of odd fields, we call u (4.2.13) the *gauge symmetry* of a graded Lagrangian L which is associated to the complete Noether identities (4.1.13). This association is unique due to the following *direct second Noether theorem* extenting Theorem 2.3.1.

Theorem 4.2.2. *The variational derivative of the equality (4.2.11) with respect to ghosts c^r leads to the equality*

$$\delta_r(u^A\mathcal{E}_A\omega) = \sum_{0\leq|\Lambda|}(-1)^{|\Lambda|}d_\Lambda(\eta(\Delta^A_r)^\Lambda\mathcal{E}_A) =$$

$$\sum_{0\leq|\Lambda|}(-1)^{|\Lambda|}\eta(\eta(\Delta^A_r))^\Lambda d_\Lambda\mathcal{E}_A = 0,$$

which reproduces the complete Noether identities (4.1.13) by means of the relation (4.2.6).

Moreover, the gauge symmetry u (4.2.13) is complete in the following sense. Let

$$\sum_{0\leq|\Xi|} C^R G_R^{r,\Xi} d_\Xi \Delta_r \omega$$

be some projective $C^\infty(X)$-module of finite rank of non-trivial Noether identities (4.1.12) parameterized by the corresponding ghosts C^R. We have the equalities

$$0 = \sum_{0\leq|\Xi|} C^R G_R^{r,\Xi} d_\Xi \left(\sum_{0\leq|\Lambda|} \Delta_r^{A,\Lambda} d_\Lambda \mathcal{E}_A \right) \omega =$$

$$\sum_{0\leq|\Lambda|} \left(\sum_{0\leq|\Xi|} \eta(G_R^r)^\Xi C_\Xi^R \right) \Delta_r^{A,\Lambda} d_\Lambda \mathcal{E}_A \omega + d_H(\sigma) =$$

$$\sum_{0\leq|\Lambda|} (-1)^{|\Lambda|} d_\Lambda \left(\Delta_r^{A,\Lambda} \sum_{0\leq|\Xi|} \eta(G_R^r)^\Xi C_\Xi^R \right) \mathcal{E}_A \omega + d_H \sigma =$$

$$\sum_{0\leq|\Lambda|} \eta(\Delta_r^A)^\Lambda d_\Lambda \left(\sum_{0\leq|\Xi|} \eta(G_R^r)^\Xi C_\Xi^R \right) \mathcal{E}_A \omega + d_H \sigma =$$

$$\sum_{0\leq|\Lambda|} u_r^{A,\Lambda} d_\Lambda \left(\sum_{0\leq|\Xi|} \eta(G_R^r)^\Xi C_\Xi^R \right) \mathcal{E}_A \omega + d_H \sigma.$$

It follows that the graded derivation

$$d_\Lambda \left(\sum_{0\leq|\Xi|} \eta(G_R^r)^\Xi C_\Xi^R \right) u_r^{A,\Lambda} \frac{\partial}{\partial s^A}$$

is a variational symmetry of a graded Lagrangian L and, consequently, its gauge symmetry parameterized by ghosts C^R. It factorizes through the gauge symmetry (4.2.13) by putting ghosts

$$c^r = \sum_{0\leq|\Xi|} \eta(G_R^r)^\Xi C_\Xi^R.$$

Thus, we come to the following definition.

Definition 4.2.1. The odd graded derivation u (4.2.13) is said to be a *complete non-trivial gauge symmetry* of a graded Lagrangian L associated to the complete Noether identities (4.1.13).

4.2. Second Noether theorems in a general setting

Remark 4.2.1. In contrast with Definitions 2.3.1 and 2.3.2, gauge symmetries in Definitions 4.2.2 and 4.2.1 are parameterized by ghosts, but not gauge parameters. Given a gauge symmetry u (2.3.3) defined as a derivation of the real ring $\mathcal{O}_\infty^0[Y \times E]$, one can associate to it the gauge symmetry

$$u = \left(\sum_{0 \le |\Lambda| \le m} u_u^{\lambda\Lambda}(x^\mu) c_\Lambda^a \right) \partial_\lambda + \left(\sum_{0 \le |\Lambda| \le m} u_a^{i\Lambda}(x^\mu, y_\Sigma^j) c_\Lambda^a \right) \partial_i, \quad (4.2.17)$$

which is an odd graded derivation of the real ring $\mathcal{S}_\infty^0[E; Y]$, and *vice versa*.

Let us note that, being a variational symmetry, a gauge symmetry u (4.2.13) defines the weak conservation law (3.5.16). Let u be an exact Lagrangian symmetry. In this case, the associated symmetry current

$$\mathcal{J}_u = -h_0(u \rfloor \Xi_L) \qquad (4.2.18)$$

is conserved. The peculiarity of gauge conservation laws always is that the symmetry current (4.2.18) is reduced to a superpotential in accordance with the following generalization of Theorem 2.4.2.

Theorem 4.2.3. *If u (4.2.13) is an exact gauge symmetry of a graded Lagrangian L, the corresponding conserved symmetry current $\mathcal{J}_u$ (4.2.18) takes the form*

$$\mathcal{J}_u = W + d_H U = (W^\mu + d_\nu U^{\nu\mu})\omega_\mu, \qquad (4.2.19)$$

where the term W vanishes on-shell, i.e., $W \approx 0$, and U is a horizontal graded $(n-2)$-form.

Proof. The proof follows that of Theorem 2.4.2. Let us note only that, if u is of N-order in jets of ghosts, the symmetry current $\mathcal{J}_u$ (4.2.18) depends on jets of ghosts whose maximal order N' exceed N in the case of a higher-order Lagrangian. Therefore, the equalities (2.4.43) are replaced with a set of equalities

$$0 = J_a^{(\mu\mu_1)\cdots\mu_{N'}},$$
$$0 = J_a^{(\mu_k\mu_{k+1})\cdots\mu_{N'}} + d_\nu J_a^{\nu\mu_k\cdots\mu_{N'}}, \qquad 1 < k < N' - N.$$
$\square$

Turn now to the relation (4.2.15). For $k = 1$, it takes the form

$$\sum_{0 \le |\Sigma|} d_\Sigma u^r \frac{\partial}{\partial c_\Sigma^r} u^A = \overline{\delta}(\alpha^A)$$

of a first-stage gauge symmetry condition on-shell which the non-trivial gauge symmetry u (4.2.13) satisfies. Therefore, one can treat the odd graded derivation

$$u^{(1)} = u^r \frac{\partial}{\partial c^r}, \qquad u^r = \sum_{0 \leq |\Lambda|} c^{r_1}_\Lambda \eta(\Delta^r_{r_1})^\Lambda,$$

as a *first-stage gauge symmetry* associated to the complete first-stage Noether identities

$$\sum_{0 \leq |\Lambda|} \Delta^{r,\Lambda}_{r_1} d_\Lambda \left(\sum_{0 \leq |\Sigma|} \Delta^{A,\Sigma}_r \overline{s}_{\Sigma A} \right) = -\overline{\delta} \left(\sum_{0 \leq |\Sigma|, |\Xi|} h^{(B,\Sigma)(A,\Xi)}_{r_1} \overline{s}_{\Sigma B} \overline{s}_{\Xi A} \right).$$

Iterating the arguments, one comes to the relation (4.2.15) which provides a k-stage gauge symmetry condition which is associated to the complete k-stage Noether identities (4.1.30).

Theorem 4.2.4. *Conversely, given the k-stage gauge symmetry condition (4.2.15), the variational derivative of the equality (4.2.14) with respect to ghosts c^{r_k} leads to the equality, reproducing the k-stage Noether identities (4.1.30) by means of the relations (4.2.6) and (4.2.6).*

This is a higher-stage extension of the direct second Noether theorem to reducible gauge symmetries.

The odd graded derivation $u_{(k)}$ (4.2.16) is called the *k-stage gauge symmetry*. It is complete as follows. Let

$$\sum_{0 \leq |\Xi|} C^{R_k} G^{r_k,\Xi}_{R_k} d_\Xi \Delta_{r_k} \omega$$

be a projective $C^\infty(X)$-module of finite rank of non-trivial k-stage Noether identities (4.1.12) factorizing through the complete ones (4.1.28) and parameterized by the corresponding ghosts C^{R_k}. One can show that it defines a k-stage gauge symmetry factorizing through $u^{(k)}$ (4.2.16) by putting k-stage ghosts

$$c^{r_k} = \sum_{0 \leq |\Xi|} \eta(G^{r_k}_{R_k})^\Xi C^{R_k}_\Xi.$$

Definition 4.2.2. The odd graded derivation $u_{(k)}$ (4.2.16) is said to be a *complete non-trivial k-stage gauge symmetry* of a Lagrangian L.

In accordance with Definitions 4.2.1 and 4.2.2, components of the ascent operator **u** (4.2.8) are complete non-trivial gauge and higher-stage gauge symmetries. Therefore, we agree to call this operator the *gauge operator*.

Remark 4.2.2. With the gauge operator (4.2.8), the extended Lagrangian L_e (4.2.9) takes the form

$$L_e = L + \mathbf{u}(\sum_{0 \leq k \leq N} c^{r_k-1}\overline{c}_{r_{k-1}})\omega + L_1^* + d_H\sigma, \qquad (4.2.20)$$

where L_1^* is a term of polynomial degree in antifields exceeding 1.

4.3 BRST operator

In contrast with the Koszul–Tate operator (4.1.27), the gauge operator $\mathbf{u}$ (4.2.7) need not be nilpotent. Following the example of Yang–Mills gauge theory (see Section 5.8), let us study its extension to a nilpotent graded derivation

$$\mathbf{b} = \mathbf{u} + \gamma = \mathbf{u} + \sum_{1 \leq k \leq N+1} \gamma^{(k)} = \mathbf{u} + \sum_{1 \leq k \leq N+1} \gamma^{r_{k-1}} \frac{\partial}{\partial c^{r_{k-1}}} \qquad (4.3.1)$$

$$= \left(u^A \frac{\partial}{\partial s^A} + \gamma^r \frac{\partial}{\partial c^r}\right) + \sum_{0 \leq k \leq N-1} \left(u^{r_k} \frac{\partial}{\partial c^{r_k}} + \gamma^{r_{k+1}} \frac{\partial}{\partial c^{r_{k+1}}}\right)$$

of ghost number 1 by means of antifield-free terms $\gamma^{(k)}$ of higher polynomial degree in ghosts c^{r_i} and their jets $c_\Lambda^{r_i}$, $0 \leq i < k$. We call $\mathbf{b}$ (4.3.1) the *BRST operator*, where k-stage gauge symmetries are extended to k-stage BRST transformations acting both on $(k-1)$-stage and k-stage ghosts [61]. If the BRST operator exists, the cochain sequence (4.2.7) is brought into the *BRST complex*

$$0 \to \mathcal{P}_\infty^{0,n}\{N\}^1 \xrightarrow{\mathbf{b}} \mathcal{P}_\infty^{0,n}\{N\}^1 \xrightarrow{\mathbf{b}} \mathcal{P}_\infty^{0,n}\{N\}^2 \xrightarrow{\mathbf{b}} \cdots.$$

There is following necessary condition of the existence of such a BRST extension.

Theorem 4.3.1. *The gauge operator (4.2.7) admits the nilpotent extension (4.3.1) only if the gauge symmetry conditions (4.2.15) and the higher-stage Noether identities (4.1.30) are satisfied off-shell.*

Proof. It is easily justified that, if the graded derivation $\mathbf{b}$ (4.3.1) is nilpotent, then the right hand sides of the equalities (4.2.15) equal zero, i.e.,

$$u^{(k+1)}(u^{(k)}) = 0, \qquad 0 \leq k \leq N-1, \qquad u^{(0)} = u. \qquad (4.3.2)$$

Using the relations (4.2.3) – (4.2.6), one can show that, in this case, the right hand sides of the higher-stage Noether identities (4.1.30) also equal zero [13]. It follows that the summand G_{r_k} of each cocycle Δ_{r_k} (4.1.28) is δ_{k-1}-closed. Then its summand h_{r_k} also is δ_{k-1}-closed and, consequently, δ_{k-2}-closed. Hence it is δ_{k-1}-exact by virtue of Condition 4.1.1. Therefore, Δ_{r_k} contains only the term G_{r_k} linear in antifields. □

It follows at once from the equalities (4.3.2) that the *higher-stage gauge operator*

$$u_{\mathrm{HS}} = \mathbf{u} - u = u^{(1)} + \cdots + u^{(N)}$$

is nilpotent, and

$$\mathbf{u}(\mathbf{u}) = u(\mathbf{u}). \qquad (4.3.3)$$

Therefore, the nilpotency condition for the BRST operator $\mathbf{b}$ (4.3.1) takes the form

$$\mathbf{b}(\mathbf{b}) = (u + \gamma)(\mathbf{u}) + (u + u_{\mathrm{HS}} + \gamma)(\gamma) = 0. \qquad (4.3.4)$$

Let us denote

$$\gamma^{(0)} = 0,$$
$$\gamma^{(k)} = \gamma^{(k)}_{(2)} + \cdots + \gamma^{(k)}_{(k+1)}, \quad k = 1, \ldots, N+1,$$
$$\gamma^{r_{k-1}}_{(i)} = \sum_{k_1 + \cdots + k_i = k+1-i} \left(\sum_{0 \leq |\Lambda_{k_j}|} \gamma^{r_{k-1}, \Lambda_{k_1}, \ldots, \Lambda_{k_i}}_{(i) r_{k_1}, \ldots, r_{k_i}} c^{r_{k_1}}_{\Lambda_{k_1}} \cdots c^{r_{k_i}}_{\Lambda_{k_i}} \right),$$
$$\gamma^{(N+2)} = 0,$$

where $\gamma^{(k)}_{(i)}$ are terms of polynomial degree $2 \leq i \leq k+1$ in ghosts. Then the nilpotent property (4.3.4) of $\mathbf{b}$ falls into a set of equalities

$$u^{(k+1)}(u^{(k)}) = 0, \quad 0 \leq k \leq N-1, \qquad (4.3.5)$$
$$(u + \gamma^{(k+1)}_{(2)})(u^{(k)}) + u_{\mathrm{HS}}(\gamma^{(k)}_{(2)}) = 0, \quad 0 \leq k \leq N+1, \qquad (4.3.6)$$
$$\gamma^{(k+1)}_{(i)}(u^{(k)}) + u(\gamma^{(k)}_{(i-1)}) + u_{\mathrm{HS}}(\gamma^{k}_{(i)}) + \qquad (4.3.7)$$
$$\sum_{2 \leq m \leq i-1} \gamma_{(m)}(\gamma^{(k)}_{(i-m+1)}) = 0, \quad i-2 \leq k \leq N+1,$$

of ghost polynomial degree 1, 2 and $3 \leq i \leq N+3$, respectively.

The equalities (4.3.5) are exactly the gauge symmetry conditions (4.3.2) in Theorem 4.3.1.

The equality (4.3.6) for $k=0$ reads

$$(u + \gamma^{(1)})(u) = 0, \quad \sum_{0 \leq |\Lambda|} (d_\Lambda(u^A) \partial^\Lambda_A u^B + d_\Lambda(\gamma^r) u^{B,\Lambda}_r) = 0. \qquad (4.3.8)$$

4.3. BRST operator

It takes the form of the Lie antibracket

$$[u,u] = -2\gamma^{(1)}(u) = -2 \sum_{0 \leq |\Lambda|} d_\Lambda(\gamma^r) u_r^{B,\Lambda} \partial_B \qquad (4.3.9)$$

of the odd gauge symmetry u. Its right-hand side is a non-linear differential operator on the module $\mathcal{C}^{(0)}$ of ghosts taking values into the real space $\mathcal{G}_L$ of variational symmetries. Following Remark 2.3.3, we treat it as a generalized gauge symmetry factorizing through the gauge symmetry u. Thus, we come to the following.

Theorem 4.3.2. *The gauge operator (4.2.7) admits the nilpotent extension (4.3.1) only if the Lie antibracket of the odd gauge symmetry u (4.2.13) is a generalized gauge symmetry factorizing through u.*

The equalities (4.3.6) – (4.3.7) for $k = 1$ take the form

$$(u + \gamma^{(2)}_{(2)})(u^{(1)}) + u^{(1)}(\gamma^{(1)}) = 0, \qquad (4.3.10)$$

$$\gamma^{(2)}_{(3)}(u^{(1)}) + (u + \gamma^{(1)})(\gamma^{(1)}) = 0. \qquad (4.3.11)$$

In particular, if a Lagrangian system is irreducible, i.e., $u^{(k)} = 0$ and $\mathbf{u} = u$, the BRST operator reads

$$\mathbf{b} = u + \gamma^{(1)} = u^A \partial_A + \gamma^r \partial_r = \sum_{0 \leq |\Lambda|} u^{A,\Lambda}_r c^r_\Lambda \partial_A + \sum_{0 \leq |\Lambda|, |\Xi|} \gamma^{r,\Lambda,\Xi}_{pq} c^p_\Lambda c^q_\Xi \partial_r.$$

In this case, the nilpotency conditions (4.3.10) - (4.3.11) are reduced to the equality

$$(u + \gamma^{(1)})(\gamma^{(1)}) = 0. \qquad (4.3.12)$$

Furthermore, let a gauge symmetry u be affine in fields s^A and their jets. It follows from the nilpotency condition (4.3.8) that the BRST term $\gamma^{(1)}$ is independent of original fields and their jets. Then the relation (4.3.12) takes the form of the Jacobi identity

$$\gamma^{(1)})(\gamma^{(1)}) = 0 \qquad (4.3.13)$$

for coefficient functions $\gamma^{r,\Lambda,\Xi}_{pq}(x)$ in the Lie antibracket (4.3.9).

The relations (4.3.9) and (4.3.13) motivate us to think of the equalities (4.3.6) – (4.3.7) in a general case of reducible gauge symmetries as being *sui generis* generalized commutation relations and Jacobi identities of gauge symmetries, respectively [61]. Based on Theorem 4.3.2, we therefore say that non-trivial gauge symmetries are *algebraically closed* (in the terminology of [63]) if the gauge operator $\mathbf{u}$ (4.2.8) admits the nilpotent BRST extension $\mathbf{b}$ (4.3.1).

Example 4.3.1. A Lagrangian system is called *Abelian* if its gauge symmetry u is Abelian and the higher-stage gauge symmetries are independent of original fields, i.e., if $u(\mathbf{u}) = 0$. It follows from the relation (4.3.3) that, in this case, the gauge operator itself is the BRST operator $\mathbf{u} = \mathbf{b}$. For instance, let a Lagrangian L be variationally trivial. Its variational derivatives $\mathcal{E}_i \equiv 0$ obey irreducible complete Noether identities

$$\overline{\delta}\Delta_i = 0, \qquad \Delta_i = \overline{s}_i. \tag{4.3.14}$$

By the formulas (2.3.6) – (2.3.7), the associated irreducible gauge symmetry is given by the gauge operator

$$\mathbf{u} = c^i \partial_i. \tag{4.3.15}$$

Thus, a Lagrangian system with a variationally trivial Lagrangian is Abelian, and $\mathbf{u}$ (4.3.15) is the BRST operator. The topological BF theory exemplifies a reducible Abelian Lagrangian system (see Section 8.3).

4.4 BRST extended Lagrangian field theory

The differential bigraded algebra $P^*_\infty\{N\}$ (4.2.2) is a particular *field-antifield theory* of the following type [9; 15; 63].

Let us consider a pull-back composite bundle

$$W = Z \underset{X}{\times} Z' \to Z \to X$$

where $Z' \to X$ is a vector bundle. Let us regard it as a graded vector bundle over Z possessing only odd part. The density-dual $\overline{VW}$ of the vertical tangent bundle VW of $W \to X$ is a graded vector bundle

$$\overline{VW} = ((\overline{Z}' \underset{Z}{\oplus} V^*Z) \underset{Z}{\otimes} \overset{n}{\wedge} T^*X) \underset{Y}{\oplus} Z'$$

over Z (cf. (4.1.5)). Let us consider the differential bigraded algebra $\mathcal{P}^*_\infty[\overline{VW}; Z]$ (4.1.6) with the local generating basis

$$(z^a, \overline{z}_a), \qquad [\overline{z}_a] = ([z^a] + 1) \bmod 2.$$

Its elements z^a and $\overline{z}_a$ are called *fields* and *antifields*, respectively.

Graded densities of this differential bigraded algebra are endowed with the *antibracket*

$$\{\mathfrak{L}\omega, \mathfrak{L}'\omega\} = \left[\frac{\overleftarrow{\delta}\mathfrak{L}}{\delta \overline{z}_a} \frac{\delta \mathfrak{L}'}{\delta z^a} + (-1)^{[\mathfrak{L}']([\mathfrak{L}']+1)} \frac{\overleftarrow{\delta}\mathfrak{L}'}{\delta \overline{z}_a} \frac{\delta \mathfrak{L}}{\delta z^a} \right] \omega. \tag{4.4.1}$$

4.4. BRST extended Lagrangian field theory

With this antibracket, one associates to any (even) Lagrangian $\mathfrak{L}\omega$ the odd vertical graded derivations

$$\upsilon_{\mathfrak{L}} = \overleftarrow{\mathcal{E}}^a \partial_a = \frac{\overleftarrow{\delta}\mathfrak{L}}{\delta \overline{z}_a} \frac{\partial}{\partial z^a}, \qquad (4.4.2)$$

$$\overline{\upsilon}_{\mathfrak{L}} = \overleftarrow{\partial}{}^a \mathcal{E}_a = \frac{\overleftarrow{\partial}}{\partial \overline{z}_a} \frac{\delta \mathfrak{L}}{\delta z^a}, \qquad (4.4.3)$$

$$\vartheta_{\mathfrak{L}} = \upsilon_{\mathfrak{L}} + \overline{\upsilon}_{\mathfrak{L}}^l = (-1)^{[a]+1}\left(\frac{\delta \mathfrak{L}}{\delta \overline{z}^a}\frac{\partial}{\partial z_a} + \frac{\delta \mathfrak{L}}{\delta z^a}\frac{\partial}{\partial \overline{z}_a}\right), \qquad (4.4.4)$$

such that

$$\vartheta_{\mathfrak{L}}(\mathfrak{L}'\omega) = \{\mathfrak{L}\omega, \mathfrak{L}'\omega\}.$$

Theorem 4.4.1. *The following conditions are equivalent.*

(i) The antibracket of a Lagrangian $\mathfrak{L}\omega$ is d_H-exact, i.e.,

$$\{\mathfrak{L}\omega, \mathfrak{L}\omega\} = 2\frac{\overleftarrow{\delta}\mathfrak{L}}{\delta \overline{z}_a}\frac{\delta \mathfrak{L}}{\delta z^a}\omega = d_H \sigma. \qquad (4.4.5)$$

(ii) The graded derivation υ (4.4.2) is a variational symmetry of a Lagrangian $\mathfrak{L}\omega$.

(iii) The graded derivation $\overline{\upsilon}$ (4.4.3) is a variational symmetry of $\mathfrak{L}\omega$.

(iv) The graded derivation $\vartheta_{\mathfrak{L}}$ (4.4.4) is nilpotent.

Proof. By virtue of the first variational formula (3.5.14), conditions (ii) and (iii) are equivalent to condition (i). The equality (4.4.5) is equivalent to that the odd density $\overleftarrow{\mathcal{E}}{}^a \mathcal{E}_a \omega$ is variationally trivial. Replacing right variational derivatives $\overleftarrow{\mathcal{E}}{}^a$ with $(-1)^{[a]+1}\mathcal{E}^a$, we obtain

$$2\sum_a (-1)^{[a]} \mathcal{E}^a \mathcal{E}_a \omega = d_H \sigma.$$

The variational operator acting on this relation results in the equalities

$$\sum_{0 \leq |\Lambda|} (-1)^{[a]+|\Lambda|} d_\Lambda(\partial_b^\Lambda(\mathcal{E}^a \mathcal{E}_a)) =$$

$$\sum_{0 \leq |\Lambda|} (-1)^{[a]}[\eta(\partial_b \mathcal{E}^a)^\Lambda \mathcal{E}_{\Lambda a} + \eta(\partial_b \mathcal{E}_a)^\Lambda \mathcal{E}_\Lambda^a)] = 0,$$

$$\sum_{0 \leq |\Lambda|} (-1)^{[a]+|\Lambda|} d_\Lambda(\partial^{\Lambda b}(\mathcal{E}^a \mathcal{E}_a)) =$$

$$\sum_{0 \leq |\Lambda|} (-1)^{[a]}[\eta(\partial^b \mathcal{E}^a)^\Lambda \mathcal{E}_{\Lambda a} + \eta(\partial^b \mathcal{E}_a)\mathcal{E}_\Lambda^a] = 0.$$

Due to the identity

$$(\delta \circ \delta)(L) = 0, \qquad \eta(\partial_B \mathcal{E}_A)^\Lambda = (-1)^{[A][B]} \partial_A^\Lambda \mathcal{E}_B,$$

we obtain

$$\sum_{0 \leq |\Lambda|} (-1)^{[a]} [(-1)^{[b]([a]+1)} \partial^{\Lambda a} \mathcal{E}_b \mathcal{E}_{\Lambda a} + (-1)^{[b][a]} \partial_a^\Lambda \mathcal{E}_b \mathcal{E}_\Lambda^a] = 0,$$

$$\sum_{0 \leq |\Lambda|} (-1)^{[a]+1} [(-1)^{([b]+1)([a]+1)} \partial^{\Lambda a} \mathcal{E}^b \mathcal{E}_{\Lambda a} + (-1)^{([b]+1)[a]} \partial_a^\Lambda \mathcal{E}^b \mathcal{E}_\Lambda^a] = 0$$

for all $\mathcal{E}_b$ and $\mathcal{E}^b$. This is exactly condition (iv). □

The equality (4.4.5) is called the *classical master equation*. For instance, any variationally trivial Lagrangian satisfies the master equation. A solution of the master equation (4.4.5) is called non-trivial if both the derivations (4.4.2) and (4.4.3) do not vanish.

Being an element of the differential bigraded algebra $P_\infty^*\{N\}$ (4.2.2), an original Lagrangian L obeys the master equation (4.4.5) and yields the graded derivations $v_L = 0$ (4.4.2) and $\overline{v}_L = \overline{\delta}$ (4.4.3), i.e., it is a trivial solution of the master equation.

The graded derivations (4.4.2) – (4.4.3) associated to the extended Lagrangian L_e (4.2.20) are extensions

$$v_e = \mathbf{u} + \frac{\overleftarrow{\delta} \mathcal{L}_1^*}{\delta \overline{s}_A} \frac{\partial}{\partial s^A} + \sum_{0 \leq k \leq N} \frac{\overleftarrow{\delta} \mathcal{L}_1^*}{\delta \overline{c}_{r_k}} \frac{\partial}{\partial c^{r_k}},$$

$$\overline{v}_e = \delta_{\mathrm{KT}} + \frac{\overleftarrow{\partial}}{\partial \overline{s}_A} \frac{\delta \mathcal{L}_1}{\delta s^A}$$

of the gauge and Koszul–Tate operators, respectively. However, the Lagrangian L_e need not satisfy the master equation. Therefore, let us consider its extension

$$L_E = L_e + L' = L + L_1 + L_2 + \cdots \tag{4.4.6}$$

by means of even densities L_i, $i \geq 2$, of zero antifield number and polynomial degree i in ghosts. The corresponding graded derivations (4.4.2) – (4.4.3) read

$$v_E = v_e + \frac{\overleftarrow{\delta} \mathcal{L}'}{\delta \overline{s}_A} \frac{\partial}{\partial s^A} + \sum_{0 \leq k \leq N} \frac{\overleftarrow{\delta} \mathcal{L}'}{\delta \overline{c}_{r_k}} \frac{\partial}{\partial c^{r_k}}, \tag{4.4.7}$$

$$\overline{v}_E = \overline{v}_e + \frac{\overleftarrow{\partial}}{\partial \overline{s}_A} \frac{\delta \mathcal{L}'}{\delta s^A} + \sum_{0 \leq k \leq N} \frac{\overleftarrow{\partial}}{\partial \overline{c}_{r_k}} \frac{\delta \mathcal{L}'}{\delta c^{r_k}}. \tag{4.4.8}$$

4.4. BRST extended Lagrangian field theory

The Lagrangian L_E (4.4.6) where $L + L_1 = L_e$ is called a *proper extension* of an original Lagrangian L. The following is a corollary of Theorem 4.4.1.

Corollary 4.4.1. *A Lagrangian L is extended to a proper solution L_E (4.4.6) of the master equation only if the gauge operator $\mathbf{u}$ (4.2.7) admits a nilpotent extension.*

By virtue of condition (iv) of Theorem 4.4.1, this nilpotent extension is the derivation $\vartheta_E = v_E + \overline{v}_E^l$ (4.4.4), called the *KT-BRST operator*. With this operator, the module of densities $P_\infty^{0,n}\{N\}$ is split into the KT-BRST complex

$$\cdots \longrightarrow P_\infty^{0,n}\{N\}_2 \longrightarrow P_\infty^{0,n}\{N\}_1 \longrightarrow P_\infty^{0,n}\{N\}_0 \longrightarrow \quad (4.4.9)$$
$$P_\infty^{0,n}\{N\}^1 \longrightarrow P_\infty^{0,n}\{N\}^2 \longrightarrow \cdots .$$

Putting all ghosts zero, we obtain a cochain morphism of this complex onto the Koszul–Tate complex, extended to $\overline{\mathcal{P}}_\infty^{0,n}\{N\}$ and reversed into the cochain one. Letting all antifields zero, we come to a cochain morphism of the KT-BRST complex (4.4.9) onto the cochain sequence (4.2.7), where the gauge operator is extended to the antifield-free part of the KT-BRST operator.

Theorem 4.4.2. *If the gauge operator $\mathbf{u}$ (4.2.7) can be extended to the BRST operator $\mathbf{b}$ (4.3.1), then the master equation has a non-trivial proper solution*

$$L_E = L_e + \sum_{1 \leq k \leq N} \gamma^{r_k-1} \overline{c}_{r_k-1} \omega = \quad (4.4.10)$$

$$L + \mathbf{b}\left(\sum_{0 \leq k \leq N} c^{r_k-1} \overline{c}_{r_k-1}\right) \omega + d_H \sigma.$$

Proof. By virtue of Theorem 4.3.1, if the BRST operator $\mathbf{b}$ (4.3.1) exists, the densities Δ_{r_k} (4.1.28) contain only the terms G_{r_k} linear in antifields. It follows that the extended Lagrangian L_e (4.2.9) and, consequently, the Lagrangian L_E (4.4.10) are affine in antifields. In this case, we have

$$u^A = \overleftarrow{\delta}{}^A(\mathcal{L}_e), \qquad u^{\prime k} = \overleftarrow{\delta}{}^{r_k}(\mathcal{L}_e)$$

for all indices A and r_k and, consequently,

$$\mathbf{b}^A = \overleftarrow{\delta}{}^A(\mathcal{L}_E), \qquad \mathbf{b}^{r_k} = \overleftarrow{\delta}{}^{r_k}(\mathcal{L}_E),$$

i.e., $\mathbf{b} = v_E$ is the graded derivation (4.4.7) defined by the Lagrangian L_E. Its nilpotency condition takes the form

$$\mathbf{b}(\overleftarrow{\delta}{}^A(\mathcal{L}_E)) = 0, \qquad \mathbf{b}(\overleftarrow{\delta}{}^{r_k}(\mathcal{L}_E)) = 0.$$

Hence, we obtain

$$\mathbf{b}(\mathcal{L}_E) = \mathbf{b}(\overleftarrow{\delta}{}^A(\mathcal{L}_E)\overline{s}_A + \overleftarrow{\delta}{}^{r_k}(\mathcal{L}_E)\overline{c}_{r_k}) = 0,$$

i.e., $\mathbf{b}$ is a variational symmetry of L_E. Consequently, L_E obeys the master equation. □

For instance, let a gauge symmetry u be Abelian, and let the higher-stage gauge symmetries be independent of original fields, i.e., $u(\mathbf{u}) = 0$. Then $\mathbf{u} = \mathbf{b}$ and $L_E = L_e$.

The proper solution L_E (4.4.10) of the master equation is called the *BRST extension* of an original Lagrangian L. As was mentioned above, it is a necessary step towards quantization of classical Lagrangian field theory in terms of functional integrals.

4.5 Appendix. Noether identities of differential operators

Noether identities of a Lagrangian system in Section 4.1 are particular Noether identities of differential operators which are described in homology terms as follows [141].

Let $E \to X$ be a vector bundle, and let $\mathcal{E}$ be a E-valued k-order differential operator on a fibre bundle $Y \to X$ in accordance with Definition 1.6.2. It is represented by a section $\mathcal{E}^a$ (1.6.5) of the pull-back bundle (1.6.4) endowed with bundle coordinates $(x^\lambda, y^j_\Sigma, \chi^a)$, $0 \leq |\Sigma| \leq k$.

Definition 4.5.1. One says that a differential operator $\mathcal{E}$ obeys *Noether identities* if there exist an r-order differential operator Φ on the pull-back bundle

$$E_Y = Y \underset{X}{\times} E \to X \tag{4.5.1}$$

such that its restriction onto E is a linear differential operator and its kernel contains $\mathcal{E}$, i.e.,

$$\Phi = \sum_{0 \leq |\Lambda|} \Phi_a^\Lambda \chi_\Lambda^a, \qquad \sum_{0 \leq |\Lambda|} \Phi_a^\Lambda \mathcal{E}_\Lambda^a = 0. \tag{4.5.2}$$

4.5. Appendix. Noether identities of differential operators

Any differential operator admits Noether identities, e.g.,

$$\Phi = \sum_{0 \leq |\Lambda|, |\Sigma|} T_{ab}^{\Lambda\Sigma} d_\Sigma \mathcal{E}^b \chi_\Lambda^a, \qquad T_{ab}^{\Lambda\Sigma} = -T_{ba}^{\Sigma\Lambda}. \qquad (4.5.3)$$

Therefore, they must be separated into the trivial and non-trivial ones.

Lemma 4.5.1. *One can associate to $\mathcal{E}$ a chain complex whose boundaries vanish on* $\operatorname{Ker}\mathcal{E}$.

Proof. Let us consider the composite graded manifold $(Y, \mathfrak{A}_{E_Y})$ modelled over the vector bundle $E_Y \to Y$. Let $\mathcal{S}_\infty^0[E_Y; Y]$ be the ring of graded functions on the infinite order jet manifold $J^\infty Y$ possessing the local generating basis (y^i, ε^a) of Grassmann parity $[\varepsilon^a] = 1$ (see Section 3.4). It is provided with the nilpotent graded derivation

$$\overline{\delta} = \overleftarrow{\partial}_a \mathcal{E}^a. \qquad (4.5.4)$$

whose definition is independent of the choice of a local basis. Then we have the chain complex

$$0 \leftarrow \operatorname{Im}\delta \xleftarrow{\overline{\delta}} \mathcal{S}_\infty^0[E_Y; Y]_1 \xleftarrow{\overline{\delta}} \mathcal{S}_\infty^0[E_Y; Y]_2 \qquad (4.5.5)$$

of graded functions of antifield number $k \leq 2$. Its one-boundaries $\overline{\delta}\Phi$, $\Phi \in \mathcal{S}_\infty^0[E_Y; Y]_2$, by very definition, vanish on $\operatorname{Ker}\mathcal{E}$. □

Every one-cycle

$$\Phi = \sum_{0 \leq |\Lambda|} \Phi_a^\Lambda \varepsilon_\Lambda^a \in \mathcal{S}_\infty^0[E_Y; Y]_1 \qquad (4.5.6)$$

of the complex (4.5.5) defines a linear differential operator on pull-back bundle E_Y (4.5.1) such that it is linear on E and its kernel contains $\mathcal{E}$, i.e.,

$$\delta\Phi = 0, \qquad \sum_{0 \leq |\Lambda|} \Phi_a^\Lambda d_\Lambda \mathcal{E}^a = 0. \qquad (4.5.7)$$

In accordance with Definition 4.5.1, the one-cycles (4.5.6) define the Noether identities (4.5.7) of a differential operator $\mathcal{E}$. These Noether identities are trivial if a cycle is a boundary, i.e., it takes the form (4.5.3). Accordingly, non-trivial Noether identities modulo the trivial ones are associated to elements of the homology $H_1(\delta)$ of the complex (4.5.6).

A differential operator is called *degenerate* if it obeys non-trivial Noether identities.

One can say something more if the $\mathcal{O}_\infty^0 Y$-module $H_1(\delta)$ is finitely generated, i.e., it possesses the following particular structure. There are elements

$\Delta \in H_1(\delta)$ making up a projective $C^\infty(X)$-module $\mathcal{C}_{(0)}$ of finite rank which, by virtue of the Serre–Swan theorem 10.9.3, is isomorphic to the module of sections of some vector bundle $E_0 \to X$. Let $\{\Delta^r\}$:

$$\Delta^r = \sum_{0 \le |\Lambda|} \Delta_a^{\Lambda r} \varepsilon_\Lambda^a, \qquad \Delta_a^{\Lambda r} \in \mathcal{O}_\infty^0 Y, \qquad (4.5.8)$$

be local bases for this $C^\infty(X)$-module. Then every element $\Phi \in H_1(\delta)$ factorizes as

$$\Phi = \sum_{0 \le |\Xi|} G_r^\Xi d_\Xi \Delta^r, \qquad G_r^\Xi \in \mathcal{O}_\infty Y, \qquad (4.5.9)$$

through elements of $\mathcal{C}_{(0)}$, i.e., any Noether identity (4.5.7) is a corollary of the Noether identities

$$\sum_{0 \le |\Lambda|} \Delta_a^{\Lambda r} d_\Lambda \mathcal{E}^a = 0, \qquad (4.5.10)$$

called complete Noether identities.

Notation 4.5.1. Given an integer $N \ge 1$, let $E_1, \ldots, E_N$ be vector bundles over X. Let us denote

$$\mathcal{P}_\infty^0\{N\} = \mathcal{S}_\infty^0[E_{N-1} \underset{X}{\oplus} \cdots \underset{X}{\oplus} E_1 \underset{X}{\oplus} E_Y; Y \underset{X}{\times} E_0 \underset{X}{\oplus} \cdots \underset{X}{\oplus} E_N]$$

if N is even and

$$\mathcal{P}_\infty^0\{N\} = \mathcal{S}_\infty^0[E_N \underset{X}{\oplus} \cdots \underset{X}{\oplus} E_1 \underset{X}{\oplus} E_Y; Y \underset{X}{\times} E_0 \underset{X}{\oplus} \cdots \underset{X}{\oplus} E_{N-1}]$$

if N is odd.

Lemma 4.5.2. *If the homology $H_1(\delta)$ of the complex (4.5.5) is finitely generated, this complex can be extended to the one-exact complex (4.5.12) with a boundary operator whose nilpotency conditions are equivalent to complete Noether identities.*

Proof. Let us consider the graded commutative ring $\mathcal{P}_\infty^0\{0\}$ (see Notation 4.5.1). It possesses the local generating basis $\{y^i, \varepsilon^a, \varepsilon^r\}$ of Grassmann parity $[\varepsilon^r] = 0$ and antifield number $\text{Ant}[\varepsilon^r] = 2$. This ring is provided with the nilpotent graded derivation

$$\delta_0 = \delta + \overleftarrow{\partial}_r \Delta^r. \qquad (4.5.11)$$

Its nilpotency conditions are equivalent to the complete Noether identities (4.5.10). Then the module $\mathcal{P}_\infty^0\{0\}_{\le 3}$ of graded functions of antifield number ≤ 3 is decomposed into the chain complex

$$0 \leftarrow \text{Im}\,\delta \xleftarrow{\delta} \mathcal{S}_\infty^0[E_Y; Y]_1 \xleftarrow{\delta_0} \mathcal{P}_\infty^0\{0\}_2 \xleftarrow{\delta_0} \mathcal{P}_\infty^0\{0\}_3. \qquad (4.5.12)$$

4.5. Appendix. Noether identities of differential operators

Let $H_*(\delta_0)$ denote its homology. We have
$$H_0(\delta_0) = H_0(\delta) = 0.$$
Furthermore, any one-cycle Φ up to a boundary takes the form (4.5.9) and, therefore, it is a δ_0-boundary
$$\Phi = \sum_{0\leq|\Xi|} G_r^\Xi d_\Xi \Delta^r = \delta_0 \left(\sum_{0\leq|\Xi|} G_r^\Xi \varepsilon_\Xi^r \right).$$
Hence, $H_1(\delta_0) = 0$, i.e., the complex (4.5.12) is one-exact. □

Let us consider the second homology $H_2(\delta_0)$ of the complex (4.5.12). Its two-chains read
$$\Phi = G + H = \sum_{0\leq|\Lambda|} G_r^\Lambda \varepsilon_\Lambda^r + \sum_{0\leq|\Lambda|,|\Sigma|} H_{ab}^{\Lambda\Sigma} \varepsilon_\Lambda^a \varepsilon_\Sigma^b. \qquad (4.5.13)$$
Its two-cycles define the first-stage Noether identities
$$\delta_0 \Phi = 0, \qquad \sum_{0\leq|\Lambda|} G_r^\Lambda d_\Lambda \Delta^r + \delta H = 0. \qquad (4.5.14)$$
Conversely, let the equality (4.5.14) hold. Then it is a cycle condition of the two-chain (4.5.13). The first-stage Noether identities (4.5.14) are trivial either if a two-cycle Φ (4.5.13) is a boundary or its summand G vanishes on $\operatorname{Ker} \mathcal{E}$.

Lemma 4.5.3. *First-stage Noether identities can be identified with non-trivial elements of the homology $H_2(\delta_0)$ if and only if any δ-cycle $\Phi \in \mathcal{S}_\infty^0[E_Y;Y]_2$ is a δ_0-boundary.*

Proof. The proof is similar to that of Lemma 4.1.3 [141]. □

A degenerate differential operator is called *reducible* if there exist non-trivial first-stage Noether identities.

If the condition of Lemma 4.5.3 is satisfied, let us assume that non-trivial first-stage Noether identities are finitely generated as follows. There exists a graded projective $C^\infty(X)$-module $\mathcal{C}_{(1)} \subset H_2(\delta_0)$ of finite rank possessing a local basis $\Delta_{(1)}$:
$$\Delta^{r_1} = \sum_{0\leq|\Lambda|} \Delta_r^{\Lambda r_1} \varepsilon_\Lambda^r + h^{r_1},$$
such that any element $\Phi \in H_2(\delta_0)$ factorizes as
$$\Phi = \sum_{0\leq|\Xi|} \Phi_{r_1}^\Xi d_\Xi \Delta^{r_1} \qquad (4.5.15)$$

through elements of $\mathcal{C}_{(1)}$. Thus, all non-trivial first-stage Noether identities (4.5.14) result from the equalities

$$\sum_{0\leq|\Lambda|} \Delta_r^{r_1\Lambda} d_\Lambda \Delta^r + \delta h^{r_1} = 0, \qquad (4.5.16)$$

called the complete first-stage Noether identities.

Lemma 4.5.4. *If non-trivial first-stage Noether identities are finitely generated, the one-exact complex (4.5.12) is extended to the two-exact one (4.5.18) with a boundary operator whose nilpotency conditions are equivalent to complete Noether and first-stage Noether identities.*

Proof. By virtue of Serre–Swan Theorem 10.9.3, the module $\mathcal{C}_{(1)}$ is isomorphic to a module of sections of some vector bundle $E_1 \to X$. Let us consider the ring $\mathcal{P}_\infty^0\{1\}$ of graded functions on $J^\infty Y$ possessing the local generating bases $\{y^i, \varepsilon^a, \varepsilon^r, \varepsilon^{r_1}\}$ of Grassmann parity $[\varepsilon^{r_1}] = 1$ and antifield number $\mathrm{Ant}[\varepsilon^{r_1}] = 3$. It can be provided with the nilpotent graded derivation

$$\delta_1 = \delta_0 + \overleftarrow{\partial}_{r_1} \Delta^{r_1}. \qquad (4.5.17)$$

Its nilpotency conditions are equivalent to the complete Noether identities (4.5.10) and the complete first-stage Noether identities (4.5.16). Then the module $\mathcal{P}_\infty^0\{1\}_{\leq 4}$ of graded functions of antifield number ≤ 4 is decomposed into the chain complex

$$0 \leftarrow \mathrm{Im}\,\delta \xleftarrow{\delta} \mathcal{S}_\infty[E_Y;Y]_1 \xleftarrow{\delta_0} \mathcal{P}_\infty^0\{0\}_2 \xleftarrow{\delta_1} \mathcal{P}_\infty^0\{1\}_3 \xleftarrow{\delta_1} \mathcal{P}_\infty^0\{1\}_4. \qquad (4.5.18)$$

Let $H_*(\delta_1)$ denote its homology. It is readily observed that

$$H_0(\delta_1) = H_0(\delta) = 0, \qquad H_1(\delta_1) = H_1(\delta_0) = 0.$$

By virtue of the expression (4.5.15), any two-cycle of the complex (4.5.18) is a boundary

$$\Phi = \sum_{0\leq|\Xi|} \Phi_{r_1}^\Xi d_\Xi \Delta^{r_1} = \delta_1 \left(\sum_{0\leq|\Xi|} \Phi_{r_1}^\Xi \varepsilon_\Xi^{r_1} \right).$$

It follows that $H_2(\delta_1) = 0$, i.e., the complex (4.5.18) is two-exact. $\square$

If the third homology $H_3(\delta_1)$ of the complex (4.5.18) is not trivial, its elements correspond to second-stage Noether identities, and so on. Iterating the arguments, we come to the following.

4.5. Appendix. Noether identities of differential operators

A degenerate differential operator $\mathcal{E}$ is called N-stage reducible if it admits finitely generated non-trivial N-stage Noether identities, but no non-trivial $(N+1)$-stage ones. It is characterized as follows [141].

• There are graded vector bundles $E_0, \ldots, E_N$ over X, and the graded commutative ring $\mathcal{S}^0_\infty[E_Y;Y]$ is enlarged to the graded commutative ring $\overline{\mathcal{P}}^0_\infty\{N\}$ with the local generating basis

$$(y^i, \varepsilon^a, \varepsilon^r, \varepsilon^{r_1}, \ldots, \varepsilon^{r_N})$$

of Grassmann parity $[\varepsilon^{r_k}] = (k+1) \bmod 2$ and antifield number $\mathrm{Ant}[\varepsilon^{r_k}_\Lambda] = k+2$.

• The graded commutative ring $\overline{\mathcal{P}}^0_\infty\{N\}$ is provided with the nilpotent right graded derivation

$$\delta_{\mathrm{KT}} = \delta_N = \delta_0 + \sum_{1 \leq k \leq N} \overleftarrow{\partial}_{r_k} \Delta^{r_k}, \qquad (4.5.19)$$

$$\Delta^{r_k} = \sum_{0 \leq |\Lambda|} \Delta^{\Lambda r_k}_{r_{k-1}} \varepsilon^{r_{k-1}}_\Lambda + \sum_{0 \leq \Sigma, 0 \leq \Xi} (h^{\Xi \Sigma r_k}_{a r_{k-2}} \varepsilon^a_\Xi \varepsilon^{r_{k-2}}_\Sigma + \ldots),$$

of antifield number -1.

• With this graded derivation, the module $\mathcal{P}^0_\infty\{N\}_{\leq N+3}$ of graded functions of antifield number $\leq (N+3)$ is decomposed into the exact Koszul–Tate complex

$$0 \leftarrow \mathrm{Im}\,\delta \xleftarrow{\delta} \mathcal{S}^0_\infty[E_Y;Y]_1 \xleftarrow{\delta_0} \mathcal{P}^0_\infty\{0\}_2 \xleftarrow{\delta_1} \mathcal{P}^0_\infty\{1\}_3 \cdots \qquad (4.5.20)$$
$$\xleftarrow{\delta_{N-1}} \mathcal{P}^0_\infty\{N-1\}_{N+1} \xleftarrow{\delta_{\mathrm{KT}}} \mathcal{P}^0_\infty\{N\}_{N+2} \xleftarrow{\delta_{\mathrm{KT}}} \mathcal{P}_\infty\{N\}_{N+3},$$

which satisfies the following homology regularity condition.

Condition 4.5.1. Any $\delta_{k<N-1}$-cycle

$$\Phi \in \mathcal{P}^0_\infty\{k\}_{k+3} \subset \mathcal{P}^0_\infty\{k+1\}_{k+3}$$

is a δ_{k+1}-boundary.

• The nilpotentness $\delta^2_{\mathrm{KT}} = 0$ of the Koszul–Tate operator (4.5.19) is equivalent to the complete non-trivial Noether identities (4.5.10) and the complete non-trivial $(k \leq N)$-stage Noether identities

$$\sum_{0 \leq |\Lambda|} \Delta^{\Lambda r_k}_{r_{k-1}} d_\Lambda \left(\sum_{0 \leq |\Sigma|} \Delta^{\Sigma r_{k-1}}_{r_{k-2}} \varepsilon^{r_{k-2}}_\Sigma \right) + \qquad (4.5.21)$$

$$\delta \left(\sum_{0 \leq \Sigma, \Xi} h^{\Xi \Sigma r_k}_{a r_{k-2}} \varepsilon^a_\Xi \varepsilon^{r_{k-2}}_\Sigma \right) = 0.$$

Let us study the following example of reducible Noether identities of a differential operator which is relevant to topological BF theory (Section 8.3).

Example 4.5.1. Let us consider the fibre bundles

$$Y = X \times \mathbb{R}, \qquad E = \overset{n-1}{\wedge} TX, \qquad 2 < n, \qquad (4.5.22)$$

coordinated by (x^λ, y) and $(x^\lambda, \chi^{\mu_1\cdots\mu_{n-1}})$, respectively. We study the E-valued differential operator

$$\mathcal{E}^{\mu_1\cdots\mu_{n-1}} = -\epsilon^{\mu\mu_1\cdots\mu_{n-1}} y_\mu, \qquad (4.5.23)$$

where ϵ is the Levi–Civita symbol. It defines the first order differential equation

$$d_H y = 0 \qquad (4.5.24)$$

on the fibre bundle Y (4.5.22).

Putting

$$E_Y = \mathbb{R} \underset{X}{\times} \overset{n-1}{\wedge} TX,$$

let us consider the graded commutative ring $\mathcal{S}^*_\infty[E_Y; Y]$ of graded functions on $J^\infty Y$. It possesses the local generating basis $(y, \varepsilon^{\mu_1\cdots\mu_{n-1}})$ of Grassmann parity $[\varepsilon^{\mu_1\cdots\mu_{n-1}}] = 1$ and antifield number $\text{Ant}[\varepsilon^{\mu_1\cdots\mu_{n-1}}] = 1$. With the nilpotent derivation

$$\overline{\delta} = \frac{\overleftarrow{\partial}}{\partial \varepsilon^{\mu_1\cdots\mu_{n-1}}} \mathcal{E}^{\mu_1\cdots\mu_{n-1}},$$

we have the complex (4.5.5). Its one-chains read

$$\Phi = \sum_{0 \leq |\Lambda|} \Phi^\Lambda_{\mu_1\cdots\mu_{n-1}} \varepsilon^{\mu_1\cdots\mu_{n-1}}_\Lambda,$$

and the cycle condition $\overline{\delta}\Phi = 0$ takes the form

$$\Phi^\Lambda_{\mu_1\cdots\mu_{n-1}} \mathcal{E}^{\mu_1\cdots\mu_{n-1}}_\Lambda = 0. \qquad (4.5.25)$$

This equality is satisfied if and only if

$$\Phi^{\lambda_1\cdots\lambda_k}_{\mu_1\cdots\mu_{n-1}} \epsilon^{\mu\mu_1\cdots\mu_{n-1}} = -\Phi^{\mu\lambda_2\cdots\lambda_k}_{\mu_1\cdots\mu_{n-1}} \epsilon^{\lambda_1\mu_1\cdots\mu_{n-1}}.$$

It follows that Φ factorizes as

$$\Phi = \sum_{0 \leq |\Xi|} G^\Xi_{\nu_2\cdots\nu_{n-1}} d_\Xi \Delta^{\nu_2\cdots\nu_{n-1}} \omega$$

4.5. Appendix. Noether identities of differential operators

through graded functions

$$\Delta^{\nu_2\ldots\nu_{n-1}} = \Delta^{\lambda,\nu_2\ldots\nu_{n-1}}_{\alpha_1\ldots\alpha_{n-1}}\varepsilon^{\alpha_1\ldots\alpha_{n-1}}_\lambda = \qquad (4.5.26)$$
$$\delta^\lambda_{\alpha_1}\delta^{\nu_2}_{\alpha_2}\cdots\delta^{\nu_{n-1}}_{\alpha_{n-1}}\varepsilon^{\alpha_1\ldots\alpha_{n-1}}_\lambda = d_{\nu_1}\varepsilon^{\nu_1\nu_2\ldots\nu_{n-1}},$$

which provide the complete Noether identities

$$d_{\nu_1}\mathcal{E}^{\nu_1\nu_2\ldots\nu_{n-1}} = 0. \qquad (4.5.27)$$

They can be written in the form

$$d_H d_H y = 0. \qquad (4.5.28)$$

The graded functions (4.5.26) form a basis for a projective $C^\infty(X)$-module of finite rank which is isomorphic to the module of sections of the vector bundle

$$E_0 = \overset{n-2}{\wedge} TX.$$

Therefore, let us extend the graded commutative ring $\mathcal{S}^0_\infty[E_Y;Y]$ to that $\mathcal{P}^*_\infty\{0\}$ (see Notation 4.5.1) possessing the local generating basis

$$(y, \varepsilon^{\mu_1\ldots\mu_{n-1}}, \varepsilon^{\mu_2\ldots\mu_{n-1}}\},$$

where $\varepsilon^{\mu_2\ldots\mu_{n-1}}$ are even antifields of antifield number 2. We have the nilpotent graded derivation

$$\delta_0 = \overline{\delta} + \frac{\overleftarrow{\partial}}{\partial\varepsilon^{\mu_2\ldots\mu_{n-1}}}\Delta^{\mu_2\ldots\mu_{n-1}}$$

of $\mathcal{P}^0_\infty\{0\}$. Its nilpotency is equivalent to the complete Noether identities (4.5.27). Then we obtain the one-exact complex (4.5.12).

Iterating the arguments, let us consider the vector bundles

$$E_k = \overset{n-k-2}{\wedge} TX, \qquad k = 1, \ldots, n-3,$$
$$E_{N=n-2} = X \times \mathbb{R}$$

and the graded commutative ring $\mathcal{P}^0_\infty\{N\}$ (see Notation 4.5.1), possessing the local generating basis

$$(y, \varepsilon^{\mu_1\ldots\mu_{n-1}}, \varepsilon^{\mu_2\ldots\mu_{n-1}}, \ldots, \varepsilon^{\mu_{n-1}}, \varepsilon)$$

of Grassmann parity

$$[\varepsilon^{\mu_{k+2}\ldots\mu_{n-1}}] = k \bmod 2, \qquad [\varepsilon] = n,$$

and of antifield number

$$\mathrm{Ant}[\varepsilon^{\mu_{k+2}\ldots\mu_{n-1}}] = k+2, \qquad \mathrm{Ant}[\varepsilon] = n.$$

It is provided with the nilpotent graded derivation

$$\delta_{\mathrm{KT}} = \delta_0 + \sum_{1 \leq k \leq n-3} \frac{\overleftarrow{\partial}}{\partial \varepsilon^{\mu_{k+2} \cdots \mu_{n-1}}} + \frac{\overleftarrow{\partial}}{\partial \varepsilon} d_{\mu_{n-1}} \varepsilon^{\mu_{n-1}}, \quad (4.5.29)$$
$$\Delta^{\mu_{k+2} \cdots \mu_{n-1}} = d_{\mu_{k+1}} \varepsilon^{\mu_{k+1} \mu_{k+2} \cdots \mu_{n-1}},$$

of antifield number -1. Its nilpotency results from the complete Noether identities (4.5.27) and the equalities

$$d_{\mu_{k+2}} \Delta^{\mu_{k+2} \cdots \mu_{n-1}} = 0, \quad k = 0, \ldots, n-3, \quad (4.5.30)$$

which are the $(k+1)$-stage Noether identities (4.5.21). Then the Koszul–Tate complex (4.5.20) reads

$$0 \leftarrow \mathrm{Im}\,\overline{\delta} \xleftarrow{\overline{\delta}} \mathcal{S}^0_\infty[E_Y;Y]_1 \xleftarrow{\delta_0} \mathcal{P}^0_\infty\{0\}_2 \xleftarrow{\delta_1} \mathcal{P}^0_\infty\{1\}_3 \cdots \quad (4.5.31)$$
$$\xleftarrow{\delta_{n-3}} \mathcal{P}^0_\infty\{n-3\}_{n-1} \xleftarrow{\delta_{\mathrm{KT}}} \mathcal{P}^0_\infty\{n-2\}_n \xleftarrow{\delta_{\mathrm{KT}}} \mathcal{P}^0_\infty\{n-2\}_{n+1}.$$

It obeys Condition 4.5.1 as follows.

Lemma 4.5.5. *Any* δ_k*-cycle* $\Phi \in \mathcal{P}^0_\infty\{k\}_{k+3}$ *up to a* δ_k*-boundary takes the form*

$$\Phi = \sum_{(k_1+\cdots+k_i+3i=k+3)} \sum_{(0 \leq |\Lambda_1|,\ldots,|\Lambda_i|)} G^{\Lambda_1 \cdots \Lambda_i}_{\mu^1_{k_1+2} \cdots \mu^1_{n-1}; \ldots; \mu^i_{k_i+2} \cdots \mu^i_{n-1}} \quad (4.5.32)$$
$$d_{\Lambda_1} \Delta^{\mu^1_{k_1+2} \cdots \mu^1_{n-1}} \cdots d_{\Lambda_i} \Delta^{\mu^i_{k_i+2} \cdots \mu^i_{n-1}}, \quad k_j = -1, 0, 1, \ldots, n-3,$$

where $k_j = -1$ *stands for* $\varepsilon^{\mu_1 \cdots \mu_{n-1}}$ *and*

$$\Delta^{\mu_1 \cdots \mu_{n-1}} = \mathcal{E}^{\mu_1 \cdots \mu_{n-1}}.$$

It follows that Φ *is a* δ_{k+1}*-boundary.*

Proof. Let us choose some basis element $\varepsilon^{\mu_{k+2} \cdots \mu_{n-1}}$ and denote it, simply, by ε. Let Φ contain a summand $\phi_1 \varepsilon$, linear in ε. Then the cycle condition reads

$$\delta_k \Phi = \delta_k(\Phi - \phi_1 \varepsilon) + (-1)^{[\varepsilon]} \delta_k(\phi_1) \varepsilon + \phi \Delta = 0, \quad \Delta = \delta_k \varepsilon.$$

It follows that Φ contains a summand $\psi \Delta$ such that

$$(-1)^{[\varepsilon]+1} \delta_k(\psi) \Delta + \phi \Delta = 0.$$

This equality implies the relation

$$\phi_1 = (-1)^{[\varepsilon]+1} \delta_k(\psi) \quad (4.5.33)$$

4.5. Appendix. Noether identities of differential operators

because the reduction conditions (4.5.30) involve total derivatives of Δ, but not Δ. Hence,

$$\Phi = \Phi' + \delta_k(\psi\varepsilon),$$

where Φ' contains no term linear in ε. Furthermore, let ε be even and Φ have a summand $\sum \phi_r \varepsilon^r$ polynomial in ε. Then the cycle condition leads to the equalities

$$\phi_r \Delta = -\delta_k \phi_{r-1}, \qquad r \geq 2.$$

Since ϕ_1 (4.5.33) is δ_k-exact, then $\phi_2 = 0$ and, consequently, $\phi_{r>2} = 0$. Thus, a cycle Φ up to a δ_k-boundary contains no term polynomial in c. It reads

$$\Phi = \sum_{(k_1+\cdots+k_i+3i=k+3)} \sum_{(0<|\Lambda_1|,\ldots,|\Lambda_i|)} G^{\Lambda_1\cdots\Lambda_i}_{\mu^1_{k_1+2}\cdots\mu^1_{n-1};\ldots;\mu^i_{k_i+2}\cdots\mu^i_{n-1}}$$
$$\varepsilon^{\mu^1_{k_1+2}\cdots\mu^1_{n-1}}_{\Lambda_1} \cdots \varepsilon^{\mu^i_{k_i+2}\cdots\mu^i_{n-1}}_{\Lambda_i}. \qquad (4.5.34)$$

However, the terms polynomial in ε may appear under general coordinate transformations

$$\varepsilon'^{\nu_{k+2}\cdots\nu_{n-1}} = \det\left(\frac{\partial x^\alpha}{\partial x'^\beta}\right) \frac{\partial x'^{\nu_{k+2}}}{\partial x^{\mu_{k+2}}} \cdots \frac{\partial x'^{\nu_{n-1}}}{\partial x^{\mu_{n-1}}} \varepsilon^{\mu_{k+2}\cdots\mu_{n-1}}$$

of a chain Φ (4.5.34). In particular, Φ contains the summand

$$\sum_{k_1+\cdots+k_i+3i=k+3} F_{\nu^1_{k_1+2}\cdots\nu^1_{n-1};\ldots;\nu^i_{k_i+2}\cdots\nu^i_{n-1}} \varepsilon'^{\nu^1_{k_1+2}\cdots\nu^1_{n-1}} \cdots \varepsilon'^{\nu^i_{k_i+2}\cdots\nu^i_{n-1}},$$

which must vanish if Φ is a cycle. This takes place only if Φ factorizes through the graded densities $\Delta^{\mu_{k+2}\cdots\mu_{n-1}}$ (4.5.29) in accordance with the expression (4.5.32). $\square$

Following the proof of Lemma 4.5.5, one also can show that any δ_k-cycle $\Phi \in \mathcal{P}^0_\infty\{k\}_{k+2}$ up to a boundary takes the form

$$\Phi = \sum_{0 \leq |\Lambda|} G^\Lambda_{\mu_{k+2}\cdots\mu_{n-1}} d_\Lambda \Delta^{\mu_{k+2}\cdots\mu_{n-1}},$$

i.e., the homology $H_{k+2}(\delta_k)$ of the complex (4.5.31) is finitely generated by the cycles $\Delta^{\mu_{k+2}\cdots\mu_{n-1}}$.

Chapter 5

Gauge theory on principal bundles

Classical gauge theory is adequately formulated as Lagrangian field theory on principal and associated bundles where gauge potentials are identified with principal connections. The reader is referred, e.g., to [38; 92; 147] for the standard exposition of geometry of principal bundles. In this Chapter, we present gauge theory on principal bundles as a particular Lagrangian field theory on fibre bundles formulated in terms of jet manifolds [112]. The main ingredient in this formulation is the bundle of principal connections $C = J^1P/G$ whose sections are principal connections on a principal bundle P with a structure group G. Its first order jet manifold J^1C plays the role of a configuration space of gauge theory.

5.1 Geometry of Lie groups

Let G be a topological group which is not reduced to the unit element **1**. Let V be a topological space. By a *continuous action of G on V on the left* is meant a continuous map

$$\zeta : G \times V \to V \qquad (5.1.1)$$

such that

$$\zeta(g'g, v) = \zeta(g', \zeta(g, v)).$$

If there is no danger of confusion, we denote $\zeta(g, v) = gv$. One says that a group G acts on V *on the right* if the map (5.1.1) obeys the relations

$$\zeta(g'g, v) = \zeta(g, \zeta(g', v)).$$

In this case, we agree to write $\zeta(g, v) = vg$.

Remark 5.1.1. Strictly speaking, by an action of a group G on V is meant a class of morphisms ζ (5.1.1) which differ from each other in inner automorphisms of G, that is,

$$\zeta'(g,v) = \zeta(g'^{-1}gg',v)$$

for some element $g' \in G$.

An action of G on V is called:
- *effective* if there is no $g \neq \mathbf{1}$ such that $\zeta(g,v) = v$ for all $v \in V$,
- *free* if, for any two elements $v,v \in V$, there exists an element $g \in G$ such that $\zeta(g,v) = v'$.
- *transitive* if there is no element $v \in V$ such that $\zeta(g,v) = v$ for all $g \in G$.

Unless otherwise stated, an action of a group is assumed to be effective. If an action ζ (5.1.1) of G on V is transitive, then V is called the *homogeneous space*, homeomorphic to the quotient $V = G/H$ of G with respect to some subgroup $H \subset G$. If an action ζ is both free and transitive, then V is homeomorphic to the group space of G. For instance, this is the case of action of G on itself by left ($\zeta = L_G$) and right ($\zeta = R_G$) multiplications.

Let G be a connected real Lie group of finite dimension $\dim G > 0$. A vector field ξ on G is called *left-invariant* if

$$\xi(g) = TL_g(\xi(\mathbf{1})), \qquad g \in G,$$

where TL_g denotes the tangent morphism to the map

$$L_g : G \to gG.$$

Accordingly, *right-invariant* vector fields ξ on G obey the condition

$$\xi(g) = TR_g(\xi(\mathbf{1})),$$

where TR_g is the tangent morphism to the map

$$T_g : G \to gG.$$

Let $\mathfrak{g}_l$ (resp. $\mathfrak{g}_r$) denote the Lie algebra of left-invariant (resp. right-invariant) vector fields on G. They are called the *left* and *right* Lie algebras of G, respectively. Every left-invariant vector field $\xi_l(g)$ (resp. a right-invariant vector field $\xi_r(g)$) can be associated to the element $v = \xi_l(\mathbf{1})$ (resp. $v = \xi_r(\mathbf{1})$) of the tangent space $T_\mathbf{1}G$ at the unit $\mathbf{1}$ of G. Accordingly, this tangent space is provided both with left and right Lie algebra structures. For instance, given $v \in T_\mathbf{1}G$, let $v_l(g)$ and $v_r(g)$ be the corresponding left-invariant and right-invariant vector fields on G, respectively. There is the relation

$$v_l(g) = (TL_g \circ TR_g^{-1})(v_r(g)).$$

5.1. Geometry of Lie groups

Let $\{\epsilon_m = \epsilon_m(1)\}$ (resp. $\{\varepsilon_m = \varepsilon_m(1)\}$) denote the basis for the left (resp. right) Lie algebra, and let c^k_{mn} be the *right structure constants*:

$$[\varepsilon_m, \varepsilon_n] = c^k_{mn}\varepsilon_k.$$

The map $g \to g^{-1}$ yields an isomorphism

$$\mathfrak{g}_l \ni \epsilon_m \to \varepsilon_m = -\epsilon_m \in \mathfrak{g}_r$$

of left and right Lie algebras.

The tangent bundle

$$\pi_G : TG \to G \qquad (5.1.2)$$

of a Lie group G is trivial because of the isomorphisms

$$\varrho_l : TG \ni q \to (g = \pi_G(q), TL_g^{-1}(q)) \in G \times \mathfrak{g}_l,$$
$$\varrho_r : TG \ni q \to (g = \pi_G(q), TR_g^{-1}(q)) \in G \times \mathfrak{g}_r.$$

Let ζ (5.1.1) be a smooth action of a Lie group G on a smooth manifold V. Let us consider the tangent morphism

$$T\zeta : TG \times TV \to TV \qquad (5.1.3)$$

to this action. Given an element $g \in G$, the restriction of $T\zeta$ (5.1.3) to $(g,0) \times TV$ is the tangent morphism $T\zeta_g$ to the map

$$\zeta_g : g \times V \to V.$$

Therefore, the restriction

$$T\zeta_G : \widehat{0}(G) \times TV \to TV \qquad (5.1.4)$$

of the tangent morphism $T\zeta$ (5.1.3) to $\widehat{0}(G) \times TV$ (where $\widehat{0}$ is the canonical zero section of $TG \to G$) is called the *tangent prolongation* of a smooth action of G on V.

In particular, the above mentioned morphisms

$$TL_g = TL_G|_{(g,0) \times TG}, \qquad TR_g = TR_G|_{(g,0) \times TG}$$

are of this type. For instance, the morphism TL_G (resp. TR_G) (5.1.4) defines the adjoint representation $g \to \mathrm{Ad}_g$ (resp. $g \to \mathrm{Ad}_{g^{-1}}$) of a group G in its right Lie algebra $\mathfrak{g}_r$ (resp. left Lie algebra $\mathfrak{g}_l$) and the identity representation in its left (resp. right) one.

Restricting $T\zeta$ (5.1.3) to $T_1G \times \widehat{0}(V)$, one obtains a homomorphism of the right (resp. left) Lie algebra of G to the Lie algebra $\mathcal{T}(V)$ of vector field on V if ζ is a left (resp. right) action. We call this homomorphism a *representation* of the Lie algebra of G in V. For instance, a vector field on a manifold V associated to a local one-parameter group G of diffeomorphisms of V (see Section 1.1.4) is exactly an image of such a homomorphism of the one-dimensional Lie algebra of G to $\mathcal{T}(V)$.

In particular, the adjoint representation Ad_g of a Lie group G in its right Lie algebra $\mathfrak{g}_r$ yields the corresponding *adjoint representation*

$$\varepsilon' : \varepsilon \to \mathrm{ad}_{\varepsilon'}(\varepsilon) = [\varepsilon', \varepsilon],$$
$$\mathrm{ad}_{\varepsilon_m}(\varepsilon_n) = c_{mn}^k \varepsilon_k, \qquad (5.1.5)$$

of the right Lie algebra $\mathfrak{g}_r$ in itself. Accordingly, the adjoint representation of the left Lie algebra $\mathfrak{g}_l$ in itself reads

$$\mathrm{ad}_{\epsilon_m}(\epsilon_n) = -c_{mn}^k \epsilon_k,$$

where c_{mn}^k are the right structure constants (5.1.5).

Remark 5.1.2. Let G be a *matrix group*, i.e., a subgroup of the algebra $M(V)$ of endomorphisms of some finite-dimensional vector space V. Then its Lie algebras are Lie subalgebras of $M(V)$. In this case, the adjoint representation Ad_g of G reads

$$\mathrm{Ad}_g(e) = g e g^{-1}, \qquad e \in \mathfrak{g}. \qquad (5.1.6)$$

Let $\mathfrak{g}^*$ be the vector space, dual of the right Lie algebra $\mathfrak{g}_r$. It is called the *Lie coalgebra*, and is provided with the dual $\{\varepsilon^m\}$ of the basis $\{\varepsilon_m\}$ for $\mathfrak{g}_r$. The group G and the right Lie algebra $\mathfrak{g}_r$ act ion $\mathfrak{g}^*$ by the *coadjoint representation*

$$(\mathrm{Ad}_g^*(\varepsilon^*), \varepsilon) = (\varepsilon^*, \mathrm{Ad}_{g^{-1}}(\varepsilon)), \qquad \varepsilon^* \in \mathfrak{g}^*, \qquad \varepsilon \in \mathfrak{g}_r, \qquad (5.1.7)$$
$$(\mathrm{ad}_{\varepsilon'}^*(\varepsilon^*), \varepsilon) = -(\varepsilon^*, [\varepsilon', \varepsilon]), \qquad \varepsilon' \in \mathfrak{g}_r,$$
$$\mathrm{ad}_{\varepsilon_m}^*(\varepsilon^n) = -c_{mk}^n \varepsilon^k.$$

An exterior form ϕ on a Lie group G is said to be *left-invariant* (resp. *right-invariant*) if $\phi(\mathbf{1}) = L_g^*(\phi(g))$ (resp. $\phi(\mathbf{1}) = R_g^*(\phi(g))$). The exterior differential of a left-invariant (resp right-invariant) form is left-invariant (resp. right-invariant). In particular, the left-invariant one-forms satisfy the *Maurer–Cartan equation*

$$d\phi(\epsilon, \epsilon') = -\frac{1}{2}\phi([\epsilon, \epsilon']), \qquad \epsilon, \epsilon' \in \mathfrak{g}_l. \qquad (5.1.8)$$

There is the canonical $\mathfrak{g}_l$-valued left-invariant one-form

$$\theta_l : T_1 G \ni \epsilon \to \epsilon \in \mathfrak{g}_l \qquad (5.1.9)$$

on a Lie group G. The components θ_l^m of its decomposition $\theta_l = \theta_l^m \epsilon_m$ with respect to the basis for the left Lie algebra $\mathfrak{g}_l$ make up the basis for the space of left-invariant exterior one-forms on G:
$$\epsilon_m \rfloor \theta_l^n = \delta_m^n.$$
The Maurer–Cartan equation (5.1.8), written with respect to this basis, reads
$$d\theta_l^m = \frac{1}{2} c_{nk}^m \theta_l^n \wedge \theta_l^k.$$

5.2 Bundles with structure groups

Principal bundles are particular bundles with a structure group. Since equivalence classes of these bundles are topological invariants (see Theorem 5.2.5), we consider continuous bundles with a structure topological group.

Let G be a topological group. Let
$$\pi : Y \to X \tag{5.2.1}$$
be a locally trivial continuous bundle (see Remark 1.1.1) whose typical fibre V is provided with a certain left action (5.1.1) of a topological group G (see Remark 5.1.1). Moreover, let Y (5.2.1) admit an atlas
$$\Psi = \{(U_\alpha, \psi_\alpha), \varrho_{\alpha\beta}\}, \qquad \psi_\alpha = \varrho_{\alpha\beta} \psi_\beta, \tag{5.2.2}$$
whose transition functions $\varrho_{\alpha\beta}$ (1.1.3) factorize as
$$\varrho_{\alpha\beta} : U_\alpha \cap U_\beta \times V \longrightarrow U_\alpha \cap U_\beta \times (G \times V) \xrightarrow{\mathrm{Id} \times \varsigma} U_\alpha \cap U_\beta \times V \tag{5.2.3}$$
through local continuous G-valued functions
$$\varrho_{\alpha\beta}^G : U_\alpha \cap U_\beta \to G \tag{5.2.4}$$
on X. This means that transition morphisms $\varrho_{\alpha\beta}(x)$ (1.1.6) are elements of G acting on V. Transition functions (5.2.3) are called G-valued.

Provided with an atlas (5.2.2) with G-valued transition functions, a locally trivial continuous bundle Y is called the *bundle with a structure group G* or, in brief, a *G-bundle*. Two G-bundles (Y, Ψ) and (Y, Ψ') are called equivalent if their atlases Ψ and Ψ' are equivalent. Atlases Ψ and Ψ' with G-valued transition functions are said to be *equivalent* if and only if, given a common cover $\{U_i\}$ of X for the union of these atlases, there exists a continuous G-valued function g_i on each U_i such that
$$\psi_i'(x) = g_i(x) \psi_i(x), \qquad x \in U_i. \tag{5.2.5}$$

Remark 5.2.1. It may happen that a bundle Y admits non-equivalent atlases Ψ and Ψ' with G-valued transition functions. Then the pairs (Y, Ψ) and (Y, Ψ') are regarded as non-equivalent G-bundles (see Remark 5.10.3).

Let $h(X, G, V)$ denote the set of equivalence classes of continuous bundles over X with a structure group G and a typical fibre V. In order to characterize this set, let us consider the presheaf $G^0_{\{U\}}$ of continuous G-valued functions on a topological space X. Let G^0_X be the sheaf of germs of these functions generated by the canonical presheaf $G^0_{\{U\}}$, and let $H^1(X; G^0_X)$ be the first cohomology of X with coefficients in G^0_X (see Remark 10.7.2). The group functions $\varrho^G_{\alpha\beta}$ (5.2.4) obey the cocycle condition

$$\varrho^G_{\alpha\beta}\varrho^G_{\beta\gamma} = \varrho^G_{\alpha\gamma}$$

on overlaps $U_\alpha \cap U_\beta \cap U_\gamma$ (cf. (10.7.12)) and, consequently, they form a one-cocycle $\{\varrho^G_{\alpha\beta}\}$ of the presheaf $G^0_{\{U\}}$. This cocycle is a representative of some element of the first cohomology $H^1(X; G^0_X)$ of X with coefficients in the sheaf G^0_X (see Remark 10.7.3).

Thus, any atlas of a G-bundle over X defines an element of the cohomology set $H^1(X; G^0_X)$. Moreover, it follows at once from the condition (5.2.5) that equivalent atlases define the same element of $H^1(X; G^0_X)$. Thus, there is an injection

$$h(X, G, V) \to H^1(X; G^0_X) \qquad (5.2.6)$$

of the set of equivalence classes of G-bundles over X with a typical fibre V to the first cohomology $H^1(X; G^0_X)$ of X with coefficients in the sheaf G^0_X. Moreover, the injection (5.2.6) is a bijection as follows [80].

Theorem 5.2.1. *There is one-to-one correspondence between the equivalence classes of G-bundles over X with a typical fibre V and the elements of the cohomology set $H^1(X; G^0_X)$.*

The bijection (5.2.6) holds for G-bundles with any typical fibre V. Two G-bundles (Y, Ψ) and (Y', Ψ') over X with different typical fibres are called *associated* if the cocycles of transition functions of their atlases Ψ and Ψ' are representatives of the same element of the cohomology set $H^1(X; G^0_X)$. Then Theorem 5.2.1 can be reformulated as follows.

Theorem 5.2.2. *There is one-to-one correspondence between the classes of associated G-bundles over X and the elements of the cohomology set $H^1(X; G^0_X)$.*

Let $f : X' \to X$ be a continuous map. Every continuous G-bundle $Y \to X$ yields the pull-back bundle $f^*Y \to X'$ (1.1.8) with the same structure group G. Therefore, f induces the map

$$[f] : H^1(X; G^0_X) \to H^1(X'; G^0_{X'}).$$

Theorem 5.2.3. *Given a continuous G-bundle Y over a paracompact base X, let f_1 and f_2 be two continuous maps of X' to X. If these maps are homotopic, the pull-back G-bundles f_1^*Y and f_2^*Y over X' are equivalent [80; 147].*

Let us return to smooth fibre bundles. Let G, $\dim G > 0$, be a real Lie group which acts on a smooth manifold V on the left. A smooth fibre bundle Y (5.2.1) is called a *bundle with a structure group G* if it is a continuous G-bundle possessing a smooth atlas Ψ (5.2.2) whose transition functions factorize as those (5.2.2) through smooth G-valued functions (5.2.4).

Example 5.2.1. Any vector (resp. affine) bundle of fibre dimension $\dim V = m$ is a bundle with a structure group which is the general linear group $GL(m, \mathbb{R})$ (resp. the general affine group $GA(m, \mathbb{R})$).

Let G_X^∞ be the sheaf of germs of smooth G-valued functions on X and $H^1(X; G_X^\infty)$ the first cohomology of a manifold X with coefficients in the sheaf G_X^∞. The following theorem is analogous to Theorem 5.2.2.

Theorem 5.2.4. *There is one-to-one correspondence between the classes of associated smooth G-bundles over X and the elements of the cohomology set $H^1(X; G_X^\infty)$.*

Moreover, since a smooth manifold is paracompact, one can show the following [80].

Theorem 5.2.5. *There is a bijection*

$$H^1(X; G_X^\infty) = H^1(X; G_X^0), \qquad (5.2.7)$$

where a Lie group G is treated as a topological group.

The bijection (5.2.7) enables one to classify smooth G-bundles as the continuous ones by means of topological invariants (see Section 8.1).

5.3 Principal bundles

Unless otherwise stated (see Section 8.1), we restrict our consideration to smooth bundles with a structure Lie group of non-zero dimension.

Given a real Lie group G, let

$$\pi_P : P \to X \qquad (5.3.1)$$

be a G-bundle whose typical fibre is the group space of G, which a group G acts on by left multiplications. It is called a *principal bundle* with a structure group G or, simply, a principal bundle if there is no danger of confusion. Equivalently, a principal G-bundle is defined as a fibre bundle P (5.3.1) which admits an *action of G on P* on the right by a fibrewise morphism

$$R_{GP} : G \underset{X}{\times} P \underset{X}{\longrightarrow} P, \qquad (5.3.2)$$

$$R_{gP} : p \to pg, \qquad \pi_P(p) = \pi_P(pg), \qquad p \in P,$$

which is free and transitive on each fibre of P. As a consequence, the quotient of P with respect to the action (5.3.2) of G is diffeomorphic to a base X, i.e., $P/G = X$.

Remark 5.3.1. The definition of a *continuous principal bundle* is a repetition of that of a smooth one, but all morphisms are continuous.

A principal G-bundle P is equipped with a bundle atlas

$$\Psi_P = \{(U_\alpha, \psi_\alpha^P), \varrho_{\alpha\beta}\} \qquad (5.3.3)$$

whose trivialization morphisms

$$\psi_\alpha^P : \pi_P^{-1}(U_\alpha) \to U_\alpha \times G$$

obey the condition

$$\psi_\alpha^P(pg) = g\psi_\alpha^P(p), \qquad g \in G.$$

Due to this property, every trivialization morphism ψ_α^P determines a unique local section $z_\alpha : U_\alpha \to P$ such that

$$(\psi_\alpha^P \circ z_\alpha)(x) = \mathbf{1}, \qquad x \in U_\alpha.$$

The transformation law for z_α reads

$$z_\beta(x) = z_\alpha(x)\varrho_{\alpha\beta}(x), \qquad x \in U_\alpha \cap U_\beta. \qquad (5.3.4)$$

Conversely, the family

$$\{(U_\alpha, z_\alpha), \varrho_{\alpha\beta}\} \qquad (5.3.5)$$

of local sections of P which obey the transformation law (5.3.4) uniquely determines a bundle atlas Ψ_P of a principal bundle P.

Assertion 5.3.1. It follows that a principal bundle admits a global section if and only if it is trivial.

5.3. Principal bundles

Example 5.3.1. Let H be a closed subgroup of a real Lie group G. Then H is a Lie group. Let G/H be the quotient of G with respect to an action of H on G by right multiplications. Then

$$\pi_{GH} : G \to G/H \qquad (5.3.6)$$

is a principal H-bundle [147]. If H is a maximal compact subgroup of G, then G/H is diffeomorphic to $\mathbb{R}^m$ and, by virtue of Theorem 1.1.7, $G \to G/H$ is a trivial bundle, i.e., G is diffeomorphic to the product $\mathbb{R}^m \times H$.

Remark 5.3.2. The pull-back f^*P (1.1.8) of a principal bundle also is a principal bundle with the same structure group.

Remark 5.3.3. Let $P \to X$ and $P' \to X'$ be principal G- and G'-bundles, respectively. A bundle morphism $\Phi : P \to P'$ is a *morphism of principal bundles* if there exists a Lie group homomorphism $\gamma : G \to G'$ such that

$$\Phi(pg) = \Phi(p)\gamma(g).$$

In particular, equivalent principal bundles are isomorphic.

Any class of associated smooth bundles on X with a structure Lie group G contains a principal bundle. In other words, any smooth bundle with a structure Lie group G is associated with some principal bundle.

Let us consider the tangent morphism

$$TR_{GP} : (G \times \mathfrak{g}_l) \underset{X}{\times} TP \to TP \qquad (5.3.7)$$

to the right action R_{GP} (5.3.2) of G on P. Its restriction to

$$T_1 G \underset{X}{\times} TP$$

provides a homomorphism

$$\mathfrak{g}_l \ni \epsilon \to \xi_\epsilon \in \mathcal{T}(P) \qquad (5.3.8)$$

of the left Lie algebra $\mathfrak{g}_l$ of G to the Lie algebra $\mathcal{T}(P)$ of vector fields on P. Vector fields ξ_ϵ (5.3.8) are obviously vertical. They are called *fundamental vector fields* [92]. Given a basis $\{\epsilon_r\}$ for $\mathfrak{g}_l$, the corresponding fundamental vector fields $\xi_r = \xi_{\epsilon_r}$ form a family of $m = \dim \mathfrak{g}_l$ nowhere vanishing and linearly independent sections of the vertical tangent bundle VP of $P \to X$. Consequently, this bundle is trivial

$$VP = P \times \mathfrak{g}_l \qquad (5.3.9)$$

by virtue of Theorem 1.1.11.

Restricting the tangent morphism TR_{GP} (5.3.7) to

$$TR_{GP} : \widehat{0}(G) \underset{X}{\times} TP \underset{X}{\longrightarrow} TP, \qquad (5.3.10)$$

we obtain the *tangent prolongation* of the structure group action R_{GP} (5.3.2). If there is no danger of confusion, it is simply called the *action of G on TP*. Since the action of G (5.3.2) on P is fibrewise, its action (5.3.10) is restricted to the vertical tangent bundle VP of P.

Taking the quotient of the tangent bundle $TP \to P$ and the vertical tangent bundle VP of P by G (5.3.10), we obtain the vector bundles

$$T_G P = TP/G, \qquad V_G P = VP/G \qquad (5.3.11)$$

over X. Sections of $T_G P \to X$ are G-invariant vector fields on P. Accordingly, sections of $V_G P \to X$ are G-invariant vertical vector fields on P. Hence, a typical fibre of $V_G P \to X$ is the right Lie algebra $\mathfrak{g}_r$ of G subject to the adjoint representation of a structure group G. Therefore, $V_G P$ (5.3.11) is called the *Lie algebra bundle*.

Given a bundle atlas Ψ_P (5.3.3) of P, there is the corresponding atlas

$$\Psi = \{(U_\alpha, \psi_\alpha), \mathrm{Ad}_{\varrho_{\alpha\beta}}\} \qquad (5.3.12)$$

of the Lie algebra bundle $V_G P$, which is provided with bundle coordinates $(U_\alpha; x^\mu, \chi^m)$ with respect to the fibre frames

$$\{e_m = \psi_\alpha^{-1}(x)(\varepsilon_m)\},$$

where $\{\varepsilon_m\}$ is a basis for the Lie algebra $\mathfrak{g}_r$. These coordinates obey the transformation rule

$$\varrho(\chi^m)\varepsilon_m = \chi^m \mathrm{Ad}_{\varrho^{-1}}(\varepsilon_m). \qquad (5.3.13)$$

A glance at this transformation rule shows that $V_G P$ is a bundle with a structure group G. Moreover, it is associated with a principal G-bundle P (see Example 5.7.1).

Accordingly, the vector bundle $T_G P$ (5.3.11) is endowed with bundle coordinates $(x^\mu, \dot{x}^\mu, \chi^m)$ with respect to the fibre frames $\{\partial_\mu, e_m\}$. Their transformation rule is

$$\varrho(\chi^m)\varepsilon_m = \chi^m \mathrm{Ad}_{\varrho^{-1}}(\varepsilon_m) + \dot{x}^\mu R_\mu^m \varepsilon_m. \qquad (5.3.14)$$

For instance, if G is a matrix group (see Remark 5.1.2), this transformation rule reads

$$\varrho(\chi^m)\varepsilon_m = \chi^m \varrho^{-1}\varepsilon_m \varrho - \dot{x}^\mu \partial_\mu(\varrho^{-1})\varrho. \qquad (5.3.15)$$

Since the second term in the right-hand sides of expressions (5.3.14) – (5.3.15) depend on derivatives of a G-valued function ϱ on X, the vector bundle $T_G P$ (5.3.11) fails to be a G-bundle.

The Lie bracket of G-invariant vector fields on P goes to the quotient by G and defines the Lie bracket of sections of the vector bundle $T_G P \to X$. This bracket reads

$$\xi = \xi^\lambda \partial_\lambda + \xi^p e_p, \qquad \eta = \eta^\mu \partial_\mu + \eta^q e_q, \qquad (5.3.16)$$

$$[\xi, \eta] = (\xi^\mu \partial_\mu \eta^\lambda - \eta^\mu \partial_\mu \xi^\lambda) \partial_\lambda + \qquad (5.3.17)$$
$$(\xi^\lambda \partial_\lambda \eta^r - \eta^\lambda \partial_\lambda \xi^r + c_{pq}^r \xi^p \eta^q) e_r.$$

Putting $\xi^\lambda = 0$ and $\eta^\mu = 0$ in the formulas (5.3.16) – (5.3.17), we obtain the Lie bracket

$$[\xi, \eta] = c_{pq}^r \xi^p \eta^q e_r \qquad (5.3.18)$$

of sections of the Lie algebra bundle $V_G P$. A glance at the expression (5.3.18) shows that sections of $V_G P$ form a finite-dimensional Lie $C^\infty(X)$-algebra, called the *gauge algebra*. Therefore, $V_G P$ also is called the *gauge algebra bundle*.

5.4 Principal connections. Gauge fields

In classical gauge theory, gauge fields are conventionally described as principal connections on principal bundles. Principal connections on a principal bundle P (5.3.1) are connections on P which are equivariant with respect to the right action (5.3.2) of a structure group G on P. In order to describe them, we follow the definition of connections on a fibre bundle $Y \to X$ as global sections of the affine jet bundle $J^1 Y \to X$ (Theorem 1.3.1).

Let $J^1 P$ be the first order jet manifold of a principal G-bundle $P \to X$ (5.3.1). Then connections on a principal bundle $P \to X$ are global sections

$$A : P \to J^1 P \qquad (5.4.1)$$

of the affine jet bundle $J^1 P \to P$ modelled over the vector bundle

$$T^* X \underset{P}{\otimes} VP = (T^* X \underset{P}{\otimes} \mathfrak{g}_l).$$

In order to define principal connections on $P \to X$, let us consider the jet prolongation

$$J^1 R_G : J^1(X \times G) \underset{X}{\times} J^1 P \to J^1 P$$

of the morphism R_{GP} (5.3.2). Restricting this morphism to

$$J^1 R_G : \widehat{0}(G) \underset{X}{\times} J^1 P \to J^1 P,$$

we obtain the *jet prolongation* of the structure group action R_{GP} (5.3.2). If there is no danger of confusion, we call it, simply, the *action of G on $J^1 P$*. It reads

$$G \ni g : j_x^1 p \to (j_x^1 p)g = j_x^1(pg). \tag{5.4.2}$$

Taking the quotient of the affine jet bundle $J^1 P$ by G (5.4.2), we obtain the affine bundle

$$C = J^1 P / G \to X \tag{5.4.3}$$

modelled over the vector bundle

$$\overline{C} = T^* X \underset{X}{\otimes} V_G P \to X.$$

Hence, there is the vertical splitting

$$VC = C \underset{X}{\otimes} \overline{C}$$

of the vertical tangent bundle VC of $C \to X$.

Remark 5.4.1. A glance at the expression (5.4.2) shows that the fibre bundle $J^1 P \to C$ is a principal bundle with the structure group G. It is canonically isomorphic to the pull-back

$$J^1 P = P_C = C \underset{X}{\times} P \to C. \tag{5.4.4}$$

Taking the quotient with respect to the action of a structure group G, one can reduce the canonical imbedding (1.2.5) (where $Y = P$) to the bundle monomorphism

$$\lambda_C : C \xrightarrow[X]{} T^* X \underset{X}{\otimes} T_G P,$$

$$\lambda_C : dx^\mu \otimes (\partial_\mu + \chi_\mu^m e_m). \tag{5.4.5}$$

It follows that, given atlases Ψ_P (5.3.3) of P and Ψ (5.3.12) of $T_G P$, the bundle of principal connections C is provided with bundle coordinates (x^λ, a_μ^m) possessing the transformation rule

$$\varrho(a_\mu^m)\varepsilon_m = (a_\nu^m \text{Ad}_{\varrho^{-1}}(\varepsilon_m) + R_\nu^m \varepsilon_m) \frac{\partial x^\nu}{\partial x'^\mu}. \tag{5.4.6}$$

If G is a matrix group, this transformation rule reads

$$\varrho(a_\mu^m)\varepsilon_m = (a_\nu^m \varrho^{-1}(\varepsilon_m)\varrho - \partial_\mu(\varrho^{-1})\varrho) \frac{\partial x^\nu}{\partial x'^\mu}. \tag{5.4.7}$$

5.4. Principal connections. Gauge fields

A glance at this expression shows that the bundle of principal connections C as like as the vector bundle $T_G P$ (5.3.11) fails to be a bundle with a structure group G.

As was mentioned above, a connection A (5.4.1) on a principal bundle $P \to X$ is called a *principal connection* if it is *equivariant* under the action (5.4.2) of a structure group G, i.e.,

$$A(pg) = A(p)g \qquad g \in G. \tag{5.4.8}$$

There is obvious one-to-one correspondence between the principal connections on a principal G-bundle P and global sections

$$A : X \to C \tag{5.4.9}$$

of the factor bundle $C \to X$ (5.4.3), called the *bundle of principal connections*.

Assertion 5.4.1. Since the bundle of principal connections $C \to X$ is affine, principal connections on a principal bundle always exist.

Due to the bundle monomorphism (5.4.5), any principal connection A (5.4.9) is represented by a $T_G P$-valued form

$$A = dx^\lambda \otimes (\partial_\lambda + A_\lambda^q e_q). \tag{5.4.10}$$

Taking the quotient with respect to the action of a structure group G, one can reduce the exact sequence (1.1.19) (where $Y = P$) to the exact sequence

$$0 \to V_G P \xrightarrow[X]{} T_G P \longrightarrow TX \to 0. \tag{5.4.11}$$

A principal connection A (5.4.10) defines a splitting of this exact sequence.

Remark 5.4.2. A principal connection A (5.4.1) on a principal bundle $P \to X$ can be represented by the vertical-valued form A (1.3.9) on P which is a $\mathfrak{g}_l$-valued form due to the trivialization (5.3.9). It is the familiar $\mathfrak{g}_l$-valued *connection form* on a principal bundle P [92]. Given a local bundle splitting (U_α, z_α) of P, this form reads

$$\overline{A} = \psi_\alpha^*(\theta_l - \overline{A}_\lambda^q dx^\lambda \otimes \epsilon_q), \tag{5.4.12}$$

where θ_l is the canonical $\mathfrak{g}_l$-valued one-form (5.1.9) on G and A_λ^p are local functions on P such that

$$\overline{A}_\lambda^q(pg)\epsilon_q = \overline{A}_\lambda^q(p)\mathrm{Ad}_{g^{-1}}(\epsilon_q).$$

The pull-back $z_\alpha^* \overline{A}$ of the connection form $\overline{A}$ (5.4.12) onto U_α is the well-known *local connection one-form*

$$A_\alpha = -A_\lambda^q dx^\lambda \otimes \epsilon_q = A_\lambda^q dx^\lambda \otimes \varepsilon_q, \tag{5.4.13}$$

where $A_\lambda^q = \overline{A}_\lambda^q \circ z_\alpha$ are local functions on X. It is readily observed that the coefficients A_λ^q of this form are exactly the coefficients of the form (5.4.10).

In classical gauge theory, coefficients of the local connection one-form (5.4.13) are treated as *gauge potentials*. We use this term in order to refer to sections A (5.4.9) of the bundle $C \to X$ of principal connections.

There are both pull-back and push-forward operations of principal connections [92].

Theorem 5.4.1. *Let P be a principal fibre bundle and f^*P (1.1.8) the pull-back principal bundle with the same structure group. Let f_P be the canonical morphism (1.1.9) of f^*P to P. If A is a principal connection on P, then the pull-back connection f^*A (1.3.12) on f^*P is a principal connection.*

Theorem 5.4.2. *Let $P' \to X$ and $P \to X$ be principle bundles with structure groups G' and G, respectively. Let $\Phi : P' \to P$ be a principal bundle morphism over X with the corresponding homomorphism $G' \to G$ (see Remark 5.3.3). For every principal connection A' on P', there exists a unique principal connection A on P such that $T\Phi$ sends the horizontal subspaces of TP' A' onto the horizontal subspaces of TP with respect to A.*

Let $P \to X$ be a principal G-bundle. The Frölicher–Nijenhuis bracket (1.1.40) on the space $\mathcal{O}^*(P) \otimes \mathcal{T}(P)$ of tangent-valued forms on P is compatible with the right action R_{GP} (5.3.2). Therefore, it induces the Frölicher–Nijenhuis bracket on the space

$$\mathcal{O}^*(X) \otimes T_G P(X)$$

of $T_G P$-valued forms on X, where $T_G P(X)$ is the vector space of sections of the vector bundle $T_G P \to X$. Note that, as it follows from the exact sequence (5.4.11), there is an epimorphism

$$T_G P(X) \to \mathcal{T}(X).$$

Let

$$A \in \mathcal{O}^1(X) \otimes T_G P(X)$$

be a principal connection (5.4.10). The associated Nijenhuis differential is

$$d_A : \mathcal{O}^r(X) \otimes T_G P(X) \to \mathcal{O}^{r+1}(X) \otimes V_G P(X),$$
$$d_A \phi = [A, \phi]_{\mathrm{FN}}, \quad \phi \in \mathcal{O}^r(X) \otimes T_G P(X). \tag{5.4.14}$$

The *strength* of a principal connection A (5.4.10) is defined as the $V_G P$-valued two-form

$$F_A = \frac{1}{2} d_A A = \frac{1}{2} [A, A]_{\mathrm{FN}} \in \mathcal{O}^2(X) \otimes V_G P(X). \tag{5.4.15}$$

Its coordinated expression

$$F_A = \frac{1}{2} F^r_{\lambda\mu} dx^\lambda \wedge dx^\mu \otimes e_r,$$
$$F^r_{\lambda\mu} = [\partial_\lambda + A^p_\lambda e_p, \partial_\mu + A^q_\mu e_q]^r = \qquad (5.4.16)$$
$$\partial_\lambda A^r_\mu - \partial_\mu A^r_\lambda + c^r_{pq} A^p_\lambda A^q_\mu,$$

results from the bracket (5.3.17).

Remark 5.4.3. However, it should be emphasized that the strength F_A (5.4.15) is not the standard curvature (1.3.23) of a principal connection because A (5.4.10) is not a tangent-valued form. The *curvature of a principal connection* A (5.4.1) on P is the VP-valued two-form R (1.3.23) on P, which can be brought into the $\mathfrak{g}_l$-valued form [92] owing to the canonical isomorphism (5.3.9).

Regarding principal connections A as gauge potentials in classical gauge theory, one calls their strength F_A (5.4.16) the *strength of a gauge field*.

Remark 5.4.4. Given a principal connection A (5.4.9), let Φ_C be a vertical principal automorphism of the bundle of principal connections C. The connection $A' = \Phi_C \circ A$ is called *conjugate* to a principal connection A. The strength forms (5.4.15) of conjugate principal connections A and A' coincide with each other, i.e., $F_A = F_{A'}$.

5.5 Canonical principal connection

Since gauge potentials are represented by sections of the bundle of principal connections $C \to X$ (5.4.3), classical gauge theory is formulated as field theory on C. In order to introduce vector fields and connections on C, one can use the canonical connection on the pull-back principal bundle $P_C \to C$ (5.4.4).

Given a principal G-bundle $P \to X$ and its jet manifold $J^1 P$, let us consider the canonical morphism $\theta_{(1)}$ (1.2.5) where $Y = P$. By virtue of Remark 1.1.2, this morphism defines the morphism

$$\theta : J^1 P \underset{P}{\times} TP \to VP.$$

Taking its quotient with respect to G, we obtain the morphism

$$C \underset{X}{\times} T_G P \xrightarrow{\theta} V_G P, \qquad (5.5.1)$$
$$\theta(\partial_\lambda) = -a^p_\lambda e_p, \qquad \theta(e_p) = e_p.$$

This means that the exact sequence (5.4.11) admits the canonical splitting over C [50].

In view of this fact, let us consider the pull-back principal G-bundle P_C (5.4.4). Since
$$V_G(C \underset{X}{\times} P) = C \underset{X}{\times} V_G P, \qquad T_G(C \underset{X}{\times} P) = TC \underset{X}{\times} T_G P, \qquad (5.5.2)$$
the exact sequence (5.4.11) for the principal bundle P_C reads
$$0 \to C \underset{X}{\times} V_G P \underset{C}{\longrightarrow} TC \underset{X}{\times} T_G P \longrightarrow TC \to 0. \qquad (5.5.3)$$
It is readily observed that the morphism (5.5.1) yields the horizontal splitting (1.3.3):
$$TC \underset{X}{\times} T_G P \longrightarrow C \underset{X}{\times} T_G P \longrightarrow C \underset{X}{\times} V_G P,$$
of the exact sequence (5.5.3) and, consequently, it defines the principal connection
$$\mathcal{A} : TC \to TC \underset{X}{\times} T_G P,$$
$$\mathcal{A} = dx^\lambda \otimes (\partial_\lambda + a^p_\lambda e_p) + da^r_\lambda \otimes \partial^\lambda_r, \qquad (5.5.4)$$
$$\mathcal{A} \in \mathcal{O}^1(C) \otimes T_G(C \underset{X}{\times} P)(X),$$
on the principal bundle
$$P_C = C \underset{X}{\times} P \to C. \qquad (5.5.5)$$
It follows that the principal bundle P_C admits the *canonical principal connection* (5.5.4).

Following the expression (5.4.15), let us define the strength
$$F_\mathcal{A} = \frac{1}{2} d_\mathcal{A} \mathcal{A} = \frac{1}{2}[\mathcal{A}, \mathcal{A}] \in \mathcal{O}^2(C) \otimes V_G P(X),$$
$$F_\mathcal{A} = (da^r_\mu \wedge dx^\mu + \frac{1}{2} c^r_{pq} a^p_\lambda a^q_\mu dx^\lambda \wedge dx^\mu) \otimes e_r, \qquad (5.5.6)$$
of the canonical principal connection $\mathcal{A}$ (5.5.4). It is called the *canonical strength* because, given a principal connection A (5.4.9) on a principal bundle $P \to X$, the pull-back
$$A^* F_\mathcal{A} = F_A \qquad (5.5.7)$$
is the strength (5.4.16) of A.

With the $V_G P$-valued two-form $F_\mathcal{A}$ (5.5.6) on C, let us define the $V_G P$-valued horizontal two-form
$$\mathcal{F} = h_0(F_\mathcal{A}) = \frac{1}{2} \mathcal{F}^r_{\lambda\mu} dx^\lambda \wedge dx^\mu \otimes \varepsilon_r,$$
$$\mathcal{F}^r_{\lambda\mu} = a^r_{\lambda\mu} - a^r_{\mu\lambda} + c^r_{pq} a^p_\lambda a^q_\mu, \qquad (5.5.8)$$

on J^1C. It is called the *strength form*. For each principal connection A (5.4.9) on P, the pull-back

$$J^1A^*\mathcal{F} = F_A \qquad (5.5.9)$$

is the strength (5.4.16) of A.

The strength form (5.5.8) yields an affine surjection

$$\mathcal{F}/2 : J^1C \xrightarrow[C]{} C \underset{X}{\times} (\overset{2}{\wedge} T^*X \otimes V_G P) \qquad (5.5.10)$$

over C of the affine jet bundle $J^1C \to C$ to the vector (and, consequently, affine) bundle

$$C \underset{X}{\times} (\overset{2}{\wedge} T^*X \otimes V_G P) \to C.$$

By virtue of Theorem 1.1.10, its kernel $C_+ = \mathrm{Ker}\,\mathcal{F}/2$ is an affine subbundle of $J^1C \to C$. Thus, we have the canonical splitting of the affine jet bundle

$$J^1C = C_+ \underset{C}{\oplus} C_- = C_+ \underset{C}{\oplus} (C \underset{X}{\times} \overset{2}{\wedge} T^*X \otimes V_G P), \qquad (5.5.11)$$

$$a^r_{\lambda\mu} = \frac{1}{2}(\mathcal{F}^r_{\lambda\mu} + \mathcal{S}^r_{\lambda\mu}) = \frac{1}{2}(a^r_{\lambda\mu} + a^r_{\mu\lambda} - c^r_{pq}a^p_\lambda a^q_\mu) + \qquad (5.5.12)$$
$$\frac{1}{2}(a^r_{\lambda\mu} - a^r_{\mu\lambda} + c^r_{pq}a^p_\lambda a^q_\mu).$$

The corresponding canonical projections are $\mathrm{pr}_2 = \mathcal{F}/2$ (5.5.10) and

$$\mathrm{pr}_1 = \mathcal{S}/2 : J^1C \to C_+. \qquad (5.5.13)$$

The jet manifold J^1C plays a role of the configuration space of classical gauge theory on principal bundles. Its splitting (5.5.11) exemplifies the splitting (2.4.65), but it is not related to a Lagrangian.

5.6 Gauge transformations

In classical gauge theory, *gauge transformations* are defined as principal automorphisms of a principal bundle P. In accordance with Remark 5.3.3, an automorphism Φ_P of a principal G-bundle P is called *principal* if it is equivariant under the right action (5.3.2) of a structure group G on P, i.e.,

$$\Phi_P(pg) = \Phi_P(p)g, \qquad g \in G, \qquad p \in P. \qquad (5.6.1)$$

In particular, every vertical principal automorphism of a principal bundle P is represented as

$$\Phi_P(p) = pf(p), \qquad p \in P, \qquad (5.6.2)$$

where f is a G-valued equivariant function on P, i.e.,

$$f(pg) = g^{-1}f(p)g, \qquad g \in G. \tag{5.6.3}$$

There is one-to-one correspondence between the equivariant functions f (5.6.3) and the global sections s of the associated group bundle

$$\pi_{P^G} : P^G \to X \tag{5.6.4}$$

whose fibres are groups isomorphic to G and whose typical fibre is the group G which acts on itself by the adjoint representation. This one-to-one correspondence is defined by the relation

$$s(\pi_P(p))p = pf(p), \qquad p \in P, \tag{5.6.5}$$

(see Example 5.7.2). The group of vertical principal automorphisms of a principal G-bundle is called the *gauge group*. It is isomorphic to the group $P^G(X)$ of global sections of the group bundle (5.6.4). Its unit element is the canonical global section $\widehat{1}$ of $P^G \to X$ whose values are unit elements of fibres of P^G.

Remark 5.6.1. Note that transition functions of atlases of a principle bundle P also are represented by local sections of the associated group bundle P^G (5.6.4).

Remark 5.6.2. Though $P^G \to X$ is not a vector bundle, one can define an appropriate Sobolev completion $\overline{P^G(X)}$ of the gauge group $P^G(X)$ if G is a matrix group [112; 116] such that $\overline{P^G(X)}$ is a Lie group. Its Lie algebra is the corresponding Sobolev completion $\overline{\mathcal{G}(X)}$ of the gauge algebra $\mathcal{G}(X)$ of global sections of the Lie algebra bundle $V_G P \to X$.

In order to describe gauge symmetries of gauge theory on a principal bundle P, let us restrict our consideration to (local) one-parameter groups of principal automorphisms of P. Their infinitesimal generators are G-invariant projectable vector fields ξ on P, and *vice versa*. We call ξ the *principal vector fields* or the *infinitesimal gauge transformations*. They are represented by sections ξ (5.3.16) of the vector bundle $T_G P$ (5.3.11). Principal vector fields constitute a real Lie algebra $T_G P(X)$ with respect to the Lie bracket (5.3.17). *Vertical principal vector fields* are the sections

$$\xi = \xi^p e_p \tag{5.6.6}$$

of the gauge algebra bundle $V_G P \to X$ (5.3.11). They form a finite-dimensional Lie $C^\infty(X)$-algebra $\mathcal{G}(X) = V_G P(X)$ (5.3.18) that has been called the *gauge algebra*.

5.6. Gauge transformations

Any (local) one-parameter group of principal automorphism Φ_P (5.6.1) of a principal bundle P admits the jet prolongation $J^1\Phi_P$ (1.2.7) to a one-parameter group of G-equivariant automorphism of the jet manifold $J^1 P$ which, in turn, yields a one-parameter group of *principal automorphisms* Φ_C of the bundle of principal connections C (5.4.3) [89; 112]. Its infinitesimal generator is a vector field on C, called the principal vector field on C and regarded as an infinitesimal gauge transformation of C. Thus, any principal vector field ξ (5.3.16) on P yields a principal vector field u_ξ on C, which can be defined as follows.

Using the morphism (5.5.1), we obtain the morphism

$$\xi \rfloor \theta : C \xrightarrow[X]{} V_G P,$$

which is a section of of the Lie algebra bundle

$$V_G(C \underset{X}{\times} P) \to C$$

in accordance with the first formula (5.5.2). Then the equation

$$u_\xi \rfloor F_A = d_A(\xi \rfloor \theta)$$

uniquely determines a desired vector field u_ξ on C. A direct computation leads to

$$u_\xi = \xi^\mu \partial_\mu + (\partial_\mu \xi^r + c^r_{pq} a^p_\mu \xi^q - a^r_\nu \partial_\mu \xi^\nu)\partial^\mu_r. \tag{5.6.7}$$

In particular, if ξ is a vertical principal field (5.6.6), we obtain the vertical vector field

$$u_\xi = (\partial_\mu \xi^r + c^r_{pq} a^p_\mu \xi^q)\partial^\mu_r. \tag{5.6.8}$$

Remark 5.6.3. The jet prolongation (1.2.8) of the vector field u_ξ (5.6.7) onto $J^1 C$ reads

$$J^1 u_\xi = u_\xi + (\partial_\lambda \partial_\mu \xi^r + c^r_{pq} a^p_\mu \partial_\lambda \xi^q + c^r_{pq} a^p_{\lambda\mu} \xi^q - \tag{5.6.9}$$
$$a^r_\nu \partial_\lambda \partial_\mu \xi^\nu - a^r_{\lambda\nu} \partial_\mu \xi^\nu - a^r_{\nu\mu} \partial_\lambda \xi^\nu)\partial^{\lambda\mu}_r.$$

Example 5.6.1. Let A (5.4.10) be a principal connection on P. For any vector field τ on X, this connection yields a section

$$\tau \rfloor A = \tau^\lambda \partial_\lambda + A^p_\lambda \tau^\lambda e_p$$

of the vector bundle $T_G P \to X$. It, in turn, defines a principal vector field (5.6.7) on the bundle of principal connection C which reads

$$\tau_A = \tau^\lambda \partial_\lambda + (\partial_\mu(A^r_\nu \tau^\nu) + c^r_{pq} a^p_\mu A^q_\nu \tau^\nu - a^r_\nu \partial_\mu \tau^\nu)\partial^\mu_r, \tag{5.6.10}$$
$$\xi^\lambda = \tau^\lambda, \qquad \xi^p = A^p_\nu \tau^\nu.$$

It is readily justified that the monomorphism

$$T_G P(X) \ni \xi \to u_\xi \in \mathcal{T}(C) \tag{5.6.11}$$

obeys the equality

$$u_{[\xi,\eta]} = [u_\xi, u_\eta], \tag{5.6.12}$$

i.e., it is a monomorphism of the real Lie algebra $T_G P(X)$ to the real Lie algebra $\mathcal{T}(C)$. In particular, the image of the gauge algebra $\mathcal{G}(X)$ in $\mathcal{T}(C)$ also is a real Lie algebra, but not the $C^\infty(X)$-one because

$$u_{f\xi} \neq f u_\xi, \qquad f \in C^\infty(X).$$

Remark 5.6.4. A glance at the expression (5.6.7) shows that the monomorphism (5.6.11) is a linear first order differential operator which sends sections of the pull-back bundle

$$C \underset{X}{\times} T_G P \to C$$

onto sections of the tangent bundle $TC \to C$. Refereing to Definition 2.3.1, we therefore can treat principal vector fields (5.6.7) as infinitesimal gauge transformations depending on gauge parameters $\xi \in T_G P(X)$.

5.7 Geometry of associated bundles. Matter fields

Given a principal G-bundle P (5.3.1), any associated G-bundle over X with a typical fibre V is equivalent to the following one.

Let us consider the quotient

$$Y = (P \times V)/G \tag{5.7.1}$$

of the product $P \times V$ by identification of elements (p, v) and $(pg, g^{-1}v)$ for all $g \in G$. Let $[p]$ denote the restriction of the canonical surjection

$$P \times V \to (P \times V)/G \tag{5.7.2}$$

to the subset $\{p\} \times V$ so that

$$[p](v) = [pg](g^{-1}v).$$

Then the map

$$Y \ni [p](V) \to \pi_P(p) \in X$$

makes the quotient Y (5.7.1) into a fibre bundle over X. This is a smooth G-bundle with the typical fibre V which is associated with the principal

5.7. Geometry of associated bundles. Matter fields

G-bundle P. For short, we call it the *P-associated bundle*. In classical gauge theory, sections of a P-associated bundle describe *matter fields*.

Remark 5.7.1. The tangent morphism to the morphism (5.7.2) and the jet prolongation of the morphism (5.7.2) lead to the bundle isomorphisms

$$TY = (TP \times TV)/G, \qquad (5.7.3)$$

$$J^1 Y = (J^1 P \times V)/G. \qquad (5.7.4)$$

The peculiarity of the P-associated bundle Y (5.7.1) is the following.

(i) Every bundle atlas $\Psi_P = \{(U_\alpha, z_\alpha)\}$ (5.3.5) of P defines a unique associated bundle atlas

$$\Psi = \{(U_\alpha, \psi_\alpha(x) = [z_\alpha(x)]^{-1})\} \qquad (5.7.5)$$

of the quotient Y (5.7.1).

Example 5.7.1. Because of the splitting (5.3.9), the Lie algebra bundle

$$V_G P = (P \times \mathfrak{g}_l)/G,$$

by definition, is of the form (5.7.1). Therefore, it is a P-associated bundle.

Example 5.7.2. The group bundle $\overline{P}$ (5.6.4) is defined as the quotient

$$P^G = (P \times G)/G, \qquad (5.7.6)$$

where the group G which acts on itself by the adjoint representation. There is the following fibre-to-fibre action of the group bundle P^G on any P-associated bundle Y (5.7.1):

$$P^G \underset{X}{\times} Y \longrightarrow Y,$$

$$((p, g)/G, (p, v)/G) \to (p, gv)/G, \qquad g \in G, \qquad v \subset V.$$

For instance, the action of P^G on P in the formula (5.6.5) is of this type.

(ii) Any principal automorphism Φ_P (5.6.1) of P yields a unique *principal automorphism*

$$\Phi_Y : (p, v)/G \to (\Phi_P(p), v)/G, \qquad p \in P, \qquad v \in V, \qquad (5.7.7)$$

of the P-associated bundle Y (5.7.1). For the sake of brevity, we agree to write

$$\Phi_Y : (P \times V)/G \to (\Phi_P(P) \times V)/G.$$

Accordingly, any (local) one-parameter group of principal automorphisms of P induces a (local) one-parameter group of automorphisms of the P-associated bundle Y (5.7.1). Passing to infinitesimal generators of these groups, we obtain that any principal vector field ξ (5.3.16) yields a vector field v_ξ on Y regarded as an infinitesimal gauge transformation of Y. Owing to the bundle isomorphism (5.7.3), we have

$$v_\xi : X \to (\xi(P) \times TV)/G \subset TY,$$
$$v_\xi = \xi^\lambda \partial_\lambda + \xi^p I_p^i \partial_i, \qquad (5.7.8)$$

where $\{I_p\}$ is a representation of the Lie algebra $\mathfrak{g}_r$ of G in V.

(iii) Any principal connection on $P \to X$ defines a unique connection on the P-associated fibre bundle Y (5.7.1) as follows. Given a principal connection A (5.4.8) on P and the corresponding horizontal distribution $HP \subset TP$, the tangent map to the canonical morphism (5.7.2) defines the horizontal splitting of the tangent bundle TY of Y (5.7.1) and the corresponding connection on $Y \to X$ [92]. Owing to the bundle isomorphism (5.7.4), we have

$$A : (P \times V)/G \to (A(P) \times V)/G \subset J^1 Y,$$
$$A = dx^\lambda \otimes (\partial_\lambda + A_\lambda^p I_p^i \partial_i), \qquad (5.7.9)$$

where $\{I_p\}$ is a representation of the Lie algebra $\mathfrak{g}_r$ of G in V [94]. The connection A (5.7.9) on Y is called the *associated principal connection* or, simply, a principal connection on $Y \to X$. The *curvature* (1.3.24) of this connection takes the form

$$R = \frac{1}{2} F_{\lambda\mu}^p I_p^i dx^\lambda \wedge dx^\mu \otimes \partial_i. \qquad (5.7.10)$$

Example 5.7.3. A principal connection A on P yields the associated connection (5.7.9) on the associated Lie algebra bundle $V_G P$ which reads

$$A = dx^\lambda \otimes (\partial_\lambda - c_{pq}^m A_\lambda^p \xi^q e_m). \qquad (5.7.11)$$

Remark 5.7.2. If an associated principal connection A is linear, one can define its *strength*

$$F_A = \frac{1}{2} F_{\lambda\mu}^p I_p dx^\lambda \wedge dx^\mu, \qquad (5.7.12)$$

where I_p are matrices of a representation of the Lie algebra $\mathfrak{g}_r$ in fibres of Y with respect to the fibre bases $\{e_i(x)\}$. They coincide with the matrices of a representation of $\mathfrak{g}_r$ in the typical fibre V of Y with respect to its fixed basis $\{e_i\}$ (see the relation (1.1.10). It follows that G-valued transition functions act on I_p by the adjoint representation. Note that, because of the canonical splitting (1.1.17), one can identify $e_i(x) = \partial_i$. Therefore, the strength form (5.7.12) can be represented as a $E \underset{X}{\otimes} E^*$-valued two-form on X.

5.7. Geometry of associated bundles. Matter fields

In view of the above mentioned properties, the P-associated bundle Y (5.7.1) is called *canonically associated* to a principal bundle P. Unless otherwise stated, only canonically associated bundles are considered, and we simply call Y (5.7.1) an *associated bundle*.

Remark 5.7.3. Since the bundle of principal connection C is not P-associated, connections on C are introduced in a different way. For this purpose, let us consider a symmetric world connection K^* (1.3.40) on the cotangent bundle $T^*X \to X$ of X and a principal connection A on $P \to X$. The latter defines the associated connection A (5.7.11) on the Lie algebra bundle $V_G P \to X$. Let us consider the tensor product connection $\overline{\Gamma}$ (1.3.38) on the bundle

$$T^*X \underset{X}{\otimes} V_G P \to X$$

induced by K (1.3.39) and A (5.7.11). Given the coordinates $(x^\lambda, \overline{a}_\mu^r, \overline{a}_{\lambda\mu}^r)$ on the jet manifold

$$J^1(T^*X \underset{X}{\otimes} V_G P),$$

we obtain

$$\overline{\Gamma} = dx^\lambda \otimes [\partial_\lambda + (-K_\lambda{}^\nu{}_\mu \overline{a}_\nu^r - c_{pq}^r \overline{a}_\mu^q A_\lambda^p)\partial_r^\mu)]. \qquad (5.7.13)$$

Given the bundle morphism

$$D_A : C \ni a_\mu^r \underset{X}{\longrightarrow} a_\mu^r - A_\mu^r \in T^*X \underset{X}{\otimes} V_G P$$

(1.1.21), the commutative diagram

$$\begin{array}{ccc} J^1 C & \xrightarrow{J^1 D_A} & J^1(T^*X \underset{X}{\otimes} V_G P) \\ {\scriptstyle \Gamma_A} \uparrow & & \uparrow {\scriptstyle \overline{\Gamma}} \\ C & \xrightarrow{D_A} & T^*X \underset{X}{\otimes} V_G P \end{array}$$

defines a section

$$a_{\lambda\mu}^r \circ \Gamma = \partial_\lambda A_\mu^r - c_{pq}^r(a_\mu^q - A_\mu^q)A_\lambda^p - K_\lambda{}^\nu{}_\mu(a_\nu^r - A_\nu^r)$$

of the affine jet bundle $J^1 C \to C$, i.e., the connection

$$\Gamma_A = dx^\lambda \otimes [\partial_\lambda + (\partial_\lambda A_\mu^r - c_{pq}^r(a_\mu^q - A_\mu^q)A_\lambda^p - \qquad (5.7.14)$$
$$K_\lambda{}^\nu{}_\mu(a_\nu^r - A_\nu^r))\partial_r^\mu]$$

on the bundle of principal connections $C \to X$. A glance at the expression (5.7.14) shows that Γ_A is an affine connection on the affine bundle $C \to X$, while the corresponding linear connection (1.3.46) is $\overline{\Gamma}$. Moreover, it is easily seen that A is an integral section of the connection Γ_A (5.7.14). The connection (5.7.14) is not a unique one defined by a symmetric world connection K and a principal connection A. The strength F_A of A can be seen as a soldering form

$$F_A = F^r_{\lambda\mu} dx^\lambda \otimes \partial^\mu_r \qquad (5.7.15)$$

on C. Then there is another connection

$$\Gamma'_A = \Gamma_A - F_A \qquad (5.7.16)$$

on $C \to X$. Let us assume that a vector field τ on X is an integral section of a symmetric world connection K (see Remark 1.3.4). Then it is readily observed that the horizontal lift $\Gamma'_A \tau$ of τ by means of the connection Γ'_A (5.7.16) coincides with the vector field $\widetilde{\tau}_A$ (5.6.10) on the fibre bundle C.

5.8 Yang–Mills gauge theory

Let us consider first order Lagrangian gauge theory on a principal bundle P. Its configuration space is the first order jet manifold $J^1 C$ of the bundle of principal connections C (5.4.3), endowed with bundle coordinates (x^μ, a^m_μ) possessing transition functions (5.4.6). Given a first order Lagrangian

$$L = \mathcal{L}\omega : J^1 C \to \overset{n}{\wedge} T^* X \qquad (5.8.1)$$

on $J^1 C$, the corresponding Euler–Lagrange operator (2.1.12) reads

$$\mathcal{E}_L = \mathcal{E}^\mu_r \theta^r_\mu \wedge \omega = (\partial^\mu_r - d_\lambda \partial^{\lambda\mu}_r)\mathcal{L}\theta^r_\mu \wedge \omega. \qquad (5.8.2)$$

Its kernel defines the Euler–Lagrange equation

$$\mathcal{E}^\mu_r = (\partial^\mu_r - d_\lambda \partial^{\lambda\mu}_r)\mathcal{L} = 0. \qquad (5.8.3)$$

5.8.1 Gauge field Lagrangian

Let us assume that a gauge theory Lagrangian L (5.8.1) on $J^1 C$ is invariant under vertical gauge transformations (or, in short, *gauge invariant*). This means that vertical principal vector fields (5.6.8):

$$u_\xi = (\partial_\mu \xi^r + c^r_{pq} a^p_\mu \xi^q)\partial^\mu_r, \qquad (5.8.4)$$

are exact symmetries of L, that is,

$$\mathbf{L}_{J^1 u_\xi} L = 0, \qquad (5.8.5)$$

5.8. Yang–Mills gauge theory

where
$$J^1 u_\xi = u_\xi + (\partial_{\lambda\mu}\xi^r + c^r_{pq} a^p_\mu \partial_\lambda \xi^q + c^r_{pq} a^p_{\lambda\mu} \xi^q)\partial_r^{\lambda\mu}$$
(see the formula (5.6.9)). Then it follows from Remark 5.6.4 that vertical principal vector fields u_ξ (5.8.4) are gauge symmetries of L whose gauge parameters are sections ξ (5.6.6) of the Lie algebra bundle $V_G P$. In this case, the first variational formula (2.4.29) for the Lie derivative (5.8.5) takes the form
$$0 = (\partial_\mu \xi^r + c^r_{pq} a^p_\mu \xi^q)\mathcal{E}^\mu_r + d_\lambda[(\partial_\mu \xi^r + c^r_{pq} a^p_\mu \xi^q)\partial_r^{\lambda\mu}\mathcal{L})]. \tag{5.8.6}$$
It leads to the gauge invariance conditions (2.4.43) – (2.4.46) which read
$$\partial_p^{\mu\lambda}\mathcal{L} + \partial_p^{\lambda\mu}\mathcal{L} = 0, \tag{5.8.7}$$
$$\mathcal{E}^\mu_r + d_\lambda \partial_r^{\lambda\mu}\mathcal{L} + c^q_{pr} a^p_\nu \partial_q^{\mu\nu}\mathcal{L} = 0, \tag{5.8.8}$$
$$c^r_{pq}(a^p_\mu \mathcal{E}^\mu_r + d_\lambda(a^p_\mu \partial_r^{\lambda\mu}\mathcal{L})) = 0. \tag{5.8.9}$$

One can regard the equalities (5.8.7) – (5.8.9) as the conditions of a Lagrangian L to be gauge invariant. They are brought into the form
$$\partial_p^{\mu\lambda}\mathcal{L} + \partial_p^{\lambda\mu}\mathcal{L} = 0. \tag{5.8.10}$$
$$\partial_q^\mu \mathcal{L} + c^r_{pq} a^p_\nu \partial_r^{\mu\nu}\mathcal{L} = 0, \tag{5.8.11}$$
$$c^r_{pq}(a^p_\mu \partial_r^\mu \mathcal{L} + a^p_{\lambda\mu} \partial_r^{\lambda\mu}\mathcal{L}) = 0. \tag{5.8.12}$$

Let us utilize the coordinates $(a^q_\mu, \mathcal{F}^r_{\lambda\mu}, \mathcal{S}^r_{\lambda\mu})$ (5.5.12) which correspond to the canonical splitting (5.5.11) of the affine jet bundle $J^1 C \to C$. With respect to these coordinates, the equation (5.8.10) reads
$$\frac{\partial \mathcal{L}}{\partial \mathcal{S}^p_{\mu\lambda}} = 0. \tag{5.8.13}$$

Then the equation (5.8.11) takes the form
$$\frac{\partial \mathcal{L}}{\partial a^q_\mu} = 0. \tag{5.8.14}$$

A glance at the equalities (5.8.13) and (5.8.14) shows that a gauge invariant Lagrangian factorizes through the strength $\mathcal{F}$ (5.5.8) of a gauge field. Then the equation (5.8.12, written as
$$c^r_{pq} \mathcal{F}^p_{\lambda\mu} \frac{\partial \mathcal{L}}{\partial \mathcal{F}^r_{\lambda\mu}} = 0,$$
shows that the gauge symmetry u_ξ of a Lagrangian L is exact. The following thus has been proved.

Theorem 5.8.1. *A gauge theory Lagrangian (5.8.1) possesses the exact gauge symmetry u_ξ (5.8.4) only if it factorizes through the strength $\mathcal{F}$ (5.5.8).*

A corollary of this result is the well-known *Utiyama theorem* [23].

Theorem 5.8.2. *There is a unique gauge invariant quadratic first order Lagrangian (with the accuracy to variationally trivial ones). It is the conventional* Yang–Mills Lagrangian

$$L_{\mathrm{YM}} = \frac{1}{4}a^G_{pq}g^{\lambda\mu}g^{\beta\nu}\mathcal{F}^p_{\lambda\beta}\mathcal{F}^q_{\mu\nu}\sqrt{|g|}\,\omega, \qquad g=\det(g_{\mu\nu}), \qquad (5.8.15)$$

where a^G is a G-invariant bilinear form on the Lie algebra $\mathfrak{g}_r$ and g is a world metric on X.

The Euler–Lagrange operator of the Yang–Mills Lagrangian L_{YM} is

$$\mathcal{E}_{\mathrm{YM}} = \mathcal{E}^\mu_r\theta^\mu_r \wedge \omega = (\delta^n_r d_\lambda + c^n_{rp}a^p_\lambda)(a^G_{nq}g^{\mu\alpha}g^{\lambda\beta}\mathcal{F}^q_{\alpha\beta}\sqrt{|g|})\theta^r_\mu \wedge \omega. \quad (5.8.16)$$

Its kernel defines the Yang–Mills equations

$$\mathcal{E}^\mu_r = (\delta^n_r d_\lambda + c^n_{rp}a^p_\lambda)(a^G_{nq}g^{\mu\alpha}g^{\lambda\beta}\mathcal{F}^q_{\alpha\beta}\sqrt{|g|}) = 0. \qquad (5.8.17)$$

We call a Lagrangian system $(\mathcal{S}^*_\infty[C], L_{\mathrm{YM}})$ the *Yang–Mills gauge theory*.

Remark 5.8.1. In classical gauge theory, there are Lagrangians, e.g., the Chern–Simons one (see Section 8.2) which do not factorize through the strength of a gauge field, and whose gauge symmetry u_ξ (5.8.4) is variational, but not exact.

5.8.2 Conservation laws

Since the gauge symmetry u_ξ of the Yang–Mills Lagrangian (5.8.15) is exact, the first variational formula (5.8.6) leads to the weak conservation law

$$0 \approx d_\lambda(-u^\mu_\xi{}_r \partial^{\lambda\mu}_r \mathcal{L}_{\mathrm{YM}}) \qquad (5.8.18)$$

of the Noether current

$$\mathcal{J}^\lambda_\xi = -(\partial_\mu\xi^r + c^r_{pq}a^p_\mu\xi^q)(a^G_{rq}g^{\mu\alpha}g^{\lambda\beta}\mathcal{F}^q_{\alpha\beta}\sqrt{|g|}). \qquad (5.8.19)$$

In accordance with Theorem 2.4.2, the Noether current (5.8.19) is brought into the superpotential form (2.4.50) which reads

$$\mathcal{J}^\lambda_\xi = \xi^r \mathcal{E}^\mu_r + d_\nu(\xi^r \partial^{[\nu\mu]}_r \mathcal{L}_{\mathrm{YM}}), \qquad (5.8.20)$$

$$U^{\nu\mu} = \xi^r a^G_{rq}g^{\nu\alpha}g^{\mu\beta}\mathcal{F}^q_{\alpha\beta}\sqrt{|g|}. \qquad (5.8.21)$$

Let us study energy-momentum conservation laws in Yang–Mills gauge theory. If a background world metric g is specified, one can find a particular

5.8. Yang–Mills gauge theory

vector field τ on X and its lift $\gamma\tau$ (2.4.39) onto C which is an exact symmetry of the Yang–Mills Lagrangian (5.8.15). Then the energy-momentum current (2.4.40) along such a symmetry is conserved. We here obtain the energy-momentum conservation law as a gauge conservation law in the case of an arbitrary world metric g and any vector field τ on X.

Given an arbitrary vector field τ on X, let A be a principal connection on P and τ_A (5.6.10) the lift of τ onto the bundle of principal connections C. Let us consider the energy-momentum current along the vector field τ_A (5.6.10) [112; 135].

Since the Yang–Mills Lagrangian (5.8.15) depends on a background world metric g, the vector field τ_A (5.6.10) is not its exact symmetry in general. Following the procedure in Section 2.4.6, let us consider the total Lagrangian

$$L = \frac{1}{4} a^G_{pq} \sigma^{\lambda\mu} \sigma^{\beta\nu} \mathcal{F}^p_{\lambda\beta} \mathcal{F}^q_{\mu\nu} \sqrt{|\sigma|} \omega, \qquad \sigma = \det(\sigma_{\mu\nu}), \qquad (5.8.22)$$

on the total configuration space

$$J^1(C \underset{X}{\times} \overset{2}{\vee} TX)$$

where the tensor bundle $\overset{2}{\vee} TX$ is provided with the holonomic fibre coordinates $(\sigma^{\mu\nu})$. Given a vector field τ on X, there exists its canonical lift (1.1.26):

$$\widetilde{\tau} = \tau^\lambda \partial_\lambda + (\partial_\nu \tau^\alpha \sigma^{\nu\beta} + \partial_\nu \tau^\beta \sigma^{\nu\alpha})\partial_{\alpha\beta},$$

onto the tensor bundle $\overset{2}{\vee} T^*X$. It is the infinitesimal generator of a local one-parameter group of general covariant transformations of $\overset{2}{\vee} T^*X$ (see Section 6.1). Thus, we have the lift

$$\widetilde{\tau}_A = \tau^\lambda \partial_\lambda + (\partial_\mu(A^r_\nu \tau^\nu) + c^r_{pq} a^p_\mu A^q_\nu \tau^\nu - a^r_\nu \partial_\mu \tau^\nu)\partial^\mu_r + \qquad (5.8.23)$$
$$(\partial_\nu \tau^\alpha \sigma^{\nu\beta} + \partial_\nu \tau^\beta \sigma^{\nu\alpha})\partial_{\alpha\beta}$$

of a vector field τ on X onto the product

$$C \underset{X}{\times} \overset{2}{\vee} T^*X.$$

Its is readily observed that the vector field $\widetilde{\tau}_A$ (5.8.23) is an exact symmetry of the total Lagrangian (5.8.22). One the shell (5.8.17), we then obtain the weak transformation law (2.4.75). This reads

$$0 \approx (\partial_\nu \tau^\alpha g^{\nu\beta} + \partial_\nu \tau^\beta g^{\nu\alpha} - \partial_\lambda g^{\alpha\beta} \tau^\lambda)\partial_{\alpha\beta}\mathcal{L} - d_\lambda \mathcal{J}^\lambda_A, \qquad (5.8.24)$$

where
$$J_A^\lambda = \partial_r^{\lambda\mu}\mathcal{L}_{\text{YM}}[\tau^\nu a_{\nu\mu}^r - \partial_\mu(A_\nu^r\tau^\nu) - c_{pq}^r a_\mu^p A_\nu^q \tau^\nu + a_\nu^r \partial_\mu \tau^\nu] - \tau^\lambda \mathcal{L}_{\text{YM}} \qquad (5.8.25)$$

is the energy-momentum current along the vector field τ_A (5.6.10).

The weak identity (5.8.24) can be written in the form
$$0 \approx \partial_\lambda \tau^\mu t_\mu^\lambda \sqrt{|g|} - \tau^\mu \{\mu^\beta{}_\lambda\} t_\beta^\lambda \sqrt{|g|} - d_\lambda J_A^\lambda, \qquad (5.8.26)$$

where $\{\mu^\beta{}_\lambda\}$ are the Christoffel symbols (1.3.44) of a world metric g and
$$t_\beta^\mu \sqrt{|g|} = 2g^{\mu\alpha}\partial_{\alpha\beta}\mathcal{L}_{\text{YM}} = (\mathcal{F}_{\beta\nu}^q \partial_q^{\mu\nu} - \delta_\beta^\mu)\mathcal{L}_{\text{YM}}$$

is the *metric energy-momentum tensor* of a gauge field. In particular, let a principal connection B be a solution of the Yang–Mills equations (5.8.17). Let us consider the lift (5.6.10) of a vector field τ on X onto C by means of the principal connection $A = B$. In this case, the energy-momentum current (5.8.25) reads
$$J_B^\lambda \circ B = \tau^\mu (t_\mu^\lambda \circ B)\sqrt{|g|}. \qquad (5.8.27)$$

Then the weak identity (5.8.26) on a solution B takes the form of the familiar *covariant conservation law*
$$\nabla_\lambda((t_\mu^\lambda \circ B)\sqrt{|g|}) = 0, \qquad (5.8.28)$$

where ∇ is the covariant derivative with respect to the Levi-Civita connection $\{\mu^\beta{}_\lambda\}$ of the background metric g.

Note that, considering a different lift $\gamma\tau$ of a vector field τ on X to a principal vector field on C, we obtain an energy-momentum current along $\gamma\tau$ which differs from J_B^λ (5.8.27) in a Noether current (5.8.20). Since such a current is reduced to a superpotential, one can always bring the energy-momentum transformation law (5.8.24) into the covariant conservation law (5.8.28).

5.8.3 BRST extension

The gauge invariance conditions (5.8.7) – (5.8.9) lead to the Noether identities which the Euler–Lagrange operator $\mathcal{E}_{\text{YM}}$ (5.8.16) of the Yang–Mills Lagrangian (5.8.15) satisfies (see Remark 2.4.6). These Noether identities are associated to the gauge symmetry u_ξ (5.8.4). By virtue of the formula (2.3.7), they read
$$c_{rq}^p a_\mu^q \mathcal{E}_p^\mu + d_\mu \mathcal{E}_r^\mu = 0. \qquad (5.8.29)$$

Lemma 5.8.1. *The Noether identities (5.8.29) are non-trivial.*

5.8. Yang–Mills gauge theory

Proof. Following the procedure in Section 4.1, let us consider the density dual

$$\overline{VC} = V^*C \underset{C}{\otimes} \overset{n}{\wedge} T^*X = (T^*X \underset{X}{\otimes} V_G P)^* \underset{C}{\otimes} \overset{n}{\wedge} T^*X \qquad (5.8.30)$$

of the vertical tangent bundle VC of $C \to X$, and let us enlarge the differential graded algebra $\mathcal{S}^*_\infty[C]$ to the differential bigraded algebra (4.1.6):

$$\mathcal{P}^*_\infty[VC; C] = \mathcal{S}^*_\infty[\overline{VC}; C],$$

possessing the local generating basis $(a^r_\mu, \overline{a}^\mu_r)$ where $\overline{a}^\mu_r$ are odd antifields. Providing this differential bigraded algebra with the nilpotent right graded derivation

$$\overline{\delta} = \frac{\overleftarrow{\partial}}{\partial \overline{a}^\mu_r} \mathcal{E}^\mu_r,$$

let us consider the chain complex (4.1.8). Its one-chains

$$\Delta_r = c^p_{rq} a^q_\mu \overline{a}^\mu_p + d_\mu \overline{a}^\mu_r \qquad (5.8.31)$$

are $\overline{\delta}$-cycles which define the Noether identities (5.8.29). Clearly, they are not $\overline{\delta}$-boundaries. Therefore, the Noether identities (5.8.29) are non-trivial.

Lemma 5.8.2. *The Noether identities (5.8.29) are complete.*

Proof. The second order Euler–Lagrange operator $\mathcal{E}_{YM}$ (5.8.16) takes its values into the space of sections of the vector bundle

$$(T^*X \underset{X}{\otimes} V_G P)^* \underset{X}{\otimes} \overset{n}{\wedge} T^*X \to X.$$

Let Φ be a first order differential operator on this vector bundle such that

$$\Phi \circ \mathcal{E}_{YM} = 0.$$

This condition holds only if the highest derivative term of the composition $\Phi^1 \circ \mathcal{E}^2_{YM}$ of the first order derivative term Φ^1 of Φ and the second order derivative term $\mathcal{E}^2_{YM}$ of $\mathcal{E}_{YM}$ vanishes. This is the case only of

$$\Phi^1 = \Delta^1_r = d_\mu \overline{a}^\mu_r. \qquad \square$$

The graded densities $\Delta_r \omega$ (5.8.31) constitute a local basis for a $C^\infty(X)$-module $\mathcal{C}_{(0)}$ isomorphic to the module $\overline{V_G P}(X)$ of sections of the density dual $\overline{V_G P}$ of the Lie algebra bundle $V_G P \to X$. Let us enlarge the differential bigraded algebra $\mathcal{P}^*_\infty[VC; C]$ to the differential bigraded algebra

$$\overline{\mathcal{P}}^*_\infty\{0\} = \mathcal{S}^*_\infty[\overline{VC}; C \underset{X}{\times} \overline{V_G P}]$$

possessing the local generating basis $(a^r_\mu, \overline{a}^\mu_r, \overline{c}_r)$ where $\overline{c}_r$ are even Noether antifields.

Lemma 5.8.3. *The Noether identities (5.8.29) are irreducible.*

Proof. Providing the differential bigraded algebra $\overline{\mathcal{P}}^*_\infty\{0\}$ with the nilpotent odd graded derivation

$$\delta_0 = \overline{\delta} + \frac{\overleftarrow{\partial}}{\partial \overline{c}_r}\Delta_r,$$

let us consider the chain complex (4.1.16). Let us assume that Φ (4.1.17) is a two-cycle of this complex, i.e., the relation (4.1.18) holds. It is readily observed that Φ obeys this relation only if its first term G is $\overline{\delta}$-exact, i.e., the first-stage Noether identities (4.1.18) are trivial. $\square$

It follows from Lemmas 5.8.1 – 5.8.3 that Yang–Mills gauge theory is an irreducible degenerate Lagrangian theory characterized by the complete Noether identities (5.8.29).

Following inverse second Noether Theorem 4.2.1, let us consider the differential bigraded algebra

$$P^*_\infty\{0\} = \mathcal{S}^*_\infty[\overline{VC} \underset{C}{\oplus} V_G P; C \underset{X}{\times} \overline{V_G P}] \qquad (5.8.32)$$

with the local generating basis $(a^r_\mu, \overline{a}^\mu_r, c^r, \overline{c}_r)$ where c_r are odd ghosts. The gauge operator $\mathbf{u}$ (4.2.8) associated to the Noether identities (5.8.29) reads

$$\mathbf{u} = u = (c^r_\mu + c^r_{pq} a^p_\mu c^q)\partial^\mu_r. \qquad (5.8.33)$$

It is an odd gauge symmetry of the Yang–Mills Lagrangian L_{YM} which can be obtained from the gauge symmetry u_ξ (5.8.4) by replacement of gauge parameters ξ^r with odd ghosts c^r (see Remark 4.2.1).

Since the gauge symmetries u_ξ form a Lie algebra (5.6.12), the gauge operator $\mathbf{u}$ (5.8.33) admits the nilpotent BRST extension

$$\mathbf{b} = (c^r_\mu + c^r_{pq} a^p_\mu c^q)\frac{\partial}{\partial a^r_\mu} - \frac{1}{2} c^r_{pq} c^p c^q \frac{\partial}{\partial c^r},$$

which is the well-known BRST operator in Yang–Mills gauge theory [63]. Then, by virtue of Theorem 4.4.2, the Yang–Mills Lagrangian L_{YM} is extended to a proper solution of the master equation

$$L_E = L_{\text{YM}} + (c^r_\mu + c^r_{pq} a^p_\mu c^q)\overline{a}^\mu_r \omega - \frac{1}{2} c^r_{pq} c^p c^q \overline{c}_r \omega.$$

5.8.4 Matter field Lagrangian

Let P be a principal bundle, and let Y (5.7.1) be a P-associated bundle coordinated by (x^μ, y^i). Treating sections of Y as matter fields, let us consider a Lagrangian system of gauge and matter fields. Its total Lagrangian L_{tot} is defined on the configuration space

$$J^1(C \underset{X}{\times} V) = J^1 C \underset{X}{\times} J^1 Y. \qquad (5.8.34)$$

5.8. Yang–Mills gauge theory

This Lagrangian is the sum

$$L_{\text{tot}} = L_{\text{YM}} + L_{\text{m}} \qquad (5.8.35)$$

where L_{YM} is the pull-back of the Yang–Mills Lagrangian (5.8.15) onto the total configuration space (5.8.34) and L_{m} is a matter field Lagrangian describing matter fields in the presence of a gauge field.

Let us assume that the total Lagrangian (5.8.34) is gauge invariant and, consequently, that a matter field Lagrangian L_{m} is separately gauge invariant. We also assume that L_{m} depends on gauge fields, but not their derivatives, i.e., L_{m} is defined on the pull-back bundle

$$C \underset{X}{\times} J^1 Y, \qquad (5.8.36)$$

coordinated by $(x^\lambda, a_\lambda^r, y^i, y_\lambda^i)$. Infinitesimal gauge transformations of this bundle are given by vector fields

$$\vartheta_\xi = u_\xi + J^1 v_\xi = (\partial_\mu \xi^r + c_{pq}^r a_\mu^p \xi^q)\partial_r^\mu + \xi^p I_p^i \partial_i + d_\lambda(\xi^p I_p^i)\partial_i^\lambda. \qquad (5.8.37)$$

The gauge invariance condition reads

$$\mathbf{L}_{\vartheta_x i} L_{\text{m}} = [(\partial_\mu \xi^r + c_{pq}^r a_\mu^p \xi^q)\partial_r^\mu L_{\text{m}} + \xi^p I_p^i \mathcal{E}_i + d_\mu(\xi^p I_p^i \partial_i^\mu L_{\text{m}})]\omega = 0.$$

This condition leads to the equality (2.4.45):

$$\partial_r^\mu L_{\text{m}} + I_r^i \partial_i^\mu L_{\text{m}} = 0,$$

which results in the following.

Assertion 5.8.1. It follows that a gauge invariant matter field Lagrangian factorizes as

$$L_{\text{m}} : C \underset{X}{\times} J^1 Y \xrightarrow{D} C \underset{X}{\times} (T^* X \underset{X}{\otimes} VY) \to \overset{n}{\wedge} T^* X$$

through the covariant differential

$$D = (y_\mu^i - a_\mu^r I_r^i)dx^\mu \otimes \partial_i \qquad (5.8.38)$$

relative to some principal connection on $Y \to X$.

Remark 5.8.2. If a matter field Lagrangian L_{m} factorizes through the covariant differential D (5.8.38), it can be regarded as a function of formal variables y^i and $k_\lambda^i = D_\lambda^i$. The infinitesimal gauge transformations (5.8.37) of these variables read

$$\vartheta_\xi = \xi^p I_p^i \partial_i + \xi^p \partial_j(I_p^i) k_\lambda^j \frac{\partial}{\partial k_\lambda^i}.$$

It is independent of derivatives $\partial_\mu \xi^p$ of gauge parameters ξ^p. Therefore, the gauge invariance condition (2.4.45) is trivially satisfied.

As a consequence, the Euler–Lagrange equation of the total Lagrangian (5.8.35) of gauge and matter fields takes the form

$$\delta_i \mathcal{L}_{\mathrm{m}} - d_\lambda \partial_i^\lambda \mathcal{L}_{\mathrm{m}} = 0, \tag{5.8.39}$$

$$(\delta_r^n d_\lambda + c_{rp}^n a_\lambda^p)(a_{nq}^G g^{\mu\alpha} g^{\lambda\beta} \mathcal{F}_{\alpha\beta}^q \sqrt{|g|}) - \partial_i^\mu \mathcal{L}_{\mathrm{m}} I_r^i = 0. \tag{5.8.40}$$

The equation (5.8.39) is an equation of motion of matter fields in the presence of a gauge fields. A glance at the equation (5.8.40) shows that matter sources of a gauge field are components

$$J_r^\mu = -\partial_i^\mu \mathcal{L}_{\mathrm{m}} I_r^i$$

of the Noether current (2.4.34):

$$\mathcal{J}^\mu = J_r^\mu \xi^r = -\partial_i^\mu \mathcal{L}_{\mathrm{m}} v_\xi^i,$$

of matter fields.

5.9 Yang–Mills supergauge theory

Yang–Mills gauge theory on $X = \mathbb{R}^n$ is straightforwardly extended to Yang–Mills supergauge theory by replacement of a Lie algebra $\mathfrak{g}_r$ with a Lie superlagebra. Yang–Mills supergauge theory exemplifies Lagrangian theory of even and odd fields.

Let

$$\mathfrak{g} = \mathfrak{g}_0 \oplus \mathfrak{g}_1$$

be a finite-dimensional real Lie superalgebra with the basis $\{\varepsilon_r\}$, $r = 1, \ldots, m$, and real structure constants c_{ij}^r. They obey the relations

$$c_{ij}^r = -(-1)^{[i][j]} c_{ji}^r, \qquad [r] = [i] + [j],$$
$$(-1)^{[i][b]} c_{ij}^r c_{ab}^j + (-1)^{[a][i]} c_{aj}^r c_{bi}^j + (-1)^{[b][a]} c_{bj}^r c_{ia}^j = 0,$$

where $[r]$ denotes the Grassmann parity of ε_r. Given the universal enveloping algebra $\overline{\mathfrak{g}}$ of $\mathfrak{g}$, we assume that there exists an even quadratic Casimir element $h^{ij}\varepsilon_i\varepsilon_j$ of $\overline{\mathfrak{g}}$ such that the matrix h^{ij} is non-degenerate. Yang–Mills supergauge theory on $X = \mathbb{R}^n$ associated to this Lie superalgebra is described by the differential bigraded algebra (4.1.6):

$$\mathcal{P}_\infty^*[F; Y] = \mathcal{S}_\infty^*[F; Y],$$

where

$$F = \mathfrak{g} \underset{X}{\otimes} T^*X, \qquad Y = \mathfrak{g}_0 \underset{X}{\otimes} T^*X.$$

5.9. Yang–Mills supergauge theory

It possesses the generating basis (a_λ^r) of Grassmann parity $[a_\lambda^r] = [r]$. First jets of its elements admit the splitting

$$a_{\lambda\mu}^r = \frac{1}{2}(\mathcal{F}_{\lambda\mu}^r + \mathcal{S}_{\lambda\mu}^r) = \qquad (5.9.1)$$
$$\frac{1}{2}(a_{\lambda\mu}^r - a_{\mu\lambda}^r + c_{ij}^r a_\lambda^i a_\mu^j) + \frac{1}{2}(a_{\lambda\mu}^r + a_{\mu\lambda}^r - c_{ij}^r a_\lambda^i a_\mu^j)$$

(cf. (5.5.12)). Given a constant metric g on $\mathbb{R}^n$, the graded Yang–Mills Lagrangian reads

$$L_{\text{YM}} = \frac{1}{4} h_{ij} g^{\lambda\mu} g^{\beta\nu} \mathcal{F}_{\lambda\beta}^i \mathcal{F}_{\mu\nu}^j \omega.$$

Its variational derivatives $\mathcal{E}_r^\lambda$ obey the irreducible complete Noether identities

$$c_{ji}^r a_\lambda^i \mathcal{E}_r^\lambda + d_\lambda \mathcal{E}_j^\lambda = 0.$$

Therefore, let us enlarge the differential bigraded algebra $\mathcal{P}_\infty^*[F;Y]$ to the differential bigraded algebra

$$P_\infty^*\{0\} = \mathcal{S}_\infty^*[F \underset{X}{\oplus} \overline{F} \underset{X}{\oplus} E_0 \underset{X}{\oplus} \overline{E}_0; Y],$$
$$\overline{F} = \mathfrak{g}^* \underset{X}{\otimes} TX \underset{X}{\otimes} \overset{n}{\wedge} T^*X,$$
$$E_0 = X \times \mathfrak{g}, \qquad \overline{E}_0 = \mathfrak{g}^* \underset{X}{\times} \overset{n}{\wedge} T^*X.$$

Its generating basis $(a_\lambda^r, \overline{a}_r^\lambda, c^r, \overline{c}_r)$ contains gauge fields a_λ^r, their antifields $\overline{a}_r^\lambda$ of Grassmann parity

$$[\overline{a}_r^\lambda] = ([r] + 1) \bmod 2,$$

the ghosts c^r of Grassmann parity

$$[c^r] = ([r] + 1) \bmod 2,$$

and the Noether antifields $\overline{c}_r$ of Grassmann parity $[\overline{c}_r] = [r]$. Then the gauge operator (4.2.8) reads

$$\mathbf{u} = (c_\lambda^r - c_{ji}^r c^j a_\lambda^i) \frac{\partial}{\partial a_\lambda^r}.$$

It admits the nilpotent BRST extension

$$\mathbf{b} = (c_\lambda^r - c_{ji}^r c^j a_\lambda^i) \frac{\partial}{\partial a_\lambda^r} - \frac{1}{2}(-1)^{[i]} c_{ij}^r c^i c^j \frac{\partial}{\partial c^r}.$$

The corresponding proper solution (4.4.10) of the master equation takes the form

$$L_E = L_{\text{YM}} + (c_\lambda^r - c_{ji}^r c^j a_\lambda^i) \overline{a}_r^\lambda \omega - \frac{1}{2}(-1)^{[i]} c_{ij}^r c^i c^j \overline{c}_r \omega.$$

5.10 Reduced structure. Higgs fields

Gauge theory deals with the three types of classical fields. These are gauge potentials, matter fields and Higgs fields. Higgs fields are responsible for spontaneous symmetry breaking. Spontaneous symmetry breaking is a quantum phenomenon, but it is characterized by a classical background Higgs field. Therefore, it also is described in classical field theory. In classical gauge theory on a principal bundle $P \to X$, *spontaneous symmetry breaking* is characterized by the reduction of a structure Lie group G of this principal bundle to a closed subgroup H of exact symmetries [83; 89; 120; 131; 155].

5.10.1 Reduction of a structure group

Let H and G be Lie groups and $\phi : H \to G$ a Lie group homomorphism. If $P_H \to X$ is a principal H-bundle, there always exists a principal G-bundle $P_G \to X$ together with the principal bundle morphism

$$\Phi : P_H \xrightarrow[X]{} P_G \qquad (5.10.1)$$

over X (see Remark 5.3.3). It is the P_H-associated bundle

$$P_G = (P_H \times G)/H$$

with the typical fibre G on which H acts on the left by the rule $h(g) = \phi(h)g$, while G acts on P_G as

$$G \ni g' : (p,g)/H \to (p,gg')/H.$$

Conversely, if $P_G \to X$ is a principal G-bundle, a problem is to find a principal H-bundle $P_H \to X$ together with a principal bundle morphism (5.10.1). If $H \to G$ is a closed subgroup, we have the *structure group reduction*. If $H \to G$ is a group epimorphism (a group extension (10.4.10)), one says that P_G *lifts* to P_H.

Here, we restrict our consideration to the reduction problem (see Lemma 7.2.1 for an example of bundle lift). In this case, the bundle monomorphism (5.10.1) is called a *reduced H-structure* [64; 93].

Remark 5.10.1. Note that, in [64; 93], the reduced structures on the principle bundle LX of linear frames in the tangent bundle TX of X only are considered, and a class of isomorphisms of such reduced structures is restricted to holonomic automorphisms of LX, i.e., the canonical lifts onto LX of diffeomorphisms of the base X (see Section 6.1).

5.10. Reduced structure. Higgs fields

Let P (5.3.1) be a principal G-bundle, and let H, $\dim H > 0$, be a closed (and, consequently, Lie) subgroup of G. Then we have the composite bundle

$$P \to P/H \to X, \qquad (5.10.2)$$

where

$$P_\Sigma = P \xrightarrow{\pi_{P\Sigma}} P/H \qquad (5.10.3)$$

is a principal bundle with a structure group H and

$$\Sigma = P/H \xrightarrow{\pi_{\Sigma X}} X \qquad (5.10.4)$$

is a P-associated bundle with the typical fibre G/H on which the structure group G acts on the left (see Example 5.3.1).

One says that a structure Lie group G of a principal bundle P is reduced to its closed subgroup H if the following equivalent conditions hold.

• A principal bundle P admits a bundle atlas Ψ_P (5.3.3) with H-valued transition functions $\varrho_{\alpha\beta}$.
• There exists a principal *reduced subbundle* P_H of P with a structure group H.

Remark 5.10.2. It is easily justified that these conditions are equivalent. If $P_H \subset P$ is a reduced subbundle, its atlas (5.3.5) given by local sections z_α of $P_H \to X$ is a desired atlas of P. Conversely, let (5.3.5):

$$\Psi_P = \{(U_\alpha, z_\alpha), \varrho_{\alpha\beta}\},$$

be an atlas of P with H-valued transition functions $\varrho_{\alpha\beta}$. For any $x \in U_\alpha \subset X$, let us define a submanifold $z_\alpha(x)H \subset P_x$. These submanifolds form a desired H-subbundle of P because

$$z_\alpha(x)H = z_\beta(x)H\varrho_{\beta\alpha}(x)$$

on the overlaps $U_\alpha \cap U_\beta$.

Theorem 5.10.1. *There is one-to-one correspondence*

$$P^h = \pi_{P\Sigma}^{-1}(h(X)) \qquad (5.10.5)$$

between the reduced principal H-subbundles $i_h : P^h \to P$ of P and the global sections h of the quotient bundle $P/H \to X$ (5.10.4) [92].

Corollary 5.10.1. *A glance at the formula (5.10.5) shows that the reduced principal H-bundle P^h is the restriction h^*P_Σ (1.4.4) of the principal H-bundle P_Σ (5.10.3) to $h(X) \subset \Sigma$.*

In classical field theory, global sections of a quotient bundle $P/H \to X$ are interpreted as *Higgs fields* [131; 143].

In general, there is topological obstruction to reduction of a structure group of a principal bundle to its subgroup.

Assertion 5.10.1. In accordance with Theorem 1.1.4, the structure group G of a principal bundle P is always reducible to its closed subgroup H, if the quotient G/H is diffeomorphic to a Euclidean space $\mathbb{R}^m$.

In particular, this is the case of a maximal compact subgroup H of a Lie group G (see Example 5.3.1). Then the following is a corollary of Assertion 5.10.1 [147].

Assertion 5.10.2. A structure group G of a principal bundle is always reducible to its maximal compact subgroup H.

As a consequence, there is a bijection between the cohomology sets

$$H^1(X; G_X^\infty) = H^1(X; H_X^\infty) \qquad (5.10.6)$$

if H is a maximal compact subgroup of G.

Example 5.10.1. For instance, this is the case of $G = GL(n, \mathbb{C})$, $H = U(n)$ and $G = GL(n, \mathbb{R})$, $H = O(n)$.

Example 5.10.2. Any affine bundle admits an atlas with linear transition functions. In accordance with Theorem 5.10.1, its structure group $GA(m, \mathbb{R})$ (see Example 5.2.1) is always reducible to the linear subgroup $GL(m, \mathbb{R})$ because

$$GA(m, \mathbb{R})/GL(m, \mathbb{R}) = \mathbb{R}^m.$$

5.10.2 Reduced subbundles

Different principal H-subbundles P^h and $P^{h'}$ of a principal G-bundle P are not isomorphic to each other in general.

Theorem 5.10.2. Let a structure Lie group G of a principal bundle be reducible to its closed subgroup H.

(i) Every vertical principal automorphism Φ of P sends a reduced principal H-subbundle P^h of P onto an isomorphic principal H-subbundle $P^{h'}$.

(ii) Conversely, let two reduced subbundles P^h and $P^{h'}$ of a principal bundle $P \to X$ be isomorphic to each other, and let $\Phi : P^h \to P^{h'}$ be their isomorphism over X. Then Φ is extended to a vertical principal automorphism of P.

5.10. Reduced structure. Higgs fields

Proof. (i) Let

$$\Psi^h = \{(U_\alpha, z_\alpha^h), \varrho_{\alpha\beta}^h\} \tag{5.10.7}$$

be an atlas of a reduced principal subbundle P^h, where z_α^h are local sections of $P^h \to X$ and $\varrho_{\alpha\beta}^h$ are the transition functions. Given a vertical automorphism Φ of P, let us provide the subbundle $P^{h'} = \Phi(P^h)$ with the atlas

$$\Psi^{h'} = \{(U_\alpha, z_\alpha^{h'}), \varrho_{\alpha\beta}^{h'}\} \tag{5.10.8}$$

given by the local sections $z_\alpha^{h'} = \Phi \circ z_\alpha^h$ of $P^{h'} \to X$. Then it is readily observed that

$$\varrho_{\alpha\beta}^{h'}(x) = \varrho_{\alpha\beta}^h(x), \qquad x \in U_\alpha \cap U_\beta. \tag{5.10.9}$$

(ii) Any isomorphism $(\Phi, \operatorname{Id} X)$ of reduced principal subbundles P^h and $P^{h'}$ of P defines an H-equivariant G-valued function f on P^h given by the relation

$$pf(p) = \Phi(p), \qquad p \in P^h.$$

Its prolongation to a G-equivariant function on P is defined as

$$f(pg) = g^{-1} f(p) g, \qquad p \in P^h, \qquad g \in G.$$

In accordance with the relation (5.6.2), this function provides a vertical principal automorphism of P whose restriction to P^h coincides with Φ. □

Theorem 5.10.3. *If the quotient G/H is homeomorphic to a Euclidean space $\mathbb{R}^m$, all principal H-subbundles of a principal G-bundle P are isomorphic to each other [147].*

Remark 5.10.3. A principal G-bundle P provided with the atlas Ψ^h (5.10.7) can be regarded as a P^h-associated bundle with a structure group H acting on its typical fibre G on the left. Endowed with the atlas $\Psi^{h'}$ (5.10.8), it is a $P^{h'}$-associated H-bundle. The H-bundles (P, Ψ^h) and $(P, \Psi^{h'})$ fail to be equivalent because their atlases Ψ^h and $\Psi^{h'}$ are not equivalent. Indeed, the union of these atlases is the atlas

$$\Psi = \{(U_\alpha, z_\alpha^h, z_\alpha^{h'}), \varrho_{\alpha\beta}^h, \varrho_{\alpha\beta}^{h'}, \varrho_{\alpha\alpha} = f(z_\alpha)\}$$

possessing transition functions

$$z_\alpha^{h'} = z_\alpha^h \varrho_{\alpha\alpha}, \qquad \varrho_{\alpha\alpha}(x) = f(z_\alpha(x)), \tag{5.10.10}$$

between the bundle charts (U_α, z_α^h) and $(U_\alpha, z_\alpha^{h'})$ of Ψ^h and $\Psi^{h'}$, respectively. However, the transition functions $\varrho_{\alpha\alpha}$ are not H-valued. At the same time, a glance at the equalities (5.10.9) shows that transition functions of both the atlases form the same cocycle. Consequently, the H-bundles (P,Ψ^h) and $(P,\Psi^{h'})$ are associated. Due to the isomorphism $\Phi:P^h\to P^{h'}$, one can write

$$P=(P^h\times G)/H=(P^{h'}\times G)/H,$$
$$(p\times g)/H=(\Phi(p)\times f^{-1}(p)g)/H.$$

For any $\rho\in H$, we have

$$(p\rho,g)/H=(\Phi(p)\rho,f^{-1}(p)g)/H=(\Phi(p),\rho f^{-1}(p)g)/H=$$
$$(\Phi(p),f^{-1}(p)\rho'g)/H,$$

where

$$\rho'=f(p)\rho f^{-1}(p). \qquad (5.10.11)$$

It follows that $(P,\Psi^{h'})$ can be regarded as a P^h-associated bundle with the same typical fibre G as that of (P,Ψ^h), but the action $g\to\rho'g$ (5.10.11) of a structure group H on the typical fibre of $(P,\Psi^{h'})$ is not equivalent to its action $g\to\rho g$ on the typical fibre of (P,Ψ^h) (see Remark 5.1.1).

5.10.3 Reducible principal connections

There are the following properties of principal connections compatible with a reduced structure [92].

Theorem 5.10.4. *Since principal connections are equivariant, every principal connection A_h on a reduced principal H-subbundle P^h of a principal G-bundle P gives rise to a principal connection on P.*

Theorem 5.10.5. *A principal connection A on a principal G-bundle P is reducible to a principal connection on a reduced principal H-subbundle P^h of P if and only if the corresponding global section h of the P-associated fibre bundle $P/H\to X$ is an integral section of the associated principal connection A on $P/H\to X$.*

Theorem 5.10.6. *Let the Lie algebra $\mathfrak{g}_l$ of G be the direct sum*

$$\mathfrak{g}_l=\mathfrak{h}_l\oplus\mathfrak{m} \qquad (5.10.12)$$

of the Lie algebra $\mathfrak{h}_l$ of H and a subspace $\mathfrak{m}$ such that $\mathrm{Ad}_g(\mathfrak{m})\subset\mathfrak{m}$, $g\in H$ (e.g., H is a Cartan subgroup of G). Let $\overline{A}$ be a $\mathfrak{g}_l$-valued connection form (5.4.12) on P. Then, the pull-back of the $\mathfrak{h}_l$-valued component of $\overline{A}$ onto a reduced principal H-subbundle P^h is a $\mathfrak{h}_l$-valued connection form of a principal connection $\overline{A}_h$ on P^h.

5.10. Reduced structure. Higgs fields

The following is a corollary of Theorem 5.4.1.

Theorem 5.10.7. *Given the composite bundle (5.10.2), let A_Σ be a principal connection on the principal H-bundle $P \to \Sigma$ (5.10.3). Then, for any reduced principal H-bundle $i_h : P^h \to P$, the pull-back connection $i_h^* A_\Sigma$ (1.4.18) is a principal connection on P^h.*

5.10.4 Associated bundles. Matter and Higgs fields

In accordance with Theorem 5.10.1, there is a bijection between the set of reduced principal H-subbundles P^h of P and the set of Higgs fields h. Given such a subbundle P^h, let

$$Y^h = (P^h \times V)/H \qquad (5.10.13)$$

be the associated vector bundle with a typical fibre V which admits a representation of the group H of exact symmetries. Its sections s_h describe matter fields in the presence of the Higgs fields h and some principal connection A_h on P^h. In general, the fibre bundle Y^h (5.10.13) fails to be associated with another principal H-subbundles $P^{h'}$ of P. It follows that, in this case, a V-valued matter field can be represented only by a pair with a certain Higgs field. The goal is to describe the totality of these pairs (s_h, h) for all Higgs fields $h \in \Sigma(X)$.

Remark 5.10.4. If reduced principal H-subbundles P^h and $P^{h'}$ of a principal G-bundle are isomorphic in accordance with Theorem 5.10.2, then the P^h-associated bundle Y^h (5.10.13) is associated as

$$Y^h = (\Phi(p) \times V)/H \qquad (5.10.14)$$

to $P^{h'}$. If a typical fibre V admits an action of the whole group G, the P^h-associated bundle Y^h (5.10.13) also is P-associated as

$$Y^h = (P^h \times V)/H = (P \times V)/G.$$

In order to describe matter fields in the presence of different Higgs fields, let us consider the composite bundle (5.10.2) and the composite bundle

$$Y \xrightarrow{\pi_{Y\Sigma}} \Sigma \xrightarrow{\pi_{\Sigma X}} X \qquad (5.10.15)$$

where $Y \to \Sigma$ is a P_Σ-associated bundle

$$Y = (P \times V)/H \qquad (5.10.16)$$

with a structure group H. Given a global section h of the fibre bundle $\Sigma \to X$ (5.10.4) and the corresponding reduced principal H subbundle $P^h = h^*P$, the P^h-associated fibre bundle (5.10.13) is the restriction

$$Y^h = h^*Y = (h^*P \times V)/H \qquad (5.10.17)$$

of the fibre bundle $Y \to \Sigma$ to $h(X) \subset \Sigma$. By virtue of Theorem 1.4.2, every global section s_h of the fibre bundle Y^h (5.10.17) is a global section of the composite bundle (5.10.15) projected onto the section $h = \pi_{Y\Sigma} \circ s$ of the fibre bundle $\Sigma \to X$. Conversely, every global section s of the composite bundle $Y \to X$ (5.10.15) projected onto a section $h = \pi_{Y\Sigma} \circ s$ of the fibre bundle $\Sigma \to X$ takes its values into the subbundle Y^hY (5.10.17). Hence, there is one-to-one correspondence between the sections of the fibre bundle Y^h (5.10.13) and the sections of the composite bundle (5.10.15) which cover h.

Thus, it is the composite bundle $Y \to X$ (5.10.15) whose sections describe the above mentioned totality of pairs (s_h, h) of matter fields and Higgs fields in classical gauge theory with spontaneous symmetry breaking [112; 143].

Lemma 5.10.1. *The composite bundle $Y \to X$ (5.10.15) is a P-associated bundle with a structure group G. Its typical fibre is the H-bundle*

$$W = (G \times V)/H \qquad (5.10.18)$$

associated with the principal H-bundle $G \to G/H$ (5.3.6).

Proof. One can consider a principal bundle $P \to X$ as the P-associated bundle

$$P = (P \times G)/G,$$
$$(pg', g) = (p, g'g), \qquad p \in P, \qquad g, g' \in G,$$

whose typical fibre is the group space of G which a group G acts on by left multiplications. Then the quotient (5.10.16) can be represented as

$$Y = (P \times (G \times V)/H)/G,$$
$$(pg', (g\rho, v)) = (pg', (g, \rho v)) = (p, g'(g, \rho v)) = (p, (g'g, \rho v)).$$

It follows that Y (5.10.16) is a P-associated bundle with the typical fibre W (5.10.18) which the structure group G acts on by the law

$$g' : (G \times V)/H \to (g'G \times V)/H. \qquad (5.10.19)$$

This is a familiar *induced representation* of G [107]. $\square$

5.10. Reduced structure. Higgs fields

Given an atlas $\{(U_a, z_a)\}$ (5.3.5) of the principal H-bundle $G \to G/H$, the induced representation (5.10.19) reads

$$g' : (\sigma, v) = (z_a(\sigma), v)/H \to (\sigma', v') = (g' z_a(\sigma), v)/H = \qquad (5.10.20)$$
$$(z_b(\pi_{GH}(g' z_a(\sigma)))\rho', v)/H = (z_b(\pi_{GH}(g' z_a(\sigma))), \rho' v)/H,$$
$$\rho' = z_b^{-1}(\pi_{GH}(g' z_a(\sigma))) g' z_a(\sigma) \in H, \quad \sigma \in U_a, \quad \pi_{GH}(g' z_a(\sigma)) \in U_b.$$

An example of induced representations is the well-known non-linear realization [30; 86] (see Section 5.11).

Lemma 5.10.2. *Given a global section h of the quotient bundle $\Sigma \to X$ (5.10.4), any atlas of a P_Σ-associated bundle $Y \to \Sigma$ defines an atlas of a P-associated bundle $Y \to X$ with H-valued transition functions. The converse need not be true.*

Proof. Any atlas $\Psi_{Y\Sigma}$ of a P_Σ-associated bundle $Y \to \Sigma$ is defined by an atlas

$$\Psi_{P\Sigma} = \{(U_{\Sigma\iota}, z_\iota), \varrho_{\iota\kappa}\} \qquad (5.10.21)$$

of the principal H-bundle P_Σ (5.10.3). Given a section h of $\Sigma \to X$, we have an atlas

$$\Psi^h = \{(\pi_{P\Sigma}(U_{\Sigma\iota}), z_\iota \circ h), \varrho_{\iota\kappa} \circ h\} \qquad (5.10.22)$$

of the reduced principal H-bundle P^h which also is an atlas of P with H-valued transition functions (see Remark 5.10.2). □

Given an atlas Ψ_P of P, the quotient bundle $\Sigma \to X$ (5.10.4) is endowed with an associated atlas (5.7.5). With this atlas and an atlas $\Psi_{Y\Sigma}$ of $Y \to \Sigma$, the composite bundle $Y \to X$ (5.10.15) is endowed with adapted bundle coordinates $(x^\lambda, \sigma^m, y^i)$ where (σ^m) are fibre coordinates on $\Sigma \to X$ and (y^i) are those on $Y \to \Sigma$.

Lemma 5.10.3. *Any principal automorphism of a principal G-bundle $P \to X$ also is a principal automorphism of a principal H-bundle $P \to \Sigma$ and, consequently, it yields an automorphism of the P_Σ-associated bundle Y (5.10.15).*

Proof. A principal automorphism of $P \to X$ is G-equivariant and, consequently, H-equivariant, i.e., it is a principal automorphism of $P \to \Sigma$. □

The converse is not true. For instance, a vertical principal automorphism of $P \to \Sigma$ is never a principal automorphism of $P \to X$.

By virtue of Lemma 5.10.3, every G-principal vector field ξ (5.3.16) on $P \to X$ also is an H-principal vector field
$$\xi_H = \xi^\lambda \partial_\lambda + \xi^p(x^\mu) J_p^m \partial_m + \vartheta_\xi^a(x^\mu, \sigma^k) e_a \qquad (5.10.23)$$
on $P \to \Sigma$ where $\{J_p\}$ is a representation of the Lie algebra $\mathfrak{g}_r$ of G in G/H and $\{\varepsilon_a = \psi_\iota(e_a)\}$ is a basis for the Lie algebra $\mathfrak{h}_r$ of H. Given different principal vector fields ξ and η (5.3.16), the Lie algebra bracket (5.3.17) leads to the relation marrtya
$$\xi^p J_p^m \partial_m \vartheta_\eta^a - \eta^p J_p^m \partial_m \vartheta_\xi^a + c_{bd}^a \vartheta_\xi^b \vartheta_\eta^d = \vartheta_{[\xi,\eta]}^a. \qquad (5.10.24)$$
Both the G-principal vector field ξ (5.3.16) and the H-principal vector field ξ_H (5.10.23) yield the infinitesimal gauge transformation v_ξ (5.7.8) of the composite bundle Y seen as a P- and P_Σ-associated bundle. It reads
$$v_\xi = \xi^\lambda \partial_\lambda + \xi^p(x^\mu) J_p^m \partial_m + \vartheta_\xi^a(x^\mu, \sigma^k) I_a^i \partial_i, \qquad (5.10.25)$$
where $\{I_a\}$ is a representation of the Lie algebra $\mathfrak{h}_r$ in V.

Because the principal vector field ξ_H (5.10.23) is never vertical with respect to the fibration $P \to \Sigma$, one also considers vertical principal vector fields
$$\zeta = \zeta^a(x^\mu, \sigma^k) e_a$$
seen as sections of the Lie algebra bundle $V_H P \to \Sigma$. These vector fields yield vertical infinitesimal gauge transformations
$$v_\zeta = \zeta^a(x^\mu, \sigma^k) I_a^i \partial_i \qquad (5.10.26)$$
of the P_Σ-associated bundle $Y \to \Sigma$.

Though Y (5.10.16) is a P-associated bundle, a principal connection on P fails to be reducible to a connection on an arbitrary principal H-subbundle P^h of P (see Theorem 5.10.5). Therefore, it can not define a connection on the corresponding subbundle Y^h (5.10.13) of $Y \to X$ in general. At the same time, all H-subbundles Y^h of $Y \to X$ can be provided with associated principal connections as follows.

Lemma 5.10.4. *Given a principal connection*
$$A_\Sigma = dx^\lambda \otimes (\partial_\lambda + A_\lambda^a e_a) + d\sigma^m \otimes (\partial_m + A_m^a e_a) \qquad (5.10.27)$$
on the principal H-bundle $P \to \Sigma$, let
$$A_{Y\Sigma} = dx^\lambda \otimes (\partial_\lambda + A_\lambda^a(x^\mu, \sigma^k) I_a^i \partial_i) + \qquad (5.10.28)$$
$$d\sigma^m \otimes (\partial_m + A_m^a(x^\mu, \sigma^k) I_a^i \partial_i)$$
be an associated principal connection on $Y \to \Sigma$. Then, for any H-subbundle $Y^h \to X$ of the composite bundle $Y \to X$, the pull-back connection (1.4.18):
$$A_h = h^* A_{Y\Sigma} = dx^\lambda \otimes [\partial_\lambda + (A_m^a(x^\mu, h^k) \partial_\lambda h^m + A_\lambda^a(x^\mu, h^k)) I_a^i \partial_i], \quad (5.10.29)$$
is a connection on Y^h associated with the pull-back principal connection $h^* A_\Sigma$ on the reduced principal H-subbundle P^h in Theorem 5.10.7.

5.10.5 *Matter field Lagrangian*

Lemmas 5.10.1 – 5.10.4 lead to the following feature of formulating Lagrangian gauge theory with spontaneous symmetry breaking.

Let $P \to X$ be a principal bundle whose structure group G is reducible to a closed subgroup H. Let Y be the P_Σ-associated bundle (5.10.16). The total configuration space of gauge theory of principal connections on P in the presence of matter and Higgs fields is

$$J^1 C \underset{X}{\times} J^1 Y \tag{5.10.30}$$

where C is the bundle of principal connections on P (5.4.3) and J^1Y is the first order jet manifold of $Y \to X$. A total Lagrangian on the configuration space (5.10.30) is a sum

$$L_{\text{tot}} = L_{\text{YM}} + L_{\text{m}} + L_\sigma \tag{5.10.31}$$

of the Yang–Mills Lagrangian L_{YM} (5.8.15), a matter field Lagrangian L_{m} and a Higgs field Lagrangian L_σ. In the case of a background Higgs field h, one considers the pull-back Lagrangian $h^* L_{\text{tot}}$ on the configuration space

$$J^1 C \underset{X}{\times} J^1 Y^h. \tag{5.10.32}$$

The total Lagrangian L_{tot} is required to be invariant with respect to vertical principal automorphisms of $P \to X$ and $P \to \Sigma$. It follows that a matter field Lagrangian L_{m} and a Higgs field Lagrangian L_σ must be separately gauge invariant. This means that

$$\mathbf{L}_{J^1 v_\xi} L_{\text{m}} = 0, \qquad \mathbf{L}_{J^1 v_\zeta} L_{\text{m}} = 0, \tag{5.10.33}$$

$$v_\xi = \xi^p J_p^m \partial_m + \vartheta_\xi^a I_a^i \partial_i, \qquad v_\zeta = \zeta^a I_a^i \partial_i, \tag{5.10.34}$$

and

$$\mathbf{L}_{J^1 v_\xi} L_\sigma = 0, \qquad v_\xi = \xi^p J_p^m \partial_m.$$

Unless a Higgs field is specified, we restrict our consideration to a matter field Lagrangian L_{m}.

In order to satisfy the conditions (5.10.33), let us consider some principal connection A_Σ (5.10.27) on the principal H-bundle $P \to \Sigma$ and the associated connection $A_{Y\Sigma}$ (5.10.28) on $Y \to \Sigma$. Let a matter field Lagrangian L_{m} factorize as

$$L_{\text{m}} : J^1 Y \xrightarrow{\widetilde{D}} T^*X \underset{Y}{\otimes} V_\Sigma Y \to \overset{n}{\wedge} T^*X$$

through the vertical covariant differential $\widetilde{D}$ (1.4.17) which reads

$$\widetilde{D} = dx^\lambda \otimes (y_\lambda^i - (A_m^a \sigma_\lambda^m + A_\lambda^a) I_a^i) \partial_i. \tag{5.10.35}$$

In this case, L_{m} can be regarded as a function $L_{\mathrm{m}}(y^i, k^i_\lambda)$ of formal variables y^i and $k^i_\lambda = \widetilde{D}^i_\lambda$. Let

$$v = v^m(\xi)\partial_m + v^i(\xi)\partial_i$$

be a vertical infinitesimal gauge transformation of $Y \to X$ generalizing the infinitesimal gauge transformations (5.10.34). The corresponding infinitesimal gauge transformation of variables (y^i, k^i_λ) reads

$$v = v^i\partial_i + \partial_j v^i k^j_\lambda \frac{\partial}{\partial k^i_\lambda}.$$

It is independent of derivatives of gauge parameters ξ. Therefore, the gauge invariance condition (2.4.45) is trivially satisfied (see Remark 5.8.2). Thus, we come to the following.

Assertion 5.10.3. In gauge theory with spontaneous symmetry breaking on a principal bundle P whose structure group G is reducible to a closed subgroup H, a matter field Lagrangian is gauge invariant only if it factorizes through the vertical covariant differential of some H-principal connection on $P \to P/H$ (cf. Assertion 5.8.1).

For instance, let the Lie algebra $\mathfrak{g}_l$ of a group G admit the decomposition (5.10.12). In this case, every principal connection A (5.4.10) on a principal G-bundle $P \to X$ induces a principal connection $\overline{A}_h$ on any reduced principal subbundle P^h of P (see Theorem 5.10.12) and, consequently, on a P^h-associated bundle $Y^h = h^*Y$. By virtue of Theorem 5.10.4, it gives rise to a principal connection on P such that h is an integral section of the associated connection

$$\overline{A}_h = dx^\lambda \otimes (\partial_\lambda + \overline{A}^p_\lambda J^m_p \partial_m)$$

on the P-associated bundle $\Sigma \to X$.

Written with respect to a bundle atlas Ψ^h (5.10.7) of P with H-valued transition functions, the Higgs field h takes its values into the center of the homogeneous space G/H and the connection $\overline{A}_h$ reads

$$\overline{A}_h = dx^\lambda \otimes (\partial_\lambda + A^a_\lambda e_a). \qquad (5.10.36)$$

We have

$$A = \overline{A}_h + \Theta = dx^\lambda \otimes (\partial_\lambda + A^a_\lambda e_a) + \Theta^b_\lambda dx^\lambda \otimes e_b, \qquad (5.10.37)$$

where $\{\varepsilon_a = \psi^h(e_a)\}$ is a basis for the Lie algebra $\mathfrak{h}_r$ and $\{\varepsilon_b = \psi^h(e_b)\}$ is that for $\mathfrak{m}_r$. Written with respect to an arbitrary atlas of P, the decomposition (5.10.37) reads

$$A = \overline{A}_h + \Theta, \qquad \Theta = \Theta^p_\lambda dx^\lambda \otimes e_p,$$

5.10. Reduced structure. Higgs fields

and obeys the relation

$$\Theta_\lambda^p J_p^m = \nabla_\lambda^A h^m,$$

where D_λ are covariant derivatives (1.3.19) relative to the associated principal connection A on $\Sigma \to X$.

Based on this fact, let consider the covariant differential

$$D = D_\lambda^m dx^\lambda \otimes \partial_m = (\sigma_\lambda^m - A_\lambda^p J_p^m) dx^\lambda \otimes \partial_m$$

relative to the associated principal connection A on $\Sigma \to X$. It can be regarded as a $V\Sigma$-valued one-form on the jet manifold $J^1\Sigma$ of $\Sigma \to X$. Since the decomposition (5.10.37) holds for any section h of $\Sigma \to X$, there exists a $V_G P$-valued one-form

$$\Theta = \Theta_\lambda^p dx^\lambda \otimes e_p$$

on $J^1\Sigma$ which obeys the equation

$$\Theta_\lambda^p J_p^m = D_\lambda^m. \tag{5.10.38}$$

Then we obtain the $V_G P$-valued one-form

$$A_H = dx^\lambda \otimes (\partial_\lambda + (A_\lambda^p - \Theta_\lambda^p) e_p)$$

on $J^1\Sigma$ whose pull-back onto each $J^1 h(X) \subset J^1\Sigma$ is the connection $\overline{A}_h$ (5.10.36) written with respect to the atlas Ψ^h (5.10.22). The decomposition (5.10.37) holds and, consequently, the equation (5.10.38) possesses a solution for each principal connection A. Therefore, there exists a $V_G P$-valued one-form

$$A_H = dx^\lambda \otimes (\partial_\lambda + (a_\lambda^p - \Theta_\lambda^p) c_p) \tag{5.10.39}$$

on the product

$$J^1\Sigma \underset{X}{\times} J^1 C$$

such that, for any principal connection A and any Higgs field h, the restriction of A_H (5.10.39) to

$$J^1 h(X) \times A(X) \subset J^1\Sigma \underset{X}{\times} J^1 C$$

is the connection $\overline{A}_h$ (5.10.36) written with respect to the atlas Ψ^h (5.10.22).

Let us now assume that, whenever A is a principal connection on a principal G-bundle $P \to X$, there exists a principal connection A_Σ (5.10.27) on a principal H-bundle $P \to \Sigma$ such that the pull-back connection $A_h =$

$h^*A_{Y\Sigma}$ (5.10.29) on Y^h coincides with $\overline{A}_h$ (5.10.36) for any $h \in \Sigma(X)$. In this case, there exists $V_\Sigma Y$-valued one-form

$$\widetilde{D} = dx^\lambda \otimes (y^i_\lambda - (\mathcal{A}^a_m \sigma^m_\lambda + \mathcal{A}^a_\lambda)I^i_a)\partial_i \qquad (5.10.40)$$

on the configuration space (5.10.30) whose components are defined as follows. Given a point

$$(x^\lambda, a^r_\mu, a^r_{\lambda\mu}, \sigma^m, \sigma^m_\lambda, y^i, y^i_\lambda) \in J^1 C \underset{X}{\times} J^1 Y, \qquad (5.10.41)$$

let h be a section of $\Sigma \to X$ whose first jet $j^1_x h$ at $x \in X$ is $(\sigma^m, \sigma^m_\lambda)$, i.e.,

$$h^m(x) = \sigma^m, \qquad \partial_\lambda h^m(x) = \sigma^m_\lambda.$$

Let the bundle of principal connections C and the Lie algebra bundle $V_G P$ be provided with the atlases associated with the atlas Ψ^h (5.10.22). Then we write

$$A_h = \overline{A}_h, \qquad \mathcal{A}^a_m \sigma^m_\lambda + \mathcal{A}^a_\lambda = a^a_\lambda - \Theta^a_\lambda. \qquad (5.10.42)$$

These equations for functions $\mathcal{A}^a_m$ and $\mathcal{A}^a_\lambda$ at the point (5.10.41) have a solution because Θ^a_λ are affine functions in the jet coordinates σ^m_λ.

Given solutions of the equations (5.10.42) at all points of the configuration space (5.10.30), we require that a matter field Lagrangian factorizes as

$$L_{\mathrm{m}} : J^1 C \underset{X}{\times} J^1 Y \xrightarrow{\widetilde{D}} T^* X \underset{Y}{\otimes} V_\Sigma Y \to \overset{n}{\wedge} T^* X \qquad (5.10.43)$$

through the form $\widetilde{D}$ (5.10.40), called the *universal covariant differential*.

It should be emphasized that the universal covariant differential $\widetilde{D}$ (5.10.40) and, consequently, a Lagrangian L_{m} depends on the local transition functions

$$\psi_\iota \circ h = \varrho_{\iota\alpha} \psi_\alpha \qquad (5.10.44)$$

where ψ_α and ψ_ι are trivialization morphisms of the atlases Ψ_P (5.3.5) of $P \to X$ and $\Psi_{P\Sigma}$ (5.10.21) of $P \to \Sigma$. One can not exclude the functions (5.10.44) from a matter field Lagrangian L_{m} because the configuration space (5.10.30) is necessarily characterized by both these atlases. Therefore, let us treat them as additional variables. These are not dynamic variables because any local function (5.10.44) cab be brought into the identity map by an appropriate choice of atlases Ψ_P and $\Psi_{P\Sigma}$, but they provide the gauge invariance of the matter field Lagrangian (5.10.43) both with respect to vertical automorphisms of $P \to X$ and $P \to \Sigma$.

5.11. Appendix. Non-linear realization of Lie algebras

As a generalization of Remark 5.6.1, the transition functions (5.10.44) can be represented by local sections of the following composite bundle. Let us consider the pull-back of the principal G-bundle $P \to X$ onto Σ and the product

$$(\Sigma \underset{X}{\times} P) \times P_\Sigma.$$

This is a principal bundle over Σ with the structure group $G \times H$. Let us consider the associated bundle

$$P_\varrho = ((P \times P) \times G)/(G \times H), \qquad (5.10.45)$$
$$((pg \times p\rho) \times g') = ((p \times p) \times \rho g' g^{-1}).$$

Transition functions ϱ (5.10.44) are sections of the composite bundle

$$P_\varrho \to \Sigma \to X.$$

Hence, the total configuration space of the matter field Lagrangian L_m (5.10.43) is

$$J^1 C \underset{X}{\times} (J^1 Y \underset{\Sigma}{\times} P_\varrho). \qquad (5.10.46)$$

Vertical automorphisms of $P \to X$ and $P \to \Sigma$ yield automorphisms (5.7.7) of the bundle P_ϱ (5.10.45) and the gauge transformations of the configuration space (5.10.46).

In Section 7.3, we apply this scheme of spontaneous symmetry breaking to describing Dirac fermion fields in gauge gravitation theory.

5.11 Appendix. Non-linear realization of Lie algebras

The well-known *non-linear realization* of a Lie group G possessing a Cartan subgroup H exemplifies the induced representation (5.10.20) [30; 86]. In fact, it is a representation of the Lie algebra of G around its origin as follows.

The Lie algebra $\mathfrak{g}_r$ of a Lie group G containing a *Cartan subgroup* H is split into the sum

$$\mathfrak{g} = \mathfrak{h}_r + \mathfrak{f}$$

of the Lie algebra $\mathfrak{h}_r$ of H and its supplement $\mathfrak{f}$ obeying the commutation relations

$$[\mathfrak{f}, \mathfrak{f}] \subset \mathfrak{h}_r, \qquad [\mathfrak{f}, \mathfrak{h}_r] \subset \mathfrak{f}.$$

In this case, there exists an open neighbourhood U of the unit $\mathbf{1} \in G$ such that any element $g \in U$ is uniquely brought into the form

$$g = \exp(F)\exp(I), \qquad F \in \mathfrak{f}, \qquad I \in \mathfrak{h}_r.$$

Let U_G be an open neighbourhood of the unit of G such that $U_G^2 \subset U$, and let U_0 be an open neighbourhood of the H-invariant center σ_0 of the quotient G/H which consists of elements

$$\sigma = g\sigma_0 = \exp(F)\sigma_0, \qquad g \in U_G.$$

Then there is a local section

$$s(g\sigma_0) = \exp(F)$$

of $G \to G/H$ over U_0. With this local section, one can define the induced representation (5.10.20) of elements $g \in U_G \subset G$ on $U_0 \times V$ given by the expressions

$$g\exp(F) = \exp(F')\exp(I'), \qquad (5.11.1)$$
$$g : (\exp(F)\sigma_0, v) \to (\exp(F')\sigma_0, \exp(I')v).$$

The corresponding representation of the Lie algebra $\mathfrak{g}_r$ of G takes the following form. Let $\{F_\alpha\}$, $\{I_a\}$ be the bases for $\mathfrak{f}$ and $\mathfrak{h}$, respectively. Their elements obey the commutation relations

$$[I_a, I_b] = c_{ab}^d I_d, \qquad [F_\alpha, F_\beta] = c_{\alpha\beta}^d I_d, \qquad [F_\alpha, I_b] = c_{\alpha b}^\beta F_\beta.$$

Then the relation (5.11.1) leads to the formulas

$$F_\alpha : F \to F' = F_\alpha + \sum_{k=1}^\infty l_{2k}[\underbrace{\ldots}_{2k}[F_\alpha, F], F], \ldots, F] - \qquad (5.11.2)$$

$$l_n \sum_{n=1}^\infty [\underbrace{\ldots}_{n}[F, I'], I'], \ldots, I'],$$

$$I' = \sum_{k=1}^\infty l_{2k-1}[\underbrace{\ldots}_{2k-1}[F_\alpha, F], F], \ldots, F], \qquad (5.11.3)$$

$$I_a : F \to F' = 2\sum_{k=1}^\infty l_{2k-1}[\underbrace{\ldots}_{2k-1}[I_a, F], F], \ldots, F], \qquad (5.11.4)$$

$$I' = I_a, \qquad (5.11.5)$$

where coefficients l_n, $n = 1, \ldots$, are obtained from the recursion relation

$$\frac{n}{(n+1)!} = \sum_{i=1}^n \frac{l_i}{(n+1-i)!}.$$

Let U_F be an open subset of the origin of the vector space $\mathfrak{f}$ such that the series (5.11.2) – (5.11.5) converge for all $F \in U_F$, $F_\alpha \in \mathfrak{f}$ and $I_a \in \mathfrak{h}_r$. Then the above mentioned non-linear realization of the Lie algebra $\mathfrak{g}_r$ in $U_F \times V$ reads

$$F_\alpha : (F, v) \to (F', I'v),$$
$$I_a : (F, v) \to (F', I'v),$$

where F' and I' are given by the expressions (5.11.2) – (5.11.4). In physical models, the coefficients σ^α of $F = \sigma^\alpha F_\alpha$ are treated as *Goldstone fields*.

Non-linear realizations of many groups especially in application to gravitation theory have been studied [82; 91; 104].

Chapter 6

Gravitation theory on natural bundles

Gravitation theory (without matter fields) can be formulated as gauge theory on natural bundles T over an oriented four-dimensional manifold X [140; 142]. It is metric-affine gravitation theory whose dynamic variables are linear world connections and pseudo-Riemannian world metrics on X. Its Lagrangians are invariant under general covariant transformations. Infinitesimal generators of local one-parameter groups of these transformations are the functorial lift (i.e., the Lie algebra monomorphism) of vector fields on X onto a natural bundle. They are infinitesimal gauge transformations whose gauge parameters are vector fields on X.

Throughout Chapters 6 and 7, by X is meant an oriented simply connected four-dimensional manifold, called a *world manifold*.

6.1 Natural bundles

Let $\pi: Y \to X$ be a smooth fibre bundle coordinated by (x^λ, y^i). Any automorphism (Φ, f) of Y, by definition, is projected as

$$\pi \circ \Phi = f \circ \pi$$

onto a diffeomorphism f of its base X. The converse is not true. A diffeomorphism of X need not give rise to an automorphism of Y, unless $Y \to X$ is a trivial bundle.

Given a one-parameter group (Φ_t, f_t) of automorphisms of Y, its infinitesimal generator is a projectable vector field

$$u = u^\lambda(x^\mu)\partial_\lambda + u^i(x^\mu, y^j)\partial_i$$

on Y. This vector field is projected as

$$\tau \circ \pi = T\pi \circ u$$

onto a vector field $\tau = u^\lambda \partial_\lambda$ on X. Its flow is the one-parameter group (f_t) of diffeomorphisms of X which are projections of autmorphisms (Φ_t, f_t) of Y. Conversely, let

$$\tau = \tau^\lambda \partial_\lambda \qquad (6.1.1)$$

be a vector field on X. There is a problem of constructing its lift to a projectable vector field

$$u = \tau^\lambda \partial_\lambda + u^i \partial_i$$

on Y projected onto τ. Such a lift always exists, but it need not be canonical. Given a connection Γ on Y, any vector field τ (6.1.1) gives rise to the horizontal vector field $\Gamma\tau$ (1.3.6) on Y. This horizontal lift $\tau \to \Gamma\tau$ yields a monomorphism of the $C^\infty(X)$-module $\mathcal{T}(X)$ of vector fields on X to the $C^\infty(Y)$-module of vector fields on Y, but this monomorphisms is not a Lie algebra morphism, unless Γ is a flat connection.

In this Chapter, we address the category of *natural bundles* $T \to X$ which admit the *functorial lift* $\widetilde{\tau}$ onto T of any vector field τ (6.1.1) on X such that $\tau \to \widetilde{\tau}$ is a monomorphism

$$\mathcal{T}(X) \to \mathcal{T}(T), \qquad [\widetilde{\tau}, \widetilde{\tau}'] = \widetilde{[\tau, \tau']},$$

of the real Lie algebra $\mathcal{T}(X)$ of vector fields on X to the real Lie algebra $\mathcal{T}(Y)$ of vector fields on T [94; 153]. One treats the functorial lift $\widetilde{\tau}$ as an *infinitesimal general covariant transformation* or, strictly speaking, an infinitesimal generator of a local one-parameter group of general covariant transformations of T.

Remark 6.1.1. It should be emphasized that, in general, there exist diffeomorphisms of X which do not belong to any one-parameter group of diffeomorphisms of X. In a general setting, one therefore considers a monomorphism $f \to \widetilde{f}$ of the group of diffeomorphisms of X to the group of bundle automorphisms of a natural bundle $T \to X$. Automorphisms $\widetilde{f}$ are called *general covariant transformations* of T. No vertical automorphism of T, unless it is the identity morphism, is a general covariant transformation. The group of automorphisms of a natural bundle is a semi-direct product of its subgroup of vertical automorphisms and the subgroup of general covariant transformations.

Natural bundles are exemplified by tensor bundles (1.1.14). For instance, the tangent and cotangent bundles TX and T^*X of X are natural bundles. Given a vector field τ (6.1.1) on X, its functorial (or canonical)

6.1. Natural bundles

lift onto the tensor bundle T (1.1.14) is given by the formula (1.1.26). Let us introduce the collective index A for the tensor bundle coordinates

$$y^A = \dot{x}^{\alpha_1\cdots\alpha_m}_{\beta_1\cdots\beta_k}.$$

In this notation, the functorial lift $\widetilde{\tau}$ (1.1.26) reads

$$\widetilde{\tau} = \tau^\lambda \partial_\lambda + u^{A\beta}_\alpha \partial_\beta \tau^\alpha \partial_A. \tag{6.1.2}$$

The expression (6.1.2) is a general form of the functorial lift of a vector field τ on X onto a natural bundle T, when this lift depends only on first derivatives of components of τ. In particular, let us recall the functorial lift (1.1.28) and (1.1.29) of τ onto the tangent bundle TX and the cotangent bundle T^*X:

$$\widetilde{\tau} = \tau^\mu \partial_\mu + \partial_\nu \tau^\alpha \dot{x}^\nu \frac{\partial}{\partial \dot{x}^\alpha}, \tag{6.1.3}$$

$$\widetilde{\tau} = \tau^\mu \partial_\mu - \partial_\beta \tau^\nu \dot{x}_\nu \frac{\partial}{\partial \dot{x}_\beta}, \tag{6.1.4}$$

respectively.

Remark 6.1.2. Any diffeomorphism f of X gives rise to the tangent automorphisms $\widetilde{f} = Tf$ of TX which is a general covariant transformation of TX as a natural bundle. Accordingly, the general covariant transformation of the cotangent bundle T^*X over a diffeomorphism f of its base X reads

$$\dot{x}'_\mu = \frac{\partial x^\nu}{\partial x'^\mu} \dot{x}_\nu.$$

Tensor bundles over a world manifold X have the structure group

$$GL_4 = GL^+(4, \mathbb{R}). \tag{6.1.5}$$

The associated principal bundle is the fibre bundle

$$\pi_{LX} : LX \to X$$

of oriented linear frames in the tangent spaces to a world manifold X. It is called the *linear frame bundle*. Its (local) sections are termed *frame fields*.

Given holonomic frames $\{\partial_\mu\}$ in the tangent bundle TX associated with the holonomic atlas Ψ_T (1.1.13), every element $\{H_a\}$ of the linear frame bundle LX takes the form

$$H_a = H^\mu_a \partial_\mu,$$

where H_a^μ is a matrix of the natural representation of the group GL_4 in $\mathbb{R}^4$. These matrices constitute the bundle coordinates

$$(x^\lambda, H_a^\mu), \qquad H_a'^\mu = \frac{\partial x'^\mu}{\partial x^\lambda} H_a^\lambda,$$

on LX associated to its holonomic atlas

$$\Psi_T = \{(U_\iota, z_\iota = \{\partial_\mu\})\} \tag{6.1.6}$$

given by the local frame fields $z_\iota = \{\partial_\mu\}$. With respect to these coordinates, the right action (5.3.2) of GL_4 on LX reads

$$R_g P : H_a^\mu \to H_b^\mu g^b{}_a, \qquad g \in GL_4.$$

The linear frame bundle LX is equipped with the canonical $\mathbb{R}^4$-valued one-form

$$\theta_{LX} = H_\mu^a dx^\mu \otimes t_a, \tag{6.1.7}$$

where $\{t_a\}$ is a fixed basis for $\mathbb{R}^4$ and H_μ^a is the inverse matrix of H_a^μ.

The linear frame bundle $LX \to X$ belongs to the category of natural bundles. Any diffeomorphism f of X gives rise to the principal automorphism

$$\widetilde{f} : (x^\lambda, H_a^\lambda) \to (f^\lambda(x), \partial_\mu f^\lambda H_a^\mu) \tag{6.1.8}$$

of LX which is its general covariant transformation (or a *holonomic automorphism*). For instance, the associated automorphism of TX is the tangent morphism Tf to f.

Given a (local) one-parameter group of diffeomorphisms of X and its infinitesimal generator τ, their lift (6.1.8) results in the functorial lift

$$\widetilde{\tau} = \tau^\mu \partial_\mu + \partial_\nu \tau^\alpha H_a^\nu \frac{\partial}{\partial H_a^\alpha} \tag{6.1.9}$$

of a vector field τ (6.1.1) on X onto LX defined by the condition

$$\mathbf{L}_{\widetilde{\tau}} \theta_{LX} = 0.$$

Every LX-associated bundle $Y \to X$ admits a lift of any diffeomorphism f of its base to the principal automorphism $\widetilde{f}_Y$ (5.7.7) of Y associated with the principal automorphism $\widetilde{f}$ (6.1.8) of the liner frame bundle LX. Thus, all bundles associated with the linear frame bundle LX are natural bundles. However, there are natural bundles which are not associated with LX.

Remark 6.1.3. In a more general setting, higher order natural bundles and gauge natural bundles are considered [37; 41; 94]. Note that the linear frame bundle LX over a manifold X is the set of first order jets of local

6.2. Linear world connections

diffeomorphisms of the vector space $\mathbb{R}^n$ to X, $n = \dim X$, at the origin of $\mathbb{R}^n$. Accordingly, one considers r-order frame bundles $L^r X$ of r-order jets of local diffeomorphisms of $\mathbb{R}^n$ to X. Furthermore, given a principal bundle $P \to X$ with a structure group G, the r-order jet bundle $J^1 P \to X$ of its sections fails to be a principal bundle. However, the product

$$W^r P = L^r X \times J^r P$$

is a principal bundle with the structure group $W_n^r G$ which is a semidirect product of the group G_n^r of invertible r-order jets of maps $\mathbb{R}^n$ to itself at its origin (e.g., $G_n^1 = GL(n, \mathbb{R})$) and the group $T_n^r G$ of r-order jets of morphisms $\mathbb{R}^n \to G$ at the origin of $\mathbb{R}^n$. Moreover, if $Y \to X$ is a P-associated bundle, the jet bundle $J^r Y \to X$ is a vector bundle associated with the principal bundle $W^r P$. It exemplifies *gauge natural bundles* which can be described as fibre bundles associated with principal bundles $W^r P$. Natural bundles are gauge natural bundles for a trivial group $G = 1$. The bundle of principal connections C (5.4.3) is a first order gauge natural bundle. This fact motivates somebody to develop generalized gauge theory on gauge natural bundles [41].

6.2 Linear world connections

Since the tangent bundle TX is associated with the linear frame bundle LX, every world connection (1.3.39):

$$\Gamma = dx^\lambda \otimes (\partial_\lambda + \Gamma_\lambda{}^\mu{}_\nu \dot{x}^\nu \dot\partial_\mu), \tag{6.2.1}$$

on a world manifold X is associated with a principal connection on LX. We agree to call Γ (6.2.1) the *linear world connection* in order to distinct it from an affine world connection in Section 6.7.

Being principal connections on the linear frame bundle LX, linear world connections are represented by sections of the quotient bundle

$$C_W = J^1 LX / GL_4, \tag{6.2.2}$$

called the *bundle of world connections*. With respect to the holonomic atlas Ψ_T (6.1.6), the bundle of world connections C_W (6.2.2) is provided with the coordinates

$$(x^\lambda, k_\lambda{}^\nu{}_\alpha), \qquad k'_\lambda{}^\nu{}_\alpha = \left[\frac{\partial x'^\nu}{\partial x^\gamma} \frac{\partial x^\beta}{\partial x'^\alpha} k_\mu{}^\gamma{}_\beta + \frac{\partial x^\beta}{\partial x'^\alpha} \frac{\partial^2 x'^\nu}{\partial x^\mu \partial x^\beta} \right] \frac{\partial x^\mu}{\partial x'^\lambda},$$

so that, for any section Γ of $C_W \to X$,

$$k_\lambda{}^\nu{}_\alpha \circ \Gamma = \Gamma_\lambda{}^\nu{}_\alpha$$

are components of the linear world connection Γ (6.2.1).

Though the bundle of world connections $C_W \to X$ (6.2.2) is not LX-associated, it is a natural bundle. It admits the lift
$$\widetilde{f}_C : J^1 LX/GL_4 \to J^1 \widetilde{f}(J^1 LX)/GL_4$$
of any diffeomorphism f of its base X and, consequently, the functorial lift
$$\widetilde{\tau}_C = \tau^\mu \partial_\mu + [\partial_\nu \tau^\alpha k_\mu{}^\nu{}_\beta - \partial_\beta \tau^\nu k_\mu{}^\alpha{}_\nu - \partial_\mu \tau^\nu k_\nu{}^\alpha{}_\beta + \partial_{\mu\beta}\tau^\alpha] \frac{\partial}{\partial k_\mu{}^\alpha{}_\beta} \quad (6.2.3)$$
of any vector field τ on X [112].

The first order jet manifold $J^1 C_W$ of the bundle of world connections admits the canonical splitting (5.5.11). In order to obtain its coordinate expression, let us consider the strength (5.7.12) of the linear world connection Γ (6.2.1). It reads
$$F_\Gamma = \frac{1}{2} F_{\lambda\mu}{}^b{}_a I_b{}^a dx^\lambda \wedge dx^\mu = \frac{1}{2} R_{\mu\nu}{}^\alpha{}_\beta dx^\lambda \wedge dx^\mu,$$
where
$$(I_b{}^a)^\alpha{}_\beta = H_b^\alpha H_\beta^a$$
are generators of the group GL_4 (6.1.5) in fibres of TX with respect to the holonomic frames, and
$$R_{\lambda\mu}{}^\alpha{}_\beta = \partial_\lambda \Gamma_\mu{}^\alpha{}_\beta - \partial_\mu \Gamma_\lambda{}^\alpha{}_\beta + \Gamma_\lambda{}^\gamma{}_\beta \Gamma_\mu{}^\alpha{}_\gamma - \Gamma_\mu{}^\gamma{}_\beta \Gamma_\lambda{}^\alpha{}_\gamma \quad (6.2.4)$$
are components if the curvature (1.3.41) of a linear world connection Γ. Accordingly, the above mentioned canonical splitting (5.5.11) of $J^1 C_W$ can be written in the form
$$k_{\lambda\mu}{}^\alpha{}_\beta = \frac{1}{2}(\mathcal{R}_{\lambda\mu}{}^\alpha{}_\beta + \mathcal{S}_{\lambda\mu}{}^\alpha{}_\beta) = \quad (6.2.5)$$
$$\frac{1}{2}(k_{\lambda\mu}{}^\alpha{}_\beta - k_{\mu\lambda}{}^\alpha{}_\beta + k_\lambda{}^\gamma{}_\beta k_\mu{}^\alpha{}_\gamma - k_\mu{}^\gamma{}_\beta k_\lambda{}^\alpha{}_\gamma) +$$
$$\frac{1}{2}(k_{\lambda\mu}{}^\alpha{}_\beta + k_{\mu\lambda}{}^\alpha{}_\beta - k_\lambda{}^\gamma{}_\beta k_\mu{}^\alpha{}_\gamma + k_\mu{}^\gamma{}_\beta k_\lambda{}^\alpha{}_\gamma).$$

It is readily observed that, if Γ is a section of $C_W \to X$, then
$$\mathcal{R}_{\lambda\mu}{}^\alpha{}_\beta \circ J^1 \Gamma = R_{\lambda\mu}{}^\alpha{}_\beta.$$

Because of the canonical vertical splitting (1.1.43) of the vertical tangent bundle VTX of TX, the curvature form (1.3.41) of a linear world connection Γ can be represented by the tangent-valued two-form
$$R = \frac{1}{2} R_{\lambda\mu}{}^\alpha{}_\beta \dot{x}^\beta dx^\lambda \wedge dx^\mu \otimes \partial_\alpha \quad (6.2.6)$$

6.2. Linear world connections

on TX. Due to this representation, the *Ricci tensor*

$$R_c = \frac{1}{2} R_{\lambda\mu}{}^\lambda{}_\beta dx^\mu \otimes dx^\beta \tag{6.2.7}$$

of a linear world connection Γ is defined.

Owing to the above mentioned vertical splitting (1.1.43) of VTX, the torsion form T (1.3.42) of Γ can be written as the tangent-valued two-form

$$T = \frac{1}{2} T_\mu{}^\nu{}_\lambda dx^\lambda \wedge dx^\mu \otimes \partial_\nu, \tag{6.2.8}$$
$$T_\mu{}^\nu{}_\lambda = \Gamma_\mu{}^\nu{}_\lambda - \Gamma_\lambda{}^\nu{}_\mu,$$

on X. The *soldering torsion form*

$$T = T_\mu{}^\nu{}_\lambda \dot{x}^\lambda dx^\mu \otimes \dot{\partial}_\nu \tag{6.2.9}$$

on TX is also defined. Then one can show the following.

- Given a linear world connection Γ (6.2.1) and its soldering torsion form T (6.2.9), the sum $\Gamma + cT$, $c \in \mathbb{R}$, is a linear world connection.
- Every linear world connection Γ defines a unique symmetric world connection

$$\Gamma' = \Gamma - \frac{1}{2} T. \tag{6.2.10}$$

- If Γ and Γ' are linear world connections, then

$$c\Gamma + (1-c)\Gamma'$$

is so for any $c \in \mathbb{R}$.

A world manifold X is said to be *flat* if it admits a flat linear world connection Γ. By virtue of Theorem 1.3.4, there exists an atlas of local constant trivializations of TX such that

$$\Gamma = dx^\lambda \otimes \partial_\lambda$$

relative to this atlas. As a consequence, the curvature form R (6.2.6) of this connection equals zero. However, such an atlas is not holonomic in general. Relative to this atlas, the canonical soldering form (1.1.39) on TX reads

$$\theta_J = H_\mu^a dx^\mu \dot{\partial}_a,$$

and the torsion form T (1.3.42) of Γ defined as the Nijenhuis differential $d_\Gamma \theta_J$ (1.3.30) need not vanish.

A world manifold X is called *parallelizable* if the tangent bundle $TX \to X$ is trivial. By virtue of Theorem 1.3.4, a parallelizable world manifold is flat. Conversely, a flat world manifold is parallelizable if it is simply connected (see Theorem 8.1.6).

Every linear world connection Γ (6.2.1) yields the horizontal lift

$$\Gamma\tau = \tau^\lambda(\partial_\lambda + \Gamma_\lambda{}^\beta{}_\alpha \dot{x}^\alpha \dot{\partial}_\beta) \qquad (6.2.11)$$

of a vector field τ on X onto the tangent bundle TX. A vector field τ on X is said to be *parallel* relative to a connection Γ if it is an integral section of Γ. Its integral curve is called the *autoparallel* of a linear world connection Γ.

Remark 6.2.1. By virtue of Theorem 1.3.2, any vector field on X is an integral section of some linear world connection. If $\tau(x) \neq 0$ at a point $x \in X$, there exists a coordinate system (q^i) on some neighbourhood U of x such that $\tau^i(x) =$const. on U. Then τ on U is an integral section of the local symmetric linear world connection

$$\Gamma_\tau(x) = dq^i \otimes \partial_i, \qquad x \in U, \qquad (6.2.12)$$

on U. In particular, the functorial lift $\widetilde{\tau}$ (6.1.3) can be obtained at each point $x \in X$ as the horizontal lift of τ by means of the local symmetric connection (6.2.12).

The horizontal lift of a vector field τ on X onto the linear frame bundle LX by means of a world connection K reads

$$\Gamma\tau = \tau^\lambda \left(\partial_\lambda + \Gamma_\lambda{}^\nu{}_\alpha H_a^\alpha \frac{\partial}{\partial H_a^\nu} \right). \qquad (6.2.13)$$

It is called *standard* if the morphism

$$u \rfloor \theta_{LX} : LX \to \mathbb{R}^4$$

is constant on LX. It is readily observed that every standard horizontal vector field on LX takes the form

$$u_v = H_b^\lambda v^b \left(\partial_\lambda + \Gamma_\lambda{}^\nu{}_\alpha H_a^\alpha \frac{\partial}{\partial H_a^\nu} \right) \qquad (6.2.14)$$

where $v = v^b t_b \in \mathbb{R}^4$. A glance at this expression shows that a standard horizontal vector field is not projectable.

Since TX is an LX-associated fibre bundle, we have the canonical morphism

$$LX \times \mathbb{R}^4 \to TX,$$
$$(H_a^\mu, v^a) \to \dot{x}^\mu = H_a^\mu v^a.$$

The tangent map to this morphism sends every standard horizontal vector field (6.2.14) on LX to the horizontal vector field

$$u = \dot{x}^\lambda(\partial_\lambda + \Gamma_\lambda{}^\nu{}_\alpha \dot{x}^\alpha \dot{\partial}_\nu) \qquad (6.2.15)$$

on the tangent bundle TX. Such a vector field on TX is called *holonomic* [112]. Given holonomic coordinates $(x^\mu, \dot{x}^\mu, \dot{x}^\mu, \ddot{x}^\mu)$ on the double tangent bundle TTX, the holonomic vector field (6.2.15) defines the second order dynamic equation

$$\ddot{x}^\nu = \Gamma_\lambda{}^\nu{}_\alpha \dot{x}^\lambda \dot{x}^\alpha \qquad (6.2.16)$$

on X which is called the *geodesic equation* with respect to a linear world connection Γ. Solutions of the geodesic equation (6.2.16), called the *geodesics* of Γ, are the projection of integral curves of the vector field (6.2.15) in TX onto X. Moreover, one can show the following [92].

Theorem 6.2.1. *The projection of an integral curve of any standard horizontal vector field (6.2.14) on LX onto X is a geodesic in X. Conversely, any geodesic in X is of this type.*

It is readily observed that, if linear world connections Γ and Γ' differ from each other only in the torsion, they define the same holonomic vector field (6.2.15) and the same geodesic equation (6.2.16).

Let τ be an integral vector field of a linear world connection Γ, i.e.,

$$\nabla^\Gamma_\mu \tau = 0.$$

Consequently, it obeys the equation

$$\tau^\mu \nabla^\Gamma_\mu \tau = 0.$$

Then one can show that any autoparallel of a linear world connection Γ is its geodesic and, conversely, a geodesic of Γ is an autoparallel of its symmetric part (6.2.10).

6.3 Lorentz reduced structure. Gravitational field

Gravitation theory on a world manifold X is classical field theory with spontaneous symmetry breaking described by different reduced structures of the linear frame bundle LX as follows [83; 142].

The geometric formulation of the *equivalence principle* states the existence of an atlas of the tangent bundle $TX \to X$ and associated bundles with transition functions taking their values into the Lorentz group [83]. In other words, the structure group GL_4 (6.1.5) of the linear frame bundle LX over a world manifold X must be reducible to the Lorentz group $SO(1,3)$. At the same time, the existence of Dirac fermion fields implies that GL_4 is reducible to the *proper Lorentz group*

$$\mathrm{L} = SO^0(1,3),$$

which is the connected component of the unit of $SO(1,3)$. If a structure group of LX is reducible to the proper Lorentz group, it also is reducible to its maximal compact subgroup $SO(3)$ that defines a space-time structure of a world manifold X (see Section 6.4).

In this and next Chapters, we deal with the following reduced Lorentz and proper Lorentz structures.

- By a *Lorentz structure* is meant a reduced principal $SO(1,3)$-subbundle L^gX, called the *Lorentz subbundle*, of the linear frame bundle LX.

- A *proper Lorentz structure* is defined as a reduced L-subbundle L^hX, called *proper Lorentz subbundle*, of the linear frame bundle LX.

Lemma 6.3.1. *There is one-to-ne correspondence between the reduced Lorentz and proper Lorentz structures.*

Proof. The Lorentz group $SO(1,3)$ is isomorphic to the group product

$$SO(1,3) = \mathbb{Z}_2 \times \mathrm{L}.$$

Since a world manifold X is simply connected, any principal $\mathbb{Z}_2$-bundle is trivial by virtue of Theorem 8.1.6. Therefore, any principal Lorentz bundle $P_{SO(1,3)}$ is isomorphic to the product

$$P_{SO(1,3)} = \mathbb{Z}_2 \underset{X}{\times} P_\mathrm{L},$$

where P_L is a principal L-bundle. $\square$

One can show that different proper Lorentz subbundles L^hX and $L^{h'}X$ of the frame bundle LX are isomorphic as principal L-bundles [77]. This means that there exists a vertical automorphism of the frame bundle LX which sends L^hX onto $L^{h'}X$ (see Theorem 5.10.2). By virtue of Lemma 6.3.1, the similar property of Lorentz subbundles also is true.

Remark 6.3.1. There is the well-known topological obstruction to the existence of a Lorentz structure on a world manifold X. All non-compact manifolds and compact manifolds whose Euler characteristic equals zero admit a reduced $SO(1,3)$-structure [34]. In gravitational models, some conditions of causality should be also satisfied [74]. A compact space-time does not possess this property. At the same time, a non-compact world manifold X has a Dirac spinor structure if and only if it is parallelizable [51; 165].

6.3. Lorentz reduced structure. Gravitational field

By virtue of Theorem 5.10.1, there is one-to-one correspondence between the principal L-subbundles $L^h X$ of the frame bundle LX and the global sections h of the quotient fibre bundle

$$\Sigma_T = LX/L, \tag{6.3.1}$$

called the *tetrad bundle*. This is an LX-associated fibre bundle with the typical fibre GL_4/L. Its global sections are called the *tetrad fields*. The fibre bundle (6.3.1) is the two-fold covering

$$\zeta : \Sigma_T \to \Sigma_{PR}$$

of the *metric bundle*

$$\Sigma_{PR} = LX/SO(1,3), \tag{6.3.2}$$

whose typical fibre is $Gl_4/SO(1,3)$ and whose global sections are *pseudo-Riemannian world metrics* g on X. In particular, every tetrad field h defines uniquely a pseudo-Riemannian metric $g = \zeta \circ h$. For the sake of convenience, one usually identifies the metric bundle (6.3.2) with an open subbundle of the tensor bundle

$$\Sigma_{PR} \subset \overset{2}{\vee} TX. \tag{6.3.3}$$

Therefore, the metric bundle Σ_{PR} (6.3.2) can be equipped with the bundle coordinates $(x^\lambda, \sigma^{\mu\nu})$.

In General Relativity, a pseudo-Riemannian world metric (or a tetrad field) describes a *gravitational field*. Therefore, the existence of a reduced Lorentz structure is part and parcel of gravitation theory.

Every tetrad field h defines an associated Lorentz bundle atlas

$$\Psi^h = \{(U_\iota, z_\iota^h = \{h_a\})\} \tag{6.3.4}$$

of the linear frame bundle LX such that the corresponding local sections z_ι^h of LX take their values into the Lorentz subbundle $L^h X$ and the transition functions of Ψ^h (6.3.4) between the frames $\{h_a\}$ are L-valued. The frames (6.3.4):

$$\{h_a = h_a^\mu(x)\partial_\mu\}, \qquad h_a^\mu = H_a^\mu \circ z_\iota^h, \qquad x \in U_\iota, \tag{6.3.5}$$

are called the *tetrad frames*. Certainly, a Lorentz bundle atlas Ψ^h is not unique.

Given a Lorentz bundle atlas Ψ^h, the pull-back

$$h = h^a \otimes t_a = z_\iota^{h*}\theta_{LX} = h_\lambda^a(x)dx^\lambda \otimes t_a \tag{6.3.6}$$

of the canonical form θ_{LX} (6.1.7) by a local section z_ι^h is called the (local) tetrad form. The tetrad form (6.3.6) determines the tetrad coframes

$$\{h^a = h_\mu^a(x)dx^\mu\}, \qquad x \in U_\iota, \tag{6.3.7}$$

in the cotangent bundle T^*X. They are the dual of the tetrad frames (6.3.5). The coefficients h_a^μ and h_μ^a of the tetrad frames (6.3.5) and coframes (6.3.7) are called the tetrad functions. They are transition functions between the holonomic atlas Ψ_T (6.1.6) and the Lorentz atlas Ψ^h (6.3.4) of the linear frame bundle LX.

With respect to the Lorentz atlas Ψ^h (6.3.4), a tetrad field h can be represented by the $\mathbb{R}^4$-valued tetrad form (6.3.6). Relative to this atlas, the corresponding pseudo-Riemannian world metric $g = \zeta \circ h$ takes the well-known form

$$g = \eta(h \otimes h) = \eta_{ab} h^a \otimes h^b, \qquad g_{\mu\nu} = h_\mu^a h_\nu^b \eta_{ab}, \tag{6.3.8}$$

where η is the Minkowski metric in $\mathbb{R}^4$ written with respect to its fixed basis $\{t_a\}$. It is readily observed that the tetrad coframes $\{h^a\}$ (6.3.7) and the tetrad frames $\{h_a\}$ (6.3.5) are orthornormal relative to the pseudo-Riemannian metric (6.3.8), namely:

$$g^{\mu\nu} h_\mu^a h_\nu^b = \eta^{ab}, \qquad g_{\mu\nu} h_a^\mu h_b^\nu = \eta_{ab}.$$

Therefore, their components h^0, h_0 and h^i, h_i, $i = 1, 2, 3$, are called time-like and spatial, respectively.

A principal connection on a proper Lorentz subbundle $L^h X$ of the frame bundle LX is called the Lorentz connection. By virtue of Theorem 5.10.4, this connection is extended to a principal connection Γ on the linear frame bundle LX. It also is called the Lorentz connection. The associated linear world connection on the tangent bundle TX with respect to a Lorentz atlas Ψ^h reads

$$\Gamma = dx^\lambda \otimes (\partial_\lambda + \frac{1}{2} A_\lambda{}^{ab} I_{ab}{}^c{}_d h_\mu^d \dot{x}^\mu h_c^\nu \dot{\partial}_\nu) \tag{6.3.9}$$

where

$$I_{ab}{}^c{}_d = \eta_{bd} \delta_a^c - \eta_{ad} \delta_b^c \tag{6.3.10}$$

are generators of the right Lie algebra $\mathfrak{g}_L$ of the proper Lorentz group L in the Minkowski space $\mathbb{R}^4$. Written relative to a holonomic atlas, the connection Γ (6.3.9) possesses the components

$$\Gamma_\lambda{}^\mu{}_\nu = h_k^\mu \partial_\lambda h_\nu^k + \eta_{ka} h_b^\mu h_\nu^k A_\lambda{}^{ab}. \tag{6.3.11}$$

6.3. Lorentz reduced structure. Gravitational field

Its holonomy group is a subgroup of the Lorentz group L. Conversely, let Γ be a linear world connection whose (abstract) holonomy group $\mathcal{K}$ is a subgroup of the Lorentz group. Since a base X is simply connected, this holonomy group coincides with restricted holonomy group (see Theorem 8.1.2). Consequently, it is connected and, therefore, is a subgroup of the proper Lorentz group L. By virtue of Theorem 8.1.4, this linear world connection Γ defines a proper Lorentz subbundle of the frame bundle LX, and it is reducible to a Lorentz connection on this subbundle. Thus, we come to the following.

Assertion 6.3.1. A linear world connection is a Lorentz connection if and only if its holonomy group is a subgroup of the proper Lorentz group L.

Given a pseudo-Riemannian metric g, every linear world connection Γ (6.2.1) admits the decomposition

$$\Gamma_{\mu\nu\alpha} = \{_{\mu\nu\alpha}\} + S_{\mu\nu\alpha} + \frac{1}{2}C_{\mu\nu\alpha} \qquad (6.3.12)$$

in the Christoffel symbols $\{_{\mu\nu\alpha}\}$ (1.3.44), the *non-metricity tensor*

$$C_{\mu\nu\alpha} = C_{\mu\alpha\nu} = \nabla^{\Gamma}_{\mu} g_{\nu\alpha} = \partial_{\mu} g_{\nu\alpha} + \Gamma_{\mu\nu\alpha} + \Gamma_{\mu\alpha\nu} \qquad (6.3.13)$$

and the *contorsion*

$$S_{\mu\nu\alpha} = -S_{\mu\alpha\nu} = \frac{1}{2}(T_{\nu\mu\alpha} + T_{\nu\alpha\mu} + T_{\mu\nu\alpha} + C_{\alpha\nu\mu} - C_{\nu\alpha\mu}), \qquad (6.3.14)$$

where $T_{\mu\nu\alpha} = -T_{\alpha\nu\mu}$ are coefficients of the torsion form (6.2.8) of Γ.

A linear world connection Γ is called a *metric connection* for a pseudo-Riemannian world metric g if g is its integral section, i.e., the *metricity condition*

$$\nabla^{\Gamma}_{\mu} g_{\nu\alpha} = 0 \qquad (6.3.15)$$

holds. A metric connection reads

$$\Gamma_{\mu\nu\alpha} = \{_{\mu\nu\alpha}\} + \frac{1}{2}(T_{\nu\mu\alpha} + T_{\nu\alpha\mu} + T_{\mu\nu\alpha}). \qquad (6.3.16)$$

For instance, the Levi–Civita connection is a torsion-free metric connection Γ where $\Gamma_{\mu\nu\alpha} = \{_{\mu\nu\alpha}\}$.

By virtue of Theorem 5.10.5, a metric connection Γ for a pseudo-Riemmanian world metric $g = \zeta \circ h$ is reducible to a Lorentz connection on the proper Lorentz subbundle $L^h X$, i.e., it is a Lorentz connection. Conversely, every Lorentz connection obeys the metricity condition (6.3.15) for some pseudo-Riemannian metric g (which is not necessarily unique [154]). Thus, the following is true.

Assertion 6.3.2. A Lorentz connection is a metric connection, and *vice versa*.

Though a linear world connection is not a Lorentz connection in general, any world connection Γ defines a Lorentz connection Γ_h on each principal L-subbundle L^hX of the frame bundle as follows.

Since the Lorentz group is a Cartan subgroup of the general linear group GL_4, the Lie algebra of the general linear group GL_4 is the direct sum

$$\mathfrak{g}_{GL_4} = \mathfrak{g}_L \oplus \mathfrak{m}$$

of the Lie algebra $\mathfrak{g}_L$ of the Lorentz group and a subspace $\mathfrak{m}$ such that

$$[\mathfrak{g}_L, \mathfrak{m}] \subset \mathfrak{m}.$$

Then this is the case of Theorem 5.10.6. Therefore, let consider the local connection one-form (5.4.13) of a connection Γ with respect to a Lorentz atlas Ψ^h of LX given by the tetrad forms h^a. It reads

$$z_\iota^{h*}\overline{\Gamma} = -\Gamma_\lambda{}^b{}_a dx^\lambda \otimes I_b{}^a,$$
$$\Gamma_\lambda{}^b{}_a = -h_\mu^b \partial_\lambda h_a^\mu + \Gamma_\lambda{}^\mu{}_\nu h_\mu^b h_a^\nu,$$

where $\{e_b^a\}$ is the basis for the Lie algebra $\mathfrak{g}_{GL_4}$. Then, the Lorentz part of this form is precisely the local connection one-form (5.4.13) of the connection Γ_h on L^hX. We have

$$z_\zeta^{h*}\overline{\Gamma}_h = -\frac{1}{2}A_\lambda{}^{ab} dx^\lambda \otimes I_{ab}, \qquad (6.3.17)$$
$$A_\lambda{}^{ab} = \frac{1}{2}(\eta^{kb}h_\mu^a - \eta^{ka}h_\mu^b)(\partial_\lambda h_k^\mu - h_k^\nu \Gamma_\lambda{}^\mu{}_\nu).$$

Then combining this expression and the expression (6.3.9) gives the connection

$$\Gamma = dx^\lambda \otimes (\partial_\lambda + \frac{1}{4}(\eta^{kb}h_\mu^a - \eta^{ka}h_\mu^b)(\partial_\lambda h_k^\mu - h_k^\nu \Gamma_\lambda{}^\mu{}_\nu)I_{ab}{}^c{}_d h_\mu^d \dot{x}^\mu h_c^\nu \dot{\partial}_\nu) \quad (6.3.18)$$

with respect to a Lorentz atlas Ψ^h and this connection

$$\Gamma_h = dx^\lambda \otimes [\partial_\lambda + \frac{1}{2}(h_\alpha^k \delta_\mu^\beta - \eta^{kc}g_{\mu\alpha}h_c^\beta)(\partial_\lambda h_k^\mu - h_k^\nu \Gamma_\lambda{}^\mu{}_\nu)\dot{x}^\alpha \partial_\beta] \quad (6.3.19)$$

relative to a holonomic atlas. If Γ is a Lorentz connection (6.3.11) extended from L^hX, then obviously $\Gamma_h = \Gamma$.

6.4 Space-time structure

If the structure group GL_4 (6.1.5) of the linear frame bundle LX is reducible to the proper Lorentz group L, it is always reducible to the maximal compact subgroup $SO(3)$ of L in accordance with Assertion 5.10.2. By virtue

6.4. Space-time structure

of Assertion 5.10.2, the structure group GL_4 of LX also is reducible to its maximal compact subgroup $SO(4)$. Thus, there is the commutative diagram

$$\begin{array}{ccc} GL_4 & \longrightarrow & SO(4) \\ \downarrow & & \downarrow \\ L & \longrightarrow & SO(3) \end{array} \qquad (6.4.1)$$

of the reduction of structure groups of the linear frame bundle LX in gravitation theory [132]. This reduction diagram results in the following.

• By virtue of Theorem 5.10.1, there is one-to-one correspondence between the reduced principal $SO(4)$-subbundles $L^{g^R}X$ of the linear frame bundle LX and the global sections of the quotient bundle

$$LX/SO(4) \to X.$$

Its global sections are *Riemannian world metrics* g^R on X. Thus, a Riemannian metric on a world manifold always exists. In fact, its existence results from paracompactness of a world manifold, but the converse also is true. One can show that a smooth manifold is paracompact if it admits a Riemannian structure [92].

• As was mentioned above, a reduction of the structure group of the linear frame bundle LX to the proper Lorentz group means the existence of a reduced Lorentz subbundle $L^hX \subset LX$ associated with a tetrad field h or a pseudo-Riemannian metric $g = \zeta \circ h$ on X.

• Since the structure group L of this reduced Lorentz bundle L^hX is reducible to the group $SO(3)$ there exists a reduced principal $SO(3)$-subbundle

$$L_0^hX \subset L^hX \subset LX, \qquad (6.4.2)$$

called the *spatial structure*. The corresponding global section of the quotient fibre bundle

$$L^hX/SO(3) \to X$$

with the typical fibre $\mathbb{R}^3$ is a one-codimensional *spatial distribution* $\mathbf{F} \subset TX$ on X. Its annihilator $\mathrm{Ann}\mathbf{F}$ is a one-dimensional codistribution $\mathbf{F}^* \subset T^*X$.

Given the spatial structure L_0^hX (6.4.2), let us consider a Lorentz bundle atlas Ψ_0^h (6.3.4) given by local sections z_ι of LX taking their values into the reduced $SO(3)$-subbundle L_0^hX. Its transition functions are $SO(3)$-valued. Thus, the following is stated.

Assertion 6.4.1. In gravitation theory on a world manifold X, one can always choose an atlas of the tangent bundle TX and associated bundles with $SO(3)$-valued transition functions. This bundle atlas, called the *spatial bundle atlas*, however need not be holonomic.

Given a spatial bundle atlas Ψ_0^h, its $SO(3)$-valued transition functions preserve the time-like component

$$h^0 = h_\lambda^0 dx^\lambda \qquad (6.4.3)$$

of local tetrad forms (6.3.6) which, therefore, is globally defined. We agree to call it the *time-like tetrad form*. Accordingly, the dual time-like vector field

$$h_0 = h_0^\mu \partial_\mu \qquad (6.4.4)$$

also is globally defined. In this case, the spatial distribution $\mathbf{F}$ is spanned by spatial components h_i, $i = 1, 2, 3$, of the tetrad frames (6.3.5), while the time-like tetrad form (6.4.3) spans the tetrad codistribution $\mathbf{F}^*$, i.e.,

$$h^0 \rfloor \mathbf{F} = 0. \qquad (6.4.5)$$

Then the tangent bundle TX of a world manifold X admits the *space-time decomposition*

$$TX = \mathbf{F} \oplus T^0 X, \qquad (6.4.6)$$

where $T^0 X \to X$ is the one-dimensional fibre bundle spanned by the time-like vector field h_0 (6.4.4).

Since the diagram (6.4.1) is commutative, the reduced spatial subbundle $L_0^h X$ (6.4.2) of a reduced Lorentz bundle $L^h X$ is a reduced subbundle of some reduced $SO(4)$-bundle $L^{g^R} X$ too, i.e.,

$$L^h X \supset L_0^h X \subset L^{g^R} X. \qquad (6.4.7)$$

Let $g = \zeta \circ h$ and g^R be the corresponding pseudo-Riemannian and Riemannian world metrics on X. Written with respect to a spatial bundle atlas Ψ_0^h, they read

$$g = \eta_{ab} h^a \otimes h^b, \qquad g_{\mu\nu} = h_\mu^a h_\nu^b \eta^{ab}, \qquad (6.4.8)$$
$$g^R = \eta_{ab}^E h^a \otimes h^b, \qquad g_{\mu\nu}^R = h_\mu^a h_\nu^b \eta_{ab}^E, \qquad (6.4.9)$$

where η^E is the Euclidean metric in $\mathbb{R}^4$. The space-time decomposition (6.4.6) is orthonormal with respect to both of the metrics (6.4.8) and (6.4.9).

The world metrics (6.4.8) and (6.4.9) satisfy the following well-known theorem [74].

6.4. Space-time structure

Theorem 6.4.1. *For any pseudo-Riemannian metric g on a world manifold X, there exist a normalized time-like one-form h^0 and a Riemannian metric g^R such that*

$$g = 2h^0 \otimes h^0 - g^R. \qquad (6.4.10)$$

Conversely, let a world manifold X admit a nowhere vanishing one-form σ (or, equivalently, a nowhere vanishing vector field). Then any Riemannian world metric g^R on X yields the pseudo-Riemannian world metric g (6.4.10) where

$$h^0 = \frac{\sigma}{\sqrt{g^R(\sigma,\sigma)}}.$$

The following is a corollary of this theorem.

Corollary 6.4.1. *A world manifold X admits a pseudo-Riemannian metric if and only if there exists a nowhere vanishing one-form (or a vector field) on X.*

Note that the condition (6.4.7) gives something more.

Theorem 6.4.2. *There is one-to-one correspondence between the reduced $SO(3)$-subbundles of the linear frame bundle LX and the triples $(g, \mathbf{F}, g^R)$ of a pseudo-Riemannian metric g, a spatial distribution $\mathbf{F}$ defined by the condition (6.4.5) and a Riemannian metric g^R which obey the relation (6.4.10).*

Proof. Given the triple (6.4.7) of the reduced subbundles, let us suppose that there exists a different reduced $SO(3)$-subbundle $L_0'^h X$ both of the reduced Lorentz subbundle $L^h X$ and the $SO(4)$-one $L^{g^R} X$. By virtue of Theorem 5.10.3, the reduced $SO(3)$-subbundles $L_0^h X$ and $L_0'^h X$ of the principal Lorentz bundle $L^h X$ are isomorphic because $L/SO(3) = \mathbb{R}^3$. Consequently, there exists an automorphism of $L^h X$ which sends $L_0^h X$ onto $L_0'^h X$ (see Theorem 5.10.2). However, no automorphism of the reduced Lorentz bundle $L^h X$ extended to an automorphism of LX preserves the reduced $SO(4)$-bundle $L^{g^R} X$. $\square$

A spatial distribution $\mathbf{F}$ and a Riemannian metric g^R in the triple $(g, \mathbf{F}, g^R)$ in Theorem 6.4.2 are called *g-compatible*.

Remark 6.4.1. A g-compatible Riemannian metric g^R in a triple $(g, \mathbf{F}, g^R)$ defines a g-compatible distance function $d(x, x')$ on a world manifold X. Such a function brings X into a metric space whose locally Euclidean topology is equivalent to a manifold topology on X. Given a gravitational field

g, the g-compatible Riemannian metrics and the corresponding distance functions are different for different spatial distributions $\mathbf{F}$ and $\mathbf{F}'$. It follows that physical observers associated with different spatial distributions $\mathbf{F}$ and $\mathbf{F}'$ perceive a world manifold X as different Riemannian spaces. The well-known relativistic changes of sizes of moving bodies exemplify this phenomenon [132]. Note that there are attempts of deriving a world topology directly from a pseudo-Riemannian structure of a world manifold (path topology, etc.) [74]. However, they are rather extraordinary in general.

From the physical viewpoint, it is natural to assume that a pseudo-Riemannian metric g on X admits an integrable g-compatible spatial distribution $\mathbf{F}$ given by the condition (6.4.5). Then $\mathbf{F}$ defines a *spatial foliation* $\mathcal{F}$ of a world manifold X whose leaves are spatial three-dimensional subspaces of X. In this case, the time-like vector field h_0 (6.4.4) is transversal to the spatial foliation $\mathcal{F}$, and the one-dimensional subbundle $T^0 X$ spanned by h_0 is the normal bundle to $\mathcal{F}$ which splits the exact sequence (1.1.31).

By virtue of Theorem 1.1.14, a spatial distribution $\mathbf{F}$ is integrable if and only if the one-form h^0 (6.4.3) is closed. In this case, a spatial distribution $\mathbf{F}$. Because a world manifold X is simply connected, its first de Rham cohomology is trivial and, therefore, a closed one-form h^0 is exact, i.e., $h^0 = df$. Consequently, a spatial distribution $\mathcal{F}$ is simple, i.e., its leaves are fibres of a fibred manifold $X \to f(X)$. Since the function f has no critical points where $df = 0$, the foliation $\mathcal{F}$ obeys the notion of stable causality by Hawking [74]. No curve transversal to leaves of such a foliation intersects each leaf more than once. It follows that a world manifold X admitting this foliation is non-compact.

These speculations motivate us restrict our consideration to gravitation theory on a non-compact world manifold. This restriction is essential for describing Dirac spinor fields because, as was mentioned in Remark 6.3.1, a non-compact world manifold X admits a Dirac spinor structure if and only if it is parallelizable.

6.5 Gauge gravitation theory

At present, Yang–Mills gauge theory on principal bundles (see Section 5.8) provides a universal description of the fundamental electroweak and strong interactions. Gauge gravitation theory from the very beginning aims to extend this description to gravity.

The first gauge model of gravity was suggested by Utiyama [158] in 1956

6.5. Gauge gravitation theory

just two years after birth of the gauge theory itself. Utiyama was first who generalized the original gauge model of Yang and Mills for $SU(2)$ to an arbitrary symmetry Lie group and, in particular, to the Lorentz group in order to describe gravity. However, he met the problem of treating general covariant transformations and a pseudo-Riemannian metric (a tetrad field) which had no partner in Yang–Mills gauge theory. To eliminate this drawback, representing tetrad fields as gauge fields of the translation group was attempted (see [76; 83; 122] for a review). Since the Poincaré group comes from the Wigner–Inönii contraction of the de Sitter groups $SO(2,3)$ and $SO(1,4)$ and it is a subgroup of the conformal group, gauge theories on fibre bundles $Y \to X$ with these structure groups were also considered [84; 156]. In a different way, gravitation theory was formulated as the gauge theory with a reduced Lorentz structure where a metric (tetrad) gravitational field was treated as the corresponding Higgs field [83; 140; 142].

Studying gauge gravitation theory, one believes reasonable to require that it incorporates Einstein's General Relativity and, in particular, admits general covariant transformations. Therefore, we formulate gauge gravitation theory as Lagrangian field theory on natural bundles over a world manifold X. It is metric-affine gravitation theory whose Lagrangian L_{MA} is invariant under general covariant transformations. In the absence of matter fields, its dynamic variables are linear world connections and pseudo-Riemannian metrics on X.

Linear world connections are represented by sections of the bundle of world connections C_W (6.2.2). Pseudo-Riemannian world metrics are described by sections of the open subbundle (6.3.3). Therefore, let us consider the bundle product

$$Y = \Sigma_{PR} \underset{X}{\times} C_W \tag{6.5.1}$$

coordinated by $(x^\lambda, \sigma^{\mu\nu}, k_\mu{}^\alpha{}_\beta)$. The configuration space of gauge gravitation theory is the jet manifold

$$J^1 Y = J^1 \Sigma_{PR} \underset{X}{\times} J^1 C_W, \tag{6.5.2}$$

where $J^1 C_W$ possesses the canonical splitting (5.5.11) given by the coordinate expression (6.2.5). Let us consider the differential graded algebra (1.7.9):

$$\mathcal{S}^*_\infty[Y] = \mathcal{O}^*_\infty Y \tag{6.5.3}$$

possessing the local generating basis $(\sigma^{\alpha\beta}, k_\mu{}^\alpha{}_\beta)$. A Lagrangian L_{MA} of gauge gravitation theory is a first order Lagrangian on the configuration space (6.5.2). Its Euler–Lagrange operator reads

$$\delta L_{\mathrm{MA}} = (\mathcal{E}_{\alpha\beta} d\sigma^{\alpha\beta} + \mathcal{E}^\mu{}_\alpha{}^\beta dk_\mu{}^\alpha{}_\beta) \wedge \omega. \tag{6.5.4}$$

The fibre bundle (6.5.1) is a natural bundle admitting the functorial lift

$$\widetilde{\tau}_{\Sigma C} = \tau^\mu \partial_\mu + (\sigma^{\nu\beta} \partial_\nu \tau^\alpha + \sigma^{\alpha\nu} \partial_\nu \tau^\beta) \frac{\partial}{\partial \sigma^{\alpha\beta}} + \tag{6.5.5}$$

$$(\partial_\nu \tau^\alpha k_\mu{}^\nu{}_\beta - \partial_\beta \tau^\nu k_\mu{}^\alpha{}_\nu - \partial_\mu \tau^\nu k_\nu{}^\alpha{}_\beta + \partial_{\mu\beta} \tau^\alpha) \frac{\partial}{\partial k_\mu{}^\alpha{}_\beta}$$

of vector fields τ (6.1.1) on X [112]. Following Definition 2.3.1, one can treat vector fields $\widetilde{\tau}_{\Sigma C}$ (6.5.5) as infinitesimal gauge transformations whose gauge parameters are vector fields τ on X.

Therefore, let us consider the pull-back bundle

$$TX \underset{X}{\times} Y = TX \underset{X}{\times} \Sigma_{\mathrm{PR}} \underset{X}{\times} C_{\mathrm{W}},$$

and let us enlarge the differential graded algebra $\mathcal{S}^*_\infty[Y]$ (6.5.3) to the differential bigraded algebra

$$\mathcal{P}^*_\infty[TX; Y] \tag{6.5.6}$$

possessing the local basis $(\sigma^{\alpha\beta}, k_\mu{}^\alpha{}_\beta, c^\mu)$ of even fields $(\sigma^{\alpha\beta}, k_\mu{}^\alpha{}_\beta)$ and odd ghosts (c^μ). Taking the vertical part of vector fields $\widetilde{\tau}_{\Sigma C}$ (6.5.5) and replacing gauge parameters τ^λ with ghosts c^λ (see Remark 4.2.1), we obtain the odd vertical graded derivation

$$u = u^{\alpha\beta} \frac{\partial}{\partial \sigma^{\alpha\beta}} + u_\mu{}^\alpha{}_\beta \frac{\partial}{\partial k_\mu{}^\alpha{}_\beta} = \tag{6.5.7}$$

$$(\sigma^{\nu\beta} c^\alpha_\nu + \sigma^{\alpha\nu} c^\beta_\nu - c^\lambda \sigma^{\alpha\beta}_\lambda) \frac{\partial}{\partial \sigma^{\alpha\beta}} +$$

$$(c^\alpha_\nu k_\mu{}^\nu{}_\beta - c^\nu_\beta k_\mu{}^\alpha{}_\nu - c^\nu_\mu k_\nu{}^\alpha{}_\beta + c^\alpha_{\mu\beta} - c^\lambda k_{\lambda\mu}{}^\alpha{}_\beta) \frac{\partial}{\partial k_\mu{}^\alpha{}_\beta}$$

of the differential bigraded algebra (6.5.6).

In metric-affine gravitation theory, all gravitation Lagrangians L_{MA}, by construction, are invariant under general covariant transformations. This means that infinitesimal gauge transformations $\widetilde{\tau}_{\Sigma C}$ (6.5.5) are exact symmetries of a Lagrangian L_{MA} (see Remarks 6.5.1 – 6.5.3). By virtue of Lemma 3.5.3, it follows that the vertical graded derivation u (6.5.7) is a variational symmetry of L_{MA} and, thus, is its gauge symmetry. Then by

6.5. Gauge gravitation theory

virtue of the formulas (2.3.6) – (2.3.7), the Euler–Lagrange operator δL_{MA} (6.5.4) of this Lagrangian obeys the complete Noether identities

$$-\sigma_\lambda^{\alpha\beta}\mathcal{E}_{\alpha\beta} - 2d_\mu(\sigma^{\mu\beta}\mathcal{E}_{\lambda\beta} - k_{\lambda\mu}{}^\alpha{}_\beta\mathcal{E}^\mu{}_\alpha{}^\beta - \quad (6.5.8)$$
$$d_\mu[(k_\nu{}^\mu{}_\beta\delta_\lambda^\alpha - k_\nu{}^\alpha{}_\lambda\delta_\beta^\mu - k_\lambda{}^\alpha{}_\beta\delta_\nu^\mu)\mathcal{E}^\nu{}_\alpha{}^\beta] + d_{\mu\beta}\mathcal{E}^\mu{}_\lambda{}^\beta = 0.$$

These Noether identities are irreducible. Therefore, the gauge operator (4.2.8) of gauge gravitation theory is $\mathbf{u} = u$. It admits the nilpotent BRST extension

$$\mathbf{b} = u + c_\mu^\lambda c^\mu \frac{\partial}{\partial c^\lambda}. \quad (6.5.9)$$

Accordingly, an original gravitation Lagrangian L_{MA} is extended to the proper solution of the master equation $L_E = L_e$ (4.2.9) which reads

$$L_E = L_{\text{MA}} + u^{\alpha\beta}\overline{\sigma}_{\alpha\beta}\omega + u_\mu{}^\alpha{}_\beta \overline{k}^\mu{}_\alpha{}^\beta\omega + c_\mu^\lambda c^\mu \overline{c}_\lambda \omega,$$

where $\overline{\sigma}_{\alpha\beta}$, $\overline{k}^\mu{}_\alpha{}^\beta$ and $\overline{c}_\lambda$ are the corresponding antifields.

Remark 6.5.1. By analogy with Theorem 5.8.1, one can show that, if a first order Lagrangian L_{MA} on the configuration space (6.5.2) does not depend on the jet coordinates $\sigma_\lambda^{\alpha\beta}$ and it possesses exact gauge symmetries (6.5.5), it factorizes through the terms $\mathcal{R}_{\lambda\mu}{}^\alpha{}_\beta$ (6.2.5).

Remark 6.5.2. The Hilbert–Einstein Lagrangian L_{HE} of General Relativity depends only on metric variables $\sigma^{\alpha\beta}$. It is a reduced second order Lagrangian which differs from the first order one L'_{HE} in a variationally trivial term (see Theorem 2.4.4). The infinitesimal gauge covariant transformations $\widetilde{\tau}_{\Sigma C}$ (6.5.5) are variational (but not exact) symmetries of the first order Lagrangian L'_{HE}, and the graded derivation u (6.5.7) is so. It reads

$$u = (\sigma^{\nu\beta}c_\nu^\alpha + \sigma^{\alpha\nu}c_\nu^\beta - c^\lambda \sigma_\lambda^{\alpha\beta})\frac{\partial}{\partial \sigma^{\alpha\beta}}.$$

Then the corresponding Noether identities (6.5.8) take the familiar form

$$\nabla_\mu \mathcal{E}_\lambda^\mu = (d_\mu + \{_\mu{}^\beta{}_\lambda\})\mathcal{E}_\beta^\mu = 0,$$

where $\mathcal{E}_\lambda^\mu = \sigma^{\mu\alpha}\mathcal{E}_{\alpha\lambda}$ and

$$\{_\mu{}^\beta{}_\lambda\} = -\frac{1}{2}\sigma^{\beta\nu}(d_\mu\sigma_{\nu\lambda} + d_\lambda\sigma_{\mu\nu} - d_\nu\sigma_{\mu\lambda}) \quad (6.5.10)$$

are the Christoffel symbols expressed into function $\sigma_{\alpha\beta}$ of $\sigma^{\mu\nu}$ given by the relations $\sigma^{\mu\alpha}\sigma_{\alpha\beta} = \delta_\beta^\mu$.

Remark 6.5.3. General covariant transformations are sufficient in order to restart both Einstein's General Relativity and metric-affine gravitation theory [140]. However, one also considers the total group of automorphisms of the linear frame bundle LX [76]. Such an automorphism is the composition of some general covariant transformation and a vertical automorphism of LX (see Remark 6.1.1). Subject to associated vertical automorphisms, the tangent bundle TX is provided with non-holonomic frames. A problem is that the most of gravitation Lagrangians, e.g., the Hilbert–Einstein Lagrangian are not invariant under vertical non-holonomic frame transformations.

6.6 Energy-momentum conservation law

Since infinitesimal general covariant transformations $\widetilde{\tau}_{\Sigma C}$ (6.5.5) are exact symmetries of a metric-affine gravitation Lagrangian, let us study the corresponding conservation laws. These are the energy-momentum conservation laws because the vector fields $\widetilde{\tau}_{\Sigma C}$ are not vertical [52; 135]. There are several approaches to discover an energy-momentum conservation law in gravitation theory. Here we treat this conservation law as a particular gauge conservation law. Accordingly, the energy-momentum of gravity is seen as a particular symmetry current (see, e.g., [8; 19; 76; 79]). Since infinitesimal general covariant transformations $\widetilde{\tau}_{\Sigma C}$ (6.5.5) are infinitesimal gauge transformations depending on derivatives of gauge parameters, the corresponding energy-momentum current reduces to a superpotential (see Theorem 2.4.2).

In view of Remark 6.5.1, let us assume that a metric-affine gravitation Lagrangian L_{MA} is independent of the derivative coordinates $\sigma_\lambda{}^{\alpha\beta}$ of a world metric and that it factorizes through the curvature terms $\mathcal{R}_{\lambda\mu}{}^\alpha{}_\beta$ (6.2.5). Then the following relations take place:

$$\pi^{\lambda\nu}{}_\alpha{}^\beta = -\pi^{\nu\lambda}{}_\alpha{}^\beta, \qquad \pi^{\lambda\nu}{}_\alpha{}^\beta = \frac{\partial \mathcal{L}_{MA}}{\partial k_{\lambda\nu}{}^\alpha{}_\beta}, \qquad (6.6.1)$$

$$\frac{\partial \mathcal{L}_{MA}}{\partial k_\nu{}^\alpha{}_\beta} = \pi^{\lambda\nu}{}_\alpha{}^\sigma k_\lambda{}^\beta{}_\sigma - \pi^{\lambda\nu}{}_\sigma{}^\beta k_\lambda{}^\sigma{}_\alpha. \qquad (6.6.2)$$

Let us follow the compact notation

$$y^A = k_\mu{}^\alpha{}_\beta,$$
$$u_\mu{}^\alpha{}_\beta{}^{\varepsilon\sigma}_\gamma = \delta^\varepsilon_\mu \delta^\sigma_\beta \delta^\alpha_\gamma,$$
$$u_\mu{}^\alpha{}_\beta{}^\varepsilon_\gamma = k_\mu{}^\varepsilon{}_\beta \delta^\alpha_\gamma - k_\mu{}^\alpha{}_\gamma \delta^\varepsilon_\beta - k_\gamma{}^\alpha{}_\beta \delta^\varepsilon_\mu.$$

6.6. Energy-momentum conservation law

Then the vector fields (6.5.5) take the form
$$\widetilde{\tau}_{\Sigma C} = \tau^\lambda \partial_\lambda + (\sigma^{\nu\beta}\partial_\nu \tau^\alpha + \sigma^{\alpha\nu}\partial_\nu \tau^\beta)\partial_{\alpha\beta} + \\ (u^{A\beta}{}_\alpha \partial_\beta \tau^\alpha + u^{A\beta\mu}{}_\alpha \partial_{\beta\mu} \tau^\alpha)\partial_A.$$

We also have the equalities
$$\pi_A^\lambda u^{A\beta\mu}{}_\alpha = \pi^{\lambda\mu}{}_\alpha{}^\beta,$$
$$\pi_A^\varepsilon u^{A\beta}{}_\alpha = -\partial^\varepsilon{}_\alpha{}^\beta \mathcal{L}_{\mathrm{MA}} - \pi^{\varepsilon\beta}{}_\sigma{}^\gamma k_\alpha{}^\sigma{}_\gamma.$$

Let a Lagrangian L_{MA} be invariant under general covariant transformations, i.e.,
$$\mathbf{L}_{J^1 \widetilde{\tau}_{\Sigma C}} L_{\mathrm{MA}} = 0.$$

Then the first variational formula (2.4.29) takes the form
$$0 = (\sigma^{\nu\beta}\partial_\nu \tau^\alpha + \sigma^{\alpha\nu}\partial_\nu \tau^\beta - \tau^\lambda \sigma^{\alpha\beta}_\lambda)\delta_{\alpha\beta}\mathcal{L}_{\mathrm{MA}} + \quad (6.6.3)$$
$$(u^{A\beta}{}_\alpha \partial_\beta \tau^\alpha + u^{A\beta\mu}{}_\alpha \partial_{\beta\mu} \tau^\alpha - \tau^\lambda y_\lambda^A)\delta_A \mathcal{L}_{\mathrm{MA}} - \\ d_\lambda[\pi_A^\lambda(y_\alpha^A \tau^\alpha - u^{A\beta}{}_\alpha \partial_\beta \tau^\alpha - u^{A\varepsilon\beta}{}_\alpha \partial_{\varepsilon\beta}\tau^\alpha) - \tau^\lambda \mathcal{L}_{\mathrm{MA}}].$$

The first variational formula (6.6.3) on-shell leads to the weak conservation law
$$0 \approx -d_\lambda[\pi_A^\lambda(y_\alpha^A \tau^\alpha - u^{A\beta}{}_\alpha \partial_\beta \tau^\alpha - u^{A\varepsilon\beta}{}_\alpha \partial_{\varepsilon\beta}\tau^\alpha) - \tau^\lambda \mathcal{L}_{\mathrm{MA}}], \quad (6.6.4)$$
where
$$\mathcal{J}_{\mathrm{MA}}{}^\lambda = \pi_A^\lambda(y_\alpha^A \tau^\alpha - u^{A\beta}{}_\alpha \partial_\beta \tau^\alpha - u^{A\varepsilon\beta}{}_\alpha \partial_{\varepsilon\beta}\tau^\alpha) - \tau^\lambda \mathcal{L}_{\mathrm{MA}} \quad (6.6.5)$$
is the energy-momentum current of the metric-affine gravity.

Remark 6.6.1. It is readily observed that, with respect to a local coordinate system where a vector field τ is constant, the energy-momentum current (6.6.5) leads to the *canonical energy-momentum tensor*
$$\mathcal{J}_{\mathrm{MA}}{}^\lambda{}_\alpha \tau^\alpha = (\pi^{\lambda\mu}{}_\beta{}^\nu k_{\alpha\mu}{}^\beta{}_\nu - \delta_\alpha^\lambda \mathcal{L}_{\mathrm{MA}})\tau^\alpha.$$
This tensor was suggested in order to describe the energy-momentum complex in the Palatini model [32; 121].

Due to the arbitrariness of gauge parameters τ^λ, the first variational formula (6.6.3) falls into the set of equalities (2.4.43) – (2.4.46) which read
$$\pi^{(\lambda\varepsilon}{}_\gamma{}^{\sigma)} = 0, \quad (6.6.6)$$
$$(u^{A\varepsilon\sigma}{}_\gamma \partial_A + u^{A\varepsilon}{}_\gamma \partial_A^\sigma)\mathcal{L}_{\mathrm{MA}} = 0, \quad (6.6.7)$$
$$\delta_\alpha^\beta \mathcal{L}_{\mathrm{MA}} + 2\sigma^{\beta\mu}\delta_{\alpha\mu}\mathcal{L}_{\mathrm{MA}} + u^{A\beta}{}_\alpha \delta_A \mathcal{L}_{\mathrm{MA}} + d_\mu(\pi_A^\mu u^{A\beta}{}_\alpha) - \quad (6.6.8)$$
$$y_\alpha^A \pi_A^\beta = 0,$$
$$\partial_\lambda \mathcal{L}_{\mathrm{MA}} = 0.$$

Remark 6.6.2. It is readily observed that the equalities (6.6.6) and (6.6.7) hold due to the relations (6.6.1) and (6.6.2), respectively.

Substituting the term $y_\alpha^A \pi_A^\beta$ from the expression (6.6.8) in the energy-momentum conservation law (6.6.4), one brings this conservation law into the form

$$0 \approx -d_\lambda[2\sigma^{\lambda\mu}\tau^\alpha \delta_{\alpha\mu}\mathcal{L}_{\text{MA}} + u_A^{A\lambda}\tau^\alpha \delta_A \mathcal{L}_{\text{MA}} - \pi_A^\lambda u_\alpha^{A\beta}\partial_\beta \tau^\alpha + \quad (6.6.9)$$
$$d_\mu(\pi^{\lambda\mu}{}_\alpha{}^\beta)\partial_\beta \tau^\alpha + d_\mu(\pi_A^\mu u_\alpha^{A\lambda})\tau^\alpha - d_\mu(\pi^{\lambda\mu}{}_\alpha{}^\beta \partial_\beta \tau^\alpha)].$$

After separating the variational derivatives, the energy-momentum conservation law (6.6.9) of the metric-affine gravity takes the superpotential form

$$0 \approx -d_\lambda[2\sigma^{\lambda\mu}\tau^\alpha \delta_{\alpha\mu}\mathcal{L}_{\text{MA}} +$$
$$(k_\mu{}^\lambda{}_\gamma \delta^\mu{}_\alpha{}^\gamma \mathcal{L}_{\text{MA}} - k_\mu{}^\sigma{}_\alpha \delta^\mu{}_\sigma{}^\lambda \mathcal{L}_{\text{MA}} - k_\alpha{}^\sigma{}_\gamma \delta^\lambda{}_\sigma{}^\gamma \mathcal{L}_{\text{MA}})\tau^\alpha +$$
$$\delta^\lambda{}_\alpha{}^\mu \mathcal{L}_{\text{MA}} \partial_\mu \tau^\alpha - d_\mu(\delta^\mu{}_\alpha{}^\lambda \mathcal{L}_{\text{MA}})\tau^\alpha + d_\mu(\pi^{\mu\lambda}{}_\alpha{}^\nu (\partial_\nu \tau^\alpha - k_\sigma{}^\alpha{}_\nu \tau^\sigma))],$$

where the energy-momentum current on-shell reduces to the *generalized Komar superpotential*

$$U_{\text{MA}}{}^{\mu\lambda} = \pi^{\mu\lambda}{}_\alpha{}^\nu(\partial_\nu \tau^\alpha - k_\sigma{}^\alpha{}_\nu \tau^\sigma) \quad (6.6.10)$$

[52; 135]. We can rewrite this superpotential as

$$U_{\text{MA}}{}^{\mu\lambda} = 2\frac{\partial \mathcal{L}_{\text{MA}}}{\partial \mathcal{R}_{\mu\lambda}{}^\alpha{}_\nu}(D_\nu \tau^\alpha + T_\nu{}^\alpha{}_\sigma \tau^\sigma),$$

where D_ν is the covariant derivative relative to the connection $k_\nu{}^\alpha{}_\sigma$ and

$$T_\nu{}^\alpha{}_\sigma = k_\nu{}^\alpha{}_\sigma - k_\sigma{}^\alpha{}_\nu \quad (6.6.11)$$

is its torsion.

Example 6.6.1. Let us consider the Hilbert–Einstein Lagrangian

$$L_{\text{HE}} = \frac{1}{2\kappa}\mathcal{R}\sqrt{-\sigma}\omega,$$
$$\mathcal{R} = \sigma^{\lambda\nu}\mathcal{R}_{\alpha\lambda}{}^\alpha{}_\nu, \qquad \sigma = \det(\sigma_{\alpha\beta}),$$

in the metric-affine gravitation model. Then the generalized Komar superpotential (6.6.10) comes to the Komar superpotential if we substitute the Levi-Civita connection $k_\nu{}^\alpha{}_\sigma = \{{}_\nu{}^\alpha{}_\sigma\}$ (6.5.10).

6.7 Appendix. Affine world connections

The tangent bundle TX of a world manifold X as like as any vector bundle possesses a natural structure of an affine bundle (see Section 1.1.3). Therefore, one can consider affine connections on TX, called *affine world connections*. Here we study them as principal connections.

6.7. Appendix. Affine world connections

Let $Y \to X$ be an affine bundle with an k-dimensional typical fibre V. It is associated with a principal bundle AY of affine frames in Y, whose structure group is the general affine group $GA(k,\mathbb{R})$. Then any affine connection on $Y \to X$ can be seen as an associated with a principal connection on $AY \to X$. These connections are represented by global sections of the affine bundle

$$J^1P/GA(k,\mathbb{R}) \to X.$$

They are always exist.

As was mentioned in Section 1.3.5, every affine connection Γ (1.3.45) on $Y \to X$ defines a unique associated linear connection $\overline{\Gamma}$ (1.3.46) on the underlying vector bundle $\overline{Y} \to X$. This connection $\overline{\Gamma}$ is associated with a linear principal connection on the principal bundle $L\overline{Y}$ of linear frames in $\overline{Y}$ whose structure group is the general linear group $GL(k,\mathbb{R})$. We have the exact sequence of groups

$$0 \to T_k \to GA(k,\mathbb{R}) \to GL(k,\mathbb{R}) \to 1, \qquad (6.7.1)$$

where T_k is the group of translations in $\mathbb{R}^k$. It is readily observed that there is the corresponding principal bundle morphism $AY \to L\overline{Y}$ over X, and the principal connection $\overline{\Gamma}$ on $L\overline{Y}$ is the image of the principal connection Γ on $AY \to X$ under this morphism in accordance with Theorem 5.4.2.

The exact sequence (6.7.1) admits a splitting

$$GL(k,\mathbb{R}) \to GA(k,\mathbb{R}),$$

but this splitting is not canonical. It depends on the morphism

$$V \ni v \to v - v_0 \in \overline{V},$$

i.e., on the choice of an origin v_0 of the affine space V. Given v_0, the image of the corresponding monomorphism

$$GL(k,\mathbb{R}) \to GA(k,\mathbb{R})$$

is a stabilizer

$$G(v_0) \subset GA(k,\mathbb{R})$$

of v_0. Different subgroups $G(v_0)$ and $G(v_0')$ are related to each other as follows:

$$G(v_0') = T(v_0' - v_0)G(v_0)T^{-1}(v_0' - v_0),$$

where $T(v_0' - v_0)$ is the translation along the vector $(v_0' - v_0) \in \overline{V}$.

Remark 6.7.1. Accordingly, the well-known morphism of a k-dimensional affine space V onto a hypersurface $\overline{y}^{k+1} = 1$ in $\mathbb{R}^{k+1}$ and the corresponding representation of elements of $GA(k,\mathbb{R})$ by particular $(k+1) \times (k+1)$-matrices also fail to be canonical. They depend on a point $v_0 \in V$ sent to vector $(0,\ldots,0,1) \in \mathbb{R}^{k+1}$.

One can say something more if $Y \to X$ is a vector bundle provided with the natural structure of an affine bundle whose origin is the canonical zero section $\widehat{0}$. In this case, we have the canonical splitting of the exact sequence (6.7.1) such that $GL(k,\mathbb{R})$ is a subgroup of $GA(k,\mathbb{R})$ and $GA(k,\mathbb{R})$ is the semidirect product of $GL(k,\mathbb{R})$ and the group $T(k,\mathbb{R})$ of translations in $\mathbb{R}^k$. Given a $GA(k,\mathbb{R})$-principal bundle $AY \to X$, its affine structure group $GA(k,\mathbb{R})$ is always reducible to the linear subgroup since the quotient $GA(k,\mathbb{R})/GL(k,\mathbb{R})$ is a vector space $\mathbb{R}^k$ provided with the natural affine structure (see Example 5.10.2). The corresponding quotient bundle is isomorphic to the vector bundle $Y \to X$. There is the canonical injection of the linear frame bundle $LY \to AY$ onto the reduced $GL(k,\mathbb{R})$-principal subbundle of AY which corresponds to the zero section $\widehat{0}$ of $Y \to X$. In this case, every principal connection on the linear frame bundle LY gives rise to a principal connection on the affine frame bundle in accordance with Theorem 5.10.4. This is equivalent to the fact that any affine connection Γ on a vector bundle $Y \to X$ defines a linear connection $\overline{\Gamma}$ on $Y \to X$ and that every linear connection on $Y \to X$ can be seen as an affine one. Then any affine connection Γ on the vector bundle $Y \to X$ is represented by the sum of the associated linear connection $\overline{\Gamma}$ and a basic soldering form σ on $Y \to X$. Due to the vertical splitting (1.1.17), this soldering form is represented by a global section of the tensor product $T^*X \otimes Y$.

Let now $Y \to X$ be the tangent bundle $TX \to X$ considered as an affine bundle. Then the relationship between affine and linear world connections on TX is the repetition of that we have said in the case of an arbitrary vector bundle $Y \to X$. In particular, any affine world connection

$$\Gamma = dx^\lambda \otimes (\partial_\lambda + \Gamma_\lambda{}^\alpha{}_\mu(x)\dot{x}^\mu + \sigma_\lambda^\alpha(x))\partial_\alpha \qquad (6.7.2)$$

on $TX \to X$ is represented by the sum of the associated linear world connection

$$\overline{\Gamma} = \Gamma_\lambda{}^\alpha{}_\mu(x)\dot{x}^\mu dx^\lambda \otimes \partial_\alpha \qquad (6.7.3)$$

on $TX \to X$ and a basic soldering form

$$\sigma = \sigma_\lambda^\alpha(x)dx^\lambda \otimes \partial_\alpha \qquad (6.7.4)$$

on $Y \to X$, which is the $(1,1)$-tensor field on X. For instance, if $\sigma = \theta_X$ (1.1.37), we have the Cartan connection (1.3.48).

It is readily observed that the soldered curvature (1.3.29) of any soldering form (6.7.4) equals zero. Then we obtain from (1.3.32) that the torsion

6.7. Appendix. Affine world connections

(1.3.47) of the affine connection Γ (6.7.2) with respect to σ (6.7.4) coincides with that of the associated linear connection $\overline{\Gamma}$ (6.7.3) and reads

$$T = \frac{1}{2}T^i_{\lambda\mu}dx^\mu \wedge dx^\lambda \otimes \partial_i,$$
$$T_\lambda{}^\lambda{}_\mu = \Gamma_\lambda{}^\alpha{}_\nu \sigma^\nu_\mu - \Gamma_\mu{}^\alpha{}_\nu \sigma^\nu_\lambda. \tag{6.7.5}$$

The relation between the curvatures of an affine world connection Γ (6.7.2) and the associated linear connection $\overline{\Gamma}$ (6.7.3) is given by the general expression (1.3.33) where $\rho = 0$ and T is (6.7.5).

Remark 6.7.2. On may think on the physical meaning of the tensor field σ (6.7.4). One can use $\sigma^\alpha_\lambda dx^\lambda$ as a non-holonomic coframes in the metric-affine gauge theory with non-holonomic GL_4 gauge transformations (see, e.g., [76]). In the gauge theory of dislocations in continuous media, the field σ is treated as an elastic distortion [87; 110; 130].

Chapter 7

Spinor fields

Classical theory of Dirac spinor fields is a theory with spontaneous symmetry breaking. The following three facts make it most interesting both from physical and mathematical viewpoints.
- Dirac fermions are unique observable matter fields.
- Dirac fermions are odd fields.
- The existence of Dirac fermion matter possessing Lorentz symmetries is the underlying physical reason of breakdown of world symmetries and, consequently, of the existence of a gravitational field.

7.1 Clifford algebras and Dirac spinors

Dirac spinors are conventionally described in the framework of formalism of Clifford algebras [103].

Let $M = \mathbb{R}^4$ be the Minkowski space equipped with the Minkowski metric

$$\eta = \mathrm{diag}(1, -1, -1, -1),$$

written with respect to a fixed basis $\{e^a\}$ for M. Let $\mathbb{C}_{1,3}$ be the complex *Clifford algebra* generated by elements of M. It is defined as the complexified quotient of the tensor algebra

$$\otimes M = \mathbb{R} \oplus M \oplus \cdots \oplus M^{\otimes k} \oplus \cdots$$

of M by the two-sided ideal generated by elements

$$e \otimes e' + e' \otimes e - 2\eta(e, e') \in \otimes M, \qquad e, e' \in M.$$

Remark 7.1.1. The complex Clifford algebra $\mathbb{C}_{1,3}$ is isomorphic to the real Clifford algebra $\mathbb{R}_{2,3}$, whose generating space is $\mathbb{R}^5$ equipped with the pseudo-Euclidean metric

$$\mathrm{diag}(1, -1, -1, -1, 1).$$

Its subalgebra generated by elements of $M \subset \mathbb{R}^5$ is the real Clifford algebra $\mathbb{R}_{1,3}$.

A *Dirac spinor space* V (or, simply, a *spinor space*) is defined as a minimal left ideal of $\mathbb{C}_{1,3}$ on which this algebra acts on the left. There is the representation

$$\gamma : M \otimes V \to V, \qquad (7.1.1)$$
$$\gamma(e^\alpha) = \gamma^\alpha,$$

of elements of the Minkowski subspace $M \subset \mathbb{C}_{1,3}$ by the Dirac γ-matrices on V. Let us mention the relations

$$(\gamma^\alpha)^+ = \eta^{\alpha\alpha}\gamma^\alpha, \qquad (\gamma^0\gamma^\alpha)^+ = \gamma^0\gamma^\alpha,$$

where the symbol $()^+$ stands for a Hermitian conjugate matrix.

Remark 7.1.2. The explicit form of the representation (7.1.1) depends on the choice of a minimal left ideal V of $\mathbb{C}_{1,3}$. Different ideals lead to equivalent representations (7.1.1). One usually considers the representation, where

$$\gamma^0 = \begin{pmatrix} 1 & 0 & 0 & 0 \\ 0 & 1 & 0 & 0 \\ 0 & 0 & -1 & 0 \\ 0 & 0 & 0 & -1 \end{pmatrix}.$$

The *Clifford group* $G_{1,3} \subset \mathbb{R}_{1,3}$ is defined to consist of the invertible elements l_s of the real Clifford algebra $\mathbb{R}_{1,3}$ such that the inner automorphisms given by these elements preserve the Minkowski space $M \subset \mathbb{R}_{1,3}$, i.e.,

$$l_s e l_s^{-1} = l(e), \qquad e \in M, \qquad (7.1.2)$$

where l is a Lorentz transformation of M. Hence, there is an epimorphism of the Clifford group $G_{1,3}$ onto the Lorentz group $O(1,3)$. However, the action (7.1.2) of the Clifford group on the Minkowski space M is not effective. Therefore, one consider its *pin* and *spin subgroups*. The subgroup $\mathrm{Pin}(1,3)$ of $G_{1,3}$ is generated by elements $e \in M$ such that $\eta(e,e) = \pm 1$. The even part of $\mathrm{Pin}(1,3)$ is the spin group $\mathrm{Spin}(1,3)$, i.e., $\eta(e,e) = 1$, $e \in \mathrm{Spin}(1,3)$. Its component of the unity

$$\mathbf{L}_s = \mathrm{Spin}^0(1,3) \simeq SL(2,\mathbb{C}) \qquad (7.1.3)$$

7.1. Clifford algebras and Dirac spinors

is the well-known two-fold universal covering group

$$z_L : L_s \to L = L_s/\mathbb{Z}_2 \tag{7.1.4}$$

of the proper Lorentz group L. We agree to call L_s (7.1.4) the *spinor Lorentz group*. Its Lie algebra $\mathfrak{g}_L$ is that of the proper Lorentz group L.

Remark 7.1.3. The generating elements $e \in M$, $\eta(e,e) = \pm 1$, of the pin group Pin(1,3) act on the Minkowski space by the adjoint representation which is the composition

$$e : v \to eve^{-1} = -v + 2\frac{\eta(e,v)}{\eta(e,e)}e, \qquad e, v \in \mathbb{R}^4,$$

of the total reflection of M and the reflection across the hyperplane

$$e^\perp = \{w \in M;\ \eta(e,w) = 0\}$$

which is perpendicular to e with respect to the metric η in M. By the well-known Cartan–Dieudonné theorem, every element of the pseudo-orthogonal group $O(p,q)$ can be written as a product of $r \leq p+q$ reflections across hyperplanes in the vector space $\mathbb{R}^{p+q}$ [103]. In particular, the spin group Spin(1,3) consists of the elements of Pin(1,3) which result from an even number of reflections of M. The epimorphism of Spin(1,3) onto the Lorentz group L and the epimorphism (7.1.4) are defined by the fact that elements e and $-e$ of M determine the same reflection of M across the hyperplane $e^\perp = (-e)^\perp$. We further consider an action of the spinor Lorentz group L_s (factorizing through that of the proper Lorentz group L) on the Minkowski space M, but it is not effective.

The Clifford group $G_{1,3}$ acts on the Dirac spinor space V by left multiplications

$$G_{1,3} \ni l_s : v \mapsto l_s v, \qquad v \in V.$$

This action preserves the representation (7.1.1), i.e.,

$$\gamma(lM \otimes l_s V) = l_s \gamma(M \otimes V).$$

The spinor Lorentz group L_s acts on the Dirac spinor space V by means of the infinitesimal generators

$$I_{ab} = \frac{1}{4}[\gamma_a, \gamma_b]. \tag{7.1.5}$$

Since

$$I^+_{ab}\gamma^0 = -\gamma^0 I_{ab},$$

the Dirac spinor space V is provided with the L_s-invariant bilinear form

$$a(v, v') = \frac{1}{2}(v^+\gamma^0 v' + v'^+\gamma^0 v), \qquad (7.1.6)$$

called the *spinor metric*.

In the framework of classical field theory on fibre bundles, classical Dirac spinor fields are described by sections of a spinor bundle S on a world manifold X whose typical fibre is the Dirac spinor space V and whose structure group is the spinor Lorentz group L_s. In order to construct the Dirac operator, one however need a fibrewise action (7.1.1) of the whole Clifford algebra $\mathbb{C}_{1,3}$ on a spinor bundle (see Remark 7.2.3). Therefore, a spinor bundle must be represented as a subbundle of the bundle in Clifford algebras [103].

Let us start with a *fibre bundle in Minkowski spaces* $MX \to X$ over a world manifold X. It is defined as a fibre bundle with the typical fibre M and the structure group L. This fibre bundle is extended to a *fibre bundle in Clifford algebras* CX whose fibres $C_x X$ are the Clifford algebras generated by the fibres $M_x X$ of the fibre bundle in Minkowski spaces MX. The fibre bundle CX possesses the structure group $\mathrm{Aut}(\mathbb{C}_{1,3})$ of inner automorphisms of the complex Clifford algebra $\mathbb{C}_{1,3}$. This structure group is reducible to the proper Lorentz group L and, certainly, the bundle in Clifford algebras CX contains the subbundle MX of the generating Minkowski spaces. However, CX need not contain a spinor subbundle because a spinor subspace V of $\mathbb{C}_{1,3}$ is not stable under inner automorphisms of $\mathbb{C}_{1,3}$. A spinor subbundle S_M of CX exists if transition functions of CX can be lifted from the Clifford group $G_{1,3}$. This condition agrees with the familiar condition of the existence of a spinor structure (see Remark 7.2.1).

The bundle MX in Minkowski spaces must be isomorphic to the cotangent bundle T^*X in order that sections of the spinor bundle S_M describe Dirac fermion fields on a world manifold X. In other words, we should consider a spinor structure on the cotangent bundle T^*X of X [103].

7.2 Dirac spinor structure

There are several almost equivalent definitions of a spinor structure on a world manifold X [7; 103]. A *Dirac spinor structure* on a world manifold X is said to be a pair (P_s, z_s) of a principal L_s-bundle $P_s \to X$ and a principal bundle morphism

$$z_s : P_s \underset{X}{\to} LX \qquad (7.2.1)$$

7.2. Dirac spinor structure

of P_s to the linear frame bundle $LX \to X$.

Since the group homomorphism $L_s \to GL_4$ factorizes through the epimorphism (7.1.4), every bundle morphism (7.2.1) factorizes through a morphism

$$z_h : P_s \to L^h X, \qquad (7.2.2)$$
$$z_h \circ R_{gP} = R_{z_L(g)P}, \qquad g \in L_s,$$

of P_s to some reduced principal Lorentz subbundle $L^h X$ of the linear frame bundle LX whose structure group is the proper Lorentz group L.

It follows that the necessary condition for the existence of a Dirac spinor structure on X is that the structure group GL_4 of LX is reducible to the proper Lorentz group L. Herewith, any Dirac spinor structure on a world manifold X is associated with some tetrad field h or a pseudo-Riemannian metric $g = \zeta \circ h$. Therefore, this Dirac spinor structure also is the *pseudo-Riemannian spinor structure* on a world manifold.

Conversely, given a reduced Lorentz structure $L^h X \subset LX$, the associated Dirac spinor structure (7.2.2) exists if the following conditions hold.

Lemma 7.2.1. *All spinor structures on a world manifold X which are related to the two-fold universal covering groups possess the following two properties [70].*

(i) *Let $P \to X$ be a principal bundle whose structure group G has the fundamental group $\pi_1(G) = \mathbb{Z}_2$. Let $\widetilde{G}$ be the universal two-fold covering group of G, i.e., $\widetilde{G}$ is the extension (10.4.10):*

$$1 \to \mathbb{Z}_2 \longrightarrow \widetilde{G} \longrightarrow G \to 1,$$

of a group G by the commutative group $\mathbb{Z}_2$. The topological obstruction to that a principal G-bundle $P \to X$ lifts to a principal $\widetilde{G}$-bundle $\widetilde{P} \to X$ is given by the Čech cohomology group $H^2(X;\mathbb{Z}_2)$ of X. Namely, a principal bundle P defines an element of $H^2(X;\mathbb{Z}_2)$ which must be zero so that $P \to X$ can give rise to $\widetilde{P} \to X$.

(ii) *Non-equivalent lifts of $P \to X$ to principal $\widetilde{G}$-bundles are classified by elements of the Čech cohomology group $H^1(X;\mathbb{Z}_2)$.*

In our case, the topological obstruction to that a reduced Lorentz structure $L^h X$ lifts to the Dirac spinor one is the second Stiefel–Whitney class $w_2(X) \in H^2(X;\mathbb{Z}_2)$ of X [103].

Remark 7.2.1. A world manifold X thus must satisfy certain topological conditions in order to admit a Dirac spinor structure. Spinor bundles S

over X with the structure group L_s (7.1.3) are classified by the Chern classes $c_i(S) \in H^{2i}(X,\mathbb{Z})$, $i = 1, 2$. Since the group L_s is reducible to its maximal compact subgroup $SU(2)$, the first Chern class $c_1(S)$ of S is trivial, while the second Chern class $c_2(S)$ is represented by the cohomology of the characteristic form $c_2(F)$ (see Example 8.1.4). Let $S = S^h$ be a Dirac spinor bundle associated with the cotangent bundle T^*X both due to the bundle morphism (7.2.2) and that T^*X is associated with reduced Lorentz bundle L^hX. Strictly speaking, T^*X is associated with the tensor product $S^h \otimes S^{h*}$. Because of the inclusion

$$GL_4 \to GL(4,\mathbb{C})$$

(see the commutative diagram (8.1.24)), the cotangent bundle T^*X can be regarded as a $GL(4,\mathbb{C})$-bundle $\varphi(T^*X)$. Consequently, there is the relation (8.1.29) between the Pontryagin and Chern classes of a world manifold. Moreover, since the fibre bundle $\varphi(T^*X)$ is associated with the tensor product $S^h \otimes S^{h*}$, we obtain

$$p_1(X) = -c_2(\varphi(T^*X)) = -4c_2(S^h).$$

One can reproduce the first relation in terms of the characteristic forms (8.1.22) and (8.1.17) if a principal connection on LX is a Lorentz connection on L^hX induced by a spinor connection on P^h (see Theorem 7.2.1). Some additional properties of a space-time structure (e.g., that a spatial distribution **F** is orientable) also are required. As a result, one can state the following [51; 165].

 • A non-compact world manifold admits a Dirac spinor structure if and only if it is parallelizable.

 • For a compact world manifold X, its Euler characteristic and the second Stiefel-Whitney class w_2 must be zero, and its first Pontryagin number must be multiple of 48.

Remark 7.2.2. Let us compare a pseudo-Riemannian spinor structure with the Riemannian one. To introduce a Riemannian spinor structure, one considers the complex Clifford algebra $\mathbb{C}_{4,0}$ which is generated by elements of the vector space $\mathbb{R}^4$ equipped with the Euclidean metric [103]. The corresponding spinor space V_E is a minimal left ideal of $\mathbb{C}_{4,0}$. The spin group is Spin(4) which is the two-fold universal covering group of the group $SO(4)$. It is isomorphic to $SU(2) \otimes SU(2)$. Let us assume that the second Stiefel–Whitney class $w_2(X)$ of X vanishes. A *Riemannian spinor structure* on a world manifold X is defined as a pair of a principal Spin(4)-bundle

7.2. Dirac spinor structure

$P_s \to X$ and a principal bundle morphism z_s of P_s to LX. Since such a morphism factorizes through a bundle morphism

$$z_{g^R} : P_s \to L^{g^R} X$$

for a Riemannian metric g^R, this spinor structure is a g_R-associated spinor structure.

Hereafter, we restrict our consideration to Dirac spinor structures on a non-compact (and, consequently, parallelizable) world manifold X. In this case, all Dirac spinor structures are isomorphic [7; 51]. Therefore, there is one-to-one correspondence

$$z_h : P_s^h \to L^h X \subset LX \qquad (7.2.3)$$

between the reduced Lorentz structures $L^h X$ and the Dirac spinor structures (P_s^h, z_h) which factorize through the corresponding $L^h X$. In particular, every Lorentz bundle atlas $\Psi^h = \{z_\iota^h\}$ (6.3.4) of $L^h X$ gives rise to an atlas

$$\overline{\Psi}^h = \{\overline{z}_\iota^h\}, \qquad z_\iota^h = z_h \circ \overline{z}_\iota^h, \qquad (7.2.4)$$

of the principal L_s-bundle P_s^h. We agree to call P_s^h the *spinor principal bundles*.

Let (P_s^h, z_h) be the Dirac spinor structure associated with a tetrad field h. Let

$$S^h = (P_s^h \times V)/L_s \to X \qquad (7.2.5)$$

be the P_s^h-associated *spinor bundle* whose typical fibre V carriers the spinor representation (7.1.5) of the spinor Lorentz group L_s. One can think of sections of S^h (7.2.5) as describing *Dirac spinor fields* in the presence of a tetrad field h.

Indeed, let us consider the $L^h X$-associated bundle in Minkowski spaces

$$M^h X = (L^h X \times M)/L = (P_s^h \times M)/L_s \qquad (7.2.6)$$

and the P_s^h-associated spinor bundle S^h (7.2.5). By virtue of Remark 5.10.4, the fibre bundle $M^h X$ (7.2.6) is isomorphic to the cotangent bundle

$$T^* X = (L^h X \times M)/L. \qquad (7.2.7)$$

Then, using the morphism (7.1.1), one can define the representation

$$\gamma_h : T^* X \otimes S^h = (P_s^h \times (M \otimes V))/L_s \to \qquad (7.2.8)$$
$$(P_s^h \times \gamma(M \otimes V))/L_s = S^h$$

of covectors to X by the Dirac γ-matrices on elements of the spinor bundle S^h. Relative to a Lorentz bundle atlas $\{z_\iota^h\}$ of LX and the corresponding atlas $\{\bar{z}_\iota\}$ (7.2.4) of the spinor principal bundle P_s^h, the representation (7.2.8) reads

$$y^A(\gamma_h(h^a(x) \otimes v)) = \gamma^{aA}{}_B y^B(v), \qquad v \in S_x^h,$$

where y^A are the associated bundle coordinates on S^h, and h^a are the tetrad coframes (6.3.7). For brevity, we write

$$\widehat{h}^a = \gamma_h(h^a) = \gamma^a,$$
$$\widehat{dx}^\lambda = \gamma_h(dx^\lambda) = h_a^\lambda(x)\gamma^a.$$

Remark 7.2.3. In fact, the spinor bundle S^h is a subbundle of the bundle in Clifford algebras generated by the fibre bundle in Minkowski spaces $M^h X$. Then the representation γ_h (7.2.8) results from the action γ (7.1.1) of M on V.

Furthermore, let

$$A_h = dx^\lambda \otimes (\partial_\lambda + \frac{1}{2} A_\lambda{}^{ab} e_{ab}) \tag{7.2.9}$$

be a principal connection on a spinor principal bundle P_s^h. It is called a *spinor connection*. The associated principal connection on the spinor bundle S^h (7.2.5) reads

$$A_h = dx^\lambda \otimes (\partial_\lambda + \frac{1}{2} A^{ab}{}_\lambda I_{ab}{}^A{}_B y^B) \partial_A, \tag{7.2.10}$$

where I_{ab} are the generators (7.1.5) of the spinor Lorentz group $\mathbf{L}_s$. Let

$$D: J^1 S^h \to T^* X \underset{S^h}{\otimes} S^h,$$

$$D = (y_\lambda^A - A^{ab}{}_\lambda I_{ab}{}^A{}_B y^B) dx^\lambda \otimes \partial_A,$$

be the corresponding covariant differential (1.3.18), where the canonical vertical splitting

$$VS^h = S^h \underset{X}{\times} S^h$$

has been used. The first order differential *Dirac operator* is defined on S^h as the composition

$$\mathcal{D}_h = \gamma_h \circ D : J^1 S^h \to T^* X \otimes S^h \to S^h, \tag{7.2.11}$$

$$y^A \circ \mathcal{D}_h = h_a^\lambda \gamma^{aA}{}_B (y_\lambda^B - \frac{1}{2} A^{ab}{}_\lambda I_{ab}{}^A{}_B y^B).$$

Theorem 7.2.1. *There is one-to-one correspondence between the spinor principal connections on a spinor principal bundle P_s^h and the Lorentz connection on the principal L-bundle $L^h X$.*

7.2. Dirac spinor structure

Proof. It follows from Theorem 5.4.2 that every principal connection on P_s^h defines a principal connection on $L^h X$ which is given by the same expression (7.2.9). Conversely, the pull-back $z_h^* \overline{A}_h$ onto P_s^h of the connection form $\overline{A}_h$ of a Lorentz connection A_h on $L^h X$ is equivariant under the action of group L_s on P_s^h and, consequently, it is a connection form of a spinor principal connection on P_s^h. □

In particular, the Levi–Civita connection of a pseudo-Riemannian metric $g = \zeta \circ h$ gives rise to a spinor connection

$$A_h = dx^\lambda \otimes [\partial_\lambda + \frac{1}{2}\eta^{kb} h_\mu^a (\partial_\lambda h_k^\mu - h_k^\nu \{\lambda^\mu{}_\nu\}) I_{ab}{}^A{}_B y^B] \partial_A \qquad (7.2.12)$$

on the h-associated spinor bundle S^h.

Moreover, every linear world connection Γ on a world manifold X defines the Lorentz connection (6.3.18) on a reduced Lorentz bundle $L^h X$ and the associated spinor connection (7.2.10):

$$A_h = dx^\lambda \otimes [\partial_\lambda + \frac{1}{4}(\eta^{kb} h_\mu^a - \eta^{ka} h_\mu^b)(\partial_\lambda h_k^\mu - h_k^\nu \Gamma_\lambda{}^\mu{}_\nu) I_{ab}{}^A{}_B y^B] \partial_A, \qquad (7.2.13)$$

on the h-associated spinor bundle S^h. Such a connection has been considered in [5; 125; 136].

Substituting the spinor connection (7.2.13) in the Dirac operator (7.2.11), we obtain a description of Dirac spinor fields in the presence of an arbitrary linear world connection on a world manifold, not only of the Lorentz type.

It should be emphasized that a spinor bundle S^h is not natural. Any connection A_h (7.2.13) defines the horizontal lift

$$A_h \tau = \tau^\lambda \partial_\lambda + \frac{1}{4}\tau^\lambda (\eta^{kb} h_\mu^a - \eta^{ka} h_\mu^b)(\partial_\lambda h_k^\mu - h_k^\nu \Gamma_\lambda{}^\mu{}_\nu) I_{ab}{}^A{}_B y^B \partial_A \qquad (7.2.14)$$

onto S^h of a vector field τ on X. Moreover, there is the canonical horizontal lift

$$\widetilde{\tau} = \tau^\lambda \partial_\lambda + \frac{1}{4}(\eta^{kb} h_\mu^a - \eta^{ka} h_\mu^b)(\tau^\lambda \partial_\lambda h_k^\mu - h_k^\nu \partial_\nu \tau^\mu) I_{ab}{}^A{}_B y^B \partial_A \qquad (7.2.15)$$

onto S^h of vector fields τ on X. However, this lift fails to be functorial.

Remark 7.2.4. In order to construct the canonical lift (7.2.15), one can write the functorial lift (6.1.9) of τ onto the linear frame bundle LX with respect to a Lorentz atlas Ψ^h and, afterwards, can take its Lorentz part. Another way is the following. Let us consider a local nowhere vanishing vector field τ and the local symmetric world connection Γ_τ (6.2.12) whose integral section is τ (see Remark 6.2.1). Let A_τ be the corresponding spinor connection (7.2.13). The horizontal lift (7.2.14) of τ by means of this connection is given by the expression (7.2.15). In a straightforward manner, one can check that (7.2.15) is a well-behaved lift of any vector field τ on X. The canonical lift (7.2.15) is brought into the form

$$\widetilde{\tau} = \tau_{\{\}} - \frac{1}{4}(\eta^{kb}h_\mu^a - \eta^{ka}h_\mu^b)h_k^\nu \nabla_\nu \tau^\mu I_{ab}{}^A{}_B y^B \partial_A,$$

where $\tau_{\{\}}$ is the horizontal lift (7.2.14) of τ by means of the spinor Levi–Civita connection (7.2.12) of a tetrad field h, and $\nabla_\nu \tau^\mu$ are the covariant derivatives of τ relative to the same Levi–Civita connection. This is precisely the Lie derivative of spinor fields described in [95].

7.3 Universal spinor structure

Dirac spinor fields in the presence of different tetrad field h and h' are described by sections of different spinor bundles S^h and $S^{h'}$. A problem is that, though the reduced Lorentz bundles $L^h X$ and $L^{h'} X$ are isomorphic, the associated structures of bundles in Minkowski spaces $M^h X$ and $M^{h'} X$ (7.2.7) on the cotangent bundle T^*X are not equivalent because of the non-equivalent actions of the Lorentz group on the typical fibre of T^*X seen as a typical fibre of $M^h X$ and that of $M^{h'} X$ (see Remark 5.10.3). As a consequence, the representations γ_h and $\gamma_{h'}$ (7.2.8) for different tetrad fields h and h' are not equivalent [132]. Indeed, let

$$t^* = t_\mu dx^\mu = t_a h^a = t'_a h'^a$$

be an element of T^*X. Its representations γ_h and $\gamma_{h'}$ (7.2.8) read

$$\gamma_h(t^*) = t_a \gamma^a = t_\mu h_a^\mu \gamma^a,$$
$$\gamma_{h'}(t^*) = t'_a \gamma^a = t_\mu h'^\mu_a \gamma^a.$$

They are not equivalent because no isomorphism Φ_s of S^h onto $S^{h'}$ can obey the condition

$$\gamma_{h'}(t^*) = \Phi_s \gamma_h(t^*) \Phi_s^{-1}, \qquad t^* \in T^*X.$$

It follows that a Dirac fermion field must be described in a pair with a certain tetrad (gravitational) field. We thus observe the phenomenon of spontaneous symmetry breaking in gauge gravitation theory which exhibits the physical nature of gravity as a Higgs field [132; 140]. In order to study this phenomenon, let us follow the general scheme of describing spontaneous symmetry breaking in Section 5.10. We are based on the fact that any Dirac

7.3. Universal spinor structure

spinor structure on a world manifold is a reduced subbundle of a so called universal spinor bundle [53; 138].

The structure group GL_4 of the linear frame bundle LX is not simply-connected. Its first homotopy group is

$$\pi_1(GL_4) = \pi_1(SO(4)) = \mathbb{Z}_2$$

[68]. Therefore, the group GL_4 admits the universal two-fold covering group $\widetilde{GL}_4$ such that the diagram

$$\begin{array}{ccc} \widetilde{GL}_4 & \longrightarrow & GL_4 \\ \uparrow & & \uparrow \\ \mathrm{Spin}(4) & \longrightarrow & \mathrm{SO}(4) \end{array} \qquad (7.3.1)$$

is commutative [76; 103; 148]. The *universal spinor structure* on a world manifold X is defined as a pair $(\widetilde{LX}, \widetilde{z})$ of a principal $\widetilde{GL}_4$-bundle $\widetilde{LX} \to X$ and a principal bundle morphism

$$\widetilde{z} : \widetilde{LX} \underset{X}{\longrightarrow} LX \qquad (7.3.2)$$

[31; 148]. Due to the commutative diagram (7.3.1), there is the commutative diagram of principal bundles

$$\begin{array}{ccc} \widetilde{LX} & \xrightarrow{\widetilde{z}} & LX \\ \uparrow & & \uparrow \\ P^{g^R} & \longrightarrow & L^{g^R}X \end{array} \qquad (7.3.3)$$

for any Riemannian metric g^R [148].

Since the group $\widetilde{GL}_4$ is homotopic to the group $\mathrm{Spin}(4)$, there is one-to-one correspondence between the non-equivalent universal spinor structures and non-equivalent Riemannian spinor structures (see Remark 7.2.2) [148]. All universal spinor structures (as like as the Riemannian ones) on a parallelizable world manifold X are equivalent, i.e., the principal $\widetilde{GL}_4$-bundle $\widetilde{LX}$ (7.3.2) is uniquely defined. It is called the *universal spinor bundle*. Then it follows from the commutative diagram (7.3.3) that any Riemannian spinor structure on a world manifold is a reduced subbundle of the universal spinor bundle $\widetilde{LX}$.

Remark 7.3.1. Though the group $\widetilde{GL}_4$ has finite-dimensional representations, its spinor representation is infinite-dimensional [76; 119]. Elements of this representation are called *world spinors*. Their field model has been developed (see [76] and references therein).

The pseudo-Riemannian spinor structures as like as the Riemannian ones are subbundles of the universal spinor bundle $\widetilde{LX}$ as follows [53; 138].

Lemma 7.3.1. *Just as the diagram (7.3.1), the diagram*

$$\begin{array}{ccc} \widetilde{GL_4} & \longrightarrow & GL_4 \\ \uparrow & & \uparrow \\ L_s & \xrightarrow{z_L} & L \end{array} \qquad (7.3.4)$$

is commutative.

Proof. The restriction of the universal covering group $\widetilde{GL_4} \to GL_4$ to the proper Lorentz group $L \subset GL_4$ is obviously a covering space of L. Let us show that this is the universal covering space. Indeed, any non-contractible cycle in GL_4 belongs to some subgroup $SO(3) \subset GL_4$, and the restriction of the covering bundle $\widetilde{GL_4} \to GL_4$ to $SO(3)$ is the universal covering of $SO(3)$. Since the proper Lorentz group is homotopic to its maximal compact subgroup $SO(3)$, its universal covering space belongs to $\widetilde{GL_4}$. □

Due to the commutative diagram (7.3.4), we have the commutative diagram of principal bundles

$$\begin{array}{ccc} \widetilde{LX} & \xrightarrow{\widetilde{z}} & LX \\ \uparrow & & \uparrow \\ P_s^h & \xrightarrow{z_h} & L^h X \end{array}$$

for any tetrad field h [47; 53; 138].

It follows that any Dirac spinor structure P_s^h (7.2.3) is a reduced L_s-subbundle of the universal spinor bundle $\widetilde{LX}$. Moreover, $\widetilde{LX}$ is a principal L_s-bundle

$$\widetilde{\pi} : \widetilde{LX} \to \Sigma_T \qquad (7.3.5)$$

over the tetrad bundle (6.3.1):

$$\Sigma_T = \widetilde{LX}/L_s = LX/L, \qquad (7.3.6)$$

such that the universal spinor structure (7.3.2) is a bundle morphism

$$\widetilde{z} : \widetilde{LX} \xrightarrow[\Sigma_T]{} LX \qquad (7.3.7)$$

over Σ_T. It is called the *universal Dirac spinor structure* on the tetrad bundle Σ_T (7.3.6). Given a tetrad field h, the restriction $h^*\widetilde{LX}$ of the principal L_s-bundle (7.3.5) to $h(X) \subset \Sigma_T$ is isomorphic to the subbundle

7.3. Universal spinor structure

P^h of the fibre bundle $\widetilde{LX} \to X$ which is the h-associated Dirac spinor structure on a world manifold.

Let us consider the spinor bundle

$$S = (\widetilde{LX} \times V)/\mathrm{L_s} \to \Sigma_\mathrm{T} \tag{7.3.8}$$

associated with the principal $\mathrm{L_s}$-bundle (7.3.5). It also is the composite bundle

$$S \to \Sigma_\mathrm{T} \to X. \tag{7.3.9}$$

This however is not a spinor bundle over X. By virtue of Theorem 5.10.1, this composite bundle is a $\widetilde{LX}$-associated bundle whose typical fibre is the spinor bundle

$$(\widetilde{GL_4} \times V)/\mathrm{L_s} \to \widetilde{GL_4}/\mathrm{L_s}.$$

Given a tetrad field h, there is the canonical isomorphism

$$i_h : S^h = (P^h \times V)/\mathrm{L_s} \to (h^*\widetilde{LX} \times V)/\mathrm{L_s}$$

of the h-associated spinor bundle S^h (7.2.5) onto the restriction h^*S of the spinor bundle $S \to \Sigma_\mathrm{T}$ to $h(X) \subset \Sigma_\mathrm{T}$ (see Theorem 1.4.1). Then every global section s_h of the spinor bundle S^h corresponds to the global section $i_h \circ s_h$ of the composite bundle (7.3.9). Conversely, every global section s of the composite bundle (7.3.9) projected onto a tetrad field h takes its values into the subbundle $i_h(S^h) \subset S$ (see Theorem 1.4.2).

Let the linear frame bundle $LX \to X$ be provided with a holonomic atlas Ψ_T (6.1.6), and let the principal bundles $\widetilde{LX} \to \Sigma_\mathrm{T}$ and $LX \to \Sigma_\mathrm{T}$ have the associated atlases $\{(U_\epsilon, z_\epsilon^s)\}$ and

$$\{(U_\epsilon, z_\epsilon = \widetilde{z} \circ z_\epsilon^s)\}, \tag{7.3.10}$$

respectively. With these atlases, the composite bundle S (7.3.9) is equipped with the bundle coordinates $(x^\lambda, \sigma_a^\mu, y^A)$, where $(x^\lambda, \sigma_a^\mu)$ are coordinates on the tetrad bundle Σ_T such that, whenever h is a tetrad field, $h_a^\mu = \sigma_a^\mu \circ h$ are the tetrad functions of the Lorentz bundle atlas

$$\Psi_h = \{h^{-1}(U_\epsilon), z_\epsilon \circ h\}$$

of the reduced Lorentz bundle $L^h X$.

Remark 7.3.2. In fact, $\sigma_a^\mu = H_a^\mu \circ z_\epsilon$ are coordinates of the image of Σ_T in LX with respect to local sections z_ϵ. They are the above mentioned transition functions (5.10.44) between a holonomic bundle atlas Ψ_T of $LX \to X$ and the atlas (7.3.10) of the bundle $LX \to \Sigma_\mathrm{T}$.

The spinor bundle $S \to \Sigma_T$ is the subbundle of the bundle in Clifford algebras which is generated by the bundle in Minkowski spaces

$$E_M = (LX \times M)/\mathrm{L} \to \Sigma_T \qquad (7.3.11)$$

associated with the principal L-bundle $LX \to \Sigma$. By virtue of Lemma 5.10.1, it is the LX-associated composite bundle

$$E_M \to \Sigma_T \to X \qquad (7.3.12)$$

whose typical fibre is the L-bundle

$$(GL_4 \times \mathbb{R}^4)/\mathrm{L}$$

associated with the principal L-bundle

$$P_{\mathrm{L}} = GL_4 \to GL_4/\mathrm{L}. \qquad (7.3.13)$$

Lemma 7.3.2. *The principal L-bundle (7.3.13) is trivial.*

Proof. In accordance with the classification theorem [147], a principal G-bundle over an n-dimensional sphere S^n is trivial if the homotopy group $\pi_{n-1}(G)$ is trivial. The base space $Z = GL_4/\mathrm{L}$ of the principal bundle (7.3.13) is homeomorphic to $S^3 \times \mathbb{R}^7$. Let us consider the morphism f_1 of S^3 into Z, $f_1(p) = (p, 0)$, and the pull-back principal L-bundle $f_1^* P_{\mathrm{L}} \to S^3$. Since L is homeomorphic to $\mathbf{RP}^3 \times \mathbb{R}^3$ and $\pi_2(\mathrm{L}) = 0$, this bundle is trivial. Let f_2 be the projection of Z onto S^3. Then, the pull-back principal L-bundle

$$f_2^*(f_1^* P_{\mathrm{L}}) \to Z \qquad (7.3.14)$$

also is trivial. Since the composition $f_1 \circ f_2$ of Z into Z is homotopic to the identity morphism of Z, the fibre bundle (7.3.14) is equivalent to the bundle P_{L} [147]. It follows that the bundle (7.3.13) also is trivial. $\square$

Since a world manifold is assumed to be parallelizable, the linear frame bundle $LX \to X$ also is trivial, and the fibre bundle $E_M \to X$ is so. Hence, it is isomorphic to the product $\Sigma_T \underset{X}{\times} T^*X$. Then there exists the representation

$$\gamma_\Sigma : T^*X \underset{\Sigma_T}{\otimes} S = (\widetilde{LX} \times (M \otimes V))/\mathrm{L_s} \to \qquad (7.3.15)$$

$$(\widetilde{LX} \times \gamma(M \otimes V))/\mathrm{L_s} = S,$$

given by the coordinate expression

$$\widehat{dx}^\lambda = \gamma_\Sigma(dx^\lambda) = \sigma_a^\lambda \gamma^a.$$

7.3. Universal spinor structure

Restricted to $h(X) \subset \Sigma_T$, this representation recovers the morphism γ_h (7.2.8).

Let Γ (6.2.1) be a linear world connection. It defines the spinor connection A_h (7.2.13) on each h-associated spinor bundle S^h (7.2.5). Following the construction in Section 5.10.5, let us consider a connection A_Σ on the spinor bundle $S \to \Sigma_T$ (7.3.8) such that, for each tetrad field h, the pullback connection $h^* A_\Sigma$ (5.10.29) on S^h coincides with the spinor connection A_h (7.2.13). Such a connection A_Σ exists [138]. It takes the form

$$A_\Sigma = dx^\lambda \otimes (\partial_\lambda + \frac{1}{2} A_\lambda{}^{ab} I_{ab}{}^A{}_B y^B \partial_A) + \qquad (7.3.16)$$

$$d\sigma_k^\mu \otimes (\partial_\mu^k + \frac{1}{2} A_\mu^{k\,ab} I_{ab}{}^A{}_B y^B \partial_A),$$

$$A_\lambda{}^{ab} = -\frac{1}{2}(\eta^{kb}\sigma_\mu^a - \eta^{ka}\sigma_\mu^b)\sigma_k^\nu \Gamma^\mu{}_{\lambda\nu},$$

$$A_\mu^{k\,ab} = \frac{1}{2}(\eta^{kb}\sigma_\mu^a - \eta^{ka}\sigma_\mu^b).$$

The connection (7.3.16) yields the first order differential operator $\widetilde{D}$ (1.4.17) on the composite bundle $S \to X$ (7.3.9) which reads

$$\widetilde{D} : J^1 S \to T^* X \underset{\Sigma_T}{\otimes} S,$$

$$\widetilde{D} = dx^\lambda \otimes [y_\lambda^A - \frac{1}{2}(A_\lambda{}^{ab} + A_\mu^{k\,ab} \sigma_{\lambda k}^\mu) I_{ab}{}^A{}_B y^B] \partial_A = \qquad (7.3.17)$$

$$dx^\lambda \otimes [y_\lambda^A - \frac{1}{4}(\eta^{kb}\sigma_\mu^a - \eta^{ka}\sigma_\mu^b)(\sigma_{\lambda k}^\mu - \sigma_k^\nu \Gamma^\mu{}_{\lambda\nu}) I_{ab}{}^A{}_B y^B] \partial_A.$$

The restriction $\widetilde{D}_h$ of the operator $\widetilde{D}$ (7.3.17) to $J^1 S^h \subset J^1 S$ recovers the familiar covariant differential on the h-associated spinor bundle S^h (7.2.5) relative to the spinor connection (7.2.13).

Combining the formulas (7.3.15) and (7.3.17) gives the first order differential operator

$$\mathcal{D} = \gamma_{\Sigma_T} \circ \widetilde{D} : J^1 S \to T^* X \underset{\Sigma_T}{\otimes} S \to S, \qquad (7.3.18)$$

$$y^B \circ \mathcal{D} = \sigma_a^\lambda \gamma^{aB}{}_A [y_\lambda^A - \frac{1}{4}(\eta^{kb}\sigma_\mu^a - \eta^{ka}\sigma_\mu^b)(\sigma_{\lambda k}^\mu - \sigma_k^\nu \Gamma^\mu{}_{\lambda\nu}) I_{ab}{}^A{}_B y^B],$$

on the composite bundle $S \to X$. One can think of the operator $\mathcal{D}$ (7.3.18) as being the *total Dirac operator* on $S \to X$ since, for every tetrad field h, the restriction of $\mathcal{D}$ to $J^1 S^h \subset J^1 S$ is exactly the Dirac operator $\mathcal{D}_h$ (7.2.11) on the spinor bundle S^h in the presence of a background tetrad field h and a linear world connection Γ.

Thus, we come to metric-affine gravitation theory in the presence of Dirac spinor fields. The total configuration space of this classical field theory is the jet manifold J^1Y of the bundle product

$$Q = (\Sigma_T \underset{X}{\times} C_W) \underset{\Sigma_T}{\times} S \qquad (7.3.19)$$

where C_W is the bundle of world connections (6.2.2). This product is coordinated by $(x^\mu, \sigma_a^\mu, k_\mu{}^\alpha{}_\beta, y^A)$. The corresponding field system algebra is the differential graded algebra (1.7.9):

$$\mathcal{S}^*_\infty[Q] = \mathcal{O}^*_\infty Q. \qquad (7.3.20)$$

A question however is that Dirac fermion fields are odd.

7.4 Dirac fermion fields

In order to describe odd Dirac fermion fields, let us regard the fibre bundle (7.3.19) as a composite bundle (3.4.1):

$$F \to Y \to X,$$

where $F \to Y$ is the vector bundle

$$F = (\Sigma_T \underset{X}{\times} C_W) \underset{\Sigma_T}{\times} (S \underset{\Sigma_T}{\oplus} S^*) \to (\Sigma_T \underset{X}{\times} C_W) = Y \qquad (7.4.1)$$

and $S^* \to \Sigma_T$ is the dual of the spinor bundle $S \to \Sigma_T$. Let us consider the composite graded manifold $(Y, \mathfrak{A}_F)$ modelled over the vector bundle $F \to Y$ (7.4.1). Then we replace the differential graded algebra (7.3.20) with the differential bigraded algebra $\mathcal{S}^*_\infty[F;Y]$ (3.4.7) possessing the local generating basis

$$(\sigma_a^\mu, k_\mu{}^\alpha{}_\beta, \psi^A, \psi^*_A) \qquad (7.4.2)$$

whose odd elements are ψ^A and ψ^*_A.

In accordance with Remark 3.3.5, the first order differential operator (the vertical covariant differential) $\widetilde{\mathcal{D}}$ (7.3.17) and the total Dirac operator $\mathcal{D}$ (7.3.18) on the composite bundle $S \to X$ yield the graded first order differential operator

$$\widetilde{\mathcal{D}} = dx^\lambda \otimes [\psi^A_\lambda - \frac{1}{4}(\eta^{kb}\sigma_\mu^a - \eta^{ka}\sigma_\mu^b)(\sigma_{\lambda k}^\mu - \sigma_k^\nu k_\lambda{}^\mu{}_\nu) I_{ab}{}^A{}_B \psi^B]\partial_A \qquad (7.4.3)$$

and the graded Dirac operator

$$\mathcal{D}\psi = \sigma_a^\lambda \gamma^{aB}{}_A[\psi^A_\lambda - \frac{1}{4}(\eta^{kb}\sigma_\mu^a - \eta^{ka}\sigma_\mu^b)(\sigma_{\lambda k}^\mu - \sigma_k^\nu k_\lambda{}^\mu{}_\nu) I_{ab}{}^A{}_B \psi^B], \qquad (7.4.4)$$

7.4. Dirac fermion fields

on the differential bigraded algebra $\mathcal{S}^*_\infty[F;Y]$.

The total Lagrangian $L \in \mathcal{S}^{(0,n)}_\infty[F;Y]$ of metric-affine gravity and fermion fields is the sum

$$L = L_{\text{MA}} + L_{\text{D}} \qquad (7.4.5)$$

of a metric-affine gravitation Lagrangian

$$L_{\text{MA}} = L_{\text{MA}}(\mathcal{R}_{\mu\lambda}{}^\alpha{}_\beta, \sigma^{\mu\nu})\omega, \qquad \sigma^{\mu\nu} = \sigma^\mu_a \sigma^\nu_b \eta^{ab}, \qquad (7.4.6)$$

on the configuration space J^1Y (see Section 6.5) and the Dirac Lagrangian L_{D}.

In accordance with Assertion 5.10.3, the Dirac Lagrangian factorizes through the vertical covariant differential $\widetilde{D}$ (7.4.3) and the graded Dirac operator (7.4.4). The Dirac Lagrangian reads

$$L_{\text{D}} = [a(i\mathcal{D}\psi, \psi) - ma(\psi,\psi)]\sigma^0 \wedge \cdots \wedge \sigma^3, \qquad \sigma^a = \sigma^a_\mu dx^\mu,$$

where $a(,)$ is the spinor metric (7.1.6). Written with respect to the local generating basis (7.4.2), the Dirac Lagrangian takes the form

$$L_{\text{D}} = \qquad (7.4.7)$$

$\{\frac{i}{2}\sigma^\lambda_q [\psi^+_A (\gamma^0\gamma^q)^A{}_B(\psi^B_\lambda - \frac{1}{4}(\eta^{kb}\sigma^a_\mu - \eta^{ka}\sigma^b_\mu)(\sigma^\mu_{\lambda k} - \sigma^\nu_k k_\lambda{}^\mu{}_\nu) I_{ab}{}^B{}_C \psi^C) -$

$(\psi^+_{\lambda A} - \frac{1}{4}(\eta^{kb}\sigma^a_\mu - \eta^{ka}\sigma^b_\mu)(\sigma^\mu_{\lambda k} - \sigma^\nu_k k_\lambda{}^\mu{}_\nu)\psi^+_C I^{+C}_{ab}{}_A (\gamma^0\gamma^q)^A{}_B \psi^B] -$

$m\psi^+_A(\gamma^0)^A{}_B \psi^B\}\sqrt{|\sigma|}\omega, \qquad \sigma = \det(\sigma_{\mu\nu}).$

It is readily observed that

$$\frac{\partial \mathcal{L}_{\text{D}}}{\partial k_\lambda{}^\mu{}_\nu} + \frac{\partial \mathcal{L}_{\text{D}}}{\partial k_\nu{}^\mu{}_\lambda} = 0, \qquad (7.4.8)$$

i.e., the Dirac Lagrangian (7.4.7) depends only on the torsion (6.6.11) of a world connection.

Given a tetrad field h and a linear world connection Γ, the pull-back $h^*\Gamma^*L_{\text{D}}$ of the Dirac Lagrangian (7.4.7) is the Dirac Lagrangian of fermion fields in the presence of a background tetrad gravitational field h and a linear world connection Γ.

Let us study gauge symmetries of the Dirac Lagrangian L_{D} (7.4.7). A metric-affine gravitation Lagrangian L_{MA} (7.4.6), by construction, is invariant under general covariant transformations (see Section 6.5).

As was mentioned above, the composite bundle $S \to X$ (7.3.9) is a $\widetilde{LX}$-associated bundle in accordance with Theorem 5.10.1. Therefore, S (7.3.9) inherits automorphisms of the universal spinor bundle $\widetilde{LX}$.

Since a world manifold X is parallelizable and the universal spinor structure is unique, the principal $\widetilde{GL}_4$-bundle $\widetilde{LX} \to X$ as well as the linear frame bundle LX admits the canonical lift of any diffeomorphism f of the base X. This lift is defined by the commutative diagram

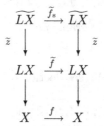

where $\widetilde{f}$ is the holonomic bundle automorphism of LX (6.1.8) induced by f [31; 77]. Consequently, the universal spinor bundle $\widetilde{LX}$ is a natural bundle, and it admits the functorial lift $\widetilde{\tau}_s$ of vector fields τ on its base X. These lifts $\widetilde{\tau}_s$ are infinitesimal general covariant transformations of $\widetilde{LX}$. Consequently, the composite bundle $S \to X$ (7.3.9) also is a natural bundle, and it possesses infinitesimal general covariant transformations [53; 138]. Let us obtain their explicit form.

By virtue of Lemma 5.10.3, infinitesimal general covariant transformations of the principal $\widetilde{GL}_4$-bundle $\widetilde{LX} \to X$ also are infinitesimal gauge transformations of the principal L_s-bundle $\widetilde{LX} \to \Sigma_T$. They take the form (5.10.23):

$$\widetilde{\tau}_s = \tau^\lambda \partial_\lambda + \partial_\nu \tau^\mu \sigma_c^\nu \frac{\partial}{\partial \sigma_c^\mu} + \frac{1}{2} \vartheta_\tau^{ab} e_{ab}, \qquad (7.4.9)$$

where the last term depends on the choice of a bundle atlas (7.3.10) and obeys the condition (5.10.24). Infinitesimal gauge transformations (7.4.9) of $\widetilde{LX} \to \Sigma_T$ depending on gauge parameters τ yield infinitesimal gauge transformations of the associated spinor bundle $S \to \Sigma_T$ given by vector fields (5.10.25):

$$\widetilde{\tau}_S = \tau^\lambda \partial_\lambda + \partial_\nu \tau^\mu \sigma_c^\nu \frac{\partial}{\partial \sigma_c^\mu} + \qquad (7.4.10)$$
$$\frac{1}{2} \vartheta_\tau^{ab} \left(-I_{ab}{}^d{}_c \sigma_d^\mu \frac{\partial}{\partial \sigma_c^\mu} + I_{ab}{}^A{}_B y^B \frac{\partial}{\partial y^A} \right),$$

where $I_{ab}{}^d{}_c$ (6.3.10) and $I_{ab}{}^A{}_B$ (7.1.5) are generators of the spinor Lorentz group L_s in the Minkowski space and the Dirac spinor space, respectively.

One can think of the vector fields (7.4.10) as being infinitesimal general covariant transformations of the natural composite bundle $S \to X$. Extended to the total bundle Q (7.3.19), these infinitesimal transformations

7.4. Dirac fermion fields

read

$$\tilde{\tau}_Q = \tau^\lambda \partial_\lambda + \partial_\nu \tau^\mu \sigma_c^\nu \frac{\partial}{\partial \sigma_c^\mu} +$$

$$[\partial_\nu \tau^\alpha k_\mu{}^\nu{}_\beta - \partial_\beta \tau^\nu k_\mu{}^\alpha{}_\nu - \partial_\mu \tau^\nu k_\nu{}^\alpha{}_\beta + \partial_{\mu\beta}\tau^\alpha]\frac{\partial}{\partial k_\mu{}^\alpha{}_\beta} +$$

$$\frac{1}{2}\vartheta_\tau^{ab}\left(-I_{ab}{}^d{}_c \sigma_d^\mu \frac{\partial}{\partial \sigma_c^\mu} + I_{ab}{}^A{}_B y^B \frac{\partial}{\partial y^A}\right).$$

Replacing the gauge parameters τ^λ with the odd ghosts c^λ and the bundle coordinates y^A with the odd elements ψ^A of the local generating basis (7.4.2), we obtain the odd graded derivation

$$u = \tilde{u} + \vartheta, \qquad (7.4.11)$$

$$\tilde{u} = \tau^\lambda \partial_\lambda + c_\nu^\mu \sigma_c^\nu \frac{\partial}{\partial \sigma_c^\mu} +$$

$$[c_\nu^\alpha k_\mu{}^\nu{}_\beta - c_\beta^\nu k_\mu{}^\alpha{}_\nu - c_\mu^\nu k_\nu{}^\alpha{}_\beta + c_{\mu\beta}^\alpha]\frac{\partial}{\partial k_\mu{}^\alpha{}_\beta},$$

$$\vartheta = \frac{1}{2}\vartheta^{ab}\left(-I_{ab}{}^d{}_c \sigma_d^\mu \frac{\partial}{\partial \sigma_c^\mu} + I_{ab}{}^A{}_B \psi^B \frac{\partial}{\partial \psi^A} + I_{ab}^{+\,A}{}_B \psi_A^+ \frac{\partial}{\partial \psi_B^+}\right),$$

of the differential bigraded algebra $\mathcal{S}^*_\infty[F;Y]$.

Besides the infinitesimal gauge transformations $\tilde{\tau}_S$ (7.4.10), one also considers vertical infinitesimal gauge transformations (5.10.26):

$$v_\zeta = \frac{1}{2}\zeta^{ab}\left(-I_{ab}{}^d{}_c \sigma_d^\mu \frac{\partial}{\partial \sigma_c^\mu} + I_{ab}{}^A{}_B y^B \frac{\partial}{\partial y^A}\right),$$

of the spinor bundle $S \to \Sigma_T$. These transformations yield the odd graded derivation

$$v = \frac{1}{2}c^{ab}\left(-I_{ab}{}^d{}_c \sigma_d^\mu \frac{\partial}{\partial \sigma_c^\mu} + I_{ab}{}^A{}_B \psi^B \frac{\partial}{\partial \psi^A} + I_{ab}^{+\,A}{}_B \psi_A^+ \frac{\partial}{\partial \psi_B^+}\right) \qquad (7.4.12)$$

(where c^{ab} are odd ghosts) of the differential bigraded algebra $\mathcal{S}^*_\infty[F;Y]$.

It is easily justified that

$$\mathbf{L}_{J^1 v} L_D = 0$$

and, obviously,

$$\mathbf{L}_v L_{\text{MA}} = 0.$$

Therefore, the graded derivation v (7.4.12) is a gauge symmetry of the total Lagrangian L (7.4.5). However, this gauge symmetry does not depend on jets of the ghosts c^{ab} and, therefore, does not lead to non-trivial Noether identities. At the same time, the corresponding Noether conservation law holds.

Since the bundle $S \to X$ is trivial and its trivialization is fixed, the decomposition (7.4.11) of the graded derivation u is global. Its second summand ϑ takes the form (7.4.12) and, consequently, it is an exact symmetry of the total Lagrangian L (7.4.5). The first summand $\widetilde{u}$ of the graded derivation (7.4.11) is independent of an atlas of the spinor bundle S and, therefore, it can be regarded as the canonical form of the infinitesimal general covariant transformation of S which differs from u in an infinitesimal vertical automorphism ϑ of S.

Since a metric-affine gravitation Lagrangian L_{MA}, by construction, is invariant under general covariant transformations, we have

$$\mathbf{L}_{J^1\widetilde{u}} L_{\text{MA}} = 0. \tag{7.4.13}$$

It is readily observed that also

$$\mathbf{L}_{J^1\widetilde{u}} L_{\text{D}} = 0. \tag{7.4.14}$$

The equalities (7.4.13) – (7.4.14) lead to the energy–momentum conservation law. Since it is a gauge conservation law, the corresponding energy–momentum current is reduced to a superpotential. One can show that it is exactly the generalized Komar superpotential (6.6.10) of metric-affine gravitation theory and that that Dirac fermion fields do not contribute to this superpotential because of the relation (7.4.8) [53; 136].

Chapter 8

Topological field theories

In classical field theory, topological field theories of Schwartz type are mainly considered [18]. They are Lagrangian field theories whose Lagrangians are independent of a world metric on a base X. Here, we are concerned with the following topological field models.

• In comparison with Yang–Mills gauge theory in Section 5.8, Chern–Simons topological field theory on a principal bundle possesses gauge symmetries which are neither vertical nor exact, and some of them become trivial if $\dim X = 3$ (Section 8.2).

• Topological BF theory in Section 8.3 exemplifies reducible degenerate Lagrangian theory.

• Since submanifolds of a smooth manifold are locally represented by sections of fibre bundles, their Lagrangian theory can be developed as the topological one (Section 8.4). For instance, classical string theory is of this type.

Note that any gauge theory of principal connections possesses characteristics which are topological invariants of a base manifold X. Section 8.1 is devoted to these characteristics.

8.1 Topological characteristics of principal connections

Theorem 8.1.1 below shows that the set $H_1(X; G_X^0)$ of equivalence classes of associated continuous G-bundles over a paracompact topological space X depends only on the homotopic class of the space X, i.e., it is a topological invariant. Due to the bijection (5.2.7), the set $H_1(X; G_X^\infty)$ of equivalence classes of associated smooth G-bundles over a manifold X also is a topological invariant.

8.1.1 Characteristic classes of principal connections

Theorem 5.2.3 leads to the following classification theorem.

Theorem 8.1.1. *For every topological group G, there exist a topological space BG, called the classifying space, and a continuous principal G-bundle $PG \to BG$ (see Remark 5.3.1), called the universal bundle, which possess the following properties.*

• *For any continuous principal G-bundle Y over a paracompact base X, there exists a continuous map $f : X \to BG$ such that Y is associated with the pull-back bundle f^*PG.*

• *If two maps f_1 and f_2 of X to BG are homotopic, then the pull-back principal bundles $f_1^*P_G$ and $f_2^*P_G$ are equivalent, and vice versa.*

Let us concentrate our attention to the most relevant physical case of bundles with the structure groups $GL(n, \mathbb{C})$ (reduced to $U(n)$) and $GL(n, \mathbb{R})$ (reduced to $O(n)$) (see Example 5.10.1). The classifying spaces for these groups are

$$BU(n) = \lim_{N \to \infty} \mathfrak{G}(n, N - n; \mathbb{C}),$$
$$BO(n) = \lim_{N \to \infty} \mathfrak{G}(n, N - n; \mathbb{R}), \qquad (8.1.1)$$

where $\mathfrak{G}(n, N - n; \mathbb{C})$ and $\mathfrak{G}(n, N - n; \mathbb{R})$ are the *Grassmann manifolds* of n-dimensional vector subspaces of $\mathbb{C}^N$ and $\mathbb{R}^N$, respectively. Then the equivalence classes of principal $U(n)$- $O(n)$-bundles over a manifold X can be represented by elements of the Čech cohomology groups $H^*(X;\mathbb{Z})$. They are called the *characteristic classes*. Due to the cohomology homomorphism (10.9.17):

$$H^*(X;\mathbb{Z}) \to H^*_{\mathrm{DR}}(X), \qquad (8.1.2)$$

these characteristic classes are given by elements of the de Rham cohomology $H^*_{\mathrm{DR}}(X)$ of X. They are cohomology classes of certain exterior forms defined as follows.

Given a principal bundle $P \to X$ with a structure Lie group G, let $C \to X$ be the bundle of principal connections (5.4.3), $\mathcal{A}$ the canonical principal connection (5.5.4) on the principal G-bundle P_C (5.5.5), and $F_{\mathcal{A}}$ (5.5.6) its strength. Let

$$I_k(\chi) = b_{r_1 \ldots r_k} \chi^{r_1} \cdots \chi^{r_k} \qquad (8.1.3)$$

be a G-invariant polynomial of degree $k > 1$ on the Lie algebra $\mathfrak{g}_r$ of G. With $F_{\mathcal{A}}$ (5.5.6), one can associate to this polynomial I_k the closed $2k$-form

$$P_{2k}(F_{\mathcal{A}}) = b_{r_1 \ldots r_k} F_{\mathcal{A}}^{r_1} \wedge \cdots \wedge F_{\mathcal{A}}^{r_k}, \qquad 2k \leq n, \qquad (8.1.4)$$

8.1. Topological characteristics of principal connections

on C which is invariant under automorphisms of C induced by vertical principal automorphisms of P. Given a section A of $C \to X$, the pull-back

$$P_{2k}(F_A) = A^* P_{2k}(F_A) \qquad (8.1.5)$$

of the form $P_{2k}(F_A)$ (8.1.4) is a closed $2k$-form on X where F_A is the strength (5.4.16) of a principal connection A. It is called the *characteristic form*. The characteristic forms (8.1.5) possess the following important properties [38; 92].

- Every characteristic form $P_{2k}(F_A)$ (8.1.5) is a closed form, i.e., $dP_{2k}(F_A) = 0$;
- The difference $P_{2k}(F_A) - P_{2k}(F_{A'})$ of characteristic forms is an exact form, whenever A and A' are different principal connections on a principal bundle P.

It follows that characteristic forms $P_{2k}(F_A)$ possesses the same de Rham cohomology class $[P_{2k}(F_A)]$ for all principal connections A on P. The association

$$I_k(\chi) \to [P_{2k}(F_A)] \in H^*_{\mathrm{DR}}(X)$$

is the well-known *Weil homomorphism*. The de Rham cohomology class $[P_{2k}(F_A)]$ is a topological invariant. Choosing a certain family of characteristic forms (8.1.5), one therefore can obtain characteristic classes of a principal bundle P.

Remark 8.1.1. If X is an oriented compact manifold of even dimension $2k$, then a value

$$C_n = \int_X P_{2k}(F_A)\omega$$

is the same for all principal connections A on P. It is called the *characteristic number* of a principal bundle P. In gauge theory, this topological number is treated as a *topological charge* [38].

However, there is problem that, if a principal bundle P admits a flat principal connection A whose strength F_A vanishes, all characteristic classes of P are trivial, but P need not be a trivial bundle (see Theorem 1.3.4). Therefore, there are topological effects related to flat principal connections, e.g., the Aharonov–Bohm effect.

8.1.2 Flat principal connections

A flat principal connection on a principal bundle $\pi_P : P \to X$, by definition, obeys the conditions of Theorem 1.3.3. It follows from Theorem 1.3.4 that a principal connection A on P represented by the $T_G P$-valued form (5.4.10) is flat if and only if its strength F_A (5.4.16) vanishes. In order to characterize flat principal connections, we however must involve the notion of a holonomy group of a principal connection [92]. For the sake of simplicity, we consider smooth paths, but everything that we say below holds true for the smooth piecewise ones.

Any principal connection A (5.4.1) on a principal G-bundle $P \to X$ is an Ehresmann connection (see Remark 1.3.2) [92]. Let

$$c : [0,1] \to X$$

be a closed path (a *loop*) through a point $x \in X$. For any point $p \in P_x = \pi_P^{-1}(x)$, there exists the horizontal lift c_p of a loop c through p such that $c_p(0) = p$. Then the map

$$\gamma_c : P_x \ni p = c_p(0) \to c_p(1) \in P_x \qquad (8.1.6)$$

defines an isomorphism g_c of the fibre P_x. This isomorphism can be seen as a *parallel displacement* of the point p along a loop c with respect to a connection A. Let us consider the group $\mathcal{C}_x$ of all loops through a point $x \in X$ and its subgroup $\mathcal{C}_x^0$ of the contractible ones. Then the set

$$\mathcal{K}_x = \{\gamma_c, c \in \mathcal{C}_x\}$$

of isomorphisms (8.1.6) and its subset

$$\mathcal{K}_x^0 = \{\gamma_c, c \in \mathcal{C}_x^0\}$$

are groups, called the *holonomy group* and the *restricted holonomy group* of a principal connection A at a point $x \in X$, respectively. Since X is assumed to be connected, the holonomy groups $\mathcal{K}_x$ for all $x \in X$ are mutually isomorphic, and one speaks on the *abstract holonomy group* $\mathcal{K}$ and its *restricted* subgroup $\mathcal{K}^0$.

There exists a monomorphism of the holonomy group $\mathcal{K}_x$ to a structure group G which, however, is not canonical. For a point $p \in \pi_P^{-1}(x)$, it is a map

$$\mathcal{K}_x \ni \gamma_c \to g_c \in \mathcal{K}_p \subset G, \qquad (8.1.7)$$

where g_c is given by the relation $\gamma_c(p) = p g_c$. The subgroup $\mathcal{K}_p$ (8.1.7) of the structure group G is called the *holonomy group at a point* $p \in P$. Accordingly, $\mathcal{K}_p^0$ denotes the *restricted holonomy group at a point* $p \in P$.

8.1. Topological characteristics of principal connections

Since a principal connection A is G-equivariant, we have
$$\gamma_c(pg) = pg_c g = pg(g^{-1} g_c g).$$
Therefore, the holonomy groups $\mathcal{K}_p$ and $\mathcal{K}_{p'}$ at different points $p, p' \in P_x$ are conjugate in G.

The forthcoming theorems summarize the main properties of holonomy groups [92].

Theorem 8.1.2. *The holonomy group $\mathcal{K}_p$ (resp. the restricted holonomy group $\mathcal{K}_p^0$) is a Lie subgroup (resp. a connected Lie subgroup) of a structure group G such that the factor group $\mathcal{K}_p/\mathcal{K}_p^0$ is countable.*

Theorem 8.1.3. *Since*
$$\mathcal{C}_x/\mathcal{C}_x^0 = \pi_1(X, x),$$
there is an epimorphism
$$\pi_1(X, x) \to \mathcal{K}_x/\mathcal{K}_x^0 \tag{8.1.8}$$
of the first homotopy group $\pi_1(X, x)$ at a point $x \in X$ onto the factor group $\mathcal{K}_x/\mathcal{K}_x^0$. In particular, if a manifold X is simply connected, then $\mathcal{K}_x = \mathcal{K}_x^0$ for all $x \in X$ and $\mathcal{K} = \mathcal{K}^0$.

Theorem 8.1.4. *Given a point $p \in P$ and the holonomy group $\mathcal{K}_p$ of a principal connection A on P, the points of P connected with p along horizontal paths form a subbundle $P(p)$ of P. It is a reduced principal bundle with the structure group $\mathcal{K}_p$, and A is reducible to a principal connection on $P(p)$.*

Theorem 8.1.5. *The values of the curvature form R of a principal connection A (5.4.1) (see Remark 5.4.3) at any point of the reduced subbundle $P(p)$ span a subalgebra of the Lie algebra $\mathfrak{g}_l$ of the group G which is isomorphic to the Lie algebra of the restricted holonomy group $\mathcal{K}_p^0$.*

The following assertion is a corollary of Theorems 8.1.3 – 8.1.5.

Theorem 8.1.6. *If a principal bundle over a simply connected base admits a flat connection, it is trivial.*

Proof. It follows from Theorem 8.1.5 that, if a principal connection A on a principal G-bundle $P \to X$ is flat, the restricted holonomy group $\mathcal{K}_p^0$ of A at any point $p \in P$ is trivial and the holonomy group $\mathcal{K}_p$ is discrete (at most countable). In accordance with Theorem 8.1.3, if this principal bundle is over a simply connected base X, then the holonomy group of this connection is trivial. Then, by virtue of Theorem 8.1.4, the principal bundle P admits a trivial reduced subbundle, i.e., it is trivial. □

The proof of Theorem 8.1.6 gives something more. If a principal connection A is flat, its holonomy group $\mathcal{K}_p$ at any point $p \in P$ is discrete. Then it follows from Theorem 8.1.4, that a principal bundle P contains a connected principal subbundle with a discrete structure group $\mathcal{K}_p$. It should be emphasized that a connected principal bundle with a discrete structure group is necessarily non-trivial.

Lemma 8.1.1. *Let K be a discrete group. A connected principal K-bundle over a connected manifold X exists if and only if there is a subgroup $N \subset \pi_1(X)$ such that $\pi_1(X)/N = K$.*

Proof. Let $\pi : Y \to X$ be a fibre bundle. Given $y \in Y$ and $x = \pi(y)$, there exists the following exact sequence of homotopy groups of $Y \to X$:

$$\cdots \to \pi_k(Y_x, y) \to \pi_k(Y, y) \to \pi_k(X, x) \to \pi_{k-1}(Y_x, y) \to \quad (8.1.9)$$
$$\cdots \to \pi_1(X, x) \to \pi_0(Y_x, y) \to \pi_0(Y, y) \to \pi_0(X, x) \to 0.$$

If Y is a connected principal bundle with a discrete structure group K, we have

$$\pi_{k>0}(Y_x, y) = \pi_0(Y, y) = \pi_0(X, x) = 0, \qquad \pi_0(Y_x, y) = K.$$

Then the exact sequence (8.1.9) is reduced to the short exact sequences

$$0 \to \pi_k(Y, y) \to \pi_k(X, x) \to 0, \quad k > 1,$$
$$0 \to \pi_1(Y, y) \to \pi_1(X, x) \to K \to 0.$$

It follows that $\pi_1(Y, y)$ is a subgroup of $\pi_1(X, x)$ and

$$\pi_1(X, x)/\pi_1(Y, y) = K. \qquad (8.1.10)$$

This is a necessary condition for a desired fibre bundle over X to exist. One can show that, since a manifold X is a locally contractible space, the condition that

$$K = \pi_1(X, .)/N$$

for some subgroup $N \subset \pi_1(X, .)$ is sufficient. □

Remark 8.1.2. The fibre bundle $Y \to X$ in Lemma 8.1.1 is called the *covering space* over X. A covering space is said to be *universal* if it is simply connected.

Now let us formulate the general result concerning flat principal connections [1; 112].

Theorem 8.1.7. *There is a bijection between the set of conjugate flat principal connections on a principal G-bundle $P \to X$ and the set $\mathrm{Hom}\,(\pi_1(X), G)/G$ of conjugate homomorphisms of the homotopy group $\pi_1(X)$ of X to the group G.*

8.1. Topological characteristics of principal connections

Proof. Given a flat principal connection A (5.4.1) on a principal G-bundle P over X, the composition of homomorphisms (8.1.8) and (8.1.7) gives the homomorphism $\pi_1(X,x) \to G$ which is not unique, but depends on a point $p \in P$ where the map (8.1.7) is defined. Therefore, every flat principal connection A on a principal G-bundle $P \to X$ defines a class of conjugate homomorphisms of the homotopy group $\pi_1(X)$ to G. Let Φ be a vertical automorphisms of a principal bundle P and A' the image of a connection A with respect to this automorphism in accordance with Theorem 5.4.2, i.e., A' is a conjugate connection to A (see Remark 5.4.4). It is obviously flat. Since a vertical automorphism of P preserves the homotopy class of curves in P, the holonomy groups $\mathcal{K}_x$ and $\mathcal{K}'_x$ of gauge conjugate connections are identically isomorphic, while $\mathcal{K}'_{\Phi(p)} = \mathcal{K}_p$.

Conversely, let $\pi_1(X) \to G$ be a homomorphism whose image is a subgroup $K \subset G$ and whose kernel is a subgroup $N \subset \pi_1(X)$. By virtue of Lemma 8.1.1, there exists a connected principal K-bundle $P_K \to X$ and, consequently, a principal G-bundle $P \to X$ which contains P_K as a subbundle, and whose structure group G is reducible to the discrete subgroup K. It follows that there exists an atlas of the principal bundle P with constant K-valued transition functions. This is an atlas of local constant trivializations. Following Theorem 1.3.4, one can define a flat connection A on $P \to X$ whose local coefficients with respect to this atlas equal zero. This is a principal connection. Certainly, the holonomy group $\mathcal{K}$ of this connection is K. □

Note that, in topological field theory, the space $\mathrm{Hom}\,(\pi_1(X), G)/G$ is treated as a moduli space of flat connections [18].

Example 8.1.1. The well-known Aharonov–Bohm effect in electromagnetic theory exemplifies phenomena related to flat principal connections. Let a Euclidean space $\mathbb{R}^3$ be equipped with the Cartesian coordinates (x, y, z) or the cylindrical ones (ρ, α, z). Its submanifold

$$X = \mathbb{R}^3 \setminus \{\rho = 0\}$$

admits an electromagnetic potential

$$A = \frac{\Phi}{2\pi} d\alpha = \frac{\Phi y}{x^2 + y^2} dx - \frac{\Phi x}{x^2 + y^2} dy, \qquad \Phi \in \mathbb{R}, \qquad (8.1.11)$$

whose strength $F = dA$ vanishes everywhere on X, i.e., the one-form A (8.1.11) is closed. However, this form is not exact. Hence, A (8.1.11) belongs to a non-vanishing element of the de Rham cohomology group $H^1_{\mathrm{DR}}(X) = \mathbb{R}$ of X. Being represented by the one-forms (8.1.11), elements of this cohomology group are indexed by real coefficients Φ in the expression (8.1.11). Let us denote them by $[\Phi] \subset H^1_{\mathrm{DR}}(X)$. Then the corresponding group operation in $H^1_{\mathrm{DR}}(X)$ reads

$$[\Phi] + [\Phi'] = [\Phi + \Phi'].$$

Let us notice that the field (8.1.11) can be extended to the whole space $\mathbb{R}^3$ in terms of generalized functions

$$A = \frac{\Phi}{2\pi\rho}\theta(\rho)d\alpha, \qquad F = \frac{\Phi}{2}\delta(\rho^2)d\rho \wedge d\alpha, \qquad (8.1.12)$$

where $\theta(\rho)$ is the step function, while $\delta(\rho^2)$ is the Dirac δ-function. The field (8.1.12) satisfies the Stokes formula

$$\int_{\partial S} A = \int_0^{2\pi} A_\alpha \rho d\alpha = \int_S F\rho d\rho d\alpha = \Phi, \qquad (8.1.13)$$

where S is a centered disc in the plane $z = 0$. Thus, we obtain the well-known *Aharonov–Bohm effect*.

One can extend the description of the Aharonov–Bohm effect in Example 8.1.1 to a non-Abelian gauge model on a principal bundle $P \to X$ which admits a flat principal connection A with a non-trivial discrete holonomy group $\mathcal{K}$. Let $Y \to X$ be a P-associated vector bundle (5.7.1). The notion of holonomy group is generalized in a straightforward manner to associated principal connections on $Y \to X$. In particular, if A is a flat principal connection with a holonomy group K on $P \to X$, the associated connection A (5.7.9) on $Y \to X$ is a flat connection with the same holonomy group. By virtue of Theorem 1.3.4, there exists an atlas $\Psi = \{U_\alpha, \varrho_{\alpha\beta}\}$ of local constant trivializations of a fibre bundle $Y \to X$ such that the connection A on $Y \to X$ takes the form $A = dx^\lambda \otimes \partial_\lambda$. Let c be a loop through a point $x \in X$ which crosses the charts $U_{\alpha_1}, \ldots, U_{\alpha_k}$ of the atlas Ψ. Then the parallel displacement of a vector $v \in Y_x$ along this curve with respect to the flat connection A reduces to the product of transition functions

$$v \to (\varrho_{\alpha_2\alpha_1} \cdots \varrho_{\alpha_1\alpha_k})(v).$$

It depends only on the homotopic class of a loop c.

8.1.3 Chern classes of unitary principal connections

Characteristic classes of principal $GL(k,\mathbb{C})$-bundles (which are always principal $U(k)$-bundles) are *Chern classes* $c_i(P) \in H^{2i}(X;\mathbb{Z})$ given by cohomology of Chern characteristic forms [38; 80].

8.1. Topological characteristics of principal connections

Let M be a complex $(k \times k)$-matrix and $r(M)$ a $GL(k,\mathbb{C})$-invariant polynomial, called the *characteristic polynomial* of components of M, i.e.,

$$r(M) = r(gMg^{-1}), \qquad \gamma \in GL(k,\mathbb{C}).$$

If a matrix M has eigenvalues $a_1, \ldots, a_k$, a characteristic polynomial $r(M)$ takes the form

$$r(M) = b_0 + b_1 S_1(a) + b_2 S_2(a) + \cdots$$

where b_i are complex numbers and

$$S_j(a) = \sum_{i_1 < \ldots < i_j} a_{i_1} \ldots a_{i_j} \qquad (8.1.14)$$

are symmetric polynomials of $a_1, \ldots, a_k$.

Example 8.1.2. An important example of a characteristic polynomial is

$$\det(\mathbf{1} + M) = 1 + S_1(a) + S_2(a) + \ldots + S_k(a),$$

where $\mathbf{1}$ denotes the unit matrix.

Let $P \to X$ be a principal $U(k)$-bundle and E the associated vector bundle with the typical fibre $\mathbb{C}^k$ which is a carrier space of the natural representation of $U(k)$.

Let F be the strength form (5.7.12) of some associated principal connection A on E. The characteristic form

$$c(F) = \det\left(\mathbf{1} + \frac{i}{2\pi} F\right) = 1 + c_1(F) + c_2(F) + \cdots \qquad (8.1.15)$$

is called the *total Chern form*, and its components $c_i(F)$ are called *Chern $2i$-forms*. For instance,

$$c_0(F) = 0,$$
$$c_1(F) = \frac{i}{2\pi} \operatorname{Tr} F, \qquad (8.1.16)$$
$$c_2(F) = \frac{1}{8\pi^2} [\operatorname{Tr}(F \wedge F) - \operatorname{Tr} F \wedge \operatorname{Tr} F]. \qquad (8.1.17)$$

All Chern forms $c_i(F)$ are closed, and their cohomology are identified with the Chern classes $c_i(E) \in H^{2i}(X;\mathbb{Z})$ of the $U(k)$-bundle E under the homomorphism (8.1.2). The total Chern form (8.1.15) corresponds to the *total Chern class*

$$c(E) = c_0(E) + c_1(E) + \cdots.$$

Example 8.1.3. Let us consider a $U(1)$-bundle $L \to X$ with the typical fibre $\mathbb{C}$ on which the group $U(1)$ acts by the generator $I = i$. It is called the *complex linear bundle*. The strength form (5.7.12) of an associated principal connection on this fibre bundle reads

$$F = \frac{i}{2} F_{\lambda\mu} dx^\lambda \wedge dx^\mu.$$

Then the total Chern form (8.1.15) of a $U(1)$-bundle is

$$c(F) = 1 + c_1(F),$$
$$c_1(F) = \frac{i}{2\pi} \operatorname{Tr} F = -\frac{1}{4\pi} F_{\lambda\mu} dx^\lambda \wedge dx^\mu.$$

Example 8.1.4. Let us consider a $SU(2)$-bundle $E \to X$. The strength form (5.7.12) of an associated principal connection on this fibre bundle is

$$F = \frac{i\sigma_a}{2} F^a,$$

where σ_a, $a = 1,2,3$ are the Pauli matrices. Then we have

$$c(F) = 1 + c_1(F) + c_2(F),$$
$$c_1(F) = 0,$$
$$c_2(F) = \frac{1}{8\pi^2} \operatorname{Tr}(F \wedge F) = -\frac{1}{(4\pi)^2} F_a \wedge F_a.$$

Using the natural properties of the Chern forms, one can obtain the following properties of Chern classes:

(i) $c_i(E) = 0$ if $2i > n = \dim X$;
(ii) $c_i(E) = 0$ if $i > k$;
(iii) $c(E \oplus E') = c(E)c(E')$;
(iv) $c_1(L \oplus L') = c_1(L) + c_1(L')$ where L and L' are complex linear bundles;
(v) if $f^*E \to X'$ is the pull-back bundle generated by a morphism $f : X' \to X$, then

$$c(f^*E) = f^*c(E),$$

where f^* also denotes the induced morphism of the cohomology groups

$$f^* : H^*(X; \mathbb{Z}) \to H^*(X'; \mathbb{Z}). \qquad (8.1.18)$$

Properties (iii) and (v) of Chern classes are utilized in the following theorem.

Theorem 8.1.8. *For any $U(k)$-bundle $E \to X$, there exists a topological space X' and a continuous morphism $f : X' \to X$ so that:*

(i) the pull-back $f^*E \to X'$ is the Whitney sum of complex linear bundles (see Example 8.1.3)
$$f^*E = L_1 \oplus \cdots \oplus L_k;$$
(ii) the induced morphism f^* (8.1.18) is an inclusion.

It follows that the total Chern class of any $U(k)$-bundle E can be seen as
$$c(E) = f^*c(E) = c(L_1 \oplus \cdots \oplus L_k) = c(L_1) \cdots c(L_k) = \quad (8.1.19)$$
$$(1 + a_1) \cdots (1 + a_k),$$
where $a_i = c_1(L_i)$ denotes the Chern class of the linear bundle L_i (see Example 8.1.3). The formula (8.1.19) is called the *splitting principle*. In particular, we have
$$c_1(E) = \sum_i a_i,$$
$$c_2(E) = \sum_{i_1 < i_2} a_{i_1} a_{i_2},$$
$$c_j(E) = \sum_{i_1 < \cdots < i_j} a_{i_1} \cdots a_{i_j}$$
(cf. (8.1.14)).

Example 8.1.5. Let E^* be the $U(k)$-bundle, dual of E. In accordance with the splitting principle, we have
$$c(E^*) = c(L_1^*) \cdots c(L_k^*) = (1 + a_1^*) \cdots (1 + a_k^*).$$
Since the generator of the dual representation of $U(1)$ is $I = -i$, then $F^* = -F$ and $a_i^* = -a_i$. It follows that
$$c_i(E^*) = (-1)^i c_i(E). \quad (8.1.20)$$

If a base X of a $U(k)$-bundle E is an oriented compact manifold of even dimension n, one can construct some exterior n-forms from the Chern forms $c_i(F)$, and can integrate them over X. Such integrals are called *Chern numbers*. Chern numbers are integer since the Chern forms belongs to the integer cohomology classes. For instance, if $n = 4$, there are the following two Chern numbers
$$C_2(E) = \int_X c_2(F),$$
$$C_1^2(E) = \int_X c_1(F) \wedge c_1(F).$$

There exist some other characteristic classes of $U(k)$-bundles which are expressed into the Chern classes. Let us mention the following two ones.

The *Chern character* $\mathrm{ch}(E)$ is given by the characteristic polynomial

$$\mathrm{ch}(M) = \mathrm{Tr}\exp\left(\frac{i}{2\pi}M\right) = \sum_{m=0}^{\infty}\frac{1}{m!}\mathrm{Tr}\left(\frac{i}{2\pi}M\right)^m.$$

It has the properties

$$\mathrm{ch}(E \oplus E') = \mathrm{ch}(E) + \mathrm{ch}(E'),$$
$$\mathrm{ch}(E \otimes E') = \mathrm{ch}(E) \cdot \mathrm{ch}(E').$$

Using the splitting principle, one can express the Chern character into the Chern classes as follows:

$$\mathrm{ch}(E) = \mathrm{ch}(L_1 \oplus \cdots \oplus L_k) = \mathrm{ch}(L_1) + \cdots + \mathrm{ch}(L_k) =$$
$$\exp a_1 + \cdots + \exp a_k = k + \sum_i a_i + \frac{1}{2}\sum_i a_i^2 + \cdots =$$
$$k + \sum_i a_i + \frac{1}{2}\left[(\sum_i a_i)^2 - 2\sum_{i_1<i_2} a_{i_1}a_{a_2}\right] + \cdots =$$
$$k + c_1(E) + \frac{1}{2}[c_1^2(E) - 2c_2(E)] + \cdots.$$

The *Todd class* is defined as

$$\mathrm{td}(E) = \sum_{i=1}^{k}\frac{a_i}{1-\exp(-a_i)} = 1 + \frac{1}{2}c_1 + \frac{1}{12}(c_1^2 + c_2) + \cdots.$$

It possesses the property

$$\mathrm{td}(E \oplus E') = \mathrm{td}(E) \cdot \mathrm{td}(E').$$

8.1.4 Characteristic classes of world connections

Characteristic classes of real $GL(k,\mathbb{R})$-bundles (which are always principal $O(k)$-bundles) are Pontryagin classes given by cohomology of the Pontryagin characteristic forms [38; 80].

Let E be a k-dimensional vector bundle with the structure group $O(k)$. Its *Pontryagin classes* in the de Rham cohomology algebra $H^*_{\mathrm{DR}}(X)$ are associated with the characteristic polynomial

$$p(F) = \mathrm{Det}\left(1 - \frac{1}{2\pi}F\right) = 1 + p_1(F) + p_2(F) + \cdots \qquad (8.1.21)$$

of the strength form F (5.7.12) of some associated principal connection A on E which takes its values into the Lie algebra $o(k)$ of the group $O(k)$. Since generators

$$I_{ab}{}^c{}_d = \eta^E_{bd}\delta^c_a - \eta^E_{ad}\delta^c_b, \qquad \operatorname{diag}\eta^E = (1,\dots,1),$$

of the group $O(k)$ in E satisfy the condition

$$(I)_b{}^a = -(I)_a{}^b,$$

the components of even degrees in F in the decomposition (8.1.21) only are different from zero, i.e., $p_i(E) \in H^{4i}_{\mathrm{DR}}(X)$.

Pontryagin classes possess the following properties:
(i) $p_i(E) = 0$ if $4i > n = \dim X$;
(ii) $p_i(E) = 0$ if $2i > k$;
(iii) $p(E \oplus E') = p(E) + p(E')$.

Note that, for the Pontriagin classes taken in the Čech cohomology $H^*(X;\mathbb{Z})$, the property (ii) is true only modulo cyclic elements of order 2.

Remark 8.1.3. Though fibre bundles with the structure groups $O(k)$ and $GL(k,\mathbb{R})$ have the same characteristic classes, their characteristic forms are different. For instance, if the strength F takes its values into the Lie algebra $gl(k,\mathbb{R})$, the characteristic polynomial $p(F)$ (8.1.21) contains the terms of odd degrees in F in general. Therefore, to construct the characteristic forms corresponding to Pontryagin classes, one should use only $O(k)$- or $O(k-m,m)$-valued strength forms F. In this case, we have

$$p_1(F) = -\frac{1}{8\pi^2}\operatorname{Tr} F \wedge F, \tag{8.1.22}$$

$$p_2(F) = \left[\frac{1}{32\pi^2}\varepsilon_{abcd}F^{ab} \wedge F^{cd}\right], \tag{8.1.23}$$

where ε_{abcd} is the skew-symmetric Levi–Civita tensor.

Remark 8.1.4. If $E = TX$ is the tangent bundle of a smooth manifold X, characteristic classes of TX are regarded as characteristic classes of a manifold X itself.

Let us consider the relations between the Pontryagin classes of $O(k)$-bundles and the Chern classes of $U(k)$-bundles. There are the following commutative diagrams of group monomorphisms

$$\begin{array}{ccc} O(k) & \longrightarrow & U(k) \\ \downarrow & & \downarrow \\ GL(k,\mathbb{R}) & \longrightarrow & GL(k,\mathbb{C}) \end{array} \tag{8.1.24}$$

$$\begin{array}{ccc} U(k) & \longrightarrow & O(2k) \\ \downarrow & & \downarrow \\ GL(k,\mathbb{C}) & \longrightarrow & GL(2k,\mathbb{R}) \end{array} \qquad (8.1.25)$$

The diagram (8.1.24) implies the inclusion
$$\varphi : H^1(X; G^\infty_{O(k)}) \to H^1(X; G^\infty_{U(k)}),$$
and one can show that
$$p_i(E) = (-1)^i c_{2i}(\varphi(E)). \qquad (8.1.26)$$
The diagram (8.1.25) yields the inclusion
$$\rho : H^1(X; G^\infty_{U(k)}) \to H^1(X; G^\infty_{O(2k)}).$$
Then we have
$$\varphi\rho : H^1(X; G^\infty_{U(k)}) \to H^1(X; G^\infty_{O(2k)}) \to H^1(X; G^\infty_{U(2k)}).$$
This means that, if E is a $U(k)$-bundle, then $\rho(E)$ is a $O(2k)$-bundle, while $\varphi\rho(E)$ is a $U(2k)$-bundle.

Remark 8.1.5. Let A be an element of $U(k)$. The group monomorphisms
$$U(k) \to O(2k) \to U(2k)$$
define the transformation of matrices
$$A \longrightarrow \begin{pmatrix} \operatorname{Re} A & -\operatorname{Im} A \\ \operatorname{Im} A & \operatorname{Re} A \end{pmatrix} \longrightarrow \begin{pmatrix} A & 0 \\ 0 & A^* \end{pmatrix}, \qquad (8.1.27)$$
written relative to complex coordinates z^i on the space $\mathbb{C}^k$, real coordinates
$$x^i = \operatorname{Re} z^i, \qquad x^{k+i} = \operatorname{Im} z^i$$
on $\mathbb{R}^{2k}$, and complex coordinates
$$z^i = x^i + ix^{k+i}, \qquad z^{k+i} = x^i - ix^{k+i}$$
on the space $\mathbb{C}^{2k}$.

A glance at the diagram (8.1.27) shows that the fibre bundle $\varphi\rho(E)$ is the Whitney sum of E and E^* and, consequently,
$$c(\varphi\rho(E)) = c(E)c(E^*).$$
Then combining (8.1.20) and (8.1.26) gives the relation
$$\sum_i (-1)^i p_i(\rho(E)) = c(\varphi\rho(E)) = c(E)c(E^*) = \qquad (8.1.28)$$
$$\left[\sum_i c_i(E)\right]\left[\sum_j (-1)^j c_j(E)\right]$$

between the Chern classes of a $U(k)$-bundle E and the Pontryagin classes of the $O(2k)$-bundle $\rho(E)$.

Example 8.1.6. In accordance with Remark 8.1.4, by *Pontryagin classes* $p_i(X)$ *of a manifold* X are meant those of the tangent bundle $T(X)$. Let a manifold X be oriented and $\dim X = 2m$. One says that a manifold X admits an *almost complex structure* if its structure group $GL(2m, \mathbb{R})$ is reducible to the image of $GL(m, \mathbb{C})$ in $GL(2m, \mathbb{R})$. By *Chern classes* $c_i(X)$ of such a manifold X are meant those of the tangent bundle $T(X)$ seen as a $GL(m, \mathbb{C})$-bundle, i.e.,

$$c_i(X) = c_i(\rho(T(X))).$$

Then the formula (8.1.28) provides the relation between the Pontryagin and Chern classes of a manifold X admitting an almost complex structure:

$$\sum_i (-1)^i p_i(X) = \left[\sum_i c_i(X)\right] \left[\sum_j (-1)^j c_j(X)\right].$$

In particular, we have

$$p_1(X) = c_1^2(X) - 2c_2(X), \tag{8.1.29}$$
$$p_2(X) = c_2^2(X) - 2c_1(X)c_3(X) + 2c_4(X).$$

If the structure group of a $O(k)$-bundle E is reduced to $SO(k)$, the *Euler class* $e(E)$ of E can be defined as an element of the Čech cohomology group $H^k(X; \mathbb{Z})$ which satisfies the conditions:

(i) $2e(E) = 0$ if k is odd;
(ii) $e(f^*E) = f^*e(E)$;
(iii) $e(E \oplus E') = e(E)e(E')$;
(iv) $e(E) = c_1(E)$ if $k = 2$ because of an isomorphism of groups $SO(2)$ and $U(1)$.

Let us consider the relationship between the Euler class and the Pontryagin ones. Let E be a $U(k)$-bundle and $\rho(E)$ the corresponding $SO(2k)$-bundle. Then, using the splitting principle and properties (iii), (iv) of the Euler class, we obtain

$$e(\rho(E)) = e(\rho(L_1) \oplus \cdots \oplus \rho(L_k)) = \tag{8.1.30}$$
$$e(\rho(L_1)) \cdots e(\rho(L_k)) =$$
$$c_1(L_1) \cdots c_1(L_k) = a_1 \cdots a_k = c_k(E).$$

At the same time, one can deduce from (8.1.28) that

$$p_k(\rho(E)) = c_k^2(E). \tag{8.1.31}$$

Combining (8.1.30) and (8.1.31) gives a desired relation

$$e(E) = [p_k(E)]^{1/2} \tag{8.1.32}$$

for any $SO(2k)$-bundle.

For instance, let X be a $2k$-dimensional oriented compact manifold. Its tangent bundle TX has the structure group $SO(2k)$. Then the integral

$$e = \int_X e(R)$$

coincides with the *Euler characteristic* of X.

In conclusion, let us also mention the *Stiefel–Whitney classes* $w_i \in H^i(X; \mathbb{Z}_2)$ of the tangent bundle TX. In particular, a manifold X is orientable if and only if $w_1 = 0$. If X admits an almost complex structure, then

$$w_{2i+1} = 0, \quad w_{2i} = c_i \bmod 2.$$

In contrast with the above mentioned characteristic classes, the Stiefel–Whitney ones are not represented by the de Rham cohomology of exterior forms.

8.2 Chern–Simons topological field theory

We consider gauge theory of principal connections on a principal bundle $P \to X$ with a structure real Lie group G. In contrast with the Yang–Mills Lagrangian L_{YM} (5.8.15), the Lagrangian L_{CS} (8.2.4) of Chern–Simons topological field theory on an odd-dimensional manifold X is independent of a world metric on X. Therefore, its non-trivial gauge symmetries are wider than those of the Yang–Mills one. However, some of them become trivial if $\dim X = 3$.

Note that one usually considers the local Chern–Simons Lagrangian which is the local Chern–Simons form derived from the local transgression formula for the Chern characteristic form. The global Chern–Simons Lagrangian is well defined, but depends on a background gauge potential [20; 42; 58].

Let I_k (8.1.3) be a G-invariant polynomial of degree $k > 1$ on the Lie algebra $\mathfrak{g}_r$ of G. Let $P_{2k}(F_A)$ (8.1.4) be the corresponding closed $2k$-form on C and $P_{2k}(F_A)$ (8.1.5) its pullback onto X by means of a section A of $C \to X$. Let the same symbol $P_{2k}(F_A)$ stand for its pull-back onto C. Since $C \to X$ is an affine bundle and, consequently, the de Rham cohomology of

8.2. Chern–Simons topological field theory

C equals that of X, the exterior forms $P_{2k}(F_\mathcal{A})$ and $P_{2k}(F_A)$ possess the same de Rham cohomology class

$$[P_{2k}(F_\mathcal{A})] = [P_{2k}(F_A)]$$

for any principal connection A. Consequently, the exterior forms $P_{2k}(F_\mathcal{A})$ and $P_{2k}(F_A)$ on C differ from each other in an exact form

$$P_{2k}(F_\mathcal{A}) - P_{2k}(F_A) = d\mathfrak{S}_{2k-1}(a, A). \tag{8.2.1}$$

This relation is called the *transgression formula* on C. Its pull-back by means of a section B of $C \to X$ gives the transgression formula on a base X:

$$P_{2k}(F_B) - P_{2k}(F_A) = d\mathfrak{S}_{2k-1}(B, A).$$

For instance, let

$$c(F_\mathcal{A}) = \det\left(1 + \frac{i}{2\pi} F_\mathcal{A}\right) = 1 + c_1(F_\mathcal{A}) + c_2(F_\mathcal{A}) + \cdots$$

be the *total Chern form* on a bundle of principal connections C. Its components $c_k(F_\mathcal{A})$ are *Chern characteristic forms* on C. If

$$P_{2k}(F_\mathcal{A}) = c_k(F_\mathcal{A})$$

is the characteristic Chern $2k$-form, then $\mathfrak{S}_{2k-1}(a, A)$ (8.2.1) is the *Chern–Simons $(2k-1)$-form*.

In particular, one can choose a local section $A = 0$. In this case, $\mathfrak{S}_{2k-1}(a, 0)$ is called the *local Chern–Simons form*. Let $\mathfrak{S}_{2k-1}(A, 0)$ be its pull-back onto X by means of a section A of $C \to X$. Then the Chern–Simons form $\mathfrak{S}_{2k-1}(a, A)$ (8.2.1) admits the decomposition

$$\mathfrak{S}_{2k-1}(a, A) = \mathfrak{S}_{2k-1}(a, 0) - \mathfrak{S}_{2k-1}(A, 0) + dK_{2k-1}. \tag{8.2.2}$$

The transgression formula (8.2.1) also yields the transgression formula

$$h_0(P_{2k}(F_\mathcal{A}) - P_{2k}(F_A)) = d_H(h_0\mathfrak{S}_{2k-1}(a, A)),$$

$$h_0\mathfrak{S}_{2k-1}(a, A) = k \int_0^1 \mathcal{P}_{2k}(t, A) dt, \tag{8.2.3}$$

$$\mathcal{P}_{2k}(t, A) = b_{r_1 \ldots r_k}(a^{r_1}_{\mu_1} - A^{r_1}_{\mu_1}) dx^{\mu_1} \wedge \mathcal{F}^{r_2}(t, A) \wedge \cdots \wedge \mathcal{F}^{r_k}(t, A),$$

$$\mathcal{F}^{r_j}(t, A) = \frac{1}{2}[ta^{r_j}_{\lambda_j \mu_j} + (1-t)\partial_{\lambda_j} A^{r_j}_{\mu_j} - ta^{r_j}_{\mu_j \lambda_j} -$$

$$(1-t)\partial_{\mu_j} A^{r_j}_{\lambda_j} + \frac{1}{2}c^{r_j}_{pq}(ta^p_{\lambda_j} + (1-t)A^p_{\lambda_j})(ta^q_{\mu_j} +$$

$$(1-t)A^q_{\mu_j}]dx^{\lambda_j} \wedge dx^{\mu_j} \otimes e_r,$$

on J^1C (where $b_{r_1...r_k}$ are coefficients of the invariant polynomial (8.1.3)).

If $2k - 1 = \dim X$, the density (8.2.3) is the global *Chern–Simons Lagrangian*

$$L_{\mathrm{CS}}(A) = h_0\mathfrak{S}_{2k-1}(a, A) \qquad (8.2.4)$$

of Chern–Simons topological field theory. It depends on a background gauge field A. The decomposition (8.2.2) induces the decomposition

$$L_{\mathrm{CS}}(A) = h_0\mathfrak{S}_{2k-1}(a, 0) - h_0\mathfrak{S}_{2k-1}(A, 0) + d_H h_0 K_{2k-1}, \qquad (8.2.5)$$

where

$$L_{\mathrm{CS}} = h_0\mathfrak{S}_{2k-1}(a, 0) \qquad (8.2.6)$$

is the local *Chern–Simons Lagrangian*.

For instance, if $\dim X = 3$, the global Chern–Simons Lagrangian (8.2.4) reads

$$L_{\mathrm{CS}}(A) = \left[\frac{1}{2}h_{mn}\varepsilon^{\alpha\beta\gamma}a_\alpha^m(\mathcal{F}_{\beta\gamma}^n - \frac{1}{3}c_{pq}^n a_\beta^p a_\gamma^q)\right]\omega - \qquad (8.2.7)$$

$$\left[\frac{1}{2}h_{mn}\varepsilon^{\alpha\beta\gamma}A_\alpha^m(F_{A\beta\gamma}^n - \frac{1}{3}c_{pq}^n A_\beta^p A_\gamma^q)\right]\omega -$$

$$d_\alpha(h_{mn}\varepsilon^{\alpha\beta\gamma}a_\beta^m A_\gamma^n)\omega,$$

where $\varepsilon^{\alpha\beta\gamma}$ is the skew-symmetric Levi–Civita tensor.

Since the density

$$-\mathfrak{S}_{2k-1}(A, 0) + d_H h_0 K_{2k-1}$$

is variationally trivial, the global Chern–Simons Lagrangian (8.2.4) possesses the same Noether identities and gauge symmetries as the local one (8.2.6). They are the following.

Recall that infinitesimal generators of local one-parameter groups of automorphisms of a principal bundle P are G-invariant projectable vector fields v_P on P. They are identified with sections (5.3.16):

$$\xi = \tau^\lambda\partial_\lambda + \xi^r e_r, \qquad (8.2.8)$$

of the vector bundle $T_G P \to X$ (5.3.11), and yield the vector fields (5.6.7):

$$v = \tau^\lambda\partial_\lambda + (-c_{pq}^r \xi^p a_\lambda^q + \partial_\lambda\xi^r - a_\mu^r\partial_\lambda\tau^\mu)\partial_r^\lambda \qquad (8.2.9)$$

on the bundle of principal connections C. Sections ξ (8.2.8) play a role of gauge parameters.

Lemma 8.2.1. *Vector fields (8.2.9) are locally variational symmetries of the global Chern–Simons Lagrangian $L_{\mathrm{CS}}(B)$ (8.2.4).*

8.2. Chern–Simons topological field theory

Proof. Since $\dim X = 2k - 1$, the transgression formula (8.2.1) takes the form

$$P_{2k}(F_A) = d\mathfrak{S}_{2k-1}(a, A).$$

The Lie derivative $\mathbf{L}_{J^1 v}$ acting on its sides results in the equality

$$0 = d(v \rfloor d\mathfrak{S}_{2k-1}(a, A)) = d(\mathbf{L}_{J^1 v}\mathfrak{S}_{2k-1}(a, A)),$$

i.e., the Lie derivative $\mathbf{L}_{J^1 v}\mathfrak{S}_{2k-1}(a, A)$ is locally d-exact. Consequently, the horizontal form $h_0 \mathbf{L}_{J^1 v}\mathfrak{S}_{2k-1}(a, A)$ is locally d_H-exact. A direct computation shows that

$$h_0 \mathbf{L}_{J^1 v}\mathfrak{S}_{2k-1}(a, A) = \mathbf{L}_{J^1 v}(h_0 \mathfrak{S}_{2k-1}(a, A)) + d_H S.$$

It follows that the Lie derivative $\mathbf{L}_{J^1 v} L_{\mathrm{CS}}(A)$ of the global Chern–Simons Lagrangian along any vector field v (8.2.9) is locally d_H-exact, i.e., this vector field is locally a variational symmetry of $L_{\mathrm{CS}}(A)$. $\square$

By virtue of Lemma 2.2.3, the vertical part

$$v_V = (-c_{pq}^r \xi^p a_\lambda^q + \partial_\lambda \xi^r - a_\mu^r \partial_\lambda \tau^\mu - \tau^\mu a_{\mu\lambda}^r)\partial_r^\lambda \tag{8.2.10}$$

of the vector field v (8.2.9) also is locally a variational symmetry of $L_{\mathrm{CS}}(A)$.

Given the fibre bundle $T_G P \to X$ (5.3.11), let the same symbol also stand for the pull-back of $T_G P$ onto C. Let us consider the differential bigraded algebra (4.1.6):

$$\mathcal{P}_\infty^*[T_G P; C] = \mathcal{S}_\infty^*[T_G P; C],$$

possessing the local generating basis $(a_\lambda^r, c^\lambda, c^r)$ of even fields a_λ^r and odd ghosts c^λ, c^r. Substituting these ghosts for gauge parameters in the vector field v (8.2.10) (see Remark 4.2.1), we obtain the odd vertical graded derivation

$$u = (-c_{pq}^r c^p a_\lambda^q + c_\lambda^r - c_\lambda^\mu a_\mu^r - c^\mu a_{\mu\lambda}^r)\partial_r^\lambda \tag{8.2.11}$$

of the differential bigraded algebra $\mathcal{P}_\infty^*[T_G P; C]$. This graded derivation as like as vector fields v_V (8.2.10) is locally a variational symmetry of the global Chern–Simons Lagrangian $L_{\mathrm{CS}}(A)$ (8.2.4), i.e., the odd density $\mathbf{L}_{J^1 u}(L_{\mathrm{CS}}(A))$ is locally d_H-exact. Hence, it is δ-closed and, consequently, d_H-exact in accordance with Corollary 3.5.1. Thus, the graded derivation u (8.2.11) is variationally trivial and, consequently, it is a gauge symmetry of the global Chern–Simons Lagrangian $L_{\mathrm{CS}}(A)$.

By virtue of the formulas (2.3.6) – (2.3.7), the corresponding Noether identities read

$$\overline{\delta}\Delta_j = -c_{ji}^r a_\lambda^i \mathcal{E}_r^\lambda - d_\lambda \mathcal{E}_j^\lambda = 0, \tag{8.2.12}$$

$$\overline{\delta}\Delta_\mu = -a_{\mu\lambda}^r \mathcal{E}_r^\lambda + d_\lambda(a_\mu^r \mathcal{E}_r^\lambda) = 0. \tag{8.2.13}$$

They are irreducible and non-trivial, unless $\dim X = 3$. Therefore, the gauge operator (4.2.8) is $\mathbf{u} = u$. It admits the nilpotent BRST extension

$$\mathbf{b} = (-c^r_{ji}c^j a^i_\lambda + c^r_\lambda - c^\mu_\lambda a^r_\mu - c^\mu a^r_{\mu\lambda})\frac{\partial}{\partial a^r_\lambda} - \qquad (8.2.14)$$

$$\frac{1}{2}c^r_{ij}c^i c^j \frac{\partial}{\partial c^r} + c^\lambda_\mu c^\mu \frac{\partial}{\partial c^\lambda}.$$

In order to include antifields $(\overline{a}^\lambda_r, \overline{c}_r, \overline{c}_\mu)$, let us enlarge the differential bigraded algebra $\mathcal{P}^*_\infty[T_GP; C]$ to the differential bigraded algebra

$$\mathcal{P}^*_\infty\{0\} = \mathcal{S}^*_\infty[\overline{VC} \underset{C}{\oplus} T_GP; C \underset{X}{\times} \overline{T_GP}]$$

where $\overline{VC}$ is the density dual (5.8.30) of the vertical tangent bundle VC of $C \to X$ and $\overline{T_GP}$ is the density dual of $T_GP \to X$ (cf. (5.8.32)). By virtue of Theorem 4.4.2, given the BRST operator $\mathbf{b}$ (8.2.14), the global Chern–Simons Lagrangian $L_{\mathrm{CS}}(A)$ (8.2.4) is extended to the proper solution (4.4.10) of the master equation which reads

$$L_E = L_{\mathrm{CS}}(A) + (-c^r_{pq}c^p a^q_\lambda + c^r_\lambda - c^\mu_\lambda a^r_\mu - c^\mu a^r_{\mu\lambda})\overline{a}^\lambda_r \omega -$$

$$\frac{1}{2}c^r_{ij}c^i c^j \overline{c}_r \omega + c^\lambda_\mu c^\mu \overline{c}_\lambda \omega.$$

If $\dim X = 3$, the global Chern–Simons Lagrangian takes the form (8.2.7). Its Euler–Lagrange operator is

$$\delta L_{\mathrm{CS}}(B) = \mathcal{E}^\lambda_r \theta^r_\lambda \wedge \omega, \qquad \mathcal{E}^\lambda_r = h_{rp}\varepsilon^{\lambda\beta\gamma}\mathcal{F}^p_{\beta\gamma}.$$

A glance at the Noether identities (8.2.12) – (8.2.13) shows that they are equivalent to the Noether identities

$$\overline{\delta}\Delta_j = -c^r_{ji}a^i_\lambda \mathcal{E}^\lambda_r - d_\lambda \mathcal{E}^\lambda_j = 0, \qquad (8.2.15)$$

$$\overline{\delta}\Delta'_\mu = \overline{\delta}\Delta_\mu + a^r_\mu \overline{\delta}\Delta_r = c^\mu \mathcal{F}^r_{\lambda\mu}\mathcal{E}^\lambda_r = 0. \qquad (8.2.16)$$

These Noether identities define the gauge symmetry u (8.2.11) written in the form

$$u = (-c^r_{pq}c'^p a^q_\lambda + c'^r_\lambda + c^\mu \mathcal{F}^r_{\lambda\mu})\partial^\lambda_r \qquad (8.2.17)$$

where $c'^r = c^r - a^r_\mu c^\mu$. It is readily observed that, if $\dim X = 3$, the Noether identities $\overline{\delta}\Delta'_\mu$ (8.2.16) are trivial. Then the corresponding part $c^\mu \mathcal{F}^r_{\lambda\mu}\partial^\lambda_r$ of the gauge symmetry u (8.2.17) also is trivial. Consequently, the non-trivial gauge symmetry of the Chern–Simons Lagrangian (8.2.7) is

$$u = (-c^r_{pq}c'^p a^q_\lambda + c'^r_\lambda)\partial^\lambda_r.$$

8.3 Topological BF theory

We address the topological BF theory of two exterior forms A and B of form degree $|A| + |B| = \dim X - 1$ on a smooth manifold X [18]. It is reducible degenerate Lagrangian theory which satisfies the homology regularity condition (Condition 4.1.1) [14]. Its dynamic variables A and B are sections of the fibre bundle

$$Y = \overset{p}{\wedge} T^*X \oplus \overset{q}{\wedge} T^*X, \qquad p + q = n - 1 > 1,$$

coordinated by

$$(x^\lambda, A_{\mu_1\ldots\mu_p}, B_{\nu_1\ldots\nu_q}).$$

Without a loss of generality, let q be even and $q \geq p$. The corresponding differential graded algebra is $\mathcal{O}^*_\infty Y$ (1.7.9).

There are the canonical p- and q-forms

$$A = A_{\mu_1\ldots\mu_p} dx^{\mu_1} \wedge \cdots \wedge dx^{\mu_p},$$
$$B = B_{\nu_1\ldots\nu_q} dx^{\nu_1} \wedge \cdots \wedge dx^{\nu_q}$$

on Y. A Lagrangian of topological BF theory reads

$$L_{\mathrm{BF}} = A \wedge d_H B = \epsilon^{\mu_1\ldots\mu_n} A_{\mu_1\ldots\mu_p} d_{\mu_{p+1}} B_{\mu_{p+2}\ldots\mu_n} \omega, \qquad (8.3.1)$$

where ϵ is the Levi–Civita symbol. It is a reduced first order Lagrangian. Its first order Euler–Lagrange operator (2.4.2) is

$$\delta L = \mathcal{E}_A^{\mu_1\ldots\mu_p} dA_{\mu_1\ldots\mu_p} \wedge \omega + \mathcal{E}_B^{\nu_{p+2}\ldots\nu_n} dB_{\nu_{p+2}\ldots\nu_n} \wedge \omega, \qquad (8.3.2)$$
$$\mathcal{E}_A^{\mu_1\ldots\mu_p} = \epsilon^{\mu_1\ldots\mu_n} d_{\mu_{p+1}} B_{\mu_{p+2}\ldots\mu_n}, \qquad (8.3.3)$$
$$\mathcal{E}_B^{\mu_{p+2}\ldots\mu_n} = -\epsilon^{\mu_1\ldots\mu_n} d_{\mu_{p+1}} A_{\mu_1\ldots\mu_p}. \qquad (8.3.4)$$

The corresponding Euler–Lagrange equations can be written in the form

$$d_H B = 0, \qquad d_H A = 0. \qquad (8.3.5)$$

They obey the Noether identities

$$d_H d_H B = 0, \qquad d_H d_H A = 0. \qquad (8.3.6)$$

One can regard the components $\mathcal{E}_A^{\mu_1\ldots\mu_p}$ (8.3.3) and $\mathcal{E}_B^{\mu_{p+2}\ldots\mu_n}$ (8.3.4) of the Euler–Lagrange operator (8.3.2) as a $(\overset{p}{\wedge} TX) \underset{X}{\otimes} (\overset{n}{\wedge} T^*X)$-valued differential operator on the fibre bundle $\overset{q}{\wedge} T^*X$ and a $(\overset{q}{\wedge} TX) \underset{X}{\otimes} (\overset{n}{\wedge} T^*X)$-valued differential operator on the fibre bundle $\overset{p}{\wedge} T^*X$, respectively. They are of the same type as the $\overset{n-1}{\wedge} TX$-valued differential operator (4.5.23) in Example 4.5.1 (cf. the equations (8.3.5) and (4.5.22)). Therefore, the analysis of the Noether identities of the differential operators (8.3.3) and (8.3.4) is a repetition of that of Noether identities of the operator (4.5.23) (cf. the Noether identities (8.3.6) and (4.5.28)).

Following Example 4.5.1, let us consider the family of vector bundles

$$E_k = \overset{p-k-1}{\wedge} T^*X \underset{X}{\times} \overset{q-k-1}{\wedge} T^*X, \qquad 0 \le k < p-1,$$

$$E_k = \mathbb{R} \underset{X}{\times} \overset{q-p}{\wedge} T^*X, \qquad k = p-1,$$

$$E_k = \overset{q-k-1}{\wedge} T^*X, \qquad p-1 < k < q-1,$$

$$E_{q-1} = X \times \mathbb{R}.$$

Let us enlarge the differential graded algebra $\mathcal{O}^*_\infty Y$ to the differential bigraded algebra $P^*_\infty\{q-1\}$ (4.2.2) which is

$$P^*_\infty\{q-1\} = \mathcal{P}^*_\infty[\overline{VY} \underset{Y}{\oplus} E_0 \underset{Y}{\oplus} \cdots \underset{Y}{\oplus} E_{q-1} \underset{Y}{\oplus} \overline{E}_0 \underset{Y}{\oplus} \cdots \oplus \overline{E}_{q-1}; Y]. \tag{8.3.7}$$

It possesses the local generating basis

$$\{A_{\mu_1\ldots\mu_p}, B_{\nu_1\ldots\nu_q}, \varepsilon_{\mu_2\ldots\mu_p}, \ldots, \varepsilon_{\mu_p}, \varepsilon, \xi_{\nu_2\ldots\nu_q}, \ldots, \xi_{\nu_q}, \xi,$$
$$\overline{A}^{\mu_1\ldots\mu_p}, \overline{B}^{\nu_1\ldots\nu_q}, \overline{\varepsilon}^{\mu_2\ldots\mu_p}, \ldots, \overline{\varepsilon}^{\mu_p}, \overline{\varepsilon}, \overline{\xi}^{\nu_2\ldots\nu_q}, \ldots, \overline{\xi}^{\nu_q}, \overline{\xi}\}$$

of Grassmann parity

$$[\varepsilon_{\mu_k\ldots\mu_p}] = [\xi_{\nu_k\ldots\nu_q}] = (k+1)\bmod 2, \qquad [\varepsilon] = p\bmod 2, \qquad [\xi] = 0,$$
$$[\overline{\varepsilon}^{\mu_k\ldots\mu_p}] = [\overline{\xi}^{\nu_k\ldots\nu_q}] = k\bmod 2, \qquad [\overline{\varepsilon}] = (p+1)\bmod 2, \qquad [\overline{\xi}] = 1,$$

of ghost number

$$\mathrm{gh}[\varepsilon_{\mu_k\ldots\mu_p}] = \mathrm{gh}[\xi_{\nu_k\ldots\nu_q}] = k, \qquad \mathrm{gh}[\varepsilon] = p+1, \qquad \mathrm{gh}[\xi] = q+1,$$

and of antifield number

$$\mathrm{Ant}[\overline{A}^{\mu_1\ldots\mu_p}] = \mathrm{Ant}[\overline{B}^{\nu_{p+1}\ldots\nu_q}] = 1,$$
$$\mathrm{Ant}[\overline{\varepsilon}^{\mu_k\ldots\mu_p}] = \mathrm{Ant}[\overline{\xi}^{\nu_k\ldots\nu_q}] = k+1,$$
$$\mathrm{Ant}[\overline{\varepsilon}] = p, \qquad \mathrm{Ant}[\overline{\xi}] = q.$$

One can show that the homology regularity condition holds (see Lemma 4.5.5) and that the differential bigraded algebra $P^*_\infty\{q-1\}$ is endowed with

8.3. Topological BF theory

the Koszul–Tate operator

$$\delta_{\mathrm{KT}} = \frac{\overleftarrow{\partial}}{\partial \overline{A}^{\mu_1\ldots\mu_p}}\mathcal{E}_A^{\mu_1\ldots\mu_p} + \frac{\overleftarrow{\partial}}{\partial \overline{B}^{\nu_1\ldots\nu_q}}\mathcal{E}_B^{\nu_1\ldots\nu_q} + \qquad (8.3.8)$$

$$\sum_{2\leq k\leq p} \frac{\overleftarrow{\partial}}{\partial \overline{\varepsilon}^{\mu_k\ldots\mu_p}}\Delta_A^{\mu_k\ldots\mu_p} + \frac{\overleftarrow{\partial}}{\partial \overline{\varepsilon}}d_{\mu_p}\overline{\varepsilon}^{\mu_p} + \qquad (8.3.9)$$

$$\sum_{2\leq k\leq q} \frac{\overleftarrow{\partial}}{\partial \overline{\xi}^{\nu_k\ldots\nu_q}}\Delta_B^{\nu_k\ldots\nu_q} + \frac{\overleftarrow{\partial}}{\partial \overline{\xi}}d_{\nu_q}\overline{\xi}^{\nu_q},$$

$$\Delta_A^{\mu_2\ldots\mu_p} = d_{\mu_1}\overline{A}^{\mu_1\ldots\mu_p}, \qquad \Delta_A^{\mu_{k+1}\ldots\mu_p} = d_{\mu_k}\overline{\varepsilon}^{\mu_k\mu_{k+1}\ldots\mu_p}, \qquad 2\leq k<p,$$
$$\Delta_B^{\nu_2\ldots\nu_q} = d_{\nu_1}\overline{B}^{\nu_1\ldots\nu_q}, \qquad \Delta_B^{\nu_{k+1}\ldots\nu_q} = d_{\nu_k}\overline{\xi}^{\nu_k\nu_{k+1}\ldots\nu_q}, \qquad 2\leq k<q.$$

Its nilpotentness provides the complete Noether identities (8.3.5):

$$d_{\mu_1}\mathcal{E}_A^{\mu_1\ldots\mu_p} = 0, \qquad d_{\nu_1}\mathcal{E}_B^{\nu_1\ldots\nu_q} = 0,$$

and the $(k-1)$-stage ones

$$d_{\mu_k}\Delta_A^{\mu_k\ldots\mu_p} = 0, \qquad k = 2,\ldots,p,$$
$$d_{\nu_k}\Delta_B^{\nu_k\ldots\nu_q} = 0, \qquad k = 2,\ldots,q,$$

(cf. (4.5.30)). It follows that the topological BF theory is $(q-1)$-reducible.

Applying inverse second Noether Theorem 4.2.1, one obtains the gauge operator (4.2.8) which reads

$$\mathbf{u} = d_{\mu_1}\varepsilon_{\mu_2\ldots\mu_p}\frac{\partial}{\partial A_{\mu_1\mu_2\ldots\mu_p}} + d_{\nu_1}\xi_{\nu_2\ldots\nu_q}\frac{\partial}{\partial B_{\nu_1\nu_2\ldots\nu_q}} + \qquad (8.3.10)$$

$$\left[d_{\mu_2}\varepsilon_{\mu_3\ldots\mu_p}\frac{\partial}{\partial \varepsilon_{\mu_2\mu_3\ldots\mu_p}} + \cdots + d_{\mu_p}\varepsilon\frac{\partial}{\partial \varepsilon_{\mu_p}}\right] +$$

$$\left[d_{\nu_2}\xi_{\nu_3\ldots\nu_q}\frac{\partial}{\partial \xi_{\nu_2\nu_3\ldots\nu_q}} + \cdots + d_{\nu_q}\xi\frac{\partial}{\partial \xi_{\nu_q}}\right].$$

In particular, the gauge symmetry of the Lagrangian L_{BF} (8.3.1) is

$$u = d_{\mu_1}\varepsilon_{\mu_2\ldots\mu_p}\frac{\partial}{\partial A_{\mu_1\mu_2\ldots\mu_p}} + d_{\nu_1}\xi_{\nu_2\ldots\nu_q}\frac{\partial}{\partial B_{\nu_1\nu_2\ldots\nu_q}}.$$

This gauge symmetry is Abelian. It also is readily observed that higher-stage gauge symmetries are independent of original fields. Consequently, topological BF theory is Abelian, and its gauge operator $\mathbf{u}$ (8.3.10) is nilpotent. Thus, it is the BRST operator $\mathbf{b} = \mathbf{u}$. As a result, the Lagrangian L_{BF} is extended to the proper solution of the master equation $L_E = L_e$ (4.2.9) which reads

$$L_e = L_{\mathrm{BF}} + \varepsilon_{\mu_2\ldots\mu_p}d_{\mu_1}\overline{A}^{\mu_1\ldots\mu_p} + \sum_{1<k<p}\varepsilon_{\mu_{k+1}\ldots\mu_p}d_{\mu_k}\overline{\varepsilon}^{\mu_k\ldots\mu_p} + \varepsilon d_{\mu_p}\overline{\varepsilon}^{\mu_p} +$$

$$\xi_{\nu_2\ldots\nu_q}d_{\nu_1}\overline{B}^{\nu_1\ldots\nu_q} + \sum_{1<k<q}\xi_{\nu_{k+1}\ldots\nu_q}d_{\nu_k}\overline{\xi}^{\nu_k\ldots\nu_q} + \xi d_{\nu_q}\overline{\xi}^{\nu_q}.$$

8.4 Lagrangian theory of submanifolds

Jets of sections of fibre bundles are particular jets of submanifolds. Namely, a space of jets of submanifolds admits a cover by charts of jets of sections [53; 96; 117]. Three-velocities in relativistic mechanics exemplify first order jets of submanifolds [111; 137]. A problem is that differential forms on jets of submanifolds fail to constitute a variational bicomplex because horizontal forms (e.g., Lagrangians) are not preserved under coordinate transformations.

We consider n-dimensional submanifolds of an m-dimensional smooth real manifold Z, and associate to them sections of the trivial fibre bundle

$$\pi : Z_Q = Q \times Z \to Q, \tag{8.4.1}$$

where Q is some n-dimensional manifold. Here, we restrict our consideration to first order jets of submanifolds, and state their relation to jets of sections of the fibre bundle (8.4.1). This relation fails to be one-to-one correspondence. The ambiguity contains, e.g., diffeomorphisms of Q. Then Lagrangian formalism on a fibre bundle $Z_Q \to Q$ is developed in a standard way, but a Lagrangian is required to be variationally invariant under the above mentioned diffeomorphisms of Q. This invariance however leads to rather restrictive Noether identities (8.4.27). If a Lagrangian is independent of a metric on Q, it is a topological field Lagrangian. For instance, this is the case of the Nambu–Goto Lagrangian (8.4.28) of classical string theory [73; 124].

Given an m-dimensional smooth real manifold Z, a *k-order jet of n-dimensional submanifolds* of Z at a point $z \in Z$ is defined as an equivalence class $j_z^k S$ of n-dimensional imbedded submanifolds of Z through z which are tangent to each other at z with order $k \geq 0$. Namely, two submanifolds

$$i_S : S \to Z, \qquad i_{S'} : S' \to Z$$

through a point $z \in Z$ belong to the same equivalence class $j_z^k S$ if and only if the images of the k-tangent morphisms

$$T^k i_S : T^k S \to T^k Z, \qquad T^k i_{S'} : T^k S' \to T^k Z$$

coincide with each other. The set

$$J_n^k Z = \bigcup_{z \in Z} j_z^k S$$

of k-order jets of submanifolds is a finite-dimensional real smooth manifold, called the *k-order jet manifold of submanifolds*. For the sake of convenience,

8.4. Lagrangian theory of submanifolds

let us denote $J_n^0 Z = Z$. If $k > 0$, let $Y \to X$ be an m-dimensional fibre bundle over an n-dimensional base X and $J^k Y$ the k-order jet manifold of sections of $Y \to X$. Given an imbedding $\Phi : Y \to Z$, there is the natural injection

$$J^k \Phi : J^k Y \to J_n^k Z,$$
$$j_x^k s \to [\Phi \circ s]_{\Phi(s(x))}^k,$$

where s are sections of $Y \to X$. This injection defines a chart on $J_n^k Z$. These charts provide a manifold atlas of $J_n^k Z$.

Let us restrict our consideration to first order jets of submanifolds. There is obvious one-to-one correspondence

$$\zeta : j_z^1 S \to V_{j^1 S} \subset T_z Z \qquad (8.4.2)$$

between the jets $j_z^1 S$ at a point $z \in Z$ and the n-dimensional vector subspaces of the tangent space $T_z Z$ of Z at z. It follows that $J_n^1 Z$ is a fibre bundle

$$\rho : J_n^1 Z \to Z \qquad (8.4.3)$$

in Grassmann manifolds. This fibre bundle possesses the following coordinate atlas.

Let $\{(U; z^\mu)\}$ be a coordinate atlas of Z. Though $J_n^0 Z = Z$, let us provide $J_m^0 Z$ with the atlas obtained by replacing every chart (U, z^A) of Z with the

$$\binom{m}{n} = \frac{m!}{n!(m-n)!}$$

charts on U which correspond to different partitions of (z^A) in collections of n and $m - n$ coordinates

$$(U; x^a, y^i), \qquad a = 1, \ldots, n, \qquad i = 1, \ldots, m - n. \qquad (8.4.4)$$

The transition functions between the coordinate charts (8.4.4) of $J_n^0 Z$ associated with a coordinate chart (U, z^A) of Z are reduced to exchange between coordinates x^a and y^i. Transition functions between arbitrary coordinate charts of the manifold $J_n^0 Z$ take the form

$$x'^a = x'^a(x^b, y^k), \qquad y'^i = y'^i(x^b, y^k). \qquad (8.4.5)$$

Given an atlas of coordinate charts (8.4.4) – (8.4.5) of the manifold $J_n^0 Z$, the first order jet manifold $J_n^1 Z$ is endowed with the coordinate charts

$$(\rho^{-1}(U) = U \times \mathbb{R}^{(m-n)n}; x^a, y^i, y_a^i), \qquad (8.4.6)$$

possessing the following transition functions. With respect to the coordinates (8.4.6) on the jet manifold $J_n^1 Z$ and the induced fibre coordinates $(\dot{x}^a, \dot{y}^i)$ on the tangent bundle TZ, the above mentioned correspondence ζ (8.4.2) reads

$$\zeta : (y_a^i) \to \dot{x}^a(\partial_a + y_a^i(j_z^1 S)\partial_i).$$

It implies the relations

$$y_a'^j = \left(\frac{\partial y'^j}{\partial y^k} y_b^k + \frac{\partial y'^j}{\partial x^b}\right)\left(\frac{\partial x^b}{\partial y'^i} y_a'^i + \frac{\partial x^b}{\partial x'^a}\right), \tag{8.4.7}$$

$$\left(\frac{\partial x^b}{\partial y'^i} y_a'^i + \frac{\partial x^b}{\partial x'^a}\right)\left(\frac{\partial x'^c}{\partial y^k} y_b^k + \frac{\partial x'^c}{\partial x^b}\right) = \delta_a^c, \tag{8.4.8}$$

which jet coordinates y_a^i must satisfy under coordinate transformations (8.4.5). Let us consider a non-degenerate $n \times n$ matrix M with the entries

$$M_b^c = \left(\frac{\partial x'^c}{\partial y^k} y_b^k + \frac{\partial x'^c}{\partial x^b}\right).$$

Then the relations (8.4.8) lead to the equalities

$$\left(\frac{\partial x^b}{\partial y'^i} y_a'^i + \frac{\partial x^b}{\partial x'^a}\right) = (M^{-1})_a^b.$$

Hence, we obtain the transformation law of first order jet coordinates

$$y_a'^j = \left(\frac{\partial y'^j}{\partial y^k} y_b^k + \frac{\partial y'^j}{\partial x^b}\right)(M^{-1})_a^b. \tag{8.4.9}$$

In particular, if coordinate transition functions x'^a (8.4.5) are independent of coordinates y^k, the transformation law (8.4.9) comes to the familiar transformations of jets of sections.

A glance at the transformations (8.4.9) shows that, in contrast with the fibre bundle of jets of sections, the fibre bundle (8.4.3) is not affine. In particular, one generalizes the notion of a connection on fibre bundles and treats global sections of the jet bundle (8.4.3) as *preconnections* [117]. However, a global section of this bundle need not exist [147].

Given a coordinate chart (8.4.6) of $J_n^1 Z$, one can regard $\rho^{-1}(U) \subset J_n^1 Z$ as the first order jet manifold $J^1 U$ of sections of the fibre bundle

$$U \ni (x^a, y^i) \to (x^a) \in U_X.$$

The graded differential algebra $\mathcal{O}^*(\rho^{-1}(U))$ of exterior forms on $\rho^{-1}(U)$ is generated by horizontal forms dx^a and contact forms $dy^i - y_a^i dx^a$. Coordinate transformations (8.4.5) and (8.4.9) preserve the ideal of contact forms,

8.4. Lagrangian theory of submanifolds

but horizontal forms are not transformed into horizontal forms, unless coordinate transition functions x'^a (8.4.5) are independent of coordinates y^k. Therefore, one can develop first order Lagrangian formalism with a Lagrangian $L = \mathcal{L}d^n x$ on a coordinate chart $\rho^{-1}(U)$, but this Lagrangian fails to be globally defined on $J_n^1 Z$.

In order to overcome this difficulty, let us consider the trivial fibre bundle $Z_Q \to Q$ (8.4.1) whose trivialization throughout holds fixed. This fibre bundle is provided with an atlas of coordinate charts

$$(U_Q \times U; q^\mu, x^a, y^i), \qquad (8.4.10)$$

where $(U; x^a, y^i)$ are the above mentioned coordinate charts (8.4.4) of the manifold $J_n^0 Z$. The coordinate charts (8.4.10) possess transition functions

$$q'^\mu = q^\mu(q^\nu), \qquad x'^a = x'^a(x^b, y^k), \qquad y'^i = y'^i(x^b, y^k). \qquad (8.4.11)$$

Let $J^1 Z_Q$ be the first order jet manifold of the fibre bundle (8.4.1). Since the trivialization (8.4.1) is fixed, it is a vector bundle

$$\pi^1 : J^1 Z_Q \to Z_Q$$

isomorphic to the tensor product

$$J^1 Z_Q = T^* Q \underset{Q \times Z}{\otimes} TZ \qquad (8.4.12)$$

of the cotangent bundle T^*Q of Q and the tangent bundle TZ of Z over Z_Q.

Given the coordinate atlas (8.4.10) - (8.4.11) of Z_Q, the jet manifold $J^1 Z_Q$ is endowed with the coordinate charts

$$((\pi^1)^{-1}(U_Q \times U) = U_Q \times U \times \mathbb{R}^{mn}; q^\mu, x^a, y^i, x_\mu^a, y_\mu^i), \qquad (8.4.13)$$

possessing transition functions

$$x'^a_\mu = \left(\frac{\partial x'^a}{\partial y^k} y_\nu^k + \frac{\partial x'^a}{\partial x^b} x_\nu^b\right) \frac{\partial q^\nu}{\partial q'^\mu}, \qquad y'^i_\mu = \left(\frac{\partial y'^i}{\partial y^k} y_\nu^k + \frac{\partial y'^i}{\partial x^b} x_\nu^b\right) \frac{\partial q^\nu}{\partial q'^\mu}. \qquad (8.4.14)$$

Relative to coordinates (8.4.13), the isomorphism (8.4.12) takes the form

$$(x_\mu^a, y_\mu^i) \to dq^\mu \otimes (x_\mu^a \partial_a + y_\mu^i \partial_i).$$

Obviously, a jet $(q^\mu, x^a, y^i, x_\mu^a, y_\mu^i)$ of sections of the fibre bundle (8.4.1) defines some jet of n-dimensional subbundles of the manifold $\{q\} \times Z$ through a point $(x^a, y^i) \in Z$ if an $m \times n$ matrix with the entries x_μ^a, y_μ^i is of maximal rank n. This property is preserved under the coordinate transformations (8.4.14). An element of $J^1 Z_Q$ is called *regular* if it possesses this property. Regular elements constitute an open subbundle of the jet bundle $J^1 Z_Q \to Z_Q$.

Since regular elements of J^1Z_Q characterize jets of submanifolds of Z, one hopes to describe the dynamics of submanifolds of a manifold Z as that of sections of the fibre bundle (8.4.1). For this purpose, let us refine the relation between elements of the jet manifolds $J_n^1 Z$ and $J^1 Z_Q$.

Let us consider the manifold product $Q \times J_n^1 Z$. It is a fibre bundle over Z_Q. Given a coordinate atlas (8.4.10) - (8.4.11) of Z_Q, this product is endowed with the coordinate charts

$$(U_Q \times \rho^{-1}(U) = U_Q \times U \times \mathbb{R}^{(m-n)n}; q^\mu, x^a, y^i, y_a^i), \tag{8.4.15}$$

possessing transition functions (8.4.9). Let us assign to an element (q^μ, x^a, y^i, y_a^i) of the chart (8.4.15) the elements $(q^\mu, x^a, y^i, x_\mu^a, y_\mu^i)$ of the chart (8.4.13) whose coordinates obey the relations

$$y_a^i x_\mu^a = y_\mu^i. \tag{8.4.16}$$

These elements make up an n^2-dimensional vector space. The relations (8.4.16) are maintained under coordinate transformations (8.4.11) and the induced transformations of the charts (8.4.13) and (8.4.15) as follows:

$$y_a'^i x_\mu'^a = (\frac{\partial y'^i}{\partial y^k} y_c^k + \frac{\partial y'^i}{\partial x^c})(M^{-1})_a^c (\frac{\partial x'^a}{\partial y^k} y_\nu^k + \frac{\partial x'^a}{\partial x^b} x_\nu^b) \frac{\partial q^\nu}{\partial q'^\mu} =$$

$$(\frac{\partial y'^i}{\partial y^k} y_c^k + \frac{\partial y'^i}{\partial x^c})(M^{-1})_a^c (\frac{\partial x'^a}{\partial y^k} y_b^k + \frac{\partial x'^a}{\partial x^b}) x_\nu^b \frac{\partial q^\nu}{\partial q'^\mu} =$$

$$(\frac{\partial y'^i}{\partial y^k} y_b^k + \frac{\partial y'^i}{\partial x^b}) x_\nu^b \frac{\partial q^\nu}{\partial q'^\mu} = (\frac{\partial y'^i}{\partial y^k} y_\nu^k + \frac{\partial y'^i}{\partial x^b} x_\nu^b) \frac{\partial q^\nu}{\partial q'^\mu} = y_\mu'^i.$$

Thus, one can associate:

$$\zeta' : (q^\mu, x^a, y^i, y_a^i) \to \{(q^\mu, x^a, y^i, x_\mu^a, y_\mu^i) \mid y_a^i x_\mu^a = y_\mu^i\}, \tag{8.4.17}$$

to each element of the manifold $Q \times J_n^1 Z$ an n^2-dimensional vector space in the jet manifold $J^1 Z_Q$. This is a subspace of elements

$$x_\mu^a dq^\mu \otimes (\partial_a + y_a^i \partial_i)$$

of a fibre of the tensor bundle (8.4.12) at a point (q^μ, x^a, y^i). This subspace always contains regular elements, e.g., whose coordinates x_μ^a form a non-degenerate $n \times n$ matrix.

Conversely, given a regular element $j_z^1 s$ of $J^1 Z_Q$, there is a coordinate chart (8.4.13) such that coordinates x_μ^a of $j_z^1 s$ constitute a non-degenerate matrix, and $j_z^1 s$ defines a unique element of $Q \times J_n^1 Z$ by the relations

$$y_a^i = y_\mu^i (x^{-1})_a^\mu. \tag{8.4.18}$$

Thus, we have shown the following. Let (q^μ, z^A) further be arbitrary coordinates on the product Z_Q (8.4.1) and (q^μ, z^A, z_μ^A) the corresponding

8.4. Lagrangian theory of submanifolds

coordinates on the jet manifold $J^1 Z_Q$. In these coordinates, an element of $J^1 Z_Q$ is regular if an $m \times n$ matrix with the entries z^A_μ is of maximal rank n.

Theorem 8.4.1. *(i) Any jet of submanifolds through a point $z \in Z$ defines some (but not unique) jet of sections of the fibre bundle Z_Q (8.4.1) through a point $q \times z$ for any $q \in Q$ in accordance with the relations (8.4.16).*

(ii) Any regular element of $J^1 Z_Q$ defines a unique element of the jet manifold $J^1_n Z$ by means of the relations (8.4.18). However, non-regular elements of $J^1 Z_Q$ can correspond to different jets of submanifolds.

(iii) Two elements (q^μ, z^A, z^A_μ) and (q^μ, z^A, z'^A_μ) of $J^1 Z_Q$ correspond to the same jet of submanifolds if

$$z'^A_\mu = M^\nu_\mu z^A_\nu,$$

where M is some matrix, e.g., it comes from a diffeomorphism of Q.

Based on this result, we can describe the dynamics of n-dimensional submanifolds of a manifold Z as that of sections of the fibre bundle Z_Q (8.4.1) for some n-dimensional manifold Q.

Let Z_Q be a fibre bundle (8.4.1) coordinated by (q^μ, z^A) with transition functions $q'^\mu(q^\nu)$ and $z'^A(z^B)$. Then the first order jet manifold $J^1 Z_Q$ of this fibre bundle is provided with coordinates (q^μ, z^A, z^A_μ) possessing transition functions

$$z'^A_\mu = \frac{\partial z'^A}{\partial z^B} \frac{\partial q^\nu}{\partial q'^\mu} z^B_\nu.$$

Let

$$L = \mathcal{L}(z^A, z^A_\mu)\omega \tag{8.4.19}$$

be a first order Lagrangian on the jet manifold $J^1 Z_Q$. The corresponding Euler–Lagrange operator (2.4.2) reads

$$\delta L = \mathcal{E}_A dz^A \wedge \omega, \qquad \mathcal{E}_A = \partial_A \mathcal{L} - d_\mu \partial^\mu_A \mathcal{L}. \tag{8.4.20}$$

It yields the Euler–Lagrange equations

$$\mathcal{E}_A = \partial_A \mathcal{L} - d_\mu \partial^\mu_A \mathcal{L} = 0. \tag{8.4.21}$$

Let

$$u = u^\mu \partial_\mu + u^A \partial_A$$

be a vector field on Z_Q. Its jet prolongation (1.2.8) onto $J^1 Z_Q$ reads

$$J^1 u = u^\mu \partial_\mu + u^A \partial_A + (d_\mu u^A - z^A_\nu d_\mu u^\nu) \partial^\mu_A. \tag{8.4.22}$$

It admits the vertical splitting (2.4.26):
$$J^1 u = u_H + u_V = u^\mu d_\mu + [(u^A - u^\nu z_\nu^A)\partial_A + d_\mu(u^A - z_\nu^A u^\nu)\partial_A^\mu]. \quad (8.4.23)$$
The Lie derivative $\mathbf{L}_u L$ of a Lagrangian L along a vector field u obeys the first variational formula (2.4.28):
$$\mathbf{L}_u L = u_V \rfloor \delta L + d_H(h_0(u \rfloor H_L)) = ((u^A - u^\mu z_\mu^A)\mathcal{E}_A - d_\mu \mathcal{J}^\mu)\omega, \quad (8.4.24)$$
where
$$H_L = L + \partial_A^\mu \mathcal{L}\theta^A \wedge \omega_\mu,$$
is the Poincaré–Cartan form (2.4.12) and
$$\mathcal{J} = \mathcal{J}^\mu \omega_\mu = (\partial_A^\mu \mathcal{L}(u^\nu z_\nu^A - u^A) - u^\mu \mathcal{L})\omega_\mu \quad (8.4.25)$$
is the symmetry current (2.4.31) along a vector field u.

For instance, let us consider a vector field $u = u^\mu \partial_\mu$ on Q. Since $Z_Q \to Q$ is a trivial bundle, this vector field gives rise to a vector field $u = u^\mu \partial_\mu$ on Z_Q. Its jet prolongation (8.4.22) onto $J^1 Z_Q$ reads
$$u = u^\mu \partial_\mu - z_\nu^A \partial_\mu u^\nu \partial_A^\mu = u^\mu d_\mu + [-u^\nu z_\nu^A \partial_A - d_\mu(u^\nu z_\nu^A)\partial_A^\mu]. \quad (8.4.26)$$
One can regard it as a generalized vector field depending on parameter functions $u^\mu(q^\nu)$. In accordance with Theorem 8.4.1, it seems reasonable to require that, in order to describe jets of submanifolds of Z, a Lagrangian L on $J^1 Z_Q$ is independent of coordinates of Q and it possesses the gauge symmetry $J^1 u$ (8.4.26) or, equivalently, its vertical part
$$u_V = -u^\nu z_\nu^A \partial_A - d_\mu(u^\nu z_\nu^A)\partial_A^\mu.$$
Then the variational derivatives of this Lagrangian obey irreducible Noether identities
$$z_\nu^A \mathcal{E}_A = 0 \quad (8.4.27)$$
(see the relations (2.3.6) – (2.3.7)). These Noether identities are rather restrictive.

For instance, let Z be a locally affine manifold, i.e., a toroidal cylinder $\mathbb{R}^{m-k} \times T^k$. Its tangent bundle can be provided with a constant non-degenerate fiber metric η_{AB}. Let Q be a two-dimensional manifold. Let us consider the 2×2 matrix with the entries
$$h_{\mu\nu} = \eta_{AB} z_\mu^A z_\nu^B.$$
Then its determinant provides a Lagrangian
$$L = (\det h)^{1/2} d^2 q = ([\eta_{AB} z_1^A z_1^B][\eta_{AB} z_2^A z_2^B] - [\eta_{AB} z_1^A z_2^B]^2)^{1/2} d^2 q \quad (8.4.28)$$
on the jet manifold $J^1 Z_Q$ (8.4.12). This is the well known *Nambu–Goto Lagrangian* of classical string theory [73; 124]. It satisfies the Noether identities (8.4.27). This Lagrangian is independent of a metric on Q and, therefore, is a topological field Lagrangian.

Chapter 9

Covariant Hamiltonian field theory

Applied to field theory, the familiar symplectic Hamiltonian technique takes the form of instantaneous Hamiltonian formalism on an infinite-dimensional phase space, where canonical coordinates are field functions at some instant of time [66]. The true Hamiltonian counterpart of classical first order Lagrangian field theory on a fibre bundle $Y \to X$ is covariant Hamiltonian formalism, where canonical momenta p_i^μ correspond to derivatives y_μ^i of field variables y^i with respect to all world coordinates x^μ. This formalism has been vigorously developed since 1970s in the Hamilton – De Donder, polysymplectic, multisymplectic and k-symplectic variants [25; 26; 36; 53; 54; 72; 90; 101; 105; 106; 113; 128; 133; 134]. If $X = \mathbb{R}$, this is the case of Hamiltonian non-relativistic time-dependent mechanics [60; 111; 137].

We follow polysymplectic Hamiltonian formalism of field theory where the Legendre bundle Π (2.4.7) plays the role of a momentum phase space of fields [53; 54; 134]. This formalism is equivalent to Lagrangian formalism in the case of hyperregular Lagrangians. However, a non-regular Lagrangian leads to constraints and requires a set of associated Hamiltonians in order to exhaust all solutions of the Euler–Lagrange equation. Moreover, the covariant Hamiltonian formulation of a non-regular Lagrangian field system contains additional gauge symmetries that is important for quantization.

9.1 Polysymplectic Hamiltonian formalism

Given a fibre bundle $Y \to X$, let

$$\pi_{\Pi X} = \pi \circ \pi_{\Pi Y} : \Pi \to Y \to X$$

be the Legendre bundle (2.4.7) endowed with the holonomic coordinates $(x^\lambda, y^i, p_i^\lambda)$. There is the canonical bundle monomorphism

$$\Theta_Y : \Pi \xrightarrow[Y]{} \overset{n+1}{\wedge} T^*Y \underset{Y}{\otimes} TX, \qquad (9.1.1)$$

$$\Theta_Y = p_i^\lambda dy^i \wedge \omega \otimes \partial_\lambda,$$

called the *tangent-valued Liouville form* on Π. It should be emphasized that the exterior differential d can not be applied to the tangent-valued form (9.1.1). At the same time, there is a unique TX-valued $(n+2)$-form

$$\Omega_Y = dp_i^\lambda \wedge dy^i \wedge \omega \otimes \partial_\lambda \qquad (9.1.2)$$

on Π such that the relation

$$\Omega_Y \rfloor \phi = d(\Theta_Y \rfloor \phi)$$

holds for an arbitrary exterior one-form ϕ on X. This form is called the *polysymplectic form*. The Legendre bundle Π endowed with the polysymplectic form (9.1.2) is said to be the *polysymplectic manifold*.

Let $J^1\Pi$ be the first order jet manifold of the fibre bundle $\Pi \to X$. It is equipped with the adapted coordinates

$$(x^\lambda, y^i, p_i^\lambda, y^i_\mu, p^\lambda_{\mu i}).$$

A connection

$$\gamma = dx^\lambda \otimes (\partial_\lambda + \gamma^i_\lambda \partial_i + \gamma^\mu_{\lambda i} \partial^i_\mu) \qquad (9.1.3)$$

on $\Pi \to X$ is called a *Hamiltonian connection* if the exterior form $\gamma \rfloor \Omega_Y$ is closed. Components of a Hamiltonian connection satisfy the conditions

$$\partial^i_\lambda \gamma^j_\mu - \partial^j_\mu \gamma^i_\lambda = 0,$$
$$\partial_i \gamma^\mu_{\mu j} - \partial_j \gamma^\mu_{\mu i} = 0, \qquad (9.1.4)$$
$$\partial_j \gamma^i_\lambda + \partial^i_\lambda \gamma^\mu_{\mu j} = 0.$$

If the form $\gamma \rfloor \Omega_Y$ is closed, there is a contractible neighbourhood U of each point of Π where the local form $\gamma \rfloor \Omega_Y$ is exact, i.e.,

$$\gamma \rfloor \Omega_Y = dH = dp_i^\lambda \wedge dy^i \wedge \omega_\lambda - (\gamma^i_\lambda dp_i^\lambda - \gamma^\lambda_{\lambda i} dy^i) \wedge \omega \qquad (9.1.5)$$

on U. It is readily observed that, by virtue of the conditions (9.1.4), the second term in the right-hand side of this equality also is an exact form on U. In accordance with the relative Poincaré lemma (see Remark 10.10.1), this term can be brought into the form $d\mathcal{H} \wedge \omega$ where $\mathcal{H}$ is a local function on U. Then the form H in the expression (9.1.5) reads

$$H = p_i^\lambda dy^i \wedge \omega_\lambda - \mathcal{H}\omega. \qquad (9.1.6)$$

9.1. Polysymplectic Hamiltonian formalism

Let us consider the homogeneous Legendre bundle Z_Y (2.4.13) and the affine bundle $Z_Y \to \Pi$ (2.4.16). This affine bundle is modelled over the pull-back vector bundle

$$\Pi \underset{X}{\times} \overset{n}{\wedge} T^*X \to \Pi$$

in accordance with the exact sequence

$$0 \longrightarrow \Pi \underset{X}{\times} \overset{n}{\wedge} T^*X \longrightarrow Z_Y \longrightarrow \Pi \longrightarrow 0. \qquad (9.1.7)$$

The homogeneous Legendre bundle Z_Y is provided with the canonical multisymplectic Liouville form Ξ_Y (2.4.23). Its exterior differential $d\Xi_Y$ is the *multisymplectic form*.

Let $h = -\mathcal{H}\omega$ be a section the affine bundle $Z_Y \to \Pi$ (2.4.16). A glance at the transformation law (2.4.15) shows that it is not a density. By analogy with Hamiltonian time-dependent mechanics [111; 137], $-h$ is said to be the *covariant Hamiltonian* of covariant Hamiltonian field theory. It defines the pull-back

$$H = h^*\Xi_Y = p_i^\lambda dy^i \wedge \omega_\lambda - \mathcal{H}\omega \qquad (9.1.8)$$

of the multisimplectic Liouville form Ξ_Y onto the Legendre bundle Π which is called the *Hamiltonian form* on Π.

The following is a straightforward corollary of this definition.

Theorem 9.1.1. *(i) Hamiltonian forms constitute a non-empty affine space modelled over the linear space of horizontal densities $\widetilde{H} = \widetilde{\mathcal{H}}\omega$ on $\Pi \to X$.*

(ii) Every connection Γ on the fibre bundle $Y \to X$ yields the splitting (1.3.8) of the exact sequence (9.1.7) and defines the Hamiltonian form

$$H_\Gamma = \Gamma^*\Xi_Y = p_i^\lambda dy^i \wedge \omega_\lambda - p_i^\lambda \Gamma_\lambda^i \omega. \qquad (9.1.9)$$

(iii) Given a connection Γ on $Y \to X$, every Hamiltonian form H admits the decomposition

$$H = H_\Gamma - \widetilde{H}_\Gamma = p_i^\lambda dy^i \wedge \omega_\lambda - p_i^\lambda \Gamma_\lambda^i \omega - \widetilde{\mathcal{H}}_\Gamma \omega. \qquad (9.1.10)$$

Remark 9.1.1. In multisymplectic formalism of field theory, the homogeneous Legendre bundle Z_Y (2.4.13) plays the role of a momentum phase space of fields. From the mathematical viewpoint, multisymplectic formalism is more convenient than the polysymplectic one, because the multisymplectic form $d\Xi_Y$ unlike the polysymplectic one Ω_Y (9.1.2) is an exterior form. However, multisymplectic formalism involves an additional non-field variable p. If $X = \mathbb{R}$, the polysymplectic and multisimplectic Hamiltonian formalisms come to Hamiltonian and homogeneous Hamiltonian formalisms of non-relativistic time-dependent mechanics on momentum phase spaces V^*Y and T^*Y, respectively. In this case, the multisymplectic form $d\Xi_Y$ is exactly the canonical symplectic form on the cotangent bundle T^*Y of Y [111; 137].

One can generalize item (ii) of Theorem 9.1.1 as follows. We agree to call any bundle morphism

$$\Phi = dx^\lambda \otimes (\partial_\lambda + \Phi^i_\lambda \partial_i) : \Pi \underset{Y}{\to} J^1 Y \qquad (9.1.11)$$

over Y the *Hamiltonian map*. In particular, let Γ be a connection on $Y \to X$. Then the composition

$$\widehat{\Gamma} = \Gamma \circ \pi_{\Pi Y} = dx^\lambda \otimes (\partial_\lambda + \Gamma^i_\lambda \partial_i) : \Pi \to Y \to J^1 Y \qquad (9.1.12)$$

is a Hamiltonian map.

Theorem 9.1.2. *Every Hamiltonian map (9.1.11) defines the Hamiltonian form*

$$H_\Phi = -\Phi \rfloor \Theta = p^\lambda_i dy^i \wedge \omega_\lambda - p^\lambda_i \Phi^i_\lambda \omega. \qquad (9.1.13)$$

Proof. Given an arbitrary connection Γ on a fibre bundle $Y \to X$, the corresponding Hamiltonian map (9.1.12) defines the form $-\widehat{\Gamma} \rfloor \Theta$ which is exactly the Hamiltonian form H_Γ (9.1.9). Since $\Phi - \widehat{\Gamma}$ is a VY-valued basic one-form on $\Pi \to X$, $H_\Phi - H_\Gamma$ is a horizontal density on Π. Then the result follows from item (i) of Theorem 9.1.1. □

Theorem 9.1.3. *Every Hamiltonian form H (9.1.8) admits a Hamiltonian connection γ_H which obeys the condition*

$$\gamma_H \rfloor \Omega_Y = dH, \qquad (9.1.14)$$

$$\gamma^i_\lambda = \partial^i_\lambda \mathcal{H}, \qquad \gamma^\lambda_{\lambda i} = -\partial_i \mathcal{H}. \qquad (9.1.15)$$

Proof. It is readily observed that the Hamiltonian form H (9.1.8) is the Poincaré–Cartan form (2.4.12) of the first order Lagrangian

$$L_H = h_0(H) = (p^\lambda_i y^i_\lambda - \mathcal{H})\omega \qquad (9.1.16)$$

on the jet manifold $J^1\Pi$. The Euler–Lagrange operator (2.1.12) associated to this Lagrangian reads

$$\mathcal{E}_H : J^1\Pi \to T^*\Pi \wedge (\overset{n}{\wedge} T^*X),$$

$$\mathcal{E}_H = [(y^i_\lambda - \partial^i_\lambda \mathcal{H})dp^\lambda_i - (p^\lambda_{\lambda i} + \partial_i \mathcal{H})dy^i] \wedge \omega. \qquad (9.1.17)$$

9.1. Polysymplectic Hamiltonian formalism

It is called the *Hamilton operator* for H. A glance at the expression (9.1.17) shows that this operator is an affine morphism over Π of constant rank. It follows that its kernel

$$y^i_\lambda = \partial^i_\lambda \mathcal{H}, \tag{9.1.18}$$

$$p^\lambda_{\lambda i} = -\partial_i \mathcal{H} \tag{9.1.19}$$

is an affine closed imbedded subbundle of the jet bundle $J^1\Pi \to \Pi$. Therefore, it admits a global section γ_H which is a first order (non-holonomic) Lagrangian connection for the Lagrangian L_H. This connection is a desired Hamiltonian connection obeying the relation (9.1.14). □

Remark 9.1.2. It follows from the expression (9.1.6) that, conversely, any Hamiltonian connection is locally associated to some Hamiltonian form.

Remark 9.1.3. In fact, the Lagrangian (9.1.16) is the pull-back onto $J^1\Pi$ of the form L_H on the product $\Pi \underset{Y}{\times} J^1 Y$.

It should be emphasized that, if $\dim X > 1$, there is a set of Hamiltonian connections associated to the same Hamiltonian form H. They differ from each other in soldering forms σ on $\Pi \to X$ which fulfill the equation $\sigma \rfloor \Omega_Y = 0$. Every Hamiltonian form H yields the Hamiltonian map

$$\widehat{H} = J^1\pi_{\Pi Y} \circ \gamma_H : \Pi \to J^1\Pi \to J^1Y, \tag{9.1.20}$$

$$y^i_\lambda \circ \widehat{H} = \partial^i_\lambda \mathcal{H},$$

which is the same for all Hamiltonian connections γ_H associated to H.

Being a closed imbedded subbundle of the jet bundle $J^1\Pi \to X$, the kernel (9.1.18) – (9.1.19) of the Euler–Lagrange operator $\mathcal{E}_H$ (9.1.17) defines the Euler–Lagrange equation on Π in accordance with Definition 1.6.1. It is a system of first order dynamic equations called the *covariant Hamilton equations*.

Every integral section

$$J^1 r = \gamma_H \circ r$$

of a Hamiltonian connection γ_H associated to a Hamiltonian form H is obviously a solution of the covariant Hamilton equations (9.1.18) – (9.1.19). By virtue of Theorem 1.1.4, if $r : X \to \Pi$ is a global solution, there exists an extension of the local section

$$J^1 r : r(X) \to \operatorname{Ker} \mathcal{E}_H$$

to a Hamiltonian connection γ_H which has r as an integral section. Substituting $J^1 r$ in (9.1.20), we obtain the equality

$$J^1(\pi_{\Pi Y} \circ r) = \widehat{H} \circ r, \tag{9.1.21}$$

which is equivalent to the covariant Hamilton equations (9.1.18) – (9.1.19).

Remark 9.1.4. Similarly to the Cartan equation (2.4.22), the covariant Hamilton equations (9.1.18) – (9.1.19) are equivalent to the condition

$$r^*(u \rfloor dH) = 0 \tag{9.1.22}$$

for any vertical vector field u on $\Pi \to X$.

9.2 Associated Hamiltonian and Lagrangian systems

Let us study the relations between first order Lagrangian and covariant Hamiltonian formalisms of field theory. We are based on the fact that any Lagrangian L on a configurations space $J^1 Y$ defines the morphism

$$J^1 \widehat{L} : J^1 J^1 Y \xrightarrow[Y]{} J^1 \Pi,$$
$$(p_i^\lambda, y_\mu^i, p_{\mu i}^\lambda) \circ J^1 \widehat{L} = (\pi_i^\lambda, \widehat{y}_\mu^i, \widehat{d}_\mu \pi_i^\lambda),$$
$$\widehat{d}_\lambda = \partial_\lambda + \widehat{y}_\lambda^j \partial_j + y_{\lambda \mu}^j \partial_j^\mu,$$

and that any Hamiltonian form H on a momentum phase space Π yields the map

$$J^1 \widehat{H} : J^1 \Pi \xrightarrow[Y]{} J^1 J^1 Y,$$
$$(y_\mu^i, \widehat{y}_\lambda^i, y_{\lambda \mu}^i) \circ J^1 \widehat{H} = (\partial_\mu^i \mathcal{H}, y_\lambda^i, d_\lambda \partial_\mu^i \mathcal{H}),$$
$$d_\lambda = \partial_\lambda + y_\lambda^j \partial_j + p_{\lambda j}^\nu \partial_\nu^j.$$

Let us start with the case of a hyperregular Lagrangian L, i.e., when the Legendre map $\widehat{L}$ is a diffeomorphism. Then $\widehat{L}^{-1}$ is a Hamiltonian map. Let us consider the Hamiltonian form

$$H = H_{\widehat{L}^{-1}} + \widehat{L}^{-1*} L, \tag{9.2.1}$$
$$\mathcal{H} = p_i^\mu \widehat{L}^{-1 i}{}_\mu - \mathcal{L}(x^\mu, y^j, \widehat{L}^{-1 j}{}_\mu),$$

where $H_{\widehat{L}^{-1}}$ is the Hamiltonian form (9.1.13) associated to the Hamiltonian map $\widehat{L}^{-1}$. Let s be a solution of the Euler–Lagrange equation (2.4.3) for the Lagrangian L. A direct computation shows that $\widehat{L} \circ J^1 s$ is a solution of the covariant Hamilton equations (9.1.18) – (9.1.18) for the Hamiltonian form

9.2. Associated Hamiltonian and Lagrangian systems

H (9.2.1). Conversely, if r is a solution of the covariant Hamilton equations (9.1.18) – (9.1.19) for the Hamiltonian form H (9.2.1), then $s = \pi_{\Pi Y} \circ r$ is a solution of the Euler–Lagrange equation (2.4.3) for L (see the equality (9.1.21)). It follows that, in the case of hyperregular Lagrangians, covariant Hamiltonian formalism is equivalent to the Lagrangian one.

Let now L be an arbitrary Lagrangian on a configuration space J^1Y. A Hamiltonian form H is said to be *associated* with a Lagrangian L if H satisfies the relations

$$\widehat{L} \circ \widehat{H} \circ \widehat{L} = \widehat{L}, \tag{9.2.2}$$

$$H = H_{\widehat{H}} + \widehat{H}^* L. \tag{9.2.3}$$

A glance at the relation (9.2.2) shows that $\widehat{L} \circ \widehat{H}$ is the projector

$$p_i^\mu(p) = \partial_i^\mu \mathcal{L}(x^\mu, y^i, \partial_\lambda^j \mathcal{H}(p)), \qquad p \in N_L, \tag{9.2.4}$$

from Π onto the Lagrangian constraint space $N_L = \widehat{L}(J^1Y)$. Accordingly, $\widehat{H} \circ \widehat{L}$ is the projector from J^1Y onto $\widehat{H}(N_L)$. A Hamiltonian form is called *weakly associated* with a Lagrangian L if the condition (9.2.3) holds on the Lagrangian constraint space N_L.

Theorem 9.2.1. *If a Hamiltonian map Φ (9.1.11) obeys the relation*

$$\widehat{L} \circ \Phi \circ \widehat{L} = \widehat{L},$$

then the Hamiltonian form

$$H = H_\Phi + \Phi^* L$$

is weakly associated to the Lagrangian L. If $\Phi = \widehat{H}$, then H is associated to L.

Theorem 9.2.2. *Any Hamiltonian form H weakly associated to a Lagrangian L fulfills the relation*

$$H|_{N_L} = \widehat{H}^* H_L|_{N_L}, \tag{9.2.5}$$

where H_L is the Poincaré–Cartan form (2.4.12).

Proof. The relation (9.2.3) takes the coordinate form

$$\mathcal{H}(p) = p_i^\mu \partial_\mu^i \mathcal{H} - \mathcal{L}(x^\mu, y^i, \partial_\lambda^j \mathcal{H}(p)), \qquad p \in N_L. \tag{9.2.6}$$

Substituting (9.2.4) and (9.2.6) in (9.1.8), we obtain the relation (9.2.5). $\square$

The difference between associated and weakly associated Hamiltonian forms lies in the following. Let H be an associated Hamiltonian form, i.e., the equality (9.2.6) holds everywhere on Π. Acting on this equality by the exterior differential, we obtain the relations

$$\partial_\mu \mathcal{H}(p) = -(\partial_\mu \mathcal{L}) \circ \widehat{H}(p), \qquad p \in N_L,$$
$$\partial_i \mathcal{H}(p) = -(\partial_i \mathcal{L}) \circ \widehat{H}(p), \qquad p \in N_L, \qquad (9.2.7)$$
$$(p_i^\mu - (\partial_i^\mu \mathcal{L})(x^\mu, y^i, \partial_\lambda^j \mathcal{H}))\partial_\mu^i \partial_\alpha^a \mathcal{H} = 0. \qquad (9.2.8)$$

The relation (9.2.8) shows that the associated Hamiltonian form (i.e., the Hamiltonian map $\widehat{H}$) is not regular outside the Lagrangian constraint space N_L. For instance, any Hamiltonian form is weakly associated to the Lagrangian $L = 0$, while the associated Hamiltonian forms are only H_Γ (9.1.9).

In particular, a hyperregular Lagrangian has a unique weakly associated Hamiltonian form (9.2.1). In the case of a regular Lagrangian L, the Lagrangian constraint space N_L is an open subbundle of the vector Legendre bundle $\Pi \to Y$. If $N_L \neq \Pi$, a weakly associated Hamiltonian form fails to be defined everywhere on Π in general. At the same time, N_L itself can be provided with the pull-back polysymplectic structure with respect to the imbedding $N_L \to \Pi$, so that one may consider Hamiltonian forms on N_L.

One can say something more in the case of semiregular Lagrangians (see Definition 2.4.1).

Lemma 9.2.1. *The Poincaré–Cartan form H_L for a semiregular Lagrangian L is constant on the connected inverse image $\widehat{L}^{-1}(p)$ of any point $p \in N_L$.*

Proof. Let u be a vertical vector field on the affine jet bundle $J^1Y \to Y$ which takes its values into the kernel of the tangent map $T\widehat{L}$ to $\widehat{L}$. Then $\mathbf{L}_u H_L = 0$. □

A corollary of Lemma 9.2.1 is the following.

Theorem 9.2.3. *All Hamiltonian forms weakly associated to a semiregular Lagrangian L coincide with each other on the Lagrangian constraint space N_L, and the Poincaré–Cartan form H_L (2.4.12) for L is the pull-back*

$$H_L = \widehat{L}^* H, \qquad (\pi_i^\lambda y_\lambda^i - \mathcal{L})\omega = \mathcal{H}(x^\mu, y^j, \pi_j^\mu)\omega, \qquad (9.2.9)$$

of any such a Hamiltonian form H.

Proof. Given a vector $v \in T_p\Pi$, the value $T\widehat{H}(v) \rfloor H_L(\widehat{H}(p))$ is the same for all Hamiltonian maps $\widehat{H}$ satisfying the relation (9.2.2). Then the result follows from the relation (9.2.5). □

9.2. Associated Hamiltonian and Lagrangian systems

Theorem 9.2.3 enables us to relate the Euler–Lagrange equation for an almost regular Lagrangian L with the covariant Hamilton equations for Hamiltonian forms weakly associated to L [53; 133; 134].

Theorem 9.2.4. *Let a section r of $\Pi \to X$ be a solution of the covariant Hamilton equations (9.1.18) – (9.1.19) for a Hamiltonian form H weakly associated to a semiregular Lagrangian L. If r lives in the Lagrangian constraint space N_L, the section $s = \pi_{\Pi Y} \circ r$ of $Y \to X$ satisfies the Euler–Lagrange equation (2.4.3).*

Proof. There is the equality
$$\overline{L} = (J^1\widehat{L})^* L_H,$$
where $\overline{L}$ is the Lagrangian (2.4.18) on J^1J^1Y and L_H is the Lagrangian (9.1.16) on $J^1\Pi$. This equality results in the relation
$$\mathcal{E}_L = (J^1\widehat{L})^*\mathcal{E}_H|_{J^2Y}. \qquad \square$$

The converse assertion is more intricate.

Theorem 9.2.5. *Given a semiregular Lagrangian L, let a section s of a fibre bundle $Y \to X$ be a solution of the Euler–Lagrange equation (2.4.3). Let H be a Hamiltonian form weakly associated to L, and let H satisfy the relation*
$$\widehat{H} \circ \widehat{L} \circ J^1 s = J^1 s. \qquad (9.2.10)$$
Then the section $r = \widehat{L} \circ J^1 s$ of the fibre bundle $\Pi \to X$ is a solution of the covariant Hamilton equations (9.1.18) – (9.1.19) for H.

We say that a set of Hamiltonian forms H weakly associated to a semiregular Lagrangian L is *complete* if, for each solution s of the Euler–Lagrange equation, there exists a solution r of the covariant Hamilton equations for a Hamiltonian form H from this set such that $s = \pi_{\Pi Y} \circ r$. By virtue of Theorem 9.2.5, a set of weakly associated Hamiltonian forms is complete if, for every solution s of the Euler-Lagrange equation for L, there is a Hamiltonian form H from this set which fulfills the relation (9.2.10).

In the case of almost regular Lagrangians (see Definition 2.4.1), one can formulate the following necessary and sufficient conditions of the existence of associated Hamiltonian forms. An immediate consequence of Theorem 9.2.1 is the following.

Theorem 9.2.6. *A Hamiltonian form H weakly associated to an almost regular Lagrangian L exists if and only if the fibred manifold $J^1Y \to N_L$ (2.4.10) admits a global section.*

In particular, on an open neighbourhood $U \subset \Pi$ of each point $p \in N_L \subset \Pi$, there exists a complete set of local Hamiltonian forms weakly associated to an almost regular Lagrangian L. Moreover, one can always construct a complete set of associated Hamiltonian forms [134].

Given a global section Ψ of the fibred manifold

$$\widehat{L} : J^1 Y \to N_L, \qquad (9.2.11)$$

let us consider the pull-back form

$$H_N = \Psi^* H_L = i_N^* H \qquad (9.2.12)$$

on N_L called the *constrained Hamiltonian form*. By virtue of Lemma 9.2.1, it does not depend on the choice of a section of the fibred manifold (9.2.11) and, consequently, $H_L = \widehat{L}^* H_N$. For sections r of the fibre bundle $N_L \to X$, one can write the *constrained Hamilton equations*

$$r^*(u_N \rfloor dH_N) = 0, \qquad (9.2.13)$$

where u_N is an arbitrary vertical vector field on $N_L \to X$. These equations possess the following important properties.

Theorem 9.2.7. *For any Hamiltonian form H weakly associated to an almost regular Lagrangian L, every solution r of the covariant Hamilton equations which lives in the Lagrangian constraint space N_L is a solution of the constrained Hamilton equations (9.2.13).*

Proof. Such a Hamiltonian form H defines the global section $\Psi = \widehat{H} \circ i_N$ of the fibred manifold (9.2.11). Since $H_N = i_N^* H$ due to the relation (9.2.9), the constrained Hamilton equations can be written as

$$r^*(u_N \rfloor di_N^* H) = r^*(u_N \rfloor dH|_{N_L}) = 0. \qquad (9.2.14)$$

Note that these equations differ from the Hamilton equations (9.1.22) restricted to N_L. These read

$$r^*(u \rfloor dH|_{N_L}) = 0, \qquad (9.2.15)$$

where r is a section of $N_L \to X$ and u is an arbitrary vertical vector field on $\Pi \to X$. A solution r of the equations (9.2.15) obviously satisfies the weaker condition (9.2.14). $\square$

Theorem 9.2.8. *The constrained Hamilton equations (9.2.13) are equivalent to the Hamilton–De Donder equation (2.4.25).*

9.2. Associated Hamiltonian and Lagrangian systems

Proof. It is readily observed that

$$\widehat{L} = \pi_{Z\Pi} \circ \widehat{H}_L.$$

Hence, the projection $\pi_{Z\Pi}$ (2.4.16) yields a surjection of Z_L onto N_L. Given a section Ψ of the fibred manifold (9.2.11), we have the morphism

$$\widehat{H}_L \circ \Psi : N_L \to Z_L.$$

By virtue of Lemma (9.2.1), this is a surjection such that

$$\pi_{Z\Pi} \circ \widehat{H}_L \circ \Psi = \operatorname{Id} N_L.$$

Hence, $\widehat{H}_L \circ \Psi$ is a bundle isomorphism over Y which is independent of the choice of a global section Ψ. Combination of (2.4.24) and (9.2.12) results in

$$H_N = (\widehat{H}_L \circ \Psi)^* \Xi_L$$

that leads to a desired equivalence. $\qquad\square$

This proof gives something more. Namely, since Z_L and N_L are isomorphic, the homogeneous Legendre map $\widehat{H}_L$ fulfils the conditions of Theorem 2.4.1. Then combining Theorem 2.4.1 and Theorem 9.2.8, we obtain the following.

Theorem 9.2.9. *Let L be an almost regular Lagrangian such that the fibred manifold (9.2.11) has a global section. A section $\overline{s}$ of the jet bundle $J^1Y \to X$ is a solution of the Cartan equation (2.4.22) if and only if $\widehat{L} \circ \overline{s}$ is a solution of the constrained Hamilton equations (9.2.13).*

Theorem 9.2.9 also is a corollary of Lemma 9.2.2 below. The constrained Hamiltonian form H_N (9.2.12) defines the *constrained Lagrangian*

$$L_N = h_0(H_N) = (J^1 i_N)^* L_H \qquad (9.2.16)$$

on the jet manifold $J^1 N_L$ of the fibre bundle $N_L \to X$.

Lemma 9.2.2. *There are the relations*

$$\overline{L} = (J^1 \widehat{L})^* L_N, \qquad L_N = (J^1 \Psi)^* \overline{L}, \qquad (9.2.17)$$

where $\overline{L}$ is the Lagrangian (2.4.18).

The Euler–Lagrange equation for the constrained Lagrangian L_N (9.2.16) is equivalent to the constrained Hamilton equations (9.2.13) and, by virtue of Lemma 9.2.2, is quasi-equivalent to the Cartan equation.

9.3 Hamiltonian conservation laws

In order to study symmetries of covariant Hamiltonian field theory, let us use the fact that a Hamiltonian form H (9.1.8) is the Poincaré–Cartan form for the Lagrangian L_H (9.1.16) and that the covariant Hamilton equations for H are the Euler–Lagrange equation for L_H. We restrict our consideration to classical symmetries defined by projectable vector fields on a fibre bundle $Y \to X$.

In accordance with the canonical lift (1.1.26), every projectable vector field u on $Y \to X$ gives rise to the vector field

$$\widetilde{u} = u^\mu \partial_\mu + u^i \partial_i + (-\partial_i u^j p_j^\lambda - \partial_\mu u^\mu p_i^\lambda + \partial_\mu u^\lambda p_i^\mu)\partial_\lambda^i \qquad (9.3.1)$$

on the Legendre bundle $\Pi \to Y$ and then to the vector field

$$J\widetilde{u} = \widetilde{u} + J^1 u \qquad (9.3.2)$$

on $\Pi \underset{Y}{\times} J^1 Y$. Then we have

$$\mathbf{L}_{\widetilde{u}} H = \mathbf{L}_{J\widetilde{u}} L_H = (-u^i \partial_i \mathcal{H} - \partial_\mu(u^\mu \mathcal{H}) - u_i^\lambda \partial_\lambda^i \mathcal{H} + p_i^\lambda \partial_\lambda u^i)\omega. \qquad (9.3.3)$$

It follows that a Hamiltonian form H and a Lagrangian L_H have the same classical symmetries.

Remark 9.3.1. Given the splitting (9.1.10) of a Hamiltonian form H, the Lie derivative (9.3.3) takes the form

$$\mathbf{L}_{\widetilde{u}} H = p_j^\lambda ([\partial_\lambda + \Gamma_\lambda^i \partial_i, u]^j - [\partial_\lambda + \Gamma_\lambda^i \partial_i, u]^\nu \Gamma_\nu^j)\omega - \qquad (9.3.4)$$
$$(\partial_\mu u^\mu \widetilde{\mathcal{H}}_\Gamma + u \rfloor d\widetilde{\mathcal{H}}_\Gamma)\omega,$$

where $[.,.]$ is the Lie bracket of vector fields.

Let us apply the first variational formula (2.4.28) to the Lie derivative $\mathbf{L}_{J\widetilde{u}} L_H$ (9.1.16) [53]. It reads

$$-u^i \partial_i \mathcal{H} - \partial_\mu(u^\mu \mathcal{H}) - u_i^\lambda \partial_\lambda^i \mathcal{H} + p_i^\lambda \partial_\lambda u^i = -(u^i - y_\mu^i u^\mu)(p_{\lambda i}^\lambda + \partial_i \mathcal{H}) +$$
$$(-\partial_i u^j p_j^\lambda - \partial_\mu u^\mu p_i^\lambda + \partial_\mu u^\lambda p_i^\mu - p_{\mu i}^\lambda u^\mu)(y_\lambda^i - \partial_\lambda^i \mathcal{H}) -$$
$$d_\lambda[p_i^\lambda(\partial_\mu^i \mathcal{H} u^\mu - u^i) - u^\lambda(p_i^\mu \partial_\mu^i \mathcal{H} - \mathcal{H})].$$

On the shell (9.1.18) – (9.1.19), this identity takes the form

$$-u^i \partial_i \mathcal{H} - \partial_\mu(u^\mu \mathcal{H}) - u_i^\lambda \partial_\lambda^i \mathcal{H} + p_i^\lambda \partial_\lambda u^i \approx - \qquad (9.3.5)$$
$$d_\lambda[p_i^\lambda(\partial_\mu^i \mathcal{H} u^\mu - u^i) - u^\lambda(p_i^\mu \partial_\mu^i \mathcal{H} - \mathcal{H})].$$

If

$$\mathbf{L}_{J^1 \widetilde{u}} L_H = 0,$$

9.3. Hamiltonian conservation laws

we obtain the weak *Hamiltonian conservation law*

$$0 \approx -d_\lambda [p_i^\lambda(u^\mu \partial_\mu^i \mathcal{H} - u^i) - u^\lambda(p_i^\mu \partial_\mu^i \mathcal{H} - \mathcal{H})] \tag{9.3.6}$$

of the *Hamiltonian symmetry current*

$$\widetilde{\mathcal{J}}_u^\lambda = p_i^\lambda(u^\mu \partial_\mu^i \mathcal{H} - u^i) - u^\lambda(p_i^\mu \partial_\mu^i \mathcal{H} - \mathcal{H}). \tag{9.3.7}$$

On solutions r of the covariant Hamilton equations (9.1.18) – (9.1.19), the weak equality (9.3.6) leads to the differential conservation law

$$\partial_\lambda (\widetilde{\mathcal{J}}_u^\lambda(r)) = 0.$$

There is the following relation between differential conservation laws in Lagrangian and Hamiltonian formalisms.

Theorem 9.3.1. *Let a Hamiltonian form H be associated to an almost regular Lagrangian L. Let r be a solution of the covariant Hamilton equations (9.1.18) – (9.1.19) for H which lives in the Lagrangian constraint space N_L. Let*

$$s = \pi_{\Pi Y} \circ r$$

be the corresponding solution of the Euler–Lagrange equation for L so that the relation (9.2.10) holds. Then, for any projectable vector field u on a fibre bundle $Y \to X$, we have

$$\widetilde{\mathcal{J}}_u(r) = \mathcal{J}_u(\pi_{\Pi Y} \circ r), \qquad \widetilde{\mathcal{J}}_u(\widehat{L} \circ J^1 s) = \mathcal{J}_u(s), \tag{9.3.8}$$

where $\mathcal{J}_u$ is the symmetry current (2.4.31) on $J^1 Y$ and $\widetilde{\mathcal{J}}_u$ is the symmetry current (9.3.7) on Π.

By virtue of Theorems 9.2.4 – 9.2.5, it follows that:

• if $\mathcal{J}_u$ in Theorem 9.3.1 is a conserved symmetry, then the symmetry current $\widetilde{\mathcal{J}}_u$ (9.3.8) is conserved on solutions of the Hamilton equations which live in the Lagrangian constraint space,

• if $\widetilde{\mathcal{J}}_u$ in Theorem 9.3.1 is a conserved symmetry current, then the symmetry current $\mathcal{J}_u$ (9.3.8) is conserved on solutions s of the Euler–Lagrange equation which obey the condition (9.2.10).

In particular, let $u = u^i \partial_i$ be a vertical vector field on $Y \to X$. Then the Lie derivative $\mathbf{L}_{\widetilde{u}} H$ (9.3.4) takes the form

$$\mathbf{L}_{\widetilde{u}} H = (p_j^\lambda \lfloor \partial_\lambda + \Gamma_\lambda^i \partial_i, u \rfloor^j - u \rfloor d\widetilde{\mathcal{H}}_\Gamma) \omega.$$

The corresponding symmetry current (9.3.7) reads

$$\widetilde{\mathcal{J}}_u^\lambda = -u^i p_i^\lambda \tag{9.3.9}$$

(cf. $\mathcal{J}_u$ (2.4.34)).

Let $\tau = \tau^\lambda \partial_\lambda$ be a vector field on X and $\Gamma\tau$ (1.3.6) its horizontal lift onto Y by means of a connection Γ on $Y \to X$. In this case, the weak identity (9.3.5) takes the form

$$-(\partial_\mu + \Gamma^j_\mu \partial_j - p^\lambda_i \partial_j \Gamma^i_\mu \partial^j_\lambda)\widetilde{\mathcal{H}}_\Gamma + p^\lambda_i R^i_{\lambda\mu} \approx -d_\lambda \widetilde{\mathcal{J}}_\Gamma{}^\lambda{}_\mu,$$

where the symmetry current (9.3.7) reads

$$\widetilde{\mathcal{J}}^\lambda_\Gamma = \tau^\mu \widetilde{\mathcal{J}}_\Gamma{}^\lambda{}_\mu = \tau^\mu (p^\lambda_i \partial^i_\mu \widetilde{\mathcal{H}}_\Gamma - \delta^\lambda_\mu (p^\nu_i \partial^i_\nu \widetilde{\mathcal{H}}_\Gamma - \widetilde{\mathcal{H}}_\Gamma)). \quad (9.3.10)$$

The relations (9.3.8) show that, on the Lagrangian constraint space N_L, the current (9.3.10) can be treated as the *Hamiltonian energy-momentum current* relative to the connection Γ.

In particular, let us consider the weak identity (9.3.5) when the vector field $\widetilde{u}$ on Π is the horizontal lift of a vector field τ on X by means of a Hamiltonian connection on $\Pi \to X$ which is associated to the Hamiltonian form H. We have

$$\widetilde{u} = \tau^\mu (\partial_\mu + \partial^i_\mu \mathcal{H} \partial_i + \gamma^\lambda_{\mu i} \partial^i_\lambda).$$

In this case, the corresponding energy-momentum current reads

$$\widetilde{\mathcal{J}}^\lambda = -\tau^\lambda (p^\mu_i \partial^i_\mu \mathcal{H} - \mathcal{H}), \quad (9.3.11)$$

and the weak identity (9.3.5) takes the form

$$-\partial_\mu \mathcal{H} + d_\lambda (p^\lambda_i \partial^i_\mu \mathcal{H}) \approx \partial_\mu (p^\lambda_i \partial^i_\lambda \mathcal{H} - \mathcal{H}). \quad (9.3.12)$$

A glance at the expression (9.3.12) shows that the energy-momentum current (9.3.11) is not conserved, the weak identity

$$-\partial_\mu \mathcal{H} + d_\lambda [p^\lambda_i \partial^i_\mu \mathcal{H} - \delta^\lambda_\mu (p^\nu_i \partial^i_\nu \mathcal{H} - \mathcal{H})] \approx 0$$

holds. This is exactly the Hamiltonian form of the canonical energy-momentum conservation law (2.4.38) in Lagrangian formalism.

9.4 Quadratic Lagrangian and Hamiltonian systems

Field theories with almost regular quadratic Lagrangians admit comprehensive Hamiltonian formulation.

Let L (2.4.52) be an almost regular quadratic Lagrangian brought into the form (2.4.66), $\sigma = \sigma_0 + \sigma_1$ a linear map (2.4.59) and Γ a connection

9.4. Quadratic Lagrangian and Hamiltonian systems

(2.4.55). Similarly to the splitting (2.4.64) of the configuration space J^1Y, we have the following decomposition of the momentum phase space:

$$\Pi = \mathcal{R}(\Pi) \underset{Y}{\oplus} \mathcal{P}(\Pi) = \operatorname{Ker}\sigma_0 \underset{Y}{\oplus} N_L, \qquad (9.4.1)$$

$$p_i^\lambda = \mathcal{R}_i^\lambda + \mathcal{P}_i^\lambda = [p_i^\lambda - a_{ij}^{\lambda\mu}\sigma_{\mu\alpha}^{jk}p_k^\alpha] + [a_{ij}^{\lambda\mu}\sigma_{\mu\alpha}^{jk}p_k^\alpha]. \qquad (9.4.2)$$

The relations (2.4.62) lead to the equalities

$$\sigma_0{}_{\mu\alpha}^{jk}\mathcal{R}_k^\alpha = 0, \qquad \sigma_1{}_{\mu\alpha}^{jk}\mathcal{P}_k^\alpha = 0, \qquad \mathcal{R}_i^\lambda \mathcal{F}_\lambda^i = 0. \qquad (9.4.3)$$

Relative to the coordinates (9.4.2), the Lagrangian constraint space N_L (2.4.53) is given by the equations

$$\mathcal{R}_i^\lambda = [p_i^\lambda - a_{ij}^{\lambda\mu}\sigma_{\mu\alpha}^{jk}p_k^\alpha] = 0. \qquad (9.4.4)$$

Let the splitting (2.4.61) be provided with the adapted fibre coordinates $(\overline{y}^a, \overline{y}^A)$ such that the matrix function a (2.4.54) is brought into a diagonal matrix with non-vanishing components a_{AA}. Then the Legendre bundle Π (9.4.1) is endowed with the dual (non-holonomic) fibre coordinates (p_a, p_A) where p_A are coordinates on the Lagrangian constraint manifold N_L, given by the equalities $p_a = 0$. Relative to these coordinates, σ_0 becomes the diagonal matrix

$$\sigma_0^{AA} = (a_{AA})^{-1}, \qquad \sigma_0^{aa} = 0, \qquad (9.4.5)$$

while

$$\sigma_1^{Aa} = \sigma_1^{AB} = 0.$$

Let us write

$$p_a = M_a{}^i_\lambda p_i^\lambda, \qquad p_A = M_A{}^i_\lambda p_i^\lambda, \qquad (9.4.6)$$

where M are the matrix functions on Y which obeys the relations

$$M_a{}^i_\lambda a_{ij}^{\lambda\mu} = 0, \qquad (M^{-1})^{a\lambda}{}_i \sigma_0{}^{ij}_{\lambda\mu} = 0, \qquad (9.4.7)$$

$$M_A{}^i_\lambda (a \circ \sigma_0)^{\lambda j}_{i\mu} = M_A{}^j_\mu, \qquad (M^{-1})^{A\mu}{}_j M_A{}^i_\lambda = a_{jk}^{\mu\nu}\sigma_0{}^{ki}_{\nu\lambda}.$$

Let us consider the affine Hamiltonian map

$$\Phi = \widehat{\Gamma} + \sigma : \Pi \to J^1Y, \qquad (9.4.8)$$

$$\Phi^i_\lambda = \Gamma^i_\lambda + \sigma^{ij}_{\lambda\mu}p^\mu_j,$$

and the Hamiltonian form

$$H(\Gamma, \sigma_1) = H_\Phi + \Phi^* L = p_i^\lambda dy^i \wedge \omega_\lambda - \qquad (9.4.9)$$

$$[\Gamma_\lambda^i p_i^\lambda + \frac{1}{2}\sigma_0{}^{ij}_{\lambda\mu}p_i^\lambda p_j^\mu + \sigma_1{}^{ij}_{\lambda\mu}p_i^\lambda p_j^\mu - c']\omega =$$

$$(\mathcal{R}_i^\lambda + \mathcal{P}_i^\lambda)dy^i \wedge \omega_\lambda -$$

$$[(\mathcal{R}_i^\lambda + \mathcal{P}_i^\lambda)\Gamma_\lambda^i + \frac{1}{2}\sigma_0{}^{ij}_{\lambda\mu}\mathcal{P}_i^\lambda \mathcal{P}_j^\mu + \sigma_1{}^{ij}_{\lambda\mu}\mathcal{R}_i^\lambda \mathcal{R}_j^\mu - c']\omega.$$

Theorem 9.4.1. *The Hamiltonian forms $H(\Gamma, \sigma_1)$ (9.4.9) parameterized by connections Γ (2.4.55) are weakly associated to the Lagrangian (2.4.52), and they constitute a complete set.*

Proof. By the very definitions of Γ and σ, the Hamiltonian map (9.4.8) satisfies the condition (9.2.2). Then $H(\Gamma, \sigma_1)$ is weakly associated to L (2.4.52) in accordance with Theorem 9.2.1. Let us write the corresponding Hamilton equations (9.1.18) for a section r of the Legendre bundle $\Pi \to X$. They are

$$J^1 s = (\widehat{\Gamma} + \sigma) \circ r, \qquad s = \pi_{\Pi Y} \circ r. \tag{9.4.10}$$

Due to the surjections $\mathcal{S}$ and $\mathcal{F}$ (2.4.64), the Hamilton equations (9.4.10) are brought into the two parts

$$\mathcal{S} \circ J^1 s = \Gamma \circ s, \tag{9.4.11}$$
$$\partial_\lambda r^i - \sigma_{0\lambda\alpha}^{ik}(a_{kj}^{\alpha\mu} \partial_\mu r^j + b_k^\alpha) = \Gamma_\lambda^i \circ s,$$
$$\mathcal{F} \circ J^1 s = \sigma \circ r, \tag{9.4.12}$$
$$\sigma_{0\lambda\alpha}^{ik}(a_{kj}^{\alpha\mu} \partial_\mu r^j + b_k^\alpha) = \sigma_{\lambda\alpha}^{ik} r_k^\alpha.$$

Let s be an arbitrary section of $Y \to X$, e.g., a solution of the Euler–Lagrange equation. There exists a connection Γ (2.4.55) such that the relation (9.4.11) holds, namely, $\Gamma = \mathcal{S} \circ \Gamma'$ where Γ' is a connection on $Y \to X$ which has s as an integral section. It is easily seen that, in this case, the Hamiltonian map (9.4.8) satisfies the relation (9.2.10) for s. Hence, the Hamiltonian forms (9.4.9) constitute a complete set. $\square$

It is readily observed that, if $\sigma_1 = 0$, then $\Phi = \widehat{H}(\Gamma)$, and the Hamiltonian forms $H(\Gamma, \sigma_1 = 0)$ (9.4.9) are associated to the Lagrangian (2.4.52).

For different σ_1, we have different complete sets of Hamiltonian forms (9.4.9). Hamiltonian forms $H(\Gamma, \sigma_1)$ and $H(\Gamma', \sigma_1)$ (9.4.9) of such a complete set differ from each other in the term $\phi_\lambda^i \mathcal{R}_i^\lambda$, where ϕ are the soldering forms (2.4.57). This term vanishes on the Lagrangian constraint space (9.4.4). Accordingly, the covariant Hamilton equations for different Hamiltonian forms $H(\Gamma, \sigma_1)$ and $H(\Gamma', \sigma_1)$ (9.4.9) differ from each other in the equations (9.4.11). These equations are independent of momenta and play the role of gauge-type conditions.

Since the Lagrangian constraint space N_L (9.4.4) is an imbedded subbundle of $\Pi \to Y$, all Hamiltonian forms $H(\Gamma, \sigma_1)$ (9.4.9) define a unique constrained Hamiltonian form H_N (9.2.12) on N_L which reads

$$H_N = i_N^* H(\Gamma, \sigma_1) = \mathcal{P}_i^\lambda dy^i \wedge \omega_\lambda - [\mathcal{P}_i^\lambda \Gamma_\lambda^i + \frac{1}{2}\sigma_{0\lambda\mu}^{ij} \mathcal{P}_i^\lambda \mathcal{P}_j^\mu - c']\omega. \tag{9.4.13}$$

9.4. Quadratic Lagrangian and Hamiltonian systems

In view of the relations (9.4.3), the corresponding constrained Lagrangian L_N (9.2.16) on $J^1 N_L$ takes the form

$$L_N = h_0(H_N) = (\mathcal{P}_i^\lambda \mathcal{F}_\lambda^i - \frac{1}{2}\sigma_0{}^{ij}_{\lambda\mu}\mathcal{P}_i^\lambda \mathcal{P}_j^\mu + c')\omega. \tag{9.4.14}$$

It is the pull-back onto $J^1 N_L$ of the Lagrangian

$$L_{H(\Gamma,\sigma_1)} = \mathcal{R}_i^\lambda(\mathcal{S}_\lambda^i - \Gamma_\lambda^i) + \mathcal{P}_i^\lambda \mathcal{F}_\lambda^i - \frac{1}{2}\sigma_0{}^{ij}_{\lambda\mu}\mathcal{P}_i^\lambda \mathcal{P}_j^\mu - \frac{1}{2}\sigma_1{}^{ij}_{\lambda\mu}\mathcal{R}_i^\lambda \mathcal{R}_j^\mu + c' \tag{9.4.15}$$

on $J^1\Pi$ for any Hamiltonian form $H(\Gamma,\sigma_1)$ (9.4.9).

In fact, the Lagrangian L_N (9.4.14) is defined on the product $N_L \times_Y J^1 Y$ (see Remark 9.1.3). Since the momentum phase space Π (9.4.1) is a trivial bundle $\mathrm{pr}_2 : \Pi \to N_L$ over the Lagrangian constraint space N_L, one can consider the pull-back

$$L_\Pi = (\mathcal{P}_i^\lambda \mathcal{F}_\lambda^i - \frac{1}{2}\sigma_0{}^{ij}_{\lambda\mu}\mathcal{P}_i^\lambda \mathcal{P}_j^\mu + c')\omega \tag{9.4.16}$$

of the constrained Lagrangian L_N (9.4.14) onto $\Pi \underset{Y}{\times} J^1 Y$.

Let us study symmetries of the Lagrangians L_N and L_Π [11]. We aim to show that, under certain conditions, they inherit symmetries of an original Lagrangian L (see Theorems 9.4.2 – 9.4.3).

Let a vertical vector field $u = u^i \partial_i$ on $Y \to X$ be a symmetry of the Lagrangian L (2.4.66), i.e.,

$$\mathbf{L}_{J^1 u} L = (u^i \partial_i + d_\lambda u^i \partial_i^\lambda)\mathcal{L}\omega = 0. \tag{9.4.17}$$

Since

$$J^1 u(y_\lambda^i - \Gamma_\lambda^i) = \partial_k u^i (y_\lambda^k - \Gamma_\lambda^k), \tag{9.4.18}$$

one easily obtains from the equality (9.4.17) that

$$u^k \partial_k a_{ij}^{\lambda\mu} + \partial_i u^k a_{kj}^{\lambda\mu} + a_{ik}^{\lambda\mu} \partial_j u^k = 0. \tag{9.4.19}$$

It follows that the summands of the Lagrangian (2.4.66) are separately invariant, i.e.,

$$J^1 u(a_{ij}^{\lambda\mu}\mathcal{F}_\lambda^i \mathcal{F}_\mu^j) = 0, \qquad J^1 u(c') = u^k \partial_k c' = 0. \tag{9.4.20}$$

The equalities (2.4.67), (9.4.18) and (9.4.19) give the transformation law

$$J^1 u(a_{ij}^{\lambda\mu}\mathcal{F}_\mu^j) = -\partial_i u^k a_{kj}^{\lambda\mu}\mathcal{F}_\mu^j. \tag{9.4.21}$$

The relations (2.4.62) and (9.4.19) lead to the equality

$$a_{ij}^{\lambda\mu}[u^k \partial_k \sigma_0{}^{jn}_{\mu\alpha} - \partial_k u^j \sigma_0{}^{kn}_{\mu\alpha} - \sigma_0{}^{jk}_{\mu\alpha}\partial_k u^n]a_{nb}^{\alpha\nu} = 0. \tag{9.4.22}$$

Let us compare symmetries of the Lagrangian L (2.4.66) and the Lagrangian L_N (9.4.14). Given the Legendre map $\widehat{L}$ (2.4.53) and the tangent morphism

$$T\widehat{L} : TJ^1Y \to TN_L,$$
$$\dot{p}_A = (\dot{y}^i \partial_i + \dot{y}^k_\nu \partial^\nu_k)(M^i_{A\lambda} a^{\lambda\mu}_{ij} \mathcal{F}^j_\mu),$$

let us consider the map

$$T\widehat{L} \circ J^1 u : J^1Y \ni (x^\lambda, y^i, y^i_\lambda) \to \qquad (9.4.23)$$
$$u^i \partial_i + (u^k \partial_k + \partial_\nu u^k \partial^\nu_k)(M^i_{A\lambda} a^{\lambda\mu}_{ij} \mathcal{F}^j_\mu)\partial^A =$$
$$u^i \partial_i + [u^k \partial_k(M^i_{A\lambda}) a^{\lambda\mu}_{ij} \mathcal{F}^j_\mu + M^i_{A\lambda} J^1 u(a^{\lambda\mu}_{ij} \mathcal{F}^j_\mu)]\partial^A =$$
$$u^i \partial_i + [u^k \partial_k(M^i_{A\lambda}) a^{\lambda\mu}_{ij} \mathcal{F}^j_\mu - M^i_{A\lambda} \partial_i u^k a^{\lambda\mu}_{kj} \mathcal{F}^j_\mu]\partial^A =$$
$$u^i \partial_i + [u^k \partial_k(a \circ \sigma_0)^{\mu i}_{j\lambda} \mathcal{P}^\lambda_i - (a \circ \sigma_0)^{\mu i}_{j\lambda} \partial_i u^k \mathcal{P}^\lambda_k]\partial^j_\mu \in TN_L,$$

where the relations (9.4.7) and (9.4.21) have been used. Let us assign to a point $(x^\lambda, y^i, \mathcal{P}^\lambda_i) \in N_L$ some point

$$(x^\lambda, y^i, y^i_\lambda) \in \widehat{L}^{-1}(x^\lambda, y^i, \mathcal{P}^\lambda_i) \qquad (9.4.24)$$

and then the image of the point (9.4.24) under the morphism (9.4.23). We obtain the map

$$v_N : (x^\lambda, y^i, \mathcal{P}^\lambda_i) \to u^i \partial_i + [u^k \partial_k(a \circ \sigma_0)^{\mu i}_{j\lambda} \mathcal{P}^\lambda_i - \qquad (9.4.25)$$
$$(a \circ \sigma_0)^{\mu i}_{j\lambda} \partial_i u^k \mathcal{P}^\lambda_k]\partial^j_\mu$$

which is independent of the choice of a point (9.4.24). Therefore, it is a vector field on the Lagrangian constraint space N_L. This vector field gives rise to the vector field

$$Jv_N = u^i \partial_i + [u^k \partial_k(a \circ \sigma_0)^{\mu i}_{j\lambda} \mathcal{P}^\lambda_i - (a \circ \sigma_0)^{\mu i}_{j\lambda} \partial_i u^k \mathcal{P}^\lambda_k]\partial^j_\mu + d_\lambda u^i \partial^\lambda_i \quad (9.4.26)$$

on $N_L \underset{Y}{\times} J^1Y$.

Theorem 9.4.2. *The Lie derivative $\mathbf{L}_{Jv_N} L_N$ of the Lagrangian L_N (9.4.14) along the vector field Jv_N (9.4.26) vanishes.*

Proof. One can show that

$$v_N(\mathcal{P}^\lambda_i) = -\partial_i u^k \mathcal{P}^\lambda_k \qquad (9.4.27)$$

on the constraint manifold $\mathcal{R}^\lambda_i = 0$. Then the invariance condition $Jv_N(\mathcal{L}_N) = 0$ falls into the three equalities

$$Jv_N(\sigma_0{}^{ij}_{\lambda\mu} \mathcal{P}^\lambda_i \mathcal{P}^\mu_j) = 0, \qquad Jv_N(\mathcal{P}^\lambda_i \mathcal{F}^i_\lambda) = 0, \qquad Jv_N(c') = 0. \qquad (9.4.28)$$

9.4. Quadratic Lagrangian and Hamiltonian systems

The latter is exactly the second equality (9.4.20). The first equality (9.4.28) is satisfied due to the relations (9.4.22) and (9.4.27). The second one takes the form

$$Jv_N(\mathcal{P}_i^\lambda(y_\lambda^i - \Gamma_\lambda^i)) = 0. \qquad (9.4.29)$$

It holds owing to the relations (9.4.18) and (9.4.27). □

Thus, any vertical symmetry u of the Lagrangian L (2.4.66) yields the symmetry v_N (9.4.25) of the Lagrangian L_N (9.4.14).

Turn now to symmetries of the Lagrangian L_Π (9.4.16). Since L_Π is the pull-back of L_N onto $\Pi \underset{Y}{\times} J^1Y$, its symmetry must be an appropriate lift of the vector field v_N (9.4.25) onto Π.

Given a vertical vector field u on $Y \to X$, let us consider its canonical lift (9.3.1):

$$\widetilde{u} = u^i \partial_i - \partial_i u^j p_j^\lambda \partial_\lambda^i, \qquad (9.4.30)$$

onto the Legendre bundle Π. It readily observed that the vector field $\widetilde{u}$ is projected onto the vector field v_N (9.4.25).

Let us additionally suppose that the one-parameter group of automorphisms of Y generated by u preserves the splitting (2.4.64), i.e., u obeys the condition

$$u^k \partial_k(\sigma_0{}_{\lambda\nu}^{im} a_{mj}^{\nu\mu}) + \sigma_0{}_{\lambda\nu}^{im} a_{mk}^{\nu\mu} \partial_j u^k - \partial_k u^i \sigma_0{}_{\lambda\nu}^{km} a_{mj}^{\nu\mu} = 0. \qquad (9.4.31)$$

The relations (9.4.18) and (9.4.31) lead to the transformation law

$$J^1 u(\mathcal{F}_\mu^i) = \partial_j u^i \mathcal{F}_\mu^j. \qquad (9.4.32)$$

Theorem 9.4.3. *If the condition (9.4.31) holds, the vector field $\widetilde{u}$ (9.4.30) is a symmetry of the Lagrangian L_Π (9.4.16) if and only if u is a symmetry of the Lagrangian L (2.4.66).*

Proof. Due to the condition (9.4.31), the vector field $\widetilde{u}$ (9.4.30) preserves the splitting (9.4.1), i.e.,

$$\widetilde{u}(\mathcal{P}_i^\lambda) = -\partial_i u^k \mathcal{P}_k^\lambda, \qquad \widetilde{u}(\mathcal{R}_i^\lambda) = -\partial_i u^k \mathcal{R}_k^\lambda. \qquad (9.4.33)$$

The vector field $\widetilde{u}$ gives rise to the vector field (9.3.2):

$$J\widetilde{u} = u^i \partial_i - \partial_i u^j p_j^\lambda \partial_\lambda^i + d_\lambda u^i \partial_i^\lambda, \qquad (9.4.34)$$

on $\Pi \underset{Y}{\times} J^1 Y$, and we obtain the Lagrangian symmetry condition

$$(u^i \partial_i - \partial_j u^i p_i^\lambda \partial_\lambda^j + d_\lambda u^i \partial_i^\lambda)\mathcal{L}_\Pi = 0. \qquad (9.4.35)$$

It is readily observed that the first and third terms of the Lagrangian L_Π are separately invariant due to the relations (9.4.20) and (9.4.32). Its second term is invariant owing to the equality (9.4.22). Conversely, let the invariance condition (9.4.35) hold. It falls into the independent equalities

$$J\widetilde{u}(\sigma_0{}^{ij}_{\lambda\mu}p_i^\lambda p_j^\mu) = 0, \qquad J\widetilde{u}(p_i^\lambda \mathcal{F}_\lambda^i) = 0, \qquad u^i \partial_i c' = 0, \qquad (9.4.36)$$

i.e., the Lagrangian L_Π is invariant if and only if its three summands are separately invariant. One obtains at once from the second condition (9.4.36) that the quantity $\mathcal{F}$ is transformed as the dual of momenta p. Then the first condition (9.4.36) shows that the quantity $\sigma_0 p$ is transformed by the same law as $\mathcal{F}$. It follows that the term $a\mathcal{F}\mathcal{F}$ in the Lagrangian L (2.4.66) is transformed as $a(\sigma_0 p)(\sigma_0 p) = \sigma_0 pp$, i.e., it is invariant. Then this Lagrangian is invariant due to the third equality (9.4.36). □

In particular, if u is a gauge symmetry of an original Lagrangian L and if it preserves the decomposition (2.4.64), then its natural lift $\widetilde{u}$ (9.4.30) onto Π is a gauge symmetry of the Lagrangian L_Π (9.4.16). However, the Lagrangian L_Π can possess additional gauge symmetries.

For instance, let us assume that $Y \to X$ is an affine bundle modelled over a vector bundle $\overline{Y} \to X$. In this case, the Legendre bundle Π (2.4.7) is isomorphic to the product

$$\Pi = Y \underset{X}{\times} (\overline{Y}^* \underset{X}{\otimes} \overset{n}{\wedge} T^*X \underset{X}{\otimes} TX)$$

such that transition functions of coordinates p_i^λ are independent of y^i. Then the splitting (9.4.1) takes the form

$$\Pi = Y \underset{X}{\times} (\overline{\mathrm{Ker}\,\sigma_0} \underset{X}{\oplus} \overline{N}_L), \qquad (9.4.37)$$

where $\overline{\mathrm{Ker}\,\sigma_0}$ and $\overline{N}_L$ are fibre bundles over X such that

$$\mathrm{Ker}\,\sigma_0 = \pi^* \overline{\mathrm{Ker}\,\sigma_0},$$

and $N_L = \pi^* \overline{N}_L$ are their pull-backs onto Y. The splitting (9.4.37) keeps the coordinate form (9.4.2). The splittings (2.4.64) and (9.4.37) lead to the decomposition

$$\Pi \underset{Y}{\times} J^1 Y = (\overline{\mathrm{Ker}\,\sigma_0} \underset{X}{\oplus} \overline{N}_L) \underset{Y}{\times} (\mathrm{Ker}\,\widehat{L} \oplus \mathrm{Im}(\sigma_0 \circ \widehat{L})). \qquad (9.4.38)$$

In view of this decomposition, let us associate to any section ξ of $\overline{\mathrm{Ker}\,\sigma_0} \to X$ the vector field

$$u_\Pi = \xi_a (M^{-1})^{a\lambda}_i \partial^i_\lambda, \qquad (9.4.39)$$

on Π. Its lift (9.3.2) onto $\Pi \underset{Y}{\times} J^1Y$ keeps the coordinate form

$$Ju_\Pi = \xi_a (M^{-1})^{a\lambda}_i \partial^i_\lambda. \tag{9.4.40}$$

It is readily observed that the Lie derivative of the Lagrangian L_Π (9.4.16) along the vector field (9.4.40) vanishes, i.e., u_Π is a symmetry of L_Π. Consequently, the vector fields (9.4.39), parameterized by sections ξ of $\overline{\text{Ker}\,\sigma_0} \to X$, is a gauge symmetry of L_Π. It does not come from gauge symmetries of an original Lagrangian L. A glance at the expression (9.4.40) shows that this gauge symmetry is independent of derivatives of gauge parameters just as that (4.3.15) of a variationally trivial Lagrangian in Example 4.3.1. The gauge symmetry (9.4.40) as like as (4.3.15) is non-trivial.

9.5 Example. Yang–Mills gauge theory

We follow the notation of Section 5.8. Let $P \to X$ be a principal bundle with a structure group G. Gauge theory of principal connections on $P \to X$ is described by the degenerate non-regular quadratic Lagrangian L_{YM} (5.8.15) on the first order jet manifold J^1C of the bundle of principal connections $C = J^1P/G$. The peculiarity of gauge theory consists in the fact that the splittings (2.4.64) and (9.4.1) of its configuration and phase spaces are canonical.

Let C and J^1C be provided with the bundle coordinates (x^λ, a^m_λ) and $(x^\lambda, a^m_\lambda, a^m_{\mu\lambda})$, respectively. As was mentioned in Section 5.5, the configuration space J^1C of gauge theory admits the canonical splitting (5.5.11):

$$J^1C = C_+ \underset{C}{\oplus} C_- = C_| \oplus (C \underset{X}{\times} \overset{2}{\wedge} T^*X \underset{X}{\otimes} V_G P), \tag{9.5.1}$$

$$a^r_{\mu\lambda} = \frac{1}{2}(a^r_{\mu\lambda} + a^r_{\lambda\mu} - c^r_{pq}a^p_\mu a^q_\lambda) + \frac{1}{2}(a^r_{\mu\lambda} - a^r_{\lambda\mu} + c^r_{pq}a^p_\mu a^q_\lambda),$$

with the corresponding projections

$$\mathcal{S} : J^1C \to C_+, \tag{9.5.2}$$

$$\mathcal{S}^r_{\mu\lambda} = a^r_{\mu\lambda} + a^r_{\lambda\mu} - c^r_{pq}a^p_\mu a^q_\lambda,$$

$$\mathcal{F} : J^1C \to C_-, \tag{9.5.3}$$

$$\mathcal{F}^r_{\mu\lambda} = a^r_{\mu\lambda} - a^r_{\lambda\mu} + c^r_{pq}a^p_\mu a^q_\lambda.$$

The Yang–Mills Lagrangian on this configuration space is

$$L_{YM} = \frac{1}{4}a^G_{pq}g^{\lambda\mu}g^{\beta\nu}\mathcal{F}^p_{\lambda\beta}\mathcal{F}^q_{\mu\nu}\sqrt{|g|}\,\omega, \qquad g = \det(g_{\mu\nu}), \tag{9.5.4}$$

where a^G is a non-degenerate G-invariant metric on the Lie algebra $\mathfrak{g}_r$ and g is a non-degenerate world metric on X.

The phase space of gauge theory is the Legendre bundle

$$\pi_{\Pi C}: \Pi \to C, \qquad \Pi = \overset{n}{\wedge} T^*X \underset{C}{\otimes} TX \underset{C}{\otimes} [C \times \overline{C}]^*, \qquad (9.5.5)$$

endowed with holonomic coordinates $(x^\lambda, a_\lambda^p, p_m^{\mu\lambda})$. The Legendre bundle Π (9.5.5) admits the canonical decomposition (9.4.1):

$$\Pi = \Pi_+ \underset{C}{\oplus} \Pi_-, \qquad (9.5.6)$$

$$p_m^{\mu\lambda} = \mathcal{R}_m^{(\mu\lambda)} + \mathcal{P}_m^{[\mu\lambda]} = p_m^{(\mu\lambda)} + p_m^{[\mu\lambda]} = \frac{1}{2}(p_m^{\mu\lambda} + p_m^{\lambda\mu}) + \frac{1}{2}(p_m^{\mu\lambda} - p_m^{\lambda\mu}).$$

The Legendre map associated to the Lagrangian (9.5.4) takes the form

$$p_m^{(\mu\lambda)} \circ \widehat{L}_{YM} = 0, \qquad (9.5.7)$$

$$p_m^{[\mu\lambda]} \circ \widehat{L}_{YM} = a_{mn}^G g^{\mu\alpha} g^{\lambda\beta} \mathcal{F}_{\alpha\beta}^n \sqrt{|g|}. \qquad (9.5.8)$$

A glance at this morphism shows that $\text{Ker}\,\widehat{L}_{YM} = C_+$, and the Lagrangian constraint space is

$$N_L = \widehat{L}_{YM}(J^1 C) = \Pi_-.$$

It is defined by the equation (9.5.7). Obviously, N_L is an imbedded submanifold of Π, and the Lagrangian L_{YM} is almost regular.

Let us consider connections Γ on the fibre bundle $C \to X$ which take their values into $\text{Ker}\,\widehat{L}$, i.e.,

$$\Gamma: C \to C_+, \qquad \Gamma_{\lambda\mu}^r - \Gamma_{\mu\lambda}^r + c_{pq}^r a_\lambda^p a_\mu^q = 0. \qquad (9.5.9)$$

Given a symmetric linear connection K on T^*X, every principal connection B on the principal bundle $P \to X$ gives rise to the connection $\Gamma_B : C \to C_+$ (5.7.14) such that

$$\Gamma_B \circ B = \mathcal{S} \circ J^1 B.$$

It reads

$$\Gamma_{B\lambda\mu}^r = \frac{1}{2}[\partial_\mu B_\lambda^r + \partial_\lambda B_\mu^r - c_{pq}^r a_\lambda^p a_\mu^q + \qquad (9.5.10)$$

$$c_{pq}^r (a_\lambda^p B_\mu^q + a_\mu^p B_\lambda^q)] - K_\lambda{}^\beta{}_\mu (a_\beta^r - B_\beta^r).$$

Given the connection (9.5.10), the corresponding Hamiltonian form (9.4.9): is

$$H_B = p_r^{\lambda\mu} da_\mu^r \wedge \omega_\lambda - p_r^{\lambda\mu} \Gamma_{B\lambda\mu}^r \omega - \widetilde{\mathcal{H}}_{YM}\omega, \qquad (9.5.11)$$

$$\widetilde{\mathcal{H}}_{YM} = \frac{1}{4} a_G^{mn} g_{\mu\nu} g_{\lambda\beta} p_m^{[\mu\lambda]} p_n^{[\nu\beta]} \sqrt{|g|},$$

9.5. Example. Yang–Mills gauge theory

is associated to the Lagrangian L_{YM}. It is the Poincaré–Cartan form of the Lagrangian

$$L_H = [p_r^{\lambda\mu}(a_{\lambda\mu}^r - \Gamma_{B\lambda\mu}^r) - \widetilde{\mathcal{H}}_{YM}]\omega \tag{9.5.12}$$

on $\Pi \underset{C}{\times} J^1C$. The pull-back of any Hamiltonian form H_B (9.5.11) onto the Lagrangian constraint space N_L is the constrained Hamiltonian form (9.2.12):

$$H_N = i_N^* H_B = p_r^{[\lambda\mu]}(da_\mu^r \wedge \omega_\lambda + \frac{1}{2}c_{pq}^r a_\lambda^p a_\mu^q \omega) - \widetilde{\mathcal{H}}_{YM}\omega. \tag{9.5.13}$$

The corresponding constrained Lagrangian L_N on

$$N_L \underset{C}{\times} J^1C$$

reads

$$L_N = (p_r^{[\lambda\mu]} \mathcal{F}_{\lambda\mu}^r - \widetilde{\mathcal{H}}_{YM})\omega. \tag{9.5.14}$$

Note that, in contrast with the Lagrangian (9.5.12), the constrained Lagrangian L_N (9.5.14) possesses gauge symmetries as follows. Gauge symmetries

$$u_\xi = (\partial_\mu \xi^r + c_{qp}^r a_\mu^q \xi^p)\partial_r^\mu$$

of the Yang–Mills Lagrangian give rise to the vector fields

$$\widetilde{u}_\xi = (\partial_\mu \xi^r + c_{qp}^r a_\mu^q \xi^p)\partial_r^\mu - c_{qp}^r \xi^p p_r^{\lambda\mu} \partial_{\lambda\mu}^q$$

on Π. Their projection onto N_L provides gauge symmetries of the Lagrangian L_N in accordance with Theorem 9.4.2.

The Hamiltonian form H_B (9.5.11) yields the covariant Hamilton equations which consist of the equations (9.5.8) and the equations

$$u_{\lambda\mu}^m + u_{\mu\lambda}^m = 2\Gamma_{B(\lambda\mu)}^m, \tag{9.5.15}$$

$$p_{\lambda r}^{\lambda\mu} = c_{pr}^q r_\lambda^p p_q^{[\lambda\mu]} - c_{rp}^q B_\lambda^p p_q^{(\lambda\mu)} + K_\lambda^{\mu\nu} p_r^{(\lambda\nu)}. \tag{9.5.16}$$

The Hamilton equations (9.5.15) and (9.5.8) are similar to the equations (9.4.11) and (9.4.12), respectively. The Hamilton equations (9.5.8) and (9.5.16) restricted to the Lagrangian constraint space (9.5.7) are precisely the constrained Hamilton equations (9.2.13) for the constrained Hamiltonian form H_N (9.5.13), and they are equivalent to the Yang-Mills equations (5.8.17) for gauge potentials $A = \pi_{\Pi C} \circ r$.

Different Hamiltonian forms H_B lead to different equations (9.5.15). This equation is independent of momenta and, thus, it exemplifies the gauge-type condition (9.4.11):

$$\Gamma_B \circ A = \mathcal{S} \circ J^1 A.$$

A glance at this condition shows that, given a solution A of the Yang–Mills equations, there always exists a Hamiltonian form H_B (e.g., $H_{B=A}$) which obeys the condition (9.2.10), i.e.,

$$\widehat{H}_B \circ \widehat{L}_{YM} \circ J^1 A = J^1 A.$$

Consequently, the Hamiltonian forms H_B (9.5.11) parameterized by principal connections B constitute a complete set.

It should be emphasized that the gauge-type condition (9.5.15) differs from familiar gauge conditions in gauge theory. If a gauge potential A is a solution of the Yang–Mills equations, there exists a gauge conjugate potential A' which also is a solution of the Yang–Mills equations and satisfies a given gauge condition. In the framework of the Hamiltonian description of quadratic Lagrangian systems, there is a complete set of gauge-type conditions in the sense that, for any solution of the Euler–Lagrange equation, there exist Hamilton equations equivalent to this Euler-Lagrange equation and a supplementary gauge-type condition which this solution satisfies.

9.6 Variation Hamilton equations. Jacobi fields

The vertical extension of covariant Hamiltonian formalism on the Legendre bundle Π (2.4.7) onto the vertical Legendre bundle Π_{VY} (2.4.82) describes Jacobi fields of solutions of the Hamilton equations. Let us utilize the compact notation $\partial_V = \dot{y}^i \partial_i + \dot{p}^\lambda_i \partial^i_\lambda$.

Due to the isomorphism (2.4.83), covariant Hamiltonian formalism on the vertical Legendre bundle Π_{VY} can be developed as the vertical extension onto $V\Pi$ of covariant Hamiltonian formalism on Π. The corresponding canonical conjugate pairs are $(y^i, \dot{p}^\lambda_i)$ and $(\dot{y}^i, p^\lambda_i)$. In particular, due to the isomorphism (2.4.83), $V\Pi$ is endowed with the canonical polysymplectic form (9.1.2) which reads

$$\Omega_{VY} = [d\dot{p}^\lambda_i \wedge dy^i + dp^\lambda_i \wedge d\dot{y}^i] \wedge \omega \otimes \partial_\lambda.$$

Let Z_{VY} be the homogeneous Legendre bundle (2.4.13) over VY with the corresponding coordinates

$$(x^\lambda, y^i, \dot{y}^i, p^\lambda_i, q^\lambda_i, p).$$

It can be endowed with the multisymplectic Liouville form Ξ_{VY} (2.4.23). Sections of the affine bundle

$$Z_{VY} \to V\Pi, \tag{9.6.1}$$

9.6. Variation Hamilton equations. Jacobi fields

by definition, provide Hamiltonian forms on $V\Pi$.

Let us consider the following particular case of these forms which are related to those on the Legendre bundle Π. Due to the fibre bundle (2.4.85):

$$\zeta : VZ_Y \to Z_{VY},$$

the vertical tangent bundle VZ_Y of $Z_Y \to X$ is provided with the exterior form

$$\Xi_V = \zeta^* \Xi_{VY} = \dot{p}\omega + (\dot{p}_i^\lambda dy^i + p_i^\lambda d\dot{y}^i) \wedge \omega_\lambda,$$

which is exactly the vertical extension (2.4.78) of the multisymplectic Liouville form Ξ on Z_Y. Given the affine bundle $Z_Y \to \Pi$ (2.4.16), we have the fibre bundle $VZ_Y \to V\Pi$ (2.4.86) where $V\pi_{Z\Pi}$ is the vertical tangent map to $\pi_{Z\Pi}$. Let h be a section of the affine bundle $Z_Y \to \Pi$ and $H = h^*\Xi$ the corresponding Hamiltonian form (9.1.8) on Π. Then a section Vh of the fibre bundle (2.4.86) and the corresponding section $\zeta \circ Vh$ of the affine bundle (9.6.1) defines the Hamiltonian form

$$H_V = (Vh)^*\Xi_V = (\dot{p}_i^\lambda dy^i + p_i^\lambda d\dot{y}^i) \wedge \omega_\lambda - \mathcal{H}_V \omega, \qquad (9.6.2)$$
$$\mathcal{H}_V = \partial_V \mathcal{H} = (\dot{y}^i \partial_i + \dot{p}_i^\lambda \partial_\lambda^i)\mathcal{H},$$

on $V\Pi$. It is called the *vertical extension* of H (or, simply, the *vertical Hamiltonian form*). In particular, given the splitting (9.1.10) of H with respect to a connection Γ on $Y \to X$, we have the corresponding splitting

$$\mathcal{H}_V = \dot{p}_i^\lambda \Gamma_\lambda^i + \dot{y}^j p_i^\lambda \partial_j \Gamma_\lambda^i + \partial_V \widetilde{\mathcal{H}}_\Gamma$$

of H_V with respect to the vertical connection $V\Gamma$ (1.4.19) on $VY \to X$.

Theorem 9.6.1. *Let γ (9.1.3) be a Hamiltonian connection on Π associated to a Hamiltonian form H. Then its vertical prolongation $V\gamma$ (1.4.19) on $V\Pi \to X$ is a Hamiltonian connection associated to the vertical Hamiltonian form H_V (9.6.2).*

Proof. The proof follows from a direct computation. We have .

$$V\gamma = \gamma + dx^\mu \otimes [\partial_V \gamma_\mu^i \dot\partial_i + \partial_V \gamma_{\mu i}^\lambda \dot\partial_\lambda^i].$$

Components of this connection obey the equations

$$\dot\gamma_\mu^i = \partial_\mu^i \mathcal{H}_V = \partial_V \partial_\mu^i \mathcal{H}, \qquad \dot\gamma_{\lambda i} = -\partial_i \mathcal{H}_V = -\partial_V \partial_i \mathcal{H} \qquad (9.6.3)$$

and the Hamilton equations (9.1.15). □

In order to clarify the physical meaning of the Hamilton equations (9.6.3), let us suppose that $Y \to X$ is a vector bundle. Given a solution r of the Hamilton equations for H, let $\overline{r}$ be a Jacobi field, i.e., $r + \varepsilon \overline{r}$ also is a solution of the same Hamilton equations modulo terms of order > 1 in ε. Then it is readily observed that the Jacobi field $\overline{r}$ satisfies the Hamilton equations (9.6.3). At the same time, the Lagrangian L_{H_V} (9.1.16) on $J^1 V\Pi$, defined by the Hamiltonian form H_V (9.6.2), takes the form

$$\mathcal{L}_{H_V} = h_0(H_V) = \dot{p}_i^\lambda(y_\lambda^i - \partial_\lambda^i \mathcal{H}) - \dot{y}^i(p_{\lambda i}^\lambda + \partial_i \mathcal{H}) + d_\lambda(p_i^\lambda \dot{y}^i), \qquad (9.6.4)$$

where $\dot{p}_i^\lambda$, $\dot{y}^i$ play the role of Lagrange multipliers.

In conclusion, let us study the relationship between the vertical extensions of Lagrangian and covariant Hamiltonian formalisms. The Hamiltonian form H_V (9.6.2) on $V\Pi$ yields the vertical Hamiltonian map

$$\widehat{H}_V = V\widehat{H} : V\Pi \underset{VY}{\to} VJ^1 Y,$$

$$y_\lambda^i = \dot{\partial}_\lambda^i \mathcal{H}_V = \partial_\lambda^i \mathcal{H}, \qquad \dot{y}_\lambda^i = \partial_V \partial_\lambda^i \mathcal{H}.$$

Theorem 9.6.2. *Let a Hamiltonian form H on Π be associated to a Lagrangian L on $J^1 Y$. Then the vertical Hamiltonian form H_V (9.6.2) is weakly associated to the Lagrangian L_V (2.4.79).*

Proof. If the morphisms $\widehat{H}$ and $\widehat{L}$ obey the relation (9.2.2), then the corresponding vertical tangent morphisms satisfy the relation

$$V\widehat{L} \circ V\widehat{H} \circ Vi_Q = Vi_Q.$$

The condition (9.2.3) for H_V reduces to the equality (9.2.7) which is fulfilled if H is associated to L. $\square$

Chapter 10

Appendixes

For the sake of convenience of the reader, several relevant mathematical topics are compiled in this Chapter.

10.1 Commutative algebra

In this Section, the relevant basics on modules over commutative algebras is summarized [102; 108].

An *algebra* $\mathcal{A}$ is an additive group which is additionally provided with distributive multiplication. All algebras throughout the book are associative, unless they are Lie algebras. A *ring* is defined to be a *unital* algebra, i.e., it contains a unit element $\mathbf{1} \neq 0$. Non-zero elements of a ring form a multiplicative monoid. A *field* is a commutative ring whose non-zero elements make up a multiplicative group.

A subset $\mathcal{I}$ of an algebra $\mathcal{A}$ is called a left (resp. right) *ideal* if it is a subgroup of the additive group $\mathcal{A}$ and $ab \in \mathcal{I}$ (resp. $ba \in \mathcal{I}$) for all $a \in \mathcal{A}$, $b \in \mathcal{I}$. If $\mathcal{I}$ is both a left and right ideal, it is called a two-sided ideal. An ideal is a subalgebra, but a *proper ideal* (i.e., $\mathcal{I} \neq \mathcal{A}$) of a ring is not a subring because it does not contain a unit element.

Let $\mathcal{A}$ be a commutative ring. Of course, its ideals are two-sided. Its proper ideal is said to be *maximal* if it does not belong to another proper ideal. A commutative ring $\mathcal{A}$ is called *local* if it has a unique maximal ideal. This ideal consists of all non-invertible elements of $\mathcal{A}$. A proper ideal $\mathcal{I}$ of a commutative ring is called *prime* if $ab \in \mathcal{I}$ implies either $a \in \mathcal{I}$ or $b \in \mathcal{I}$. Any maximal ideal is prime.

Given an ideal $\mathcal{I} \subset \mathcal{A}$, the additive factor group $\mathcal{A}/\mathcal{I}$ is an algebra, called the *factor algebra*. If $\mathcal{A}$ is a ring, then $\mathcal{A}/\mathcal{I}$ is so. If $\mathcal{I}$ is a maximal ideal, the factor ring $\mathcal{A}/\mathcal{I}$ is a field.

Given an algebra $\mathcal{A}$, an additive group P is said to be a left (resp. right) $\mathcal{A}$-*module* if it is provided with distributive multiplication $\mathcal{A} \times P \to P$ by elements of $\mathcal{A}$ such that $(ab)p = a(bp)$ (resp. $(ab)p = b(ap)$) for all $a, b \in \mathcal{A}$ and $p \in P$. If $\mathcal{A}$ is a ring, one additionally assumes that $\mathbf{1}p = p = p\mathbf{1}$ for all $p \in P$. Left and right module structures are usually written by means of left and right multiplications $(a, p) \to ap$ and $(a, p) \to pa$, respectively. If P is both a left module over an algebra $\mathcal{A}$ and a right module over an algebra $\mathcal{A}'$, it is called an $(\mathcal{A} - \mathcal{A}')$-*bimodule* (an $\mathcal{A}$-bimodule if $\mathcal{A} = \mathcal{A}'$). If $\mathcal{A}$ is a commutative algebra, an $\mathcal{A}$-bimodule P is said to be *commutative* if $ap = pa$ for all $a \in \mathcal{A}$ and $p \in P$. Any left or right module over a commutative algebra $\mathcal{A}$ can be brought into a commutative bimodule. Therefore, unless otherwise stated, any module over a commutative algebra $\mathcal{A}$ is called an $\mathcal{A}$-module. A module over a field is called a *vector space*.

If an algebra $\mathcal{A}$ is a module over a commutative ring $\mathcal{K}$, it is said to be a $\mathcal{K}$-*algebra*. Any algebra can be seen as a $\mathbb{Z}$-algebra. Any ideal of an algebra $\mathcal{A}$ is an $\mathcal{A}$-module.

Hereafter, all associative algebras are assumed to be commutative.

The following are standard constructions of new modules from old ones.

• The *direct sum* $P_1 \oplus P_2$ of $\mathcal{A}$-modules P_1 and P_2 is the additive group $P_1 \times P_2$ provided with the $\mathcal{A}$-module structure

$$a(p_1, p_2) = (ap_1, ap_2), \qquad p_{1,2} \in P_{1,2}, \qquad a \in \mathcal{A}.$$

Let $\{P_i\}_{i \in I}$ be a set of modules. Their direct sum $\oplus P_i$ consists of elements $(\ldots, p_i, \ldots)$ of the Cartesian product $\prod P_i$ such that $p_i \neq 0$ at most for a finite number of indices $i \in I$.

• Given a submodule Q of an $\mathcal{A}$-module P, the quotient P/Q of the additive group P with respect to its subgroup Q also is provided with an $\mathcal{A}$-module structure. It is called a *factor module*.

• The set $\mathrm{Hom}\,_{\mathcal{A}}(P, Q)$ of $\mathcal{A}$-linear morphisms of an $\mathcal{A}$-module P to an $\mathcal{A}$-module Q is naturally an $\mathcal{A}$-module. The $\mathcal{A}$-module

$$P^* = \mathrm{Hom}\,_{\mathcal{A}}(P, \mathcal{A})$$

is called the *dual* of an $\mathcal{A}$-module P. There is a natural monomorphism $P \to P^{**}$.

• The *tensor product* $P \otimes Q$ of $\mathcal{A}$-modules P and Q is an additive group which is generated by elements $p \otimes q$, $p \in P$, $q \in Q$, obeying the relations

$$(p + p') \otimes q = p \otimes q + p' \otimes q,$$
$$p \otimes (q + q') = p \otimes q + p \otimes q',$$
$$pa \otimes q = p \otimes aq, \qquad p \in P, \qquad q \in Q, \qquad a \in \mathcal{A}.$$

10.1. Commutative algebra

It is provided with the $\mathcal{A}$-module structure

$$a(p \otimes q) = (ap) \otimes q = p \otimes (qa) = (p \otimes q)a.$$

In particular, we have the following.

(i) If a ring $\mathcal{A}$ is treated as an $\mathcal{A}$-module, the tensor product $\mathcal{A} \otimes_\mathcal{A} Q$ is canonically isomorphic to Q via the assignment

$$\mathcal{A} \otimes_\mathcal{A} Q \ni a \otimes q \leftrightarrow aq \in Q.$$

(ii) The *tensor product of Abelian groups* G and G' is defined as their tensor product $G \otimes G'$ as $\mathbb{Z}$-modules. For instance, if one of them is a finite group, then $G \otimes G' = 0$.

(iii) The *tensor product of commutative algebras* $\mathcal{A}$ and $\mathcal{A}'$ is defined as their tensor product $\mathcal{A} \otimes \mathcal{A}'$ as modules provided with the multiplication

$$(a \otimes a')(b \otimes b') = (aa') \otimes bb'.$$

An $\mathcal{A}$-module P is called *free* if it has a *basis*, i.e., a linearly independent subset $I \subset P$ spanning P such that each element of P has a unique expression as a linear combination of elements of I with a finite number of non-zero coefficients from an algebra $\mathcal{A}$. Any vector space is free. Any module is isomorphic to a quotient of a free module. A module is said to be *finitely generated* (or of *finite rank*) if it is a quotient of a free module with a finite basis.

One says that a module P is *projective* if it is a direct summand of a free module, i.e., there exists a module Q such that $P \oplus Q$ is a free module. A module P is projective if and only if $P = \mathbf{p}S$ where S is a free module and $\mathbf{p}$ is a projector of S, i.e., $\mathbf{p}^2 = \mathbf{p}$. If P is a projective module of finite rank over a ring, then its dual P^* is so, and P^{**} is isomorphic to P.

Theorem 10.1.1. *Any projective module over a local ring is free.*

Now we focus on exact sequences, direct and inverse limits of modules [108; 114]. A composition of module morphisms

$$P \xrightarrow{i} Q \xrightarrow{j} T$$

is said to be *exact* at Q if $\operatorname{Ker} j = \operatorname{Im} i$. A composition of module morphisms

$$0 \to P \xrightarrow{i} Q \xrightarrow{j} T \to 0 \tag{10.1.1}$$

is called a *short exact sequence* if it is exact at all the terms P, Q, and T. This condition implies that: (i) i is a monomorphism, (ii) $\operatorname{Ker} j = \operatorname{Im} i$, and (iii) j is an epimorphism onto the quotient $T = Q/P$.

Theorem 10.1.2. *Given an exact sequence of modules (10.1.1) and another $\mathcal{A}$-module R, the sequence of modules*

$$0 \to \operatorname{Hom}_{\mathcal{A}}(T, R) \xrightarrow{j^*} \operatorname{Hom}_{\mathcal{A}}(Q, R) \xrightarrow{i^*} \operatorname{Hom}(P, R)$$

is exact at the first and second terms, i.e., j^* is a monomorphism, but i^* need not be an epimorphism.

One says that the exact sequence (10.1.1) is *split* if there exists a monomorphism $s : T \to Q$ such that $j \circ s = \operatorname{Id} T$ or, equivalently,

$$Q = i(P) \oplus s(T) \cong P \oplus T.$$

Theorem 10.1.3. *The exact sequence (10.1.1) is always split if T is a projective module.*

A *directed set* I is a set with an order relation $<$ which satisfies the following three conditions: (i) $i < i$, for all $i \in I$; (ii) if $i < j$ and $j < k$, then $i < k$; (iii) for any $i, j \in I$, there exists $k \in I$ such that $i < k$ and $j < k$. It may happen that $i \neq j$, but $i < j$ and $j < i$ simultaneously.

A family of modules $\{P_i\}_{i \in I}$ (over the same algebra), indexed by a directed set I, is called a *direct system* if, for any pair $i < j$, there exists a morphism $r^i_j : P_i \to P_j$ such that

$$r^i_i = \operatorname{Id} P_i, \qquad r^i_j \circ r^j_k = r^i_k, \qquad i < j < k.$$

A direct system of modules admits a *direct limit*. This is a module P_∞ together with morphisms $r^i_\infty : P_i \to P_\infty$ such that $r^i_\infty = r^j_\infty \circ r^i_j$ for all $i < j$. The module P_∞ consists of elements of the direct sum $\oplus_I P_i$ modulo the identification of elements of P_i with their images in P_j for all $i < j$. An example of a direct system is a *direct sequence*

$$P_0 \longrightarrow P_1 \longrightarrow \cdots P_i \xrightarrow{r^i_{i+1}} \cdots, \qquad I = \mathbb{N}. \qquad (10.1.2)$$

Note that direct limits also exist in the categories of commutative and graded commutative algebras and rings, but not in categories containing non-Abelian groups.

Theorem 10.1.4. *Direct limits commute with direct sums and tensor products of modules. Namely, let $\{P_i\}$ and $\{Q_i\}$ be two direct systems of modules over the same algebra which are indexed by the same directed set I, and let P_∞ and Q_∞ be their direct limits. Then the direct limits of the direct systems $\{P_i \oplus Q_i\}$ and $\{P_i \otimes Q_i\}$ are $P_\infty \oplus Q_\infty$ and $P_\infty \otimes Q_\infty$, respectively.*

10.1. Commutative algebra

Theorem 10.1.5. *A morphism of a direct system $\{P_i, r_j^i\}_I$ to a direct system $\{Q_{i'}, \rho_{j'}^{i'}\}_{I'}$ consists of an order preserving map $f : I \to I'$ and morphisms*

$$F_i : P_i \to Q_{f(i)}$$

which obey the compatibility conditions

$$\rho_{f(j)}^{f(i)} \circ F_i = F_j \circ r_j^i.$$

If P_∞ and Q_∞ are limits of these direct systems, there exists a unique morphism

$$F_\infty : P_\infty \to Q_\infty$$

such that

$$\rho_\infty^{f(i)} \circ F_i = F_\infty \circ r_\infty^i.$$

Theorem 10.1.6. *Direct limits preserve monomorphisms and epimorphisms, i.e., if all*

$$F_i : P_i \to Q_{f(i)}$$

are monomorphisms or epimorphisms, so is

$$\Phi_\infty : P_\infty \to Q_\infty.$$

Let short exact sequences

$$0 \to P_i \xrightarrow{F_i} Q_i \xrightarrow{\Phi_i} T_i \to 0 \qquad (10.1.3)$$

for all $i \in I$ define a short exact sequence of direct systems of modules $\{P_i\}_I$, $\{Q_i\}_I$, and $\{T_i\}_I$ which are indexed by the same directed set I. Then their direct limits form a short exact sequence

$$0 \to P_\infty \xrightarrow{F_\infty} Q_\infty \xrightarrow{\Phi_\infty} T_\infty \to 0. \qquad (10.1.4)$$

In particular, the direct limit of factor modules Q_i/P_i is the factor module Q_∞/P_∞. By virtue of Theorem 10.1.4, if all the exact sequences (10.1.3) are split, the exact sequence (10.1.4) is well.

Remark 10.1.1. Let P be an $\mathcal{A}$-module. We denote

$$P^{\otimes k} = \overset{k}{\otimes} P.$$

Let us consider the direct system of $\mathcal{A}$-modules with respect to monomorphisms

$$\mathcal{A} \longrightarrow (\mathcal{A} \oplus P) \longrightarrow \cdots (\mathcal{A} \oplus P \oplus \cdots \oplus P^{\otimes k}) \longrightarrow \cdots .$$

Its direct limit
$$\otimes P = \mathcal{A} \oplus P \oplus \cdots \oplus P^{\otimes k} \oplus \cdots \qquad (10.1.5)$$
is an N-graded $\mathcal{A}$-algebra with respect to the tensor product $\otimes$. It is called the *tensor algebra* of a module P. Its quotient with respect to the ideal generated by elements
$$p \otimes p' + p' \otimes p, \qquad p, p' \in P,$$
is an N-graded commutative algebra, called the *exterior algebra* of a module P.

Given an *inverse sequences* of modules
$$P^0 \longleftarrow P^1 \longleftarrow \cdots P^i \overset{\pi_i^{i+1}}{\longleftarrow} \cdots, \qquad (10.1.6)$$
its *inductive limit* is a module P^∞ together with morphisms $\pi_i^\infty : P^\infty \to P^i$ such that
$$\pi_i^\infty = \pi_i^j \circ \pi_j^\infty$$
for all $i < j$. It consists of elements $(\ldots, p^i, \ldots)$, $p^i \in P^i$, of the Cartesian product $\prod P^i$ such that $p^i = \pi_i^j(p^j)$ for all $i < j$.

Theorem 10.1.7. *Inductive limits preserve monomorphisms, but not epimorphisms. Let exact sequences*
$$0 \to P^i \xrightarrow{F^i} Q^i \xrightarrow{\Phi^i} T^i, \qquad i \in \mathbb{N},$$
for all $i \in \mathbb{N}$ define an exact sequence of inverse systems of modules $\{P^i\}$, $\{Q^i\}$ and $\{T^i\}$. Then their inductive limits form an exact sequence
$$0 \to P^\infty \xrightarrow{F^\infty} Q^\infty \xrightarrow{\Phi^\infty} T^\infty.$$

In contrast with direct limits, the inductive ones exist in the category of groups which are not necessarily commutative.

10.2 Differential operators on modules

This Section addresses the notion of a linear differential operator on a module over a commutative ring [71; 96].

Let $\mathcal{K}$ be a commutative ring and $\mathcal{A}$ a commutative $\mathcal{K}$-ring. Let P and Q be $\mathcal{A}$-modules. The $\mathcal{K}$-module $\mathrm{Hom}\,_\mathcal{K}(P,Q)$ of $\mathcal{K}$-module homomorphisms $\Phi : P \to Q$ can be endowed with the two different $\mathcal{A}$-module structures
$$(a\Phi)(p) = a\Phi(p), \qquad (\Phi \bullet a)(p) = \Phi(ap), \qquad a \in \mathcal{A}, \quad p \in P. \qquad (10.2.1)$$

10.2. Differential operators on modules

For the sake of convenience, we refer to the second one as the $\mathcal{A}^\bullet$-module structure. Let us put

$$\delta_a \Phi = a\Phi - \Phi \bullet a, \qquad a \in \mathcal{A}. \tag{10.2.2}$$

Definition 10.2.1. An element $\Delta \in \mathrm{Hom}_{\mathcal{K}}(P, Q)$ is called a Q-valued differential operator of order s on P if

$$\delta_{a_0} \circ \cdots \circ \delta_{a_s} \Delta = 0$$

for any tuple of $s+1$ elements $a_0, \ldots, a_s$ of $\mathcal{A}$. The set $\mathrm{Diff}_s(P, Q)$ of these operators inherits the $\mathcal{A}$- and $\mathcal{A}^\bullet$-module structures (10.2.1).

In particular, zero order differential operators obey the condition

$$\delta_a \Delta(p) = a\Delta(p) - \Delta(ap) = 0, \qquad a \in \mathcal{A}, \qquad p \in P,$$

and, consequently, they coincide with $\mathcal{A}$-module morphisms $P \to Q$. A first order differential operator Δ satisfies the condition

$$\delta_b \circ \delta_a \Delta(p) = ba\Delta(p) - b\Delta(ap) - a\Delta(bp) + \Delta(abp) = 0, \quad a, b \in \mathcal{A}. \tag{10.2.3}$$

The following fact reduces the study of Q-valued differential operators on an $\mathcal{A}$-module P to that of Q-valued differential operators on the ring $\mathcal{A}$.

Theorem 10.2.1. *Let us consider the $\mathcal{A}$-module morphism*

$$h_s : \mathrm{Diff}_s(\mathcal{A}, Q) \to Q, \qquad h_s(\Delta) = \Delta(1). \tag{10.2.4}$$

Any Q-valued s-order differential operator $\Delta \in \mathrm{Diff}_s(P, Q)$ on P uniquely factorizes as

$$\Delta : P \xrightarrow{\mathfrak{f}_\Delta} \mathrm{Diff}_s(\mathcal{A}, Q) \xrightarrow{h_s} Q \tag{10.2.5}$$

through the morphism h_s (10.2.4) and some homomorphism

$$\mathfrak{f}_\Delta : P \to \mathrm{Diff}_s(\mathcal{A}, Q), \qquad (\mathfrak{f}_\Delta p)(a) = \Delta(ap), \qquad a \in \mathcal{A}, \tag{10.2.6}$$

of the $\mathcal{A}$-module P to the $\mathcal{A}^\bullet$-module $\mathrm{Diff}_s(\mathcal{A}, Q)$. The assignment $\Delta \to \mathfrak{f}_\Delta$ defines the isomorphism

$$\mathrm{Diff}_s(P, Q) = \mathrm{Hom}_{\mathcal{A}-\mathcal{A}^\bullet}(P, \mathrm{Diff}_s(\mathcal{A}, Q)). \tag{10.2.7}$$

Let $P = \mathcal{A}$. Any zero order Q-valued differential operator Δ on $\mathcal{A}$ is defined by its value $\Delta(1)$. Then there is an isomorphism

$$\mathrm{Diff}_0(\mathcal{A}, Q) = Q$$

via the association

$$Q \ni q \to \Delta_q \in \mathrm{Diff}_0(\mathcal{A}, Q),$$

where Δ_q is given by the equality $\Delta_q(1) = q$. A first order Q-valued differential operator Δ on $\mathcal{A}$ fulfils the condition

$$\Delta(ab) = b\Delta(a) + a\Delta(b) - ba\Delta(1), \qquad a, b \in \mathcal{A}.$$

It is called a Q-valued *derivation* of $\mathcal{A}$ if $\Delta(1) = 0$, i.e., the *Leibniz rule*

$$\Delta(ab) = \Delta(a)b + a\Delta(b), \qquad a, b \in \mathcal{A}, \qquad (10.2.8)$$

holds. One obtains at once that any first order differential operator on $\mathcal{A}$ falls into the sum

$$\Delta(a) = a\Delta(1) + [\Delta(a) - a\Delta(1)]$$

of the zero order differential operator $a\Delta(1)$ and the derivation $\Delta(a) - a\Delta(1)$. If ∂ is a Q-valued derivation of $\mathcal{A}$, then $a\partial$ is well for any $a \in \mathcal{A}$. Hence, Q-valued derivations of $\mathcal{A}$ constitute an $\mathcal{A}$-module $\mathfrak{d}(\mathcal{A}, Q)$, called the *derivation module*. There is the $\mathcal{A}$-module decomposition

$$\text{Diff}_1(\mathcal{A}, Q) = Q \oplus \mathfrak{d}(\mathcal{A}, Q). \qquad (10.2.9)$$

If $P = Q = \mathcal{A}$, the derivation module $\mathfrak{d}\mathcal{A}$ of $\mathcal{A}$ also is a Lie $\mathcal{K}$-algebra with respect to the Lie bracket

$$[u, u'] = u \circ u' - u' \circ u, \qquad u, u' \in \mathcal{A}. \qquad (10.2.10)$$

Accordingly, the decomposition (10.2.9) takes the form

$$\text{Diff}_1(\mathcal{A}) = \mathcal{A} \oplus \mathfrak{d}\mathcal{A}. \qquad (10.2.11)$$

Definition 10.2.2. A *connection* on an $\mathcal{A}$-module P is an $\mathcal{A}$-module morphism

$$\mathfrak{d}\mathcal{A} \ni u \to \nabla_u \in \text{Diff}_1(P, P) \qquad (10.2.12)$$

such that the first order differential operators ∇_u obey the *Leibniz rule*

$$\nabla_u(ap) = u(a)p + a\nabla_u(p), \qquad a \in \mathcal{A}, \quad p \in P. \qquad (10.2.13)$$

Though ∇_u (10.2.12) is called a connection, it in fact is a *covariant differential* on a module P.

Let P be a commutative $\mathcal{A}$-ring and $\mathfrak{d}P$ the derivation module of P as a $\mathcal{K}$-ring. The $\mathfrak{d}P$ is both a P- and $\mathcal{A}$-module. Then Definition 10.2.2 is modified as follows.

Definition 10.2.3. A *connection* on an $\mathcal{A}$-ring P is an $\mathcal{A}$-module morphism

$$\mathfrak{d}\mathcal{A} \ni u \to \nabla_u \in \mathfrak{d}P \subset \text{Diff}_1(P, P), \qquad (10.2.14)$$

which is a connection on P as an $\mathcal{A}$-module, i.e., obeys the Leinbniz rule (10.2.13).

10.3 Homology and cohomology of complexes

This Section summarizes the relevant basics on complexes of modules over a commutative ring [108; 114].

Let $\mathcal{K}$ be a commutative ring. A sequence

$$0 \leftarrow B_0 \xleftarrow{\partial_1} B_1 \xleftarrow{\partial_2} \cdots B_p \xleftarrow{\partial_{p+1}} \cdots \qquad (10.3.1)$$

of $\mathcal{K}$-modules B_p and homomorphisms ∂_p is said to be a *chain complex* if

$$\partial_p \circ \partial_{p+1} = 0, \qquad p \in \mathbb{N},$$

i.e., $\operatorname{Im} \partial_{p+1} \subset \operatorname{Ker} \partial_p$. Homomorphisms ∂_p are called *boundary operators*. Elements of a module B_p, its submodules $\operatorname{Ker} \partial_p \subset B_p$ and $\operatorname{Im} \partial_{p+1} \subset \operatorname{Ker} \partial_p$ are called *p-chains*, *p-cycles* and *p-boundaries*, respectively. The *p-th homology group* of the chain complex B_* (10.3.1) is defined as the factor module

$$H_p(B_*) = \operatorname{Ker} \partial_p / \operatorname{Im} \partial_{p+1}.$$

It is a $\mathcal{K}$-module. In particular, we have

$$H_0(B_*) = B_0 / \operatorname{Im} \partial_1.$$

The chain complex (10.3.1) is exact at a term B_p if and only if $H_p(B_*) = 0$. This complex is said to be *k-exact* if its homology groups $H_{p \leq k}(B_*)$ are trivial. It is called *exact* if all its homology groups are trivial, i.e., it is an exact sequence.

A sequence

$$0 \to B^0 \xrightarrow{\delta^0} B^1 \xrightarrow{\delta^1} \cdots B^p \xrightarrow{\delta^p} \cdots \qquad (10.3.2)$$

of modules B^p and their homomorphisms δ^p is said to be a *cochain complex* (or, simply, a *complex*) if

$$\delta^{p+1} \circ \delta^p = 0, \qquad p \in \mathbb{N},$$

i.e., $\operatorname{Im} \delta^p \subset \operatorname{Ker} \delta^{p+1}$. The homomorphisms δ^p are called *coboundary operators*. For the sake of convenience, let us denote

$$\delta^{p=-1} : 0 \to B^0.$$

Elements of a module B^p, its submodules $\operatorname{Ker} \delta^p \subset B^p$ and $\operatorname{Im} \delta^{p-1}$ are called *p-cochains*, *p-cocycles* and *p-coboundaries*, respectively. The *p-th cohomology group* of the complex B^* (10.3.2) is the factor module

$$H^p(B^*) = \operatorname{Ker} \delta^p / \operatorname{Im} \delta^{p-1}.$$

It is a $\mathcal{K}$-module. In particular,
$$H^0(B^*) = \operatorname{Ker} \delta^0.$$
The complex (10.3.2) is exact at a term B^p if and only if $H^p(B^*) = 0$. This complex is an exact sequence if all its cohomology groups are trivial.

Remark 10.3.1. Given a chain complex B_* (10.3.1), let $B^p = B_p^*$ be the $\mathcal{K}$-duals of B_p. Let us define the $\mathcal{K}$-module homomorphisms $\delta^p : B^p \to B^{p+1}$ as
$$\delta^p b^p = b^p \circ \partial_{p+1} : B_{p+1} \to \mathcal{K}, \qquad b^p \in B^p. \qquad (10.3.3)$$
It is readily observed that $\delta^{p+1} \circ \delta^p = 0$. Then $\{B^p, \delta^p\}$ is the *dual complex* of the chain complex B_*. Let us note that, if the chain complex B_* is exact, the dual complex need not be so (see Theorem 10.1.2).

A complex (B^*, δ^*) is called *acyclic* if its cohomology groups $H^{0<p}(B^*)$ are trivial. It is acyclic if there exists a *homotopy operator* $\mathbf{h}$, defined as a set of module morphisms
$$\mathbf{h}^{p+1} : B^{p+1} \to B^p, \qquad p \in \mathbb{N},$$
such that
$$\mathbf{h}^{p+1} \circ \delta^p + \delta^{p-1} \circ \mathbf{h}^p = \operatorname{Id} B^p, \qquad p \in \mathbb{N}_+.$$
Indeed, if $\delta^p b^p = 0$, then
$$b^p = \delta^{p-1}(\mathbf{h}^p b^p),$$
and $H^{p>0}(B^*) = 0$. A complex (B^*, δ^*) is said to be a *resolution* of a module B if it is acyclic and
$$H^0(B^*) = \operatorname{Ker} \delta^0 = B.$$

The following are the standard constructions of new complexes from old ones.

• Given complexes (B_1^*, δ_1^*) and (B_2^*, δ_2^*), their *direct sum* $B_1^* \oplus B_2^*$ is a complex of modules
$$(B_1^* \oplus B_2^*)^p = B_1^p \oplus B_2^p$$
with respect to the coboundary operators
$$\delta_\oplus^p(b_1^p + b_2^p) = \delta_1^p b_1^p + \delta_2^p b_2^p.$$

• Given a subcomplex (C^*, δ^*) of a complex (B^*, δ^*), the *factor complex* B^*/C^* is defined as a complex of factor modules B^p/C^p provided with the coboundary operators
$$\delta^p[b^p] = [\delta^p b^p],$$

10.3. Homology and cohomology of complexes

where $[b^p] \in B^p/C^p$ denotes the coset of an element b^p.

- Given complexes (B_1^*, δ_1^*) and (B_2^*, δ_2^*), their *tensor product* $B_1^* \otimes B_2^*$ is a complex of modules

$$(B_1^* \otimes B_2^*)^p = \bigoplus_{k+r=p} B_1^k \otimes B_2^r$$

with respect to the coboundary operators

$$\delta_\otimes^p (b_1^k \otimes b_2^r) = (\delta_1^k b_1^k) \otimes b_2^r + (-1)^k b_1^k \otimes (\delta_2^r b_2^r).$$

A *cochain morphism* of complexes

$$\gamma : B_1^* \to B_2^* \tag{10.3.4}$$

is defined as a family of degree-preserving homomorphisms

$$\gamma^p : B_1^p \to B_2^p, \qquad p \in \mathbb{N},$$

such that

$$\delta_2^p \circ \gamma^p = \gamma^{p+1} \circ \delta_1^p, \qquad p \in \mathbb{N}.$$

It follows that if $b^p \in B_1^p$ is a cocycle or a coboundary, then $\gamma^p(b^p) \in B_2^p$ is so. Therefore, the cochain morphism of complexes (10.3.4) yields an induced homomorphism of their cohomology groups

$$[\gamma]^* : H^*(B_1^*) \to H^*(B_2^*).$$

Let short exact sequences

$$0 \to C^p \xrightarrow{\gamma_p} B^p \xrightarrow{\zeta_p} F^p \to 0$$

for all $p \in \mathbb{N}$ define a short exact sequence of complexes

$$0 \to C^* \xrightarrow{\gamma} B^* \xrightarrow{\zeta} F^* \to 0, \tag{10.3.5}$$

where γ is a cochain monomorphism and ζ is a cochain epimorphism onto the quotient $F^* = B^*/C^*$.

Theorem 10.3.1. *The short exact sequence of complexes (10.3.5) yields the long exact sequence of their cohomology groups*

$$0 \to H^0(C^*) \xrightarrow{[\gamma]^0} H^0(B^*) \xrightarrow{[\zeta]^0} H^0(F^*) \xrightarrow{\tau^0} H^1(C^*) \longrightarrow \cdots \tag{10.3.6}$$

$$\longrightarrow H^p(C^*) \xrightarrow{[\gamma]^p} H^p(B^*) \xrightarrow{[\zeta]^p} H^p(F^*) \xrightarrow{\tau^p} H^{p+1}(C^*) \longrightarrow \cdots .$$

Theorem 10.3.2. *A direct sequence of complexes*

$$B_0^* \longrightarrow B_1^* \longrightarrow \cdots B_k^* \xrightarrow{\gamma_{k+1}^k} B_{k+1}^* \longrightarrow \cdots \tag{10.3.7}$$

admits a direct limit B_∞^* which is a complex whose cohomology $H^*(B_\infty^*)$ is a direct limit of the direct sequence of cohomology groups

$$H^*(B_0^*) \longrightarrow H^*(B_1^*) \longrightarrow \cdots H^*(B_k^*) \xrightarrow{[\gamma_{k+1}^k]} H^*(B_{k+1}^*) \longrightarrow \cdots .$$

10.4 Cohomology of groups

Cohomology of groups that we here describe characterize extension of these groups by a commutative group [108].

One can associate to any set Z the following chain complex. Let Z_k be a free $\mathbb{Z}$-module whose basis is the Cartesian product $\overset{k+1}{\times} Z$. In particular, Z_0 is a free $\mathbb{Z}$-module whose basis is Z. Let us define $\mathbb{Z}$-linear homomorphisms

$$\partial_0 : Z_0 \ni m_i(z_0^i) \to \sum_i m_i \in \mathbb{Z}, \qquad m_i \in \mathbb{Z}, \tag{10.4.1}$$

$$\partial_{k+1} : Z_{k+1} \to Z_k, \qquad k \in \mathbb{N},$$

$$\partial_{k+1}(z_0, \ldots, z_{k+1}) = \sum_{j=0}^{k+1} (-1)^j (z_0, \ldots, \widehat{z_j}, \ldots, z_{k+1}), \tag{10.4.2}$$

where the caret $\widehat{}$ denotes omission. It is readily observed that $\partial_k \circ \partial_{k+1} = 0$ for all $k \in \mathbb{N}$. Thus, we obtain the chain complex

$$0 \xleftarrow{} \mathbb{Z} \xleftarrow{\partial_0} Z_0 \xleftarrow{\partial_1} Z_1 \xleftarrow{} \cdots Z_k \xleftarrow{\partial_{k+1}} \cdots,$$

called the *standard chain complex* of a set Z. This complex is exact.

Let $Z = G$ be a group and G_* the standard chain complex

$$0 \xleftarrow{} \mathbb{Z} \xleftarrow{\partial_0} G_0 \xleftarrow{\partial_1} G_1 \xleftarrow{} \cdots G_p \xleftarrow{\partial_{p+1}} \cdots. \tag{10.4.3}$$

A $\mathbb{Z}$-module G_0 becomes a ring with respect to the multiplication

$$(m_r g^r)(n_k g^k) = m_r n_k (g^r g^k), \qquad m_r, n_k \in \mathbb{Z}, \qquad g^r, g^k \in G.$$

It is called the *group ring* $\mathbb{Z}G$. Accordingly, a set G_k, $k > 0$, is brought into a free left $\mathbb{Z}G$-module by letting

$$g(g_0, \ldots, g_k) = (gg_0, \ldots, gg_k), \qquad g \in G. \tag{10.4.4}$$

Its basis consists of $(k+1)$-tuples $(1, g_1, \ldots, g_k)$. For the sake of simplicity, left $\mathbb{Z}G$-modules are usually called group G-modules. It is readily observed that the boundary operators ∂_0 (10.4.1) and $\partial_{k>0}$ (10.4.2) are G-module morphisms, where $\mathbb{Z}$ is regarded as a trivial G-module.

Let us consider an isomorphic chain complex $\mathcal{G}_*$ of G-modules. Its term $\mathcal{G}_{k>0}$ is a free G-module whose basis consists of the k-tuples $[g_1, \ldots, g_k]$, while elements of $\mathcal{G}_0 = \mathbb{Z}G$ are $m_r g^r [\ ]$. The boundary operators of the chain complex $\mathcal{G}_*$ are given by the formula

$$\partial_{k>0}[g_1, \ldots, g_k] = g_1[g_2, \ldots, g_k] + \tag{10.4.5}$$
$$\sum_j (-1)^j [g_1, \ldots, g_j g_{j+1}, \ldots, g_k] + (-1)^k [g_1, \ldots, g_{k-1}],$$

10.4. Cohomology of groups

and $\partial_0(g[\ ]) = 1$ is a group module morphism from $\mathbb{Z}G$ to $\mathbb{Z}$. The chain complexes $\mathcal{G}_*$ and G_* are isomorphic via the association

$$[g_1, \ldots, g_k] \to (1, g_1, g_1g_2, \ldots, g_1 \cdots g_k).$$

Let A be a left G-module. Given the chain complex $\mathcal{G}_*$, let us consider the cochain complex whose terms

$$\mathcal{G}^k = \mathrm{Hom}\,_G(\mathcal{G}_k, A)$$

are left G-modules of group module morphisms $f^k : \mathcal{G}_k \to A$. These morphisms also can be seen as A-valued functions $f^k(g_1, \ldots, g_k)$ of k arguments on a group G. In accordance with the formula (10.3.3), the coboundary operators of the complex $\mathcal{G}^*$ are defined as

$$(\delta^k f^k)(g_1, \ldots, g_{k+1}) = f^k(\partial_{k+1}[g_1, \ldots, g_{k+1}]) = \qquad (10.4.6)$$
$$g_1 f^k(g_2, \ldots, g_{k+1}) + \sum_j (-1)^j f^k(g_1, \ldots, g_j g_{j+1}, \ldots, g_{k+1})$$
$$+ (-1)^{k+1} f^k(g_1, \ldots, g_k), \qquad k \in \mathbb{N}.$$

In particular, the module $\mathcal{G}^0$ is isomorphic to A via the association

$$A \ni a \to f_a^0 \in \mathcal{G}^0, \qquad f_a^0([\ ]) = a.$$

For instance, we have

$$\delta^0 f_a^0(g) = ga - a, \qquad g \in G, \qquad (10.4.7)$$
$$\delta^1 f^1(g_1, g_2) = g_1 f^1(g_2) - f^1(g_1 g_2) + f^1(g_1), \qquad (10.4.8)$$
$$\delta^2 f^2(g_1, g_2, g_3) = g_1 f^2(g_2, g_3) - f^2(g_1 g_2, g_3) + \qquad (10.4.9)$$
$$f^2(g_1, g_2 g_3) - f^2(g_1, g_2).$$

Cohomology $H^*(G, A)$ of the complex $\mathcal{G}^*$ is called *cohomology of the group G with coefficients in a G-module A*. In particular, the expression (10.4.7) shows that

$$H^0(G, A) = \mathrm{Ker}\,\delta^0$$

is isomorphic to the additive subgroup of G-invariant elements of A. By cohomology of a group G with coefficients in a G-module A also is meant the cohomology $H_0^*(G, A)$ of the subcomplex $\mathcal{G}_0^*$ of the complex $\mathcal{G}^*$ whose k-cochains are A-valued functions of k arguments from G which vanish whenever one of the arguments is equal to 1. It is easily verified that $\delta^k \mathcal{G}_0^k \subset \mathcal{G}_0^{k+1}$.

Let us show that the cohomology group $H_0^2(G, A)$ classifies the extensions of a group G by an additive group A. Such an extension is defined as a sequence

$$0 \to A \xrightarrow{i} W \xrightarrow{\pi} G \to 1 \tag{10.4.10}$$

of group homomorphisms, where i is a monomorphism onto a normal subgroup of W and π is an epimorphism onto the factor group $G = W/A$, i.e., $\operatorname{Im} i = \pi^{-1}(\mathbf{1})$. By analogy with sequences of additive groups, the sequence (10.4.10) is said to be exact. For the sake of simplicity, let us identify A with its image $i(A) \subset W$, and let us write the group operation in W in the additive form.

Any extension (10.4.10) yields a homomorphism of G to the group of automorphisms of A as follows. Let $w(g)$ be representatives in W of elements $g \in G$. Then any element $w \in W$ is uniquely written in the form

$$w = a_w + w(g), \qquad a_w \in A.$$

Let us consider the automorphism

$$\phi_g : a \to ga = w(g) + a - w(g), \qquad a \in A. \tag{10.4.11}$$

Certainly, this automorphism depends only on an element $g \in G$, but not on its representative in W. The association $g \to \phi_g$ defines a desired homomorphism

$$\phi : G \to \operatorname{Aut} A. \tag{10.4.12}$$

This homomorphism ϕ makes A to a G-module denoted by A_ϕ.

Conversely, the homomorphism ϕ (10.4.12) corresponds to some extension (10.4.10) of the group G. Among these extensions, there is the semidirect product $W = A \times_\phi G$ with the group operation

$$(a, g) + (a', g') = (a + ga', gg'), \qquad ga' = \phi_g(a').$$

Theorem 10.4.1. *There is one-to-one correspondence between the classes of isomorphic extensions (10.4.10) of a group G by a commutative group A associated to the same homomorphism ϕ (10.4.12) and the elements of the cohomology group $H_0^2(G, A_\phi)$ [108].*

In particular, the semidirect product $A \times_\phi G$ corresponds to 0 of the cohomology group $H_0^2(G, A_\phi)$.

10.5 Cohomology of Lie algebras

Let $\mathfrak{g}$ be a Lie algebra over a commutative ring $\mathcal{K}$. Let $\mathfrak{g}$ act on a $\mathcal{K}$-module P on the left such that

$$[\varepsilon, \varepsilon']p = (\varepsilon \circ \varepsilon' - \varepsilon' \circ \varepsilon)p, \qquad \varepsilon, \varepsilon' \in \mathfrak{g}.$$

Then one calls P the Lie algebra $\mathfrak{g}$-*module*. Let us consider $\mathcal{K}$-multilinear skew-symmetric maps

$$c^k : \overset{k}{\times} \mathfrak{g} \to P.$$

They form a $\mathfrak{g}$-module $C^k[\mathfrak{g}; P]$. Let us put $C^0[\mathfrak{g}; P] = P$. We obtain the cochain complex

$$0 \to P \xrightarrow{\delta^0} C^1[\mathfrak{g}; P] \xrightarrow{\delta^1} \cdots C^k[\mathfrak{g}; P] \xrightarrow{\delta^k} \cdots \qquad (10.5.1)$$

with respect to the *Chevalley–Eilenberg coboundary operators*

$$\delta^k c^k(\varepsilon_0, \ldots, \varepsilon_k) = \sum_{i=0}^{k}(-1)^i \varepsilon_i c^k(\varepsilon_0, \ldots, \widehat{\varepsilon}_i, \ldots, \varepsilon_k) + \qquad (10.5.2)$$

$$\sum_{1 \le i < j \le k} (-1)^{i+j} c^k([\varepsilon_i, \varepsilon_j], \varepsilon_0, \ldots, \widehat{\varepsilon}_i, \ldots, \widehat{\varepsilon}_j, \ldots, \varepsilon_k),$$

where the caret $\widehat{}$ denotes omission [46]. For instance, we have

$$\delta^0 p(\varepsilon_0) = \varepsilon_0 p, \qquad (10.5.3)$$
$$\delta^1 c^1(\varepsilon_0, \varepsilon_1) = \varepsilon_0 c^1(\varepsilon_1) - \varepsilon_1 c^1(\varepsilon_0) - c^1([\varepsilon_0, \varepsilon_1]). \qquad (10.5.4)$$

The complex (10.5.1) is called the *Chevalley–Eilenberg complex*, and its cohomology $H^*(\mathfrak{g}, P)$ is the *Chevalley–Eilenberg cohomology* of a Lie algebra $\mathfrak{g}$ with coefficients in P.

In particular, let $P = \mathcal{K}$ and $\mathfrak{g} : \mathcal{K} \to 0$. Then the Chevalley–Eilenberg complex $C^*[\mathfrak{g}; \mathcal{K}]$ is the exterior algebra $\wedge \mathfrak{g}^*$ of the Lie coalgebra $\mathfrak{g}^*$. The Chevalley–Eilenberg coboundary operators (10.5.2) on this algebra read

$$\delta^k c^k(\varepsilon_0, \ldots, \varepsilon_k) = \sum_{i<j}^{k}(-1)^{i+j} c^k([\varepsilon_i, \varepsilon_j], \varepsilon_0, \ldots, \widehat{\varepsilon}_i, \ldots, \widehat{\varepsilon}_j, \ldots, \varepsilon_k).$$
$$(10.5.5)$$

In particular, we have

$$\delta^0 c^0(\varepsilon_0) = 0, \qquad c^0 \in \mathcal{K},$$
$$\delta^1 c^1(\varepsilon_0, \varepsilon_1) = -c^1([\varepsilon_0, \varepsilon_1]), \qquad c^1 \in \mathfrak{g}^*, \qquad (10.5.6)$$
$$\delta^2 c^2(\varepsilon_0, \varepsilon_1, \varepsilon_2) = -c^2([\varepsilon_0, \varepsilon_1], \varepsilon_2) + c^2([\varepsilon_0, \varepsilon_2], \varepsilon_1) - c^2([\varepsilon_1, \varepsilon_2], \varepsilon_0).$$

Cohomology $H^*(\mathfrak{g}, \mathcal{K})$ of the complex $C^*[\mathfrak{g}; \mathcal{K}]$ is called the *Chevalley–Eilenberg cohomology of a Lie algebra* $\mathcal{A}$.

For instance, let $\mathfrak{g}$ be the right Lie algebra of a finite-dimensional real Lie group G. There is a monomorphism of the Chevalley–Eilenberg complex $C^*[\mathfrak{g}; \mathbb{R}]$ onto the subcomplex of right-invariant exterior forms of the de Rham complex of exterior forms on G. In particular, the relation (10.5.6) is the Maurer–Cartan equation. The above mentioned monomorphism induces an isomorphism of the Chevalley–Eilenberg cohomology $H^*(\mathfrak{g}, \mathbb{R})$ of $\mathfrak{g}$ to the de Rham cohomology of G [46]. For instance, if G is semisimple, then

$$H^1(\mathfrak{g}, \mathbb{R}) = H^2(\mathfrak{g}, \mathbb{R}) = 0.$$

10.6 Differential calculus over a commutative ring

Let $\mathcal{A}$ be a commutative $\mathcal{K}$-ring. Since the derivation module $\partial\mathcal{A}$ of $\mathcal{A}$ is a Lie $\mathcal{K}$-algebra, one can associate to $\mathcal{A}$ the Chevalley–Eilenberg complex $C^*[\partial\mathcal{A}; \mathcal{A}]$. Its subcomplex of $\mathcal{A}$-multilinear maps is a differential graded algebra, also called the *differential calculus over* $\mathcal{A}$. By a gradation throughout this Section is meant the $\mathbb{N}$-gradation.

A *graded algebra* Ω^* over a commutative ring $\mathcal{K}$ is defined as a direct sum

$$\Omega^* = \bigoplus_k \Omega^k$$

of $\mathcal{K}$-modules Ω^k, provided with an associative multiplication law $\alpha \cdot \beta$, $\alpha, \beta \in \Omega^*$, such that $\alpha \cdot \beta \in \Omega^{|\alpha|+|\beta|}$, where $|\alpha|$ denotes the degree of an element $\alpha \in \Omega^{|\alpha|}$. In particular, it follows that Ω^0 is a (non-commutative) $\mathcal{K}$-algebra $\mathcal{A}$, while $\Omega^{k>0}$ are $\mathcal{A}$-bimodules and Ω^* is an $(\mathcal{A} - \mathcal{A})$-algebra. A graded algebra is said to be *graded commutative* if

$$\alpha \cdot \beta = (-1)^{|\alpha||\beta|} \beta \cdot \alpha, \qquad \alpha, \beta \in \Omega^*.$$

A graded algebra Ω^* is called the *differential graded algebra* or the *differential calculus* over $\mathcal{A}$ if it is a cochain complex of $\mathcal{K}$-modules

$$0 \to \mathcal{K} \longrightarrow \mathcal{A} \xrightarrow{\delta} \Omega^1 \xrightarrow{\delta} \cdots \Omega^k \xrightarrow{\delta} \cdots \qquad (10.6.1)$$

with respect to a coboundary operator δ which obeys the *graded Leibniz rule*

$$\delta(\alpha \cdot \beta) = \delta\alpha \cdot \beta + (-1)^{|\alpha|}\alpha \cdot \delta\beta. \qquad (10.6.2)$$

10.6. Differential calculus over a commutative ring

In particular, $\delta : \mathcal{A} \to \Omega^1$ is a Ω^1-valued derivation of a $\mathcal{K}$-algebra $\mathcal{A}$. The cochain complex (10.6.1) is said to be the *abstract de Rham complex* of the differential graded algebra (Ω^*, δ). Cohomology $H^*(\Omega^*)$ of the complex (10.6.1) is called the *abstract de Rham cohomology*. It is a graded algebra with respect to the *cup-product*

$$[\alpha] \smile [\beta] = [\alpha \cdot \beta], \qquad (10.6.3)$$

where $[\alpha]$ denotes the de Rham cohomology class of elements $\alpha \in \Omega^*$.

A morphism γ between two differential graded algebras (Ω^*, δ) and (Ω'^*, δ') is defined as a cochain morphism, i.e.,

$$\gamma \circ \delta = \gamma \circ \delta'.$$

It yields the corresponding morphism of the abstract de Rham cohomology groups of these algebras.

One considers the minimal differential graded subalgebra $\Omega^*\mathcal{A}$ of the differential graded algebra Ω^* which contains $\mathcal{A}$. Seen as an $(\mathcal{A}-\mathcal{A})$-algebra, it is generated by the elements δa, $a \in \mathcal{A}$, and consists of monomials

$$\alpha = a_0 \delta a_1 \cdots \delta a_k, \qquad a_i \in \mathcal{A},$$

whose product obeys the *juxtaposition rule*

$$(a_0 \delta a_1) \cdot (b_0 \delta b_1) = a_0 \delta(a_1 b_0) \cdot \delta b_1 - a_0 a_1 \delta b_0 \cdot \delta b_1$$

in accordance with the equality (10.6.2). The differential graded algebra $(\Omega^*\mathcal{A}, \delta)$ is called the *minimal differential calculus* over $\mathcal{A}$.

Let now $\mathcal{A}$ be a commutative $\mathcal{K}$-ring possessing a non-trivial Lie algebra $\partial\mathcal{A}$ of derivations. Let us consider the extended Chevalley–Eilenberg complex

$$0 \to \mathcal{K} \xrightarrow{\text{in}} C^*[\partial\mathcal{A}; \mathcal{A}]$$

of the Lie algebra $\partial\mathcal{A}$ with coefficients in the ring $\mathcal{A}$, regarded as a $\partial\mathcal{A}$-module. It is easily justified that this complex contains a subcomplex $\mathcal{O}^*[\partial\mathcal{A}]$ of $\mathcal{A}$-multilinear skew-symmetric maps

$$\phi^k : \overset{k}{\times} \partial\mathcal{A} \to \mathcal{A} \qquad (10.6.4)$$

with respect to the Chevalley–Eilenberg coboundary operator

$$d\phi(u_0, \ldots, u_k) = \sum_{i=0}^{k} (-1)^i u_i(\phi(u_0, \ldots, \widehat{u_i}, \ldots, u_k)) + \qquad (10.6.5)$$
$$\sum_{i<j} (-1)^{i+j} \phi([u_i, u_j], u_0, \ldots, \widehat{u_i}, \ldots, \widehat{u_j}, \ldots, u_k).$$

In particular, we have

$$(da)(u) = u(a), \quad a \in \mathcal{A}, \quad u \in \mathfrak{d}\mathcal{A},$$
$$(d\phi)(u_0, u_1) = u_0(\phi(u_1)) - u_1(\phi(u_0)) - \phi([u_0, u_1]), \quad \phi \in \mathcal{O}^1[\mathfrak{d}\mathcal{A}],$$
$$\mathcal{O}^0[\mathfrak{d}\mathcal{A}] = \mathcal{A},$$
$$\mathcal{O}^1[\mathfrak{d}\mathcal{A}] = \operatorname{Hom}_{\mathcal{A}}(\mathfrak{d}\mathcal{A}, \mathcal{A}) = \mathfrak{d}\mathcal{A}^*.$$

It follows that $d(1) = 0$ and d is a $\mathcal{O}^1[\mathfrak{d}\mathcal{A}]$-valued derivation of $\mathcal{A}$.

The graded module $\mathcal{O}^*[\mathfrak{d}\mathcal{A}]$ is provided with the structure of a graded $\mathcal{A}$-algebra with respect to the exterior product

$$\phi \wedge \phi'(u_1, ..., u_{r+s}) = \qquad (10.6.6)$$
$$\sum_{i_1<\cdots<i_r; j_1<\cdots<j_s} \operatorname{sgn}^{i_1\cdots i_r j_1\cdots j_s}_{1\cdots r+s} \phi(u_{i_1}, \ldots, u_{i_r}) \phi'(u_{j_1}, \ldots, u_{j_s}),$$
$$\phi \in \mathcal{O}^r[\mathfrak{d}\mathcal{A}], \quad \phi' \in \mathcal{O}^s[\mathfrak{d}\mathcal{A}], \quad u_k \in \mathfrak{d}\mathcal{A},$$

where $\operatorname{sgn}^{\cdots}_{\cdots}$ is the sign of a permutation. This product obeys the relations

$$d(\phi \wedge \phi') = d(\phi) \wedge \phi' + (-1)^{|\phi|} \phi \wedge d(\phi'), \quad \phi, \phi' \in \mathcal{O}^*[\mathfrak{d}\mathcal{A}],$$
$$\phi \wedge \phi' = (-1)^{|\phi||\phi'|} \phi' \wedge \phi. \qquad (10.6.7)$$

By virtue of the first one, $\mathcal{O}^*[\mathfrak{d}\mathcal{A}]$ is a differential graded $\mathcal{K}$-algebra, called the *Chevalley–Eilenberg differential calculus* over a $\mathcal{K}$-ring $\mathcal{A}$. The relation (10.6.7) shows that $\mathcal{O}^*[\mathfrak{d}\mathcal{A}]$ is a graded commutative algebra.

The *minimal Chevalley–Eilenberg differential calculus* $\mathcal{O}^*\mathcal{A}$ over a ring $\mathcal{A}$ consists of the monomials

$$a_0 da_1 \wedge \cdots \wedge da_k, \quad a_i \in \mathcal{A}.$$

Its complex

$$0 \to \mathcal{K} \longrightarrow \mathcal{A} \xrightarrow{d} \mathcal{O}^1\mathcal{A} \xrightarrow{d} \cdots \mathcal{O}^k\mathcal{A} \xrightarrow{d} \cdots \qquad (10.6.8)$$

is said to be the *de Rham complex* of a $\mathcal{K}$-ring $\mathcal{A}$, and its cohomology $H^*(\mathcal{A})$ is called the *de Rham cohomology* of $\mathcal{A}$. This cohomology is a graded commutative algebra with respect to the cup-product (10.6.3) induced by the exterior product $\wedge$ of elements of $\mathcal{O}^*\mathcal{A}$ so that

$$[\phi] \smile [\phi'] = [\phi \wedge \phi']. \qquad (10.6.9)$$

10.7 Sheaf cohomology

Throughout this Section, we follow the terminology of [22; 80].

A *sheaf* on a topological space X is a continuous fibre bundle $\pi : S \to X$ in modules over a commutative ring $\mathcal{K}$, where the surjection π is a local homeomorphism and fibres S_x, $x \in X$, called the *stalks*, are provided with the discrete topology. Global sections of a sheaf S make up a $\mathcal{K}$-module $S(X)$, called the *structure module* of S.

Any sheaf is generated by a presheaf. A *presheaf* $S_{\{U\}}$ on a topological space X is defined if a module S_U over a commutative ring $\mathcal{K}$ is assigned to every open subset $U \subset X$ ($S_\emptyset = 0$) and if, for any pair of open subsets $V \subset U$, there exists the restriction morphism $r_V^U : S_U \to S_V$ such that

$$r_U^U = \operatorname{Id} S_U, \qquad r_W^U = r_W^V r_V^U, \qquad W \subset V \subset U.$$

Every presheaf $S_{\{U\}}$ on a topological space X yields a sheaf on X whose stalk S_x at a point $x \in X$ is the direct limit of the modules S_U, $x \in U$, with respect to the restriction morphisms r_V^U. It means that, for each open neighborhood U of a point x, every element $s \in S_U$ determines an element $s_x \in S_x$, called the *germ* of s at x. Two elements $s \in S_U$ and $s' \in S_V$ belong to the same germ at x if and only if there exists an open neighborhood $W \subset U \cap V$ of x such that $r_W^U s = r_W^V s'$.

Example 10.7.1. Let $C_{\{U\}}^0$ be the presheaf of continuous real functions on a topological space X. Two such functions s and s' define the same germ s_x if they coincide on an open neighborhood of x. Hence, we obtain the sheaf C_X^0 of continuous functions on X. Similarly, the sheaf C_X^∞ of smooth functions on a smooth manifold X is defined. Let us also mention the presheaf of real functions which are constant on connected open subsets of X. It generates the *constant sheaf* on X denoted by $\mathbb{R}$.

Different presheaves may generate the same sheaf. Conversely, every sheaf S defines a presheaf $S(\{U\})$ of modules $S(U)$ of its local sections. It is called the *canonical presheaf* of the sheaf S. If a sheaf S is constructed from a presheaf $S_{\{U\}}$, there are natural module morphisms

$$S_U \ni s \to s(U) \in S(U), \qquad s(x) = s_x, \quad x \in U,$$

which are neither monomorphisms nor epimorphisms in general. For instance, it may happen that a non-zero presheaf defines a zero sheaf. The sheaf generated by the canonical presheaf of a sheaf S coincides with S.

A direct sum and a tensor product of presheaves (as families of modules) and sheaves (as fibre bundles in modules) are naturally defined. By virtue of Theorem 10.1.4, a direct sum (resp. a tensor product) of presheaves generates a direct sum (resp. a tensor product) of the corresponding sheaves.

Remark 10.7.1. In the terminology of [152], a sheaf is introduced as a presheaf which satisfies the following additional axioms.

(S1) Suppose that $U \subset X$ is an open subset and $\{U_\alpha\}$ is its open cover. If $s, s' \in S_U$ obey the condition
$$r^U_{U_\alpha}(s) = r^U_{U_\alpha}(s')$$
for all U_α, then $s = s'$.

(S2) Let U and $\{U_\alpha\}$ be as in previous item. Suppose that we are given a family of presheaf elements $\{s_\alpha \in S_{U_\alpha}\}$ such that
$$r^{U_\alpha}_{U_\alpha \cap U_\lambda}(s_\alpha) = r^{U_\lambda}_{U_\alpha \cap U_\lambda}(s_\lambda)$$
for all U_α, U_λ. Then there exists a presheaf element $s \in S_U$ such that $s_\alpha = r^U_{U_\alpha}(s)$.

Canonical presheaves are in one-to-one correspondence with presheaves obeying these axioms. For instance, presheaves of continuous, smooth and locally constant functions in Example 10.7.1 satisfy the axioms (S1) – (S2).

Remark 10.7.2. The notion of a sheaf can be extended to sets, but not to non-commutative groups. One can consider a presheaf of such groups, but it generates a sheaf of sets because a direct limit of non-commutative groups need not be a group. The first (but not higher) cohomology of X with coefficients in this sheaf is defined [80].

There is a useful construction of a sheaf on a topological space X from sheaves on open subsets which make up a cover of X.

Theorem 10.7.1. *Let $\{U_\zeta\}$ be an open cover of a topological space X and S_ζ a sheaf on U_ζ for every U_ζ. Let us suppose that, if $U_\zeta \cap U_\xi \neq \emptyset$, there is a sheaf isomorphism*
$$\varrho_{\zeta\xi} : S_\xi|_{U_\zeta \cap U_\xi} \to S_\zeta|_{U_\zeta \cap U_\xi}$$
and, for every triple $(U_\zeta, U_\xi, U_\iota)$, these isomorphisms fulfil the cocycle condition
$$\varrho_{\xi\zeta} \circ \varrho_{\zeta\iota}(S_\iota|_{U_\zeta \cap U_\xi \cap U_\iota}) = \varrho_{\xi\iota}(S_\iota|_{U_\zeta \cap U_\xi \cap U_\iota}).$$
Then there exists a sheaf S on X together with the sheaf isomorphisms
$$\phi_\zeta : S|_{U_\zeta} \to S_\zeta$$
such that
$$\phi_\zeta|_{U_\zeta \cap U_\xi} = \varrho_{\zeta\xi} \circ \phi_\xi|_{U_\zeta \cap U_\xi}.$$

10.7. Sheaf cohomology

A *morphism of a presheaf* $S_{\{U\}}$ to a presheaf $S'_{\{U\}}$ on the same topological space X is defined as a set of module morphisms $\gamma_U : S_U \to S'_U$ which commute with restriction morphisms. A morphism of presheaves yields a *morphism of sheaves* generated by these presheaves. This is a bundle morphism over X such that $\gamma_x : S_x \to S'_x$ is the direct limit of morphisms γ_U, $x \in U$. Conversely, any morphism of sheaves $S \to S'$ on a topological space X yields a morphism of canonical presheaves of local sections of these sheaves. Let $\mathrm{Hom}\,(S|_U, S'|_U)$ be the commutative group of sheaf morphisms $S|_U \to S'|_U$ for any open subset $U \subset X$. These groups are assembled into a presheaf, and define the sheaf $\mathrm{Hom}\,(S, S')$ on X. There is a monomorphism

$$\mathrm{Hom}\,(S, S')(U) \to \mathrm{Hom}\,(S(U), S'(U)), \qquad (10.7.1)$$

which need not be an isomorphism.

By virtue of Theorem 10.1.6, if a presheaf morphism is a monomorphism or an epimorphism, so is the corresponding sheaf morphism. Furthermore, the following holds.

Theorem 10.7.2. *A short exact sequence*

$$0 \to S'_{\{U\}} \to S_{\{U\}} \to S''_{\{U\}} \to 0 \qquad (10.7.2)$$

of presheaves on the same topological space yields the short exact sequence of sheaves generated by these presheaves

$$0 \to S' \to S \to S'' \to 0, \qquad (10.7.3)$$

where the factor sheaf $S'' = S/S'$ is isomorphic to that generated by the factor presheaf

$$S''_{\{U\}} = S_{\{U\}}/S'_{\{U\}}.$$

If the exact sequence of presheaves (10.7.2) is split, i.e.,

$$S_{\{U\}} \cong S'_{\{U\}} \oplus S''_{\{U\}},$$

the corresponding splitting

$$S \cong S' \oplus S''$$

of the exact sequence of sheaves (10.7.3) holds.

The converse is more intricate. A sheaf morphism induces a morphism of the corresponding canonical presheaves. If $S \to S'$ is a monomorphism,

$$S(\{U\}) \to S'(\{U\})$$

also is a monomorphism. However, if $S \to S'$ is an epimorphism,
$$S(\{U\}) \to S'(\{U\})$$
need not be so. Therefore, the short exact sequence (10.7.3) of sheaves yields the exact sequence of the canonical presheaves
$$0 \to S'(\{U\}) \to S(\{U\}) \to S''(\{U\}), \qquad (10.7.4)$$
where $S(\{U\}) \to S''(\{U\})$ is not necessarily an epimorphism. At the same time, there is the short exact sequence of presheaves
$$0 \to S'(\{U\}) \to S(\{U\}) \to S''_{\{U\}} \to 0, \qquad (10.7.5)$$
where the factor presheaf
$$S''_{\{U\}} = S(\{U\})/S'(\{U\})$$
generates the factor sheaf $S'' = S/S'$, but need not be its canonical presheaf.

Let us turn now to sheaf cohomology. We follow its definition in [80]. In the case of paracompact topological spaces, it coincides with a different definition of sheaf cohomology based on the canonical flabby resolution (Remark 10.7.5). Note that only proper covers are considered.

Let $S_{\{U\}}$ be a presheaf of modules on a topological space X, and let $\mathfrak{U} = \{U_i\}_{i \in I}$ be an open cover of X. One constructs a cochain complex where a p-cochain is defined as a function s^p which associates an element
$$s^p(i_0, \ldots, i_p) \in S_{U_{i_0} \cap \cdots \cap U_{i_p}} \qquad (10.7.6)$$
to each $(p+1)$-tuple $(i_0, \ldots, i_p)$ of indices in I. These p-cochains are assembled into a module $C^p(\mathfrak{U}, S_{\{U\}})$. Let us introduce the coboundary operator
$$\delta^p : C^p(\mathfrak{U}, S_{\{U\}}) \to C^{p+1}(\mathfrak{U}, S_{\{U\}}),$$
$$\delta^p s^p(i_0, \ldots, i_{p+1}) = \sum_{k=0}^{p+1} (-1)^k r_W^{W_k} s^p(i_0, \ldots, \widehat{i_k}, \ldots, i_{p+1}), \qquad (10.7.7)$$
$$W = U_{i_0} \cap \ldots \cap U_{i_{p+1}}, \qquad W_k = U_{i_0} \cap \cdots \cap \widehat{U}_{i_k} \cap \cdots \cap U_{i_{p+1}}.$$
One can easily check that $\delta^{p+1} \circ \delta^p = 0$. Thus, we obtain the cochain complex of modules
$$0 \to C^0(\mathfrak{U}, S_{\{U\}}) \xrightarrow{\delta^0} \cdots C^p(\mathfrak{U}, S_{\{U\}}) \xrightarrow{\delta^p} C^{p+1}(\mathfrak{U}, S_{\{U\}}) \longrightarrow \cdots .$$
$$(10.7.8)$$
Its cohomology groups
$$H^p(\mathfrak{U}; S_{\{U\}}) = \operatorname{Ker} \delta^p / \operatorname{Im} \delta^{p-1}$$
are modules. Of course, they depend on an open cover $\mathfrak{U}$ of X.

10.7. Sheaf cohomology

Let $\mathfrak{U}'$ be a refinement of the cover $\mathfrak{U}$. Then there is a morphism of cohomology groups

$$H^*(\mathfrak{U}; S_{\{U\}}) \to H^*(\mathfrak{U}'; S_{\{U\}}). \qquad (10.7.9)$$

Let us take the direct limit of cohomology groups $H^*(\mathfrak{U}; S_{\{U\}})$ with respect to these morphisms, where $\mathfrak{U}$ runs through all open covers of X. This limit $H^*(X; S_{\{U\}})$ is called the cohomology of X with coefficients in the presheaf $S_{\{U\}}$.

Remark 10.7.3. The cohomology $H^*(X; S_{\{U\}})$ consists of elements of the cohomology groups $H^*(\mathfrak{U}; S_{\{U\}})$ modulo the morphisms (10.7.9). It follows that any cocycle s^p (10.7.6) is a representative of some element of $H^p(X; S_{\{U\}})$.

Let S be a sheaf on a topological space X. *Cohomology of X with coefficients in S* or, simply, *sheaf cohomology* of X is defined as cohomology

$$H^*(X; S) = H^*(X; S(\{U\}))$$

with coefficients in the canonical presheaf $S(\{U\})$ of the sheaf S.

In this case, a p-cochain $s^p \in C^p(\mathfrak{U}, S(\{U\}))$ is a collection

$$s^p = \{s^p(i_0, \ldots, i_p)\}$$

of local sections $s^p(i_0, \ldots, i_p)$ of the sheaf S over $U_{i_0} \cap \cdots \cap U_{i_p}$ for each $(p+1)$-tuple $(U_{i_0}, \ldots, U_{i_p})$ of elements of the cover $\mathfrak{U}$. The coboundary operator (10.7.7) reads

$$\delta^p s^p(i_0, \ldots, i_{p+1}) = \sum_{k=0}^{p+1} (-1)^k s^p(i_0, \ldots, \widehat{i_k}, \ldots, i_{p+1})|_{U_{i_0} \cap \cdots \cap U_{i_{p+1}}}.$$

For instance, we have

$$\delta^0 s^0(i, j) = [s^0(j) - s^0(i)]|_{U_i \cap U_j}, \qquad (10.7.10)$$

$$\delta^1 s^1(i, j, k) = [s^1(j, k) - s^1(i, k) + s^1(i, j)]|_{U_i \cap U_j \cap U_k}. \qquad (10.7.11)$$

A glance at the expression (10.7.10) shows that a zero-cocycle is a collection $s = \{s(i)\}_I$ of local sections of the sheaf S over $U_i \in \mathfrak{U}$ such that $s(i) = s(j)$ on $U_i \cap U_j$. It follows from the axiom (S2) in Remark 10.7.1 that s is a global section of the sheaf S, while each $s(i)$ is its restriction $s|_{U_i}$ to U_i. Consequently, the cohomology group $H^0(\mathfrak{U}; S(\{U\}))$ is isomorphic to the structure module $S(X)$ of global sections of the sheaf S. A one-cocycle is a collection $\{s(i, j)\}$ of local sections of the sheaf S over overlaps $U_i \cap U_j$ which satisfy the *cocycle condition*

$$[s(j, k) - s(i, k) + s(i, j)]|_{U_i \cap U_j \cap U_k} = 0. \qquad (10.7.12)$$

If X is a paracompact space, the study of its sheaf cohomology is essentially simplified due to the following fact.

Theorem 10.7.3. *Cohomology of a paracompact space X with coefficients in a sheaf S coincides with cohomology of X with coefficients in any presheaf generating the sheaf S.*

Remark 10.7.4. We follow the definition of a *paracompact topological space* in [80] as a Hausdorff space such that any its open cover admits a *locally finite* open refinement, i.e., any point has an open neighborhood which intersects only a finite number of elements of this refinement. A topological space X is paracompact if and only if any cover $\{U_\xi\}$ of X admits a subordinate *partition of unity* $\{f_\xi\}$, i.e.:
  (i) f_ξ are real positive continuous functions on X;
  (ii) $\operatorname{supp} f_\xi \subset U_\xi$;
  (iii) each point $x \in X$ has an open neighborhood which intersects only a finite number of the sets $\operatorname{supp} f_\xi$;
  (iv) $\sum_\xi f_\xi(x) = 1$ for all $x \in X$.

The key point of the analysis of sheaf cohomology is that short exact sequences of sheaves yield long exact sequences of their cohomology groups.

Let $S_{\{U\}}$ and $S'_{\{U\}}$ be presheaves on the same topological space X. It is readily observed that, given an open cover $\mathfrak{U}$ of X, any morphism $S_{\{U\}} \to S'_{\{U\}}$ yields a cochain morphism of complexes

$$C^*(\mathfrak{U}, S_{\{U\}}) \to C^*(\mathfrak{U}, S'_{\{U\}})$$

and the corresponding morphism

$$H^*(\mathfrak{U}; S_{\{U\}}) \to H^*(\mathfrak{U}; S'_{\{U\}})$$

of cohomology groups of these complexes. Passing to the direct limit through all refinements of $\mathfrak{U}$, we come to a morphism of cohomology groups

$$H^*(X; S_{\{U\}}) \to H^*(X; S'_{\{U\}})$$

of X with coefficients in the presheaves $S_{\{U\}}$ and $S'_{\{U\}}$. In particular, any sheaf morphism $S \to S'$ yields a morphism of canonical presheaves

$$S(\{U\}) \to S'(\{U\})$$

and the corresponding cohomology morphism

$$H^*(X; S) \to H^*(X; S').$$

10.7. Sheaf cohomology

By virtue of Theorems 10.3.1 and 10.3.2, every short exact sequence

$$0 \to S'_{\{U\}} \longrightarrow S_{\{U\}} \longrightarrow S''_{\{U\}} \to 0 \qquad (10.7.13)$$

of presheaves on the same topological space X and the corresponding exact sequence of complexes (10.7.8) yield the long exact sequence

$$0 \to H^0(X; S'_{\{U\}}) \longrightarrow H^0(X; S_{\{U\}}) \longrightarrow H^0(X; S''_{\{U\}}) \longrightarrow \quad (10.7.14)$$
$$H^1(X; S'_{\{U\}}) \longrightarrow \cdots H^p(X; S'_{\{U\}}) \longrightarrow H^p(X; S_{\{U\}}) \longrightarrow$$
$$H^p(X; S''_{\{U\}}) \longrightarrow H^{p+1}(X; S'_{\{U\}}) \longrightarrow \cdots$$

of the cohomology groups of X with coefficients in these presheaves. This result however is not extended to an exact sequence of sheaves, unless X is a paracompact space. Let

$$0 \to S' \longrightarrow S \longrightarrow S'' \to 0 \qquad (10.7.15)$$

be a short exact sequence of sheaves on X. It yields the short exact sequence of presheaves (10.7.5) where the presheaf $S''_{\{U\}}$ generates the sheaf S''. If X is paracompact,

$$H^*(X; S''_{\{U\}}) = H^*(X; S'')$$

in accordance with Theorem 10.7.3, and we have the exact sequence of sheaf cohomology

$$0 \to H^0(X; S') \longrightarrow H^0(X; S) \longrightarrow H^0(X; S'') \longrightarrow \quad (10.7.16)$$
$$H^1(X; S') \longrightarrow \cdots H^p(X; S') \longrightarrow H^p(X; S) \longrightarrow$$
$$H^p(X; S'') \longrightarrow H^{p+1}(X; S') \longrightarrow \cdots .$$

Let us turn now to the abstract de Rham theorem which provides a powerful tool of studying algebraic systems on paracompact spaces.

Let us consider an exact sequence of sheaves

$$0 \to S \xrightarrow{h} S_0 \xrightarrow{h^0} S_1 \xrightarrow{h^1} \cdots S_p \xrightarrow{h^p} \cdots . \qquad (10.7.17)$$

It is said to be a *resolution* of the sheaf S if each sheaf $S_{p\geq 0}$ is acyclic, i.e., its cohomology groups $H^{k>0}(X; S_p)$ vanish.

Any exact sequence of sheaves (10.7.17) yields the sequence of their structure modules

$$0 \to S(X) \xrightarrow{h_*} S_0(X) \xrightarrow{h^0_*} S_1(X) \xrightarrow{h^1_*}, \cdots S_p(X) \xrightarrow{h^p_*} \cdots \qquad (10.7.18)$$

which is always exact at terms $S(X)$ and $S_0(X)$ (see the exact sequence (10.7.4)). The sequence (10.7.18) is a cochain complex because

$$h^{p+1}_* \circ h^p_* = 0.$$

If X is a paracompact space and the exact sequence (10.7.17) is a resolution of S, the forthcoming abstract de Rham theorem establishes an isomorphism of cohomology of the complex (10.7.18) to cohomology of X with coefficients in the sheaf S [80].

Theorem 10.7.4. *Given a resolution (10.7.17) of a sheaf S on a paracompact topological space X and the induced complex (10.7.18), there are isomorphisms*

$$H^0(X;S) = \operatorname{Ker} h^0_*, \qquad H^q(X;S) = \operatorname{Ker} h^q_* / \operatorname{Im} h^{q-1}_*, \qquad q > 0. \tag{10.7.19}$$

We refer to the following minor modification of Theorem 10.7.4 [56; 150].

Theorem 10.7.5. *Let*

$$0 \to S \xrightarrow{h} S_0 \xrightarrow{h^0} S_1 \xrightarrow{h^1} \cdots \xrightarrow{h^{p-1}} S_p \xrightarrow{h^p} S_{p+1}, \qquad p > 1, \tag{10.7.20}$$

be an exact sequence of sheaves on a paracompact topological space X, where the sheaves S_q, $0 \leq q < p$, are acyclic, and let

$$0 \to S(X) \xrightarrow{h_*} S_0(X) \xrightarrow{h^0_*} S_1(X) \xrightarrow{h^1_*} \cdots \tag{10.7.21}$$
$$\xrightarrow{h^{p-1}_*} S_p(X) \xrightarrow{h^p_*} S_{p+1}(X)$$

be the corresponding cochain complex of structure modules of these sheaves. Then the isomorphisms (10.7.19) hold for $0 \leq q \leq p$.

Any sheaf on a topological space admits the canonical resolution by flabby sheaves as follows. A sheaf S on a topological space X is called flabby (or flasque in the terminology of [152]), if the restriction morphism $S(X) \to S(U)$ is an epimorphism for any open $U \subset X$, i.e., if any local section of the sheaf S can be extended to a global section. A flabby sheaf is acyclic. Indeed, given an arbitrary cover $\mathfrak{U}$ of X, let us consider the complex $C^*(\mathfrak{U}, S(\{U\}))$ (10.7.8) for its canonical presheaf $S(\{U\})$. Since S is flabby, one can define a morphism

$$h : C^p(\mathfrak{U}, S(\{U\})) \to C^{p-1}(\mathfrak{U}, S(\{U\})), \qquad p > 0,$$
$$hs^p(i_0, \ldots, i_{p-1}) = j^* s^p(i_0, \ldots, i_{p-1}, j), \tag{10.7.22}$$

where U_j is a fixed element of the cover $\mathfrak{U}$ and $j^* s^p$ is an extension of $s^p(i_0, \ldots, i_{p-1}, j)$ onto $U_{i_0} \cap \cdots \cap U_{i_{p-1}}$. A direct verification shows that

10.7. Sheaf cohomology

h (10.7.22) is a homotopy operator for the complex $C^*(\mathfrak{U}, S(\{U\}))$ and, consequently,
$$H^{p>0}(\mathfrak{U}; S(\{U\})) = 0.$$

Given an arbitrary sheaf S on a topological space X, let $S_F^0(\{U\})$ denote the presheaf of all (not-necessarily continuous) sections of the sheaf S. It generates a sheaf S_F^0 on X, and coincides with the canonical presheaf of this sheaf. There are the natural monomorphisms
$$S(\{U\}) \to S_F^0(\{U\}) \qquad S \to S_F^0.$$
It is readily observed that the sheaf S_F^0 is flabby. Let us take the quotient S_F^0/S and construct the flabby sheaf
$$S_F^1 = (S_F^0/S)_F^0.$$
Continuing the procedure, we obtain the exact sequence of sheaves
$$0 \to S \longrightarrow S_F^0 \longrightarrow S_F^1 \longrightarrow \cdots , \tag{10.7.23}$$
which is a resolution of S since all sheaves are flabby and, consequently, acyclic. It is called the *canonical flabby resolution* of the sheaf S. The exact sequence of sheaves (10.7.23) yields the complex of structure modules of these sheaves
$$0 \to S(X) \longrightarrow S_F^0(X) \longrightarrow S_F^1(X) \longrightarrow \cdots . \tag{10.7.24}$$
If X is paracompact, the cohomology of X with coefficients in the sheaf S coincides with that of the complex (10.7.24) by virtue of Theorem 10.7.4.

Remark 10.7.5. An important peculiarity of flabby sheaves is that a short exact sequence of flabby sheaves on an arbitrary topological space provides the short exact sequence of their structure modules. Therefore, there is a different definition of sheaf cohomology. Cohomology of a topological space X with coefficients in a sheaf S is defined directly as cohomology of the complex (10.7.24) [22]. For a paracompact space, this definition coincides with the above mentioned one due to Theorem 10.7.4.

In the sequel, we also refer to a *fine resolution* of sheaves, i.e., a resolution by fine sheaves.

A sheaf S on a paracompact space X is called *fine* if, for each locally finite open cover $\mathfrak{U} = \{U_i\}_{i \in I}$ of X, there exists a system $\{h_i\}$ of endomorphisms $h_i : S \to S$ such that:

(i) there is a closed subset $V_i \subset U_i$ and $h_i(S_x) = 0$ if $x \notin V_i$,

(ii) $\sum_{i \in I} h_i$ is the identity map of S.

Theorem 10.7.6. *A fine sheaf on a paracompact space is acyclic.*

There is the following important example of fine sheaves.

Theorem 10.7.7. *Let X be a paracompact topological space which admits a partition of unity performed by elements of the structure module $\mathfrak{A}(X)$ of some sheaf $\mathfrak{A}$ of real functions on X. Then any sheaf S of $\mathfrak{A}$-modules on X, including $\mathfrak{A}$ itself, is fine.*

In particular, the sheaf C_X^0 of continuous functions on a paracompact topological space is fine, and so is any sheaf of C_X^0-modules.

We complete our exposition of sheaf cohomology with the following useful theorem [10].

Theorem 10.7.8. *Let $f : X \to X'$ be a continuous map and S a sheaf on X. Let either f be a closed immersion or every point $x' \in X'$ have a base of open neighborhoods $\{U\}$ such that the sheaves $S \vert_{f^{-1}(U)}$ are acyclic. Then the cohomology groups $H^*(X; S)$ and $H^*(X'; f_*S)$ are isomorphic.*

10.8 Local-ringed spaces

Local-ringed spaces are sheafs of local rings. For instance, smooth manifolds, represented by sheaves of real smooth functions, make up a subcategory of the category of local-ringed spaces (Section 10.9).

A sheaf $\mathfrak{R}$ on a topological space X is said to be a *ringed space* if its stalk $\mathfrak{R}_x$ at each point $x \in X$ is a real commutative ring [152]. A ringed space is often denoted by a pair $(X, \mathfrak{R})$ of a topological space X and a sheaf $\mathfrak{R}$ of rings on X. They are called the *body* and the *structure sheaf* of a ringed space, respectively.

A ringed space is said to be a *local-ringed space* (a *geometric space* in the terminology of [152]) if it is a sheaf of local rings.

For instance, the sheaf C_X^0 of continuous real functions on a topological space X is a local-ringed space. Its stalk C_x^0, $x \in X$, contains the unique maximal ideal of germs of functions vanishing at x.

Morphisms of local-ringed spaces are defined as those of sheaves on different topological spaces as follows.

Let $\varphi : X \to X'$ be a continuous map. Given a sheaf S on X, its *direct image* φ_*S on X' is generated by the presheaf of assignments

$$X' \supset U' \to S(\varphi^{-1}(U'))$$

for any open subset $U' \subset X'$. Conversely, given a sheaf S' on X', its *inverse image* φ^*S' on X is defined as the pull-back onto X of the continuous fibre

10.8. Local-ringed spaces

bundle S' over X', i.e., $\varphi^* S'_x = S_{\varphi(x)}$. This sheaf is generated by the presheaf which associates to any open $V \subset X$ the direct limit of modules $S'(U)$ over all open subsets $U \subset X'$ such that $V \subset f^{-1}(U)$.

Remark 10.8.1. Let $i : X \to X'$ be a closed subspace of X'. Then $i_* S$ is a unique sheaf on X' such that

$$i_* S|_X = S, \qquad i_* S|_{X' \setminus X} = 0.$$

Indeed, if $x' \in X \subset X'$, then

$$i_* S(U') = S(U' \cap X)$$

for any open neighborhood U of this point. If $x' \notin X$, there exists its neighborhood U' such that $U' \cap X$ is empty, i.e., $i_* S(U') = 0$. The sheaf $i_* S$ is called the *trivial extension* of the sheaf S.

By a *morphism of ringed spaces*

$$(X, \mathfrak{R}) \to (X', \mathfrak{R}')$$

is meant a pair $(\varphi, \widehat{\varphi})$ of a continuous map $\varphi : X \to X'$ and a sheaf morphism $\widehat{\varphi} : \mathfrak{R}' \to \varphi_* \mathfrak{R}$ or, equivalently, a sheaf morphisms $\varphi^* \mathfrak{R}' \to \mathfrak{R}$. Restricted to each stalk, a sheaf morphism Φ is assumed to be a ring homomorphism. A morphism of ringed spaces is said to be:
- a monomorphism if φ is an injection and Φ is an epimorphism,
- an epimorphism if φ is a surjection, while Φ is a monomorphism.

Let $(X, \mathfrak{R})$ be a local-ringed space. By a *sheaf $\mathfrak{dR}$ of derivations* of the sheaf $\mathfrak{R}$ is meant a subsheaf of endomorphisms of $\mathfrak{R}$ such that any section u of $\mathfrak{dR}$ over an open subset $U \subset X$ is a derivation of the real ring $\mathfrak{R}(U)$. It should be emphasized that, since the monomorphism (10.7.1) is not necessarily an isomorphism, a derivation of the ring $\mathfrak{R}(U)$ need not be a section of the sheaf $\mathfrak{dR}|_U$. Namely, it may happen that, given open sets $U' \subset U$, there is no restriction morphism

$$\mathfrak{d}(\mathfrak{R}(U)) \to \mathfrak{d}(\mathfrak{R}(U')).$$

Given a local-ringed space $(X, \mathfrak{R})$, a sheaf P on X is called a *sheaf of $\mathfrak{R}$-modules* if every stalk P_x, $x \in X$, is an $\mathfrak{R}_x$-module or, equivalently, if $P(U)$ is an $\mathfrak{R}(U)$-module for any open subset $U \subset X$. A sheaf of $\mathfrak{R}$-modules P is said to be *locally free* if there exists an open neighborhood U of every point $x \in X$ such that $P(U)$ is a free $\mathfrak{R}(U)$-module. If all these free modules are of finite rank (resp. of the same finite rank), one says that P is of *finite type* (resp. of *constant rank*). The structure module of a locally free sheaf is called a *locally free module*.

The following is a generalization of Theorem 10.7.7 [80].

Theorem 10.8.1. *Let X be a paracompact space which admits a partition of unity by elements of the structure module $S(X)$ of some sheaf S of real functions on X. Let P be a sheaf of S-modules. Then P is fine and, consequently, acyclic.*

10.9 Cohomology of smooth manifolds

It should be emphasized that cohomology of smooth manifolds are their *topological invariants* in the sense that they are the same for homotopic topological spaces. The following two facts enable one to define them for smooth manifolds.

(i) Smooth manifolds have the homotopy type of CW-complexes [35].

(ii) Assumed to be paracompact, a smooth manifold X admits a partition of unity performed by smooth real functions. It follows that the sheaf C_X^∞ of smooth real functions on X is fine, and so is any sheaf of C_X^∞-modules, e.g., the sheaves of sections of smooth vector bundles over X.

Similarly to the sheaf C_X^0 of continuous functions, the sheaf C_X^∞ of smooth real functions on a smooth manifold X is a local-ringed spaces. Its stalk C_x^∞ at a point $x \in X$ has a unique maximal ideal μ_x of germs of smooth functions vanishing at x. Though the sheaf C_X^∞ is defined on a topological space X, it fixes a unique smooth manifold structure on X as follows.

Theorem 10.9.1. *Let X be a paracompact topological space and $(X, \mathfrak{R})$ a local-ringed space. Let X admit an open cover $\{U_i\}$ such that the sheaf $\mathfrak{R}$ restricted to each U_i is isomorphic to the local-ringed space $(\mathbb{R}^n, C_{\mathbb{R}^n}^\infty)$. Then X is an n-dimensional smooth manifold together with a natural isomorphism of local-ringed spaces $(X, \mathfrak{R})$ and (X, C_X^∞).*

One can think of this result as being an alternative definition of smooth real manifolds in terms of local-ringed spaces. A smooth manifold X also is algebraically reproduced as a certain subspace of the spectrum of the real ring $C^\infty(X)$ of smooth real functions on X as follows [6].

Let $\operatorname{Spec} \mathcal{A}$ be the set of prime ideals of a commutative ring $\mathcal{A}$. It is called the *spectrum* of $\mathcal{A}$. Let us assign to each ideal $\mathcal{I}$ of $\mathcal{A}$ the set

$$V(\mathcal{I}) = \{\, x \in \operatorname{Spec} \mathcal{A} : \mathcal{I} \subset x \,\}. \tag{10.9.1}$$

10.9. Cohomology of smooth manifolds

These sets possess the properties

$$V(\{0\}) = \operatorname{Spec} \mathcal{A}, \qquad V(\mathcal{A}) = \emptyset,$$
$$\bigcap_i V(\mathcal{I}_i) = V(\oplus_i \mathcal{I}_i), \qquad V(\mathcal{I}) \cup V(\mathcal{I}') = V(\mathcal{I}\mathcal{I}').$$

In view of these properties, one can regard the sets (10.9.1) as closed sets of some topology on the spectrum $\operatorname{Spec} \mathcal{A}$. It is called the *Zariski topology*. A base for this topology consists of the closed sets

$$U(a) = \{x \in \operatorname{Spec} \mathcal{A} : a \notin x\} = \operatorname{Spec} \mathcal{A} \setminus V(\mathcal{A}a) \qquad (10.9.2)$$

as a runs through $\mathcal{A}$. In particular, the set of closed points is the set $\operatorname{Specm} \mathcal{A}$ of all maximal ideals of $\mathcal{A}$. Endowed with the relative Zariski topology, it is called the *maximal spectrum* of $\mathcal{A}$. A ring morphism $\zeta : \mathcal{A} \to \mathcal{A}'$ induces the continuous map

$$\zeta^\natural : \operatorname{Spec} \mathcal{A}' \ni x' \to \zeta^{-1}(x') \in \operatorname{Spec} \mathcal{A}. \qquad (10.9.3)$$

Let $\mathcal{A}$ be a real commutative ring. The *real spectrum* of $\mathcal{A}$ is the subspace $\operatorname{Spec}_\mathbb{R} \mathcal{A} \subset \operatorname{Specm} \mathcal{A}$ of the maximal ideals $\mathcal{I}$ such that the quotients $\mathcal{A}/\mathcal{I}$ are isomorphic to $\mathbb{R}$. It is endowed with the relative Zariski topology. There is the bijection between the set of $\mathbb{R}$-algebra morphisms of $\mathcal{A}$ to the field $\mathbb{R}$ and the real spectrum of $\mathcal{A}$, namely,

$$\operatorname{Hom}_\mathbb{R}(\mathcal{A}, \mathbb{R}) \ni \phi \to \operatorname{Ker} \phi \in \operatorname{Spec}_\mathbb{R} \mathcal{A},$$
$$\operatorname{Spec}_\mathbb{R} \mathcal{A} \ni x \to \pi_x \in \operatorname{Hom}_\mathbb{R}(\mathcal{A}, \mathbb{R}), \qquad \pi_x : \mathcal{A} \to \mathcal{A}/x \cong \mathbb{R}.$$

Any element $a \in \mathcal{A}$ induces a real function

$$f_a : \operatorname{Spec}_\mathbb{R} \mathcal{A} \ni x \to \pi_x(a)$$

on the real spectrum $\operatorname{Spec}_\mathbb{R} \mathcal{A}$. This function need not be continuous with respect to the Zariski topology, but one can provide $\operatorname{Spec}_\mathbb{R} \mathcal{A}$ with another topology, called the *Gel'fand topology*, which is the coarsest topology which makes all such functions continuous.

Theorem 10.9.2. *If $\mathcal{A} = C^\infty(X)$ is the real ring of smooth real functions on a manifold X, the Zariski and Gel'fand topologies on its real spectrum $\operatorname{Spec}_\mathbb{R} C^\infty(X)$ coincide with each other. Therefore, there is a homeomorphism*

$$\chi_X : X \ni x \to \mu_x \in \operatorname{Spec}_\mathbb{R} C^\infty(X). \qquad (10.9.4)$$

Any smooth map $\gamma : X \to X'$ induces the $\mathbb{R}$-ring morphism

$$\gamma^* : C^\infty(X') \to C^\infty(X)$$

which associates to a function f on X' the pull-back function

$$\gamma^* f = f \circ \gamma$$

on X. Conversely, each $\mathbb{R}$-ring morphism

$$\zeta : C^\infty(X') \to C^\infty(X)$$

yields the continuous map $\zeta^\natural$ (10.9.3) which sends

$$\mathrm{Spec}_\mathbb{R} C^\infty(X) \subset \mathrm{Spec}\, C^\infty(X)$$

to

$$\mathrm{Spec}_\mathbb{R} C^\infty(X') \subset \mathrm{Spec}\, C^\infty(X')$$

so that the induced map

$$\chi_{X'}^{-1} \zeta^\natural \circ \chi_X : X \to X'$$

is smooth. Thus, there is one-to-one correspondence between smooth manifold morphisms $X \to X'$ and the $\mathbb{R}$-ring morphisms $C^\infty(X') \to C^\infty(X)$.

Remark 10.9.1. Let $X \times X'$ be a manifold product. The ring $C^\infty(X \times X')$ is constructed from the rings $C^\infty(X)$ and $C^\infty(X')$ as follows. Throughout the book, by a topology on the ring $C^\infty(X)$ is meant the topology of compact convergence for all derivatives [129]. With this topology, $C^\infty(X)$ is a *Fréchet ring*, i.e., a complete metrizable locally convex topological vector space. There is an isomorphism of Fréchet rings

$$C^\infty(X) \widehat{\otimes} C^\infty(X') = C^\infty(X \times X'), \qquad (10.9.5)$$

where the left-hand side, called the *topological tensor product*, is the completion of $C^\infty(X) \otimes C^\infty(X')$ with respect to Grothendieck's topology, defined as follows. If E_1 and E_2 are locally convex topological vector spaces, *Grothendieck's topology* is the finest locally convex topology on $E_1 \otimes E_2$ such that the canonical mapping of $E_1 \times E_2$ to $E_1 \otimes E_2$ is continuous [129]. Furthermore, for any two open subsets $U \subset X$ and $U' \subset X'$, let us consider the topological tensor product of rings $C^\infty(U) \widehat{\otimes} C^\infty(U')$. These tensor products define a local-ringed space $(X \times X', C_X^\infty \widehat{\otimes} C_{X'}^\infty)$. Due to the isomorphism (10.9.5) written for all $U \subset X$ and $U' \subset X'$, we obtain the sheaf isomorphism

$$C_X^\infty \widehat{\otimes} C_{X'}^\infty = C_{X \times X'}^\infty. \qquad (10.9.6)$$

10.9. Cohomology of smooth manifolds

Since a smooth manifold admits a partition of unity by smooth functions, it follows from Theorem 10.8.1 that any sheaf of C_X^∞-modules on X is fine and, consequently, acyclic.

For instance, let $Y \to X$ be a smooth vector bundle. The germs of its sections make up a sheaf of C_X^∞-modules, called the *structure sheaf S_Y* of a vector bundle $Y \to X$. The sheaf S_Y is fine. The structure module of this sheaf coincides with the structure module $Y(X)$ of global sections of a vector bundle $Y \to X$. The following *Serre–Swan theorem* shows that these modules exhaust all projective modules of finite rank over $C^\infty(X)$. Originally proved for bundles over a compact base X, this theorem has been extended to an arbitrary X [60; 127].

Theorem 10.9.3. *Let X be a smooth manifold. A $C^\infty(X)$-module P is isomorphic to the structure module of a smooth vector bundle over X if and only if it is a projective module of finite rank.*

Proof. Let
$$\{(U_\xi, \psi_\xi), \varrho_{\xi\zeta}\}, \qquad \xi, \zeta = 1, \ldots, k,$$
be a finite atlas of a smooth vector bundle $Y \to X$ of fibre dimension m in accordance with Theorem 1.1.9. Given a smooth partition of unity $\{f_\xi\}$ subordinate to the cover $\{U_\xi\}$, let us set
$$l_\xi = f_\xi(f_1^2 + \cdots + f_k^2)^{-1/2}.$$
It is readily observed that $\{l_\xi^2\}$ also is a partition of unity subordinate to $\{U_\xi\}$. Every element $s \in Y(X)$ defines an element of a free $C^\infty(X)$-module of rank $m + k$ as follows. Let us put $s_\xi = \psi_\xi \circ s|_{U_\xi}$. It fulfils the relation
$$s_\xi = \sum_\zeta \varrho_{\xi\zeta} l_\zeta^2 s_\zeta. \tag{10.9.7}$$
There are a module monomorphism
$$F : Y(X) \ni s \to (l_1 s_1, \ldots, l_k s_k) \in \overset{m+k}{\oplus} C^\infty(X)$$
and a module epimorphism
$$\Phi : \overset{m+k}{\oplus} C^\infty(X) \ni (t_1, \ldots, t_k) \to (\widetilde{s}_1, \ldots, \widetilde{s}_k) \in Y(X),$$
$$\widetilde{s}_i = \sum_j \varrho_{ij} l_j t_j.$$
In view of the relation (10.9.7),
$$\Phi \circ F = \operatorname{Id} Y(X),$$
i.e., $Y(X)$ is a projective module of finite rank. The converse assertion is proved similarly to that in [160]. □

This theorem states the categorial equivalence between the vector bundles over a smooth manifold X and projective modules of finite rank over the ring $C^\infty(X)$ of smooth real functions on X. The following are corollaries of this equivalence

- The structure module $Y^*(X)$ of the dual $Y^* \to X$ of a vector bundle $Y \to X$ is the $C^\infty(X)$-dual $Y(X)^*$ of the structure module $Y(X)$ of $Y \to X$.
- Any exact sequence of vector bundles

$$0 \to Y \longrightarrow Y' \longrightarrow Y'' \to 0 \qquad (10.9.8)$$

over the same base X yields the exact sequence

$$0 \to Y(X) \longrightarrow Y'(X) \longrightarrow Y''(X) \to 0 \qquad (10.9.9)$$

of their structure modules, and *vice versa*. In accordance with Theorem 1.1.12, the exact sequence (10.9.8) is always split. Every its splitting defines that of the exact sequence (10.9.9), and *vice versa*.

- The derivation module of the real ring $C^\infty(X)$ coincides with the $C^\infty(X)$-module $\mathcal{T}(X)$ of vector fields on X, i.e., with the structure module of the tangent bundle TX of X. Hence, it is a projective $C^\infty(X)$-module of finite rank. It is the $C^\infty(X)$-dual $\mathcal{T}(X) = \mathcal{O}^1(X)^*$ of the structure module $\mathcal{O}^1(X)$ of the cotangent bundle T^*X of X which is the module of differential one-forms on X and, conversely,

$$\mathcal{O}^1(X) = \mathcal{T}(X)^*.$$

- Therefore, if P is a $C^\infty(X)$-module, one can reformulate Definition 10.2.2 of a connection on P as follows. A connection on a $C^\infty(X)$-module P is a $C^\infty(X)$-module morphism

$$\nabla : P \to \mathcal{O}^1(X) \otimes P, \qquad (10.9.10)$$

which satisfies the Leibniz rule

$$\nabla(fp) = df \otimes p + f\nabla(p), \qquad f \in C^\infty(X), \qquad p \in P.$$

It associates to any vector field $\tau \in \mathcal{T}(X)$ on X a first order differential operator ∇_τ on P which obeys the Leibniz rule

$$\nabla_\tau(fp) = (\tau \rfloor df)p + f\nabla_\tau p. \qquad (10.9.11)$$

In particular, let $Y \to X$ be a vector bundle and $Y(X)$ its structure module. The notion of a connection on the structure module $Y(X)$ is equivalent to the standard geometric notion of a connection on a vector bundle $Y \to X$ [112].

Since the derivation module of the real ring $C^\infty(X)$ is the $C^\infty(X)$-module $\mathcal{T}(X)$ of vector fields on X and

$$\mathcal{O}^1(X) = \mathcal{T}(X)^*,$$

the Chevalley–Eilenberg differential calculus over the real ring $C^\infty(X)$ is exactly the differential graded algebra $(\mathcal{O}^*(X), d)$ of exterior forms on X, where the Chevalley–Eilenberg coboundary operator d (10.6.5) coincides with the exterior differential. Moreover, one can show that $(\mathcal{O}^*(X), d)$ is a minimal differential calculus, i.e., the $C^\infty(X)$-module $\mathcal{O}^1(X)$ is generated by elements df, $f \in C^\infty(X)$. Therefore, the de Rham complex (10.6.8) of the real ring $C^\infty(X)$ is the *de Rham complex*

$$0 \to \mathbb{R} \longrightarrow C^\infty(X) \xrightarrow{d} \mathcal{O}^1(X) \xrightarrow{d} \cdots \mathcal{O}^k(X) \xrightarrow{d} \cdots \qquad (10.9.12)$$

of exterior forms on a manifold X.

The de Rham cohomology of the complex (10.9.12) is called the *de Rham cohomology* $H^*_{\mathrm{DR}}(X)$ of X. To describe them, let us consider the de Rham complex

$$0 \to \mathbb{R} \longrightarrow C^\infty_X \xrightarrow{d} \mathcal{O}^1_X \xrightarrow{d} \cdots \mathcal{O}^k_X \xrightarrow{d} \cdots \qquad (10.9.13)$$

of sheaves $\mathcal{O}^k_X$, $k \in \mathbb{N}_+$, of germs of exterior forms on X. These sheaves are fine. Due to the *Poincaré lemma*, the complex (10.9.13) is exact and, thereby, is a fine resolution of the constant sheaf $\mathbb{R}$ on a manifold X. Then a corollary of Theorem 10.7.4 is the classical *de Rham theorem*.

Theorem 10.9.4. *There is an isomorphism*

$$H^k_{\mathrm{DR}}(X) = H^k(X; \mathbb{R}) \qquad (10.9.14)$$

of the de Rham cohomology $H^*_{\mathrm{DR}}(X)$ of a manifold X to cohomology of X with coefficients in the constant sheaf $\mathbb{R}$.

Cohomology $H^k(X; \mathbb{R})$ in turn is related to cohomology of other types as follows. Let us consider the short exact sequence of constant sheaves

$$0 \to \mathbb{Z} \longrightarrow \mathbb{R} \longrightarrow U(1) \to 0, \qquad (10.9.15)$$

where $U(1) = \mathbb{R}/\mathbb{Z}$ is the circle group of complex numbers of unit modulus. This exact sequence yields the long exact sequence of the sheaf cohomology groups

$$0 \to \mathbb{Z} \longrightarrow \mathbb{R} \longrightarrow U(1) \longrightarrow H^1(X; \mathbb{Z}) \longrightarrow H^1(X; \mathbb{R}) \longrightarrow \cdots$$
$$H^p(X; \mathbb{Z}) \longrightarrow H^p(X; \mathbb{R}) \longrightarrow H^p(X; U(1)) \longrightarrow H^{p+1}(X; \mathbb{Z}) \longrightarrow \cdots,$$

where
$$H^0(X; \mathbb{Z}) = \mathbb{Z}, \qquad H^0(X; \mathbb{R}) = \mathbb{R}$$
and
$$H^0(X; U(1)) = U(1).$$

This exact sequence defines the homomorphism
$$H^*(X; \mathbb{Z}) \to H^*(X; \mathbb{R}) = H^*(X; \mathbb{Z}) \otimes \mathbb{R} \qquad (10.9.16)$$
of cohomology with coefficients in the constant sheaf $\mathbb{Z}$ to that with coefficients in $\mathbb{R}$. Combining the isomorphism (10.9.14) and the homomorphism (10.9.16) leads to the cohomology homomorphism
$$H^*(X; \mathbb{Z}) \to H^*_{\mathrm{DR}}(X). \qquad (10.9.17)$$
Its kernel contains all cyclic elements of cohomology groups $H^k(X; \mathbb{Z})$.

Since smooth manifolds are assumed to be paracompact, there is the following isomorphism between their sheaf cohomology and singular cohomology.

Theorem 10.9.5. *The sheaf cohomology $H^*(X; \mathbb{Z})$ (resp. $H^*(X; \mathbb{Q})$, $H^*(X; \mathbb{R})$) of a paracompact topological space X with coefficients in the constant sheaf $\mathbb{Z}$ (resp. $\mathbb{Q}$, $\mathbb{R}$) is isomorphic to the singular cohomology of X with coefficients in the ring $\mathbb{Z}$ (resp. $\mathbb{Q}$, $\mathbb{R}$) [22; 146].*

Note that singular cohomology of paracompact topological spaces coincides with the Čech and Alexandery ones. Since singular cohomology is a topological invariant [146], the sheaf cohomology groups $H^*(X; \mathbb{Z})$, $H^*(X; \mathbb{Q})$, $H^*(X; \mathbb{R})$ and, consequently, de Rham cohomology of smooth manifolds also are topological invariants.

10.10 Leafwise and fibrewise cohomology

Let $\mathcal{F}$ be a (regular) foliation of a k-dimensional manifold Z provided with the adapted coordinate atlas (1.1.30). The real Lie algebra $\mathcal{T}(\mathcal{F})$ of global sections of the tangent bundle $T\mathcal{F} \to Z$ to $\mathcal{F}$ is a $C^\infty(Z)$-submodule of the derivation module of the real ring $C^\infty(Z)$. Its kernel $S_\mathcal{F}(Z) \subset C^\infty(Z)$ consists of functions constant on leaves of $\mathcal{F}$. Therefore, $\mathcal{T}(\mathcal{F})$ is the Lie $S_\mathcal{F}(Z)$-algebra of derivations of $C^\infty(Z)$, regarded as a $S_\mathcal{F}(Z)$-ring. Then

10.10. Leafwise and fibrewise cohomology

one can introduce the *leafwise differential calculus* [57; 75] as the Chevalley–Eilenberg differential calculus over the $S_{\mathcal{F}}(Z)$-ring $C^{\infty}(Z)$. It is defined as a subcomplex

$$0 \to S_{\mathcal{F}}(Z) \longrightarrow C^{\infty}(Z) \xrightarrow{\widetilde{d}} \mathfrak{F}^{1}(Z) \cdots \xrightarrow{\widetilde{d}} \mathfrak{F}^{\dim \mathcal{F}}(Z) \to 0 \qquad (10.10.1)$$

of the Chevalley–Eilenberg complex of the Lie $S_{\mathcal{F}}(Z)$-algebra $\mathcal{T}(\mathcal{F})$ with coefficients in $C^{\infty}(Z)$ which consists of $C^{\infty}(Z)$-multilinear skew-symmetric maps

$$\overset{r}{\times} \mathcal{T}(\mathcal{F}) \to C^{\infty}(Z), \qquad r = 1, \ldots, \dim \mathcal{F}.$$

These maps are global sections of exterior products $\overset{r}{\wedge} T\mathcal{F}^{*}$ of the dual $T\mathcal{F}^{*} \to Z$ of $T\mathcal{F} \to Z$. They are called the *leafwise forms* on a foliated manifold $(Z, \mathcal{F})$, and are given by the coordinate expression

$$\phi = \frac{1}{r!} \phi_{i_{1} \ldots i_{r}} \widetilde{dz}^{i_{1}} \wedge \cdots \wedge \widetilde{dz}^{i_{r}},$$

where $\{\widetilde{dz}^{i}\}$ are the duals of the holonomic fibre bases $\{\partial_{i}\}$ for $T\mathcal{F}$ and w is the exterior product (10.6.6). Then one can think of the Chevalley–Eilenberg coboundary operator

$$\widetilde{d}\phi = \widetilde{dz}^{k} \wedge \partial_{k}\phi = \frac{1}{r!} \partial_{k} \phi_{i_{1} \ldots i_{r}} \widetilde{dz}^{k} \wedge \widetilde{dz}^{i_{1}} \wedge \cdots \wedge \widetilde{dz}^{i_{r}} \qquad (10.10.2)$$

as being the *leafwise exterior differential*. The Chevalley–Eilenberg differential calculus $\mathfrak{F}^{*}(Z)$ as like as $\mathcal{O}^{*}(Z)$ is minimal. Accordingly, the complex (10.10.1) is called the *leafwise de Rham complex* (or the *tangential de Rham complex* in the terminology of [75]). This is the complex $(\mathcal{A}^{0,*}, d_{f})$ in [159]. Its cohomology $H_{\mathcal{F}}^{*}(Z)$, called the *leafwise de Rham cohomology*, equals the cohomology $H^{*}(Z; S_{\mathcal{F}})$ of Z with coefficients in the sheaf $S_{\mathcal{F}}$ of germs of elements of $S_{\mathcal{F}}(Z)$ [88].

In order to relate the leafwise de Rham cohomology $H_{\mathcal{F}}^{*}(Z)$ with the de Rham cohomology of Z, let us consider the exact sequence (1.1.32) of vector bundles over Z. Since it admits a splitting, the epimorphism $i_{\mathcal{F}}^{*}$ yields that of the algebra $\mathcal{O}^{*}(Z)$ of exterior forms on Z to the algebra $\mathfrak{F}^{*}(Z)$ of leafwise forms. It obeys the condition

$$i_{\mathcal{F}}^{*} \circ d = \widetilde{d} \circ i_{\mathcal{F}}^{*},$$

and provides the cochain morphism

$$i_{\mathcal{F}}^{*} : (\mathbb{R}, \mathcal{O}^{*}(Z), d) \to (S_{\mathcal{F}}(Z), \mathcal{F}^{*}(Z), \widetilde{d}), \qquad (10.10.3)$$
$$dz^{\lambda} \to 0, \quad dz^{i} \to \widetilde{dz}^{i},$$

of the de Rham complex of Z to the leafwise de Rham complex (10.10.1) and the corresponding homomorphism

$$[i_\mathcal{F}^*]^* : H_{\text{DR}}^*(Z) \to H_\mathcal{F}^*(Z) \qquad (10.10.4)$$

of the de Rham cohomology of Z to the leafwise one. Let us note that $[i_\mathcal{F}^*]^{r>0}$ need not be epimorphisms [159].

Given a leaf $i_F : F \to Z$ of $\mathcal{F}$, we have the pull-back homomorphism

$$(\mathbb{R}, \mathcal{O}^*(Z), d) \to (\mathbb{R}, \mathcal{O}^*(F), d) \qquad (10.10.5)$$

of the de Rham complex of Z to that of F and the corresponding homomorphism of the de Rham cohomology groups

$$H_{\text{DR}}^*(Z) \to H_{\text{DR}}^*(F). \qquad (10.10.6)$$

Theorem 10.10.1. *The homomorphisms (10.10.5) – (10.10.6) factorize through the homomorphisms (10.10.3) – (10.10.4).*

Proof. It is readily observed that the pull-back bundles $i_F^* T\mathcal{F}$ and $i_F^* T\mathcal{F}^*$ over F are isomorphic to the tangent and the cotangent bundles of F, respectively. Moreover, a direct computation shows that

$$i_F^*(\widetilde{d}\phi) = d(i_F^*\phi)$$

for any leafwise form ϕ. It follows that the cochain morphism (10.10.5) factorizes through the cochain morphism (10.10.3) and the cochain morphism

$$i_F^* : (S_\mathcal{F}(Z), \mathfrak{F}^*(Z), \widetilde{d}) \to (\mathbb{R}, \mathcal{O}^*(F), d), \qquad \widetilde{d}z^i \to dz^i, \qquad (10.10.7)$$

of the leafwise de Rham complex of $(Z, \mathcal{F})$ to the de Rham complex of F. Accordingly, the cohomology morphism (10.10.6) factorizes through the leafwise cohomology

$$H_{\text{DR}}^*(Z) \xrightarrow{[i_\mathcal{F}^*]} H_\mathcal{F}^*(Z) \xrightarrow{[i_F^*]} H_{\text{DR}}^*(F). \qquad (10.10.8)$$
□

Let $Y \to X$ be a fibred manifold endowed with fibred coordinates (x^λ, y^i). Treating it as a particular foliated manifold, we come to the notion of the *fibrewise de Rham complex* and the *fibrewise de Rham cohomology*. This complex $V^*(Y)$ consists of sections of the exterior bundle $\wedge V^*Y \to Y$ where V^*Y is the vertical cotangent bundle of Y provided with linear bundle coordinates $(x^\lambda, y^i, \dot{y}_i)$ relative to coframes $\{\overline{d}y^i\}$. These sections, called the *fibrewise forms*, are given by the coordinate expression

$$\phi = \frac{1}{r!}\phi_{i_1\ldots i_r}\overline{d}y^{i_1} \wedge \cdots \wedge \overline{d}y^{i_r}.$$

10.10. Leafwise and fibrewise cohomology

The $C^\infty(X)$-linear fibrewise exterior differential (10.10.2) acting on these forms reads

$$\overline{d}\phi = \overline{d}y^k \wedge \partial_k \phi = \frac{1}{r!}\partial_k \phi_{i_1 \ldots i_r} \overline{d}y^k \wedge \overline{d}y^{i_1} \wedge \cdots \wedge \overline{d}y^{i_r}.$$

In particular, there is the cochain morphism (10.10.7) $\overline{d}y^i \to dy^i$ of the fibrewise de Rham complex $(C^\infty(X), V^*(Y), \overline{d})$ to the de Rham complex $(\mathbb{R}, \mathcal{O}^*(V_x), d)$ of a fibre V_x, $x \in X$, of a fibred manifold $Y \to X$. This cochain morphism yields the corresponding cohomology morphism

$$H^*_V(Y) \to H^*_{\mathrm{DR}}(V_x) \qquad (10.10.9)$$

of the fibrewise cohomology to the de Rham cohomology of V_x.

For instance, let $Y \to X$ be an affine bundle. Then

$$H^{0<k}_{\mathrm{DR}}(V_x) = 0$$

for any $x \in X$. Let us show the following.

Theorem 10.10.2. *If $Y \to X$ is an affine bundle, its fibrewise de Rham complex is acyclic, i.e., cohomology $H^{0<k}_V(Y)$ is trivial.*

Proof. Let an affine bundle $Y \to X$ be provided with bundle coordinates (x^λ, y^i) possessing linear transition functions (see Theorem 1.1.13). Given a coordinate chart $(U \times V; x^\lambda, y^i)$ of Y over a domain $U \subset X$, let us consider the homotopy operator

$$\mathbf{h} : V^k(Y) \to V^{k-1}(Y),$$

$$\mathbf{h}(\phi) = \int_0^1 \left[\frac{1}{(k-1)!} t^{k-1} y^{i_1} \phi_{i_1 i_2 \ldots i_k}(x^\lambda, ty^j) \overline{d}y^{i_2} \wedge \cdots \wedge \overline{d}y^{i_k} \right] dt.$$

This operator is globally defined due to linear transition functions of fibre coordinates y^i. $\square$

Remark 10.10.1. If

$$Y = \mathbb{R}^n \times \mathbb{R}^m \to \mathbb{R}^n,$$

Theorem 10.10.2 reproduces the so called *relative Poincaré lemma* [53]. For instance, let $\phi = \varphi \wedge \omega$ be an exact $(r+n)$-form on $\mathbb{R}^n \times \mathbb{R}^m$. Then, ϕ is brought into the form $\phi = d\sigma \wedge \omega$ where σ is an $(r-1)$-form on $\mathbb{R}^n \times \mathbb{R}^m$.

Bibliography

[1] Albertin, U. (1991). The diffeomorphism group and flat principal bundles, *J. Math. Phys.* **32**, 1975.
[2] Almorox, A. (1987). Supergauge theories in graded manifolds, In *Differential Geometric Methods in Mathematical Physics*, Lect. Notes in Math. **1251** (Springer-Verlag, Berlin) p. 114.
[3] Anderson, I. and Duchamp, T. (1980). On the existence of global variational principles, *Amer. J. Math.* **102**, 781.
[4] Anderson, I. (1992). Introduction to the variational bicomplex, *Contemp. Math.*, **132**, 51.
[5] Aringazin, A. and Mikhailov, A. (1991). Matter fields in spacetime with vector non-metricity, *Class. Quant. Grav.* **8**, 1685.
[6] Atiyah, M. and Macdonald, I. (1969). *Introduction to Commutative Algebra* (Addison-Wesley, London).
[7] Avis, S. and Isham, C. (1980). Generalized spin structure on four dimensional space-times, *Commun. Math. Phys.* **72**, 103.
[8] Bak, D., Cangemi, D. and Jackiw, R. (1994). Energy-momentum conservation in gravity theories, *Phys. Rev. D* **49**, 5173.
[9] Barnich, G., Brandt, F. and Henneaux, M. (2000). Local BRST cohomology in gauge theories, *Phys. Rep.* **338**, 439.
[10] Bartocci, C., Bruzzo, U. and Hernández Ruipérez, D.(1991). *The Geometry of Supermanifolds* (Kluwer, Dordrecht).
[11] Bashkirov, D. and Sardanashvily, G. (2005), On the BV quantization of gauge gravitation theory, *Int. J. Geom. Methods Mod. Phys.* **2**, 203.
[12] Bashkirov, D., Giachetta, G., Mangiarotti, L. and Sardanashvily, G. (2005). Noether's second theorem in a general setting. Reducible gauge theory, *J. Phys. A* **38**, 5329.
[13] Bashkirov, D., Giachetta, G., Mangiarotti, L. and Sardanashvily, G. (2005). Noether's second theorem for BRST symmetries, *J. Math. Phys.* **46**, 053517.
[14] Bashkirov, D., Giachetta, G., Mangiarotti, L. and Sardanashvily, G. (2005). The antifield Koszul–Tate complex of reducible Noether identities, *J. Math. Phys.* **46**, 103513.

[15] Bashkirov, D., Giachetta, G., Mangiarotti, L. and Sardanashvily, G. (2008). The KT-BRST complex of degenerate Lagrangian systems, *Lett. Math. Phys.* **83**, 237.
[16] Batchelor, M. (1979). The structure of supermanifolds, *Trans. Amer. Math. Soc.* **253**, 329.
[17] Bauderon, M. (1985). Differential geometry and Lagrangian formalism in the calculus of variations, in *Differential Geometry, Calculus of Variations, and their Applications*, Lecture Notes in Pure and Applied Mathematics **100** (Dekker, New York) p. 67.
[18] Birmingham, D. and Blau, M. (1991). Topological field theory, *Phys. Rep.* **209**, 129.
[19] Borowiec, A., Ferraris, M., Francaviglia, M. and Volovich, I. (1994). Energy-momentum complex for nonlinear gravitational Lagrangians in the first-order formalism, *Gen. Rel. Grav.* **26**, 637.
[20] Borowiec, A., Ferraris, M. and Francaviglia, M. (2003). A covariant formalism for Chern–Simons gravity. *J. Phys. A* **36**, 2589.
[21] Brandt, F. (2001). Jet coordinates for local BRST cohomology, *Lett. Math. Phys.* **55**, 149.
[22] Bredon, G. (1967). *Sheaf theory* (McGraw-Hill, New York).
[23] Bruzzo, U. (1987). The global Utiyama theorem in Einstein–Cartan theory, *J. Math. Phys.* **28**, 2074.
[24] Bryant, R., Chern, S., Gardner, R., Goldschmidt, H. and Griffiths, P. (1991). *Exterior Differential Systems* (Springer-Verlag, Berlin).
[25] Cantrijn, F., Ibort, A. and De Leon, M. (1999). On the geometry of multisymplectic manifolds, *J. Austral. Math. Soc. Ser. A* **66**, 303.
[26] Cariñena, J., Crampin, M. and Ibort, L. (1991). On the multisymplectic formalism for first order field theories, *Diff. Geom. Appl.* **1**, 345.
[27] Cariñena, J. and Figueroa, H. (2003). Singular Lagrangian in supermechanics, *Diff. J. Geom. Appl.* **18**, 33.
[28] Cianci, R. (1990) *Introduction to Supermanifolds* (Bibliopolis, Naples).
[29] Cianci, R., Francaviglia, M. and Volovich, I. (1995). Variational calculus and Poincaré–Cartan formalism in supermanifolds, *J. Phys. A.* **28**, 723.
[30] Coleman, S., Wess J. and Zumino, B.(1969). Structure of phenomenological Lagrangians, I, II, *Phys. Rev.* **177**, 2239.
[31] Dabrowski, L. and Percacci, R. (1986). Spinors and diffeomorphisms, *Commun. Math. Phys.* **106**, 691.
[32] Dick, R. (1993). Covariant conservation laws from the Palatini formalism, *Int. J. Theor. Phys.* **32**, 109.
[33] Dittrich, W. and Reuter, M. (1994). *Classical and Quantum Dynamics* (Springer-Verlag, Berlin).
[34] Dodson, C. (1980) *Categories, Bundles and Spacetime Topology* (Shiva Publishing Limited, Orpington).
[35] Dold, A. (1972). *Lectures on Algebraic Topology* (Springer-Verlag, Berlin).
[36] Echeverría-Enríquez, A., Muñoz-Lecanda, M. and Roman-Roy, N. (2002). Geometry of multisymplectic Hamiltonian first-order field theories, *J. Math. Phys.* **41**, 7402.

[37] Eck, D. (1981). Gauge-natural bundles and generalized gauge theories, *Mem. Amer. Math. Soc.* **247**, 1.

[38] Eguchi, T., Gilkey, P. and Hanson, A. (1980). Gravitation, gauge theories and differential geometry, *Phys. Rep.* **66**, 213.

[39] Fatibene L., Ferraris, M. and Francaviglia, M. (1994). Nöther formalism for conserved quantities in classical gauge field theories. *J. Math. Phys.* **35**, 1644.

[40] Fatibene, L., Ferraris, M., Francaviglia, M. and McLenaghan, R. (2002). Generalized symmetries in mechanics and field theories. *J. Math. Phys.* **43**, 3147.

[41] Fatibene, L. and Francaviglia, M. (2003). *Natural and Gauge Natural Formalism for Classical Field Theories. A Geometric Perspective Including Spinors and Gauge Theories* (Kluwer, Dordrecht).

[42] Fatibene, L., Francaviglia, M. and Mercadante, S. (2005). Covariant formulation of Chern–Simons theories. *Int. J. Geom. Methods Mod. Phys.* **2**, 993.

[43] Ferraris, M. and Francaviglia, M. (1985). Energy-momentum tensors and stress tensors in geometric field theories, *J. Math. Phys.* **26**, 1243.

[44] Fisch, J. and Henneaux, M. (1990). Homological perturbation theory and algebraic structure of the antifield-antibracket formalism for gauge theories, *Commun. Math. Phys.* **128**, 627.

[45] Franco, D. and Polito, C. (2004). Supersymmetric field-theoretic models on a supermanifold, *J. Math. Phys.* **45**, 1447.

[46] Fuks, D. (1986) *Cohomology of Infinite-Dimensional Lie Algebras* (Consultants Bureau, New York).

[47] Fulp, R., Lawson, J. and Norris, L. (1994). Geometric prequantization on the spin bundle based on n-symplectic geometry: the Dirac equation, *Int. J. Theor. Phys.* **33**, 1011.

[48] Fulp, R., Lada, T. and Stasheff, J. (2002). Sh-Lie algebras induced by gauge transformations, *Commun. Math. Phys.* **231**, 25.

[49] Fulp, R., Lada, T. and Stasheff, J. (2003). Noether variational Theorem II and the BV formalism, *Rend. Circ. Mat. Palermo (2) Suppl.* No. 71, 115.

[50] García, P. (1977). Gauge algebras, curvature and symplectic structure, *J. Diff. Geom.* **12**, 209.

[51] Geroch, R. (1968). Spinor structure of space-time in general relativity, *J. Math. Phys.* **9**, 1739.

[52] Giachetta, G. and Sardanashvily, G. (1996). Stress-energy-momentum of affine-metric gravity. Generalized Komar superportential, *Class. Quant. Grav.* **13**, L67.

[53] Giachetta, G., Mangiarotti, L. and Sardanashvily, G. (1997). *New Lagrangian and Hamiltonian Methods in Field Theory* (World Scientific, Singapore).

[54] Giachetta, G., Mangiarotti, L. and Sardanashvily, G. (1999). Covariant Hamilton equations for field theory, *J. Phys. A* **32**, 6629.

[55] Giachetta, G., Mangiarotti, L. and Sardanashvily, G. (2000). Iterated BRST cohomology. *Lett. Math. Phys.* **53**, 143.

[56] Giachetta, G., Mangiarotti, L. and Sardanashvily, G. (2001). Cohomology of the infinite-order jet space and the inverse problem, *J. Math. Phys.* **42**, 4272.

[57] Giachetta, G., Mangiarotti, L. and Sardanashvily, G. (2002). Geometric quantization of mechanical systems with time-dependent parameters, *J. Math. Phys* **43**, 2882.

[58] Giachetta, G., Mangiarotti, L. and Sardanashvily, G. (2003). Noether conservation laws in higher-dimensional Chern–Simons theory. *Mod. Phys. Lett. A* **18**, 2645.

[59] Giachetta, G., Mangiarotti, L. and Sardanashvily, G. (2005). Lagrangian supersymmetries depending on derivatives. Global analysis and cohomology, *Commun. Math. Phys.*, **259**, 103.

[60] Giachetta, G., Mangiarotti, L. and Sardanashvily, G. (2005). *Geometric and Topological Algebraic Methods in Quantum Mechanics* (World Scientific, Singapore).

[61] Giachetta, G., Mangiarotti, L. and Sardanashvily, G. (2008). On the notion of gauge symmetries of generic Lagrangian field theory, *J. Math. Phys* **50**, 012903.

[62] Godement, R. (1964). *Théorie des Faisceaux* (Hermann, Paris).

[63] Gomis, J., París, J. and Samuel, S. (1995). Antibracket, antifields and gauge theory quantization, *Phys. Rep.* **295**, 1.

[64] Gordejuela, F. and Masqué, J. (1995). Gauge group and G-structures, *J. Phys. A* **28**, 497.

[65] Gotay, M. (1991). A multisymplectic framework for classical field theory and the calculus of variations, in *Mechanics, Analysis and Geometry: 200 Years after Lagrange* (North Holland, Amsterdam) p. 203.

[66] Gotay, M. (1991). A multisymplectic framework for classical field theory and the calculus of variations. II. Space + time decomposition, *Diff. Geom. Appl.* **1**, 375.

[67] Gotay, M. and Marsden, J. (1992). Stress-energy-momentum tensors and the Belinfante–Rosenfeld formula, *Contemp. Math.* **132**, 367.

[68] Greenberg, M. (1971). *Lectures on Algebraic Topology* (W.A. Benjamin, Inc., Menlo Park).

[69] Greub, W., Halperin, S. and Vanstone, R. (1972). *Connections, Curvature and Cohomology* (Academic Press, New York).

[70] Greub, W. and Petry, H.-R. (1978). On the lifting of structure groups, *Differential Geometric Methods in Mathematical Physics II*, Lect. Notes in Mathematics, **676** (Springer-Verlag, Berlin) p. 217.

[71] Grothendieck, A. (1967). *Eléments de Géométrie Algébrique IV*, Publ. Math. **32** (IHES, Paris).

[72] Günther, C. (1987). The polysymplectic Hamiltonian formalism in field theory and calculus of variations, I: The local case, *J. Diff. Geom.* **25**, 23.

[73] Hatfield, B. (1992). *Quantum Field Theory of Point Particles and Strings* (Addison–Willey Publ., Redwood City, CA).

[74] Hawking, S. and Ellis, G. (1973) *The Large Scale Structure of a Space-Time* (Cambr. Univ. Press, Cambridge).

[75] Hector, G., Macías, E. and Saralegi, M. (1989). Lemme de Moser feuilleté et classification des variétés de Poisson régulières, *Publ. Mat.* **33**, 423.
[76] Hehl, F., McCrea, J., Mielke, E. and Ne'eman, Y. (1995). Metric-affine gauge theory of gravity: field equations, Noether identities, world spinors, and breaking of dilaton invariance, *Phys. Rep.* **258**, 1.
[77] Henninig, J. and Jadczyk, A. (1988). Spinors and isomorphisms of Lorentz structures, *Preprint of University of Wroclaw, ITP UWr 88/695*.
[78] Hernández Ruipérez, D. and Muñoz Masqué, J. (1984). Global variational calculus on graded manifolds, *J. Math. Pures Appl.* **63**, 283.
[79] Van der Heuvel, B. (1994). Energy-momentum conservation in gauge theories, *J. Math. Phys.* **35**, 1668.
[80] Hirzebruch, F. (1966) *Topological Methods in Algebraic Geometry* (Springer-Verlag, Berlin).
[81] Ibragimov, N. (1985). *Transformation Groups Applied to Mathematical Physics* (Riedel, Boston).
[82] Isham, C., Salam, A. and Strathdee, J. (1971). Nonlinear realizations of space-time symmetries. Scalar and tensor gravity, *Ann. Phys.* **62**, 98.
[83] Ivanenko, D. and Sardanashvily, G. (1983). The gauge treatment of gravity, *Phys. Rep.* **94**, 1.
[84] Ivanov, E. and Niederle, J. (1982). Gauge formulations of gravitation theories: I. The Poincaré, de Sitter, and conformal cases, *Phys. Rev. D* **25**, 976.
[85] Jadczyk, A. and Pilch, K. (1981). Superspaces and Supersymmetries, *Commun. Math. Phys.* **78**, 391.
[86] Joseph, A. and Solomon, A. (1970). Global and infinitesimal nonlinear chiral transformations, *J. Math. Phys.* **11**, 748.
[87] Kadić, A. and Edelen, D. (1983). *A Gauge Theory of Dislocations and Disclinations* (Springer-Verlag, Berlin).
[88] El Kasimi-Alaoui, A. (1983). Sur la cohomologie feuilletée, *Compositio Math.* **49**, 195.
[89] Keyl, M. (1991). About the geometric structure of symmetry breaking, *J. Math. Phys.* **32**, 1065.
[90] Kijowski, J. and Tulczyjew, W. (1979). *A Symplectic Framework for Field Theories* (Springer-Verlag, Berlin).
[91] Kirsch, I. (2005). A Higgs mechanism for gravity, *Phys. Rev. D* **72**, 024001.
[92] Kobayashi, S. and Nomizu, K. (1963) *Foundations of Differential Geometry* (John Wiley, New York - Singapore).
[93] Kobayashi, S. (1972). *Transformation Groups in Differential Geometry* (Springer-Verlag, Berlin).
[94] Kolář, I., Michor, P. and Slovák, J. (1993) *Natural Operations in Differential Geometry* (Springer-Verlag, Berlin).
[95] Kosmann, Y. (1972). Dérivées de Lie des spineurs, *Ann. di Matem. Pura ed Appl.* **91**, 317.
[96] Krasil'shchik, I., Lychagin, V. and Vinogradov, A. (1985). *Geometry of Jet Spaces and Nonlinear Partial Differential Equations* (Gordon and Breach, Glasgow).

[97] Krupka, D. (1990) Variational sequences on finite order jet spaces, in *Differential Geometry and its Applications* (World Scientific, Singapore) p. 236.
[98] Krupka, D. and Musilova, J. (1998). Trivial Lagrangians in field theory, *Diff. Geom. Appl.* **9**, 293.
[99] Krupka, D. (2007). Global variational theory in fibred spaces, in *Handbook of Global Analysis* (Elsevier, Amsterdam) p. 773.
[100] Krupkova, O. and Smetanova, D. (2001). Legendre transformations for regularizable Lagrangian in field theory, *Lett. Math. Phys.* **58**, 189.
[101] Krupkova, O. (2002). Hamiltonian field theory, *J. Geom. Phys.* **43**, 93.
[102] Lang, S. (1993). *Algebra* (Addison–Wisley, New York).
[103] Lawson, H. and Michelsohn, M.-L. (1989). *Spin Geometry* (Princeton Univ. Press, Princeton).
[104] Leclerk, M. (2006). The Higgs sector of gravitational gauge theories, *Ann. Phys.* **321**, 708.
[105] De Leon, M., Martín de Diego, D. and Santamaría-Merini, A. (2004). Symmetries in classical field theory, *Int. J. Geom. Methods Mod. Phys.* **1**, 651.
[106] De Leon, M., Martín de Diego, D., Saldago, M. and Vilariño, L. (2008). Nonholonomic constraints in k-symplectic classical field theory, *Int. J. Geom. Methods Mod. Phys.* **5**, 799.
[107] Mackey, G. (1968). *Induced Representations of Groups and Quantum mechanics* (W.A. Benjamin, New York).
[108] Mac Lane, S. (1967). *Homology* (Springer-Verlag, Berlin).
[109] B.Malgrange, B. (1972). Equationes de Lie I, II, *J. Diff. Geom.* **6** 503; **7**, 117.
[110] Malyshev, C. (2000). The dislocation stress functions from the double curl $T(3)$-gauge equations: Linearity and look beyond, *Ann. Phys.* **286**, 249.
[111] Mangiarotti, L. and Sardanashvily, G. (1998). *Gauge Mechanics* (World Scientific, Singapore).
[112] Mangiarotti, L. and Sardanashvily, G. (2000). *Connections in Classical and Quantum Field Theory* (World Scientific, Singapore).
[113] Marsden, J., Patrick, G. and Shkoller, S. (1998). Multisymplectic geometry, variational integrators and nonlinear PDEs, *Commun. Math. Phys.* **199**, 351.
[114] Massey, W. (1978). *Homology and Cohomology Theory* (Marcel Dekker, Inc., New York).
[115] Meigniez, G. (2002). Submersions, fibrations and bundles, *Trans. Amer. Math. Soc.* **354**, 3771.
[116] Mitter, P. and Viallet, C. (1981). On the bundle of connections and the gauge orbit manifold in Yang–Mills theory, *Commun. Math. Phys.* **79**, 457.
[117] Modugno, M. and Vinogradov, A. (1994). Some variations on the notion of connections, *Ann. Matem. Pura ed Appl.* **CLXVII**, 33.
[118] Monterde, J., Masqué, J. and Vallejo, J. (2006). The Poincaré–Cartan form in superfield theory, *Int. J. Geom. Methods Mod. Phys.* **3**, 775.
[119] Ne'eman, Y. and Šijački, Dj. (1979). Unified affine gauge theory of gravity and strong interactions with finite and infinite $\overline{GL}(4,\mathbb{R})$ spinor fields, *Ann.*

Phys. **120**, 292.
[120] Nikolova, L. and Rizov, V. (1984). Geometrical approach to the reduction of gauge theories with spontaneous broken symmetries, *Rep. Math. Phys.* **20**, 287.
[121] Novotný, J. (1993). Energy-momentum complex of gravitational field in the Palatini formalism, *Int. J. Theor. Phys.* **32**, 1033.
[122] Obukhov, Yu. (2006). Poincaré gauge theories: selected topics, *Int. J. Geom. Methods Mod. Phys.* **3**, 95.
[123] Olver, P. (1986). *Applications of Lie Groups to Differential Equations* (Springer-Verlag, Berlin).
[124] Polchinski, J. (1998). *String Theory* (Cambr. Univ. Press, Cambridge).
[125] Ponomarev, V. and Obukhov, Yu. (1982). Generalized Einstein-Maxwell theory, *Gen. Rel. Grav.* **14**, 309.
[126] Reinhart, B. (1983). *Differential Geometry and Foliations* (Springer-Verlag, Berlin).
[127] Rennie, A. (2003). Smoothness and locality for nonunital spectral triples, *K-Theory* **28**, 127.
[128] Rey, A., Roman-Roy, N. and Saldago, M. (2005). Gunther's formalism (k-symplectic formalism) in classical field theory: Skinner-Rusk approach and the evolution operator, *J. Math. Phys.* **46**, 052901.
[129] Robertson, A. and Robertson, W. (1973). *Topological Vector Spaces* (Cambridge Univ. Press., Cambridge).
[130] Sardanashvily, G. and Gogberashvily, M. (1987). The dislocation treatment of gauge fields of space-time translations, *Int. Mod. Lett. Phys. A* **2**, 609.
[131] Sardanashvily, G. (1992). On the geometry of spontaneous symmetry breaking, *J. Math. Phys.* **33**, 1546.
[132] G.Sardanashvily, G. and Zakharov, O. (1992). *Gauge Gravitation Theory* (World Scientific, Singapore).
[133] Sardanashvily, G. and Zakharov, O. (1993). On application of the Hamilton formalism in fibred manifolds to field theory, *Diff. Geom. Appl.* **3**, 245.
[134] Sardanashvily, G. (1995). *Generalized Hamiltonian Formalism for Field Theory. Constraint Systems.* (World Scientific, Singapore).
[135] Sardanashvily, G. (1997). Stress-energy-momentum tensors in constraint field theories, *J. Math. Phys.* **38**, 847.
[136] Sardanashvily, G. (1997). Stress-energy-momentum conservation law in gauge gravitation theory, *Class. Quant. Grav.* **14**, 1371.
[137] Sardanashvily, G. (1998). Hamiltonian time-dependent mechanics, *J. Math. Phys.* **39**, 2714.
[138] Sardanashvily, G. (1998). Covariant spin structure, *J. Math. Phys.* **39**, 4874.
[139] Sardanashvily, G. (2002). Cohomology of the variational complex in the class of exterior forms of finite jet order. *Int. J. Math. and Math. Sci.* **30**, 39.
[140] Sardanashvily, G. (2002). Classical gauge theory of gravity, *Theor. Math. Phys.* **132**, 1163.
[141] Sardanashvily, G. (2005). Noether identities of a differential operator. The Koszul–Tate complex, *Int. J. Geom. Methods Mod. Phys.* **2**, 873.

[142] Sardanashvily, G. (2006). Gauge gravitation theory from geometric viewpoint, *Int. J. Geom. Methods Mod. Phys.* **3** No. 1, p. v.
[143] Sardanashvily, G. (2006). Geometry of classical Higgs fields, *Int. J. Geom. Methods Mod. Phys.* **3**, 139.
[144] Sardanashvily, G. (2007). Graded infinite order jet manifolds, *Int. J. Geom. Methods Mod. Phys.* **4**, 1335.
[145] Saunders, D. (1989). *The Geometry of Jet Bundles* (Cambridge Univ. Press, Cambridge).
[146] Spanier, E. (1966). *Algebraic Topology* (McGraw-Hill, New York).
[147] Steenrod, N. (1972). *The Topology of Fibre Bundles* (Princeton Univ. Press, Princeton).
[148] Switt, S. (1993). Natural bundles. II. Spin and the diffeomorphism group, *J. Math. Phys.* **34**, 3825.
[149] Takens, F. (1977). Symmetries, conservation laws and variational principles, in *Geometry and Topology*, Lect. Notes in Mathematics, **597** (Springer-Verlag, Berlin) p. 581.
[150] Takens, F. (1979) A global version of the inverse problem of the calculus of variations, *J. Diff. Geom.* **14**, 543.
[151] Tamura, I. (1992). *Topology of Foliations: An Introduction*, Transl. Math. Monographs **97** (AMS, Providence).
[152] Tennison, B. (1975) *Sheaf Theory* (Cambridge Univ. Press, Cambridge).
[153] Terng, C. (1978). Natural vector bundles and natural differential operators, *American J. Math.* **100**, 775.
[154] Thompson, G. (1993). Non-uniqueness of metrics compatible with a symmetric connection, *Class. Quant. Grav.* **10**, 2035.
[155] Trautman, A. (1984). *Differential Geometry for Physicists* (Bibliopolis, Naples).
[156] Tseytlin, A. (1982). Poincaré and de Sitter gauge theories of gravity with propagation torsion, *Phys. Rev. D* **26**, 3327.
[157] Tulczyiew, W. (1980). The Euler–Lagrange resolution, in *Differential Geometric Methods in Mathematical Physics (Proc. Conf., Aix-en-Provence/Salamanca, 1979)* Lecture Notes in Math. **836** (Springer-Verlag, Berlin) p. 22.
[158] Utiyama, R. (1956). Invariant theoretical interpretation of interaction, *Phys. Rev.* **101**, 1597.
[159] Vaisman, I. (1973). *Cohomology and Differential Forms* (Marcel Dekker, New York).
[160] Várilly, J. and Gracia-Bondia, J. (1993). Connes' noncommutative differential geometry and the Standard Model, *J. Geom. Phys.* **12**, 223.
[161] Vinogradov, A. (1984). The C-spectral sequence, Lagrangian formalism, and conservation laws. II. The nonlinear theory, *J. Math. Anal. Appl.* **100**, 41.
[162] Vitolo, R. (1999). Finite order variational bicomplex, *Math. Proc. Cambridge Philos. Soc.* **125**, 321.
[163] Vitolo, R. (2007). Variational sequences, in *Handbook of Global Analysis* (Elsevier, Amsterdam) p. 1115.

[164] Warner, F. (1983). *Foundations of Differential Manifolds and Lie Groups* (Springer-Verlag, Berlin).
[165] Wiston, G. (1974). Topics on space-time topology, *Int. J. Theor. Phys.* **11**, 341.

Index

$B^{n,\mu}$ 102
CX, 246
C_X^∞, 337
C_W, 219
D_Γ, 33
G-bundle, 169
  principal, 172
  smooth, 171
$GL(n,\mu;\Lambda)$ 103
GL_4, 217
$G_{1,3}$, 244
HY, 30
$H^1(X;G_X^0)$, 170
H_L, 75
H_N, 302
H_Γ, 295
$J^1 J^1 Y$, 28
$J^1 Y$, 26
$J^1 Z_Q$, 289
$J^1 \Phi$, 28
$J_\Sigma^1 Y$, 43
$J^1 s$, 27
$J^1 u$, 28
$J^2 Y$, 29
$J^\infty Y$, 55
$J^k u$, 49
$J_n^k Z$, 286
$J^r Y$, 46
$J^r \Phi$, 48
$J^r s$, 48
LX, 217
L_1^*, 147

$L_0^h X$, 229
L_E, 153
L_G, 166
L_H, 296
L_Π, 309
L_e, 141
L_g, 166
L_{YM}, 190
MX, 246
N_L, 74
P^G, 182
$P^{\otimes k}$, 323
P_Σ, 199
$P_{2k}(F_A)$, 265
R_G, 166
R_g, 166
R_{GP}, 172
R_{gP}, 172
TZ, 6
T^*Z, 6
$T_G P$, 174
Tf, 6
VY, 15
$V\Gamma$, 45
$V\Phi$, 15
V^*Y, 16
$V^*\Gamma$, 46
$V_G P$, 174
$V_\Sigma Y$, 44
$V_\Sigma^* Y$, 44
$Y(X)$, 13
Y^h, 42

Z_Y, 75
$[A]$, 119
$\Gamma\tau$, 31
Γ^*, 37
$\mathbf{L}_u$, 23
$\mathbf{1}$, 165
$\mathbf{b}$, 147
$\mathbf{\Omega}_Y$, 294
$\mathbf{\Theta}_Y$, 294
$\mathbf{u}$, 141
$\mathcal{A}_E$, 109
$\mathcal{D}_h$, 250
$\mathcal{E}_H$, 297
$\mathcal{E}_L$, 63
$\mathcal{E}_i$, 63
$\mathcal{F}$, 180
$\mathfrak{g}^*$, 168
$\mathfrak{g}_L$, 226
$\mathfrak{g}_l$, 166
$\mathfrak{g}_r$, 166
$\mathcal{J}_\upsilon$, 69
$\mathcal{K}_p$, 266
$\mathcal{K}_x$, 266
$\mathcal{K}_x^0$, 266
$\mathcal{O}^*(Z)$, 21
$\mathcal{O}^*[\partial\mathcal{A}]$, 335
$\mathcal{O}^*\mathcal{A}$, 336
$\mathcal{O}_\infty^* Y$, 56
$\mathcal{O}_\infty^*$, 56
$\mathcal{O}_\infty^{k,m}$, 58
$\mathcal{O}_k^*$, 48
$\mathcal{P}_\infty^*$, 66
$\mathcal{P}_\infty^*[E;Y]$, 131
$\mathcal{Q}_\infty^*$, 57
$\mathcal{Q}_\infty^*[F;Y]$, 118
$\mathcal{S}^*[E;Z]$, 113
$\mathcal{S}_\infty^*[F;X]$, 116
$\mathcal{S}_\infty^*[F;Y]$, 117
$\mathcal{S}_r^*[F;Y]$, 116
$\mathcal{T}(Z)$, 18
$\mathcal{V}_E$, 111
$\mathcal{G}(X)$, 182
$\mathcal{G}_L$, 124
δ_N, 138
δ_{KT}, 137
$\dot{\partial}_\lambda$, 19

ϵ_m, 167
$\mathfrak{A}(Z)$, 108
$\mathfrak{A}_E$, 108
$\mathfrak{Q}_\infty^*[F;Y]$, 118
γ_H, 296
$\partial\mathcal{A}$, 326
$\partial\mathcal{A}_E$, 112
$\mathbb{C}_{1,3}$, 243
$\mathbb{R}_{1,3}$, 244
$\mathbb{R}_{2,3}$, 243
∇^Γ, 33
$\overline{E}$, 131
$\overline{dy}^i$, 16
$\overline{\delta}$, 132
ω, 22
ω_λ, 22
$\overleftarrow{v}$, 125
$\otimes P$, 324
π^1, 27
π_0^1, 27
π_r^∞, 55
π_i^λ, 73
π_k^r, 47
π_{11}, 28
$\pi_{Y\Sigma}$, 42
$\pi_{Z\Pi}$, 76
$\pi_{\Sigma X}$, 42
ψ_ξ, 8
$\mathbf{L}_s$, 245
θ^i, 27
θ_Λ^i, 50
θ_Z, 23
θ_{LX}, 218
ε_m, 167
$\varrho_{\xi\zeta}$, 8
$\wedge Y$, 14
$\widehat{0}$, 13
$\widehat{H}$, 297
$\widehat{H}_L$, 75
$\widehat{J}^2 Y$, 29
$\widehat{L}$, 74
$\widehat{d}_\lambda$, 28
$\widehat{dx}^\lambda$, 250
$\widehat{h}^a$, 250
$\widehat{\Gamma}$, 296
$\widehat{\otimes}$, 350

Index

$\widetilde{D}$, 44
$\widetilde{\Gamma}$, 41
$\widetilde{\tau}$, 18
$\overline{LX}$, 253
$\{\lambda^\nu{}_\mu\}$, 38
$c_i(F)$, 271
$c_k(F_\mathcal{A})$, 279
d_H, 59
d_V, 59
d_λ, 27
f^*Y, 11
$f^*\Gamma$, 31
$f^*\phi$, 22
h^a, 226
h_0, 51
$j_z^k S$, 286
$u \rfloor \phi$, 22
u_ξ, 183
Ad_g, 167
$\mathrm{Pin}(1,3)$, 244
$\mathrm{Spin}(1,3)$, 244
$\mathrm{Spin}^0(1,3)$, 245

action of a group, 165
  effective, 166
  free, 166
  on the left, 165
  on the right, 165
  transitive, 166
action of a structure group
  on J^1P, 176
  on P, 172
  on TP, 174
adjoint representation
  of a Lie algebra, 168
  of a Lie group, 167
affine bundle, 16
  morphism, 17
Aharonov–Bohm effect, 270
algebra, 319
  $\mathcal{O}_\infty^* Y$, 56
  $\mathcal{O}_\infty^*$, 56
  $\mathcal{P}_\infty^*[E;Y]$, 131
  $\mathcal{S}_\infty^*[F;Y]$, 117
  $\mathbb{Z}_2$-graded, 99

  commutative, 100
  differential bigraded, 106
  differential graded, 334
  graded, 334
    commutative, 334
  unital, 319
algebraic Poincaré lemma, 92
almost complex structure, 277
annihilator of a distribution, 19
antibracket, 150
antifield, 132
  k-stage, 137
  first-stage, 136
  Noether, 133
associated bundles, 170
automorphism
  associated, 185
  principal, 181
autoparallel, 222

background field, 88
base of a fibred manifold, 7
basic form, 22
basis
  for a graded manifold, 109
  for a module, 321
  generating, 118
Batchelor theorem, 108
Bianchi identity
  first, 35
  second, 35
bigraded
  de Rham complex, 107
  exterior algebra, 101
bimodule, 320
  commutative, 320
  graded, 100
body
  of a graded manifold, 108
  of a ringed space, 346
body map, 102
boundary, 327
boundary operator, 327
BRST complex, 147
BRST extension of a Lagrangian, 154
BRST operator, 147

bundle
- P-associated, 185
- affine, 16
- associated, 187
  - canonically, 187
- atlas, 9
  - associated, 185
  - holonomic, 15
  - Lorentz, 225
  - of constant local trivializations, 40
  - spatial, 230
- automorphism, 11
  - vertical, 11
- composite, 42
- continuous, 9
  - locally trivial, 9
- coordinates, 10
  - affine, 16
  - linear, 12
- cotangent, 6
- density-dual, 131
- epimorphism, 11
- exterior, 14
- gauge natural, 219
- in Clifford algebras, 246
- in Minkowski spaces, 246
- isomorphism, 11
- lift, 198
- locally trivial, 9
- monomorphism, 11
- morphism, 11
  - affine, 17
  - linear, 13
  - of principal bundles, 173
- natural, 216
- normal, 20
- of principal connections, 177
- of world connections, 219
- principal, 172
- product, 12
- smooth, 9
- spinor, 249
- tangent, 6
  - affine, 17
  - vertical, 15
- universal, 264
- with a structure group, 169
  - principal, 172
  - smooth, 171

canonical principal connection, 180
Cartan connection, 39
Cartan equation, 77
Cartan subgroup, 211
chain, 327
characteristic
- class, 264
- form, 265
- number, 265
- polynomial, 271

Chern character, 274
Chern class, 270
- of a manifold, 277
Chern form, 271
- on a bundle of principal connections, 279
- total, 271
  - on a bundle of principal connections, 279
Chern number, 273
Chern–Simons form, 279
- local, 279
Chern–Simons Lagrangian, 280
- local, 280
Chevalley–Eilenberg
- coboundary operator, 333
  - graded, 106
- cohomology, 333
  - of a Lie algebra, 334
- complex, 333
- differential calculus, 336
  - Grassmann-graded, 106
  - minimal, 336
Christoffel symbols, 38
classical solution, 52
classification theorem, 264
classifying space, 264
Clifford algebra, 243
Clifford group, 244
closed map, 8
coadjoint representation, 168

Index 373

coboundary, 327
coboundary operator, 327
    Chevalley–Eilenberg, 333
cochain, 327
cochain morphism, 329
cocycle, 327
cocycle condition, 9
    for a sheaf, 341
codistribution, 19
coframe, 6
cohomology, 327
    Čech, 354
    Alexandery, 354
    Chevalley–Eilenberg, 333
        of a Lie algebra, 334
    de Rham
        abstract, 335
        of a manifold, 353
    of a group, 331
    singular, 354
    with coefficients in a sheaf, 341
complete set of Hamiltonian forms, 301
complex, 327
    k-exact, 327
    acyclic, 328
    chain, 327
        standard, 330
    Chevalley–Eilenberg, 333
    cochain, 327
    de Rham
        abstract, 335
    dual, 328
    exact, 327
    variational, 62
complex linear bundle, 272
component of a connection, 30
composite bundle, 42
composite connection, 44
configuration space, 73
    vertical, 90
connection, 30
    affine, 38
    composite, 44
    covertical, 46
    dual, 36

flat, 39
Lagrangian, 74
linear, 36
    world, 219
on a graded commutative ring, 107
on a graded manifold, 112
on a graded module, 107
on a module, 326
on a ring, 326
principal, 177
    associated, 186
    canonical, 180
projectable, 43
reducible, 34
second order, 41
    holonomic, 41
vertical, 45
connection form, 30
    of a principal connection, 177
        local, 177
    vertical, 31
conservation law
    covariant, 192
    differential, 78
    gauge, 80
    Hamiltonian, 305
    integral, 78
    Noether, 79
    weak, 69
constrained Hamilton equations, 302
constrained Lagrangian, 303
contact derivation, 66
    graded, 122
    projectable, 67
    vertical, 67
contact form, 51
    graded, 119
    local, 27
    of higher jet order, 50
contorsion, 227
contraction, 22
cotangent bundle, 6
    vertical, 16
covariant derivative, 33
covariant differential, 33
    on a module, 326

universal, 210
vertical, 45
covariant Hamilton equations, 297
constrained, 302
covariant Hamiltonian, 295
covering space, 268
universal, 268
cup-product, 335
curvature, 34
of a principal connection, 179
of a world connection, 37
of an associated principal
connection, 186
soldered, 35
curvature-free connection, 39
curve integral, 18
cycle, 327

De Donder form, 77
de Rham cohomology
abstract, 335
fibrewise, 356
leafwise, 355
of a graded manifold, 114
of a manifold, 353
of a ring, 336
de Rham complex
abstract, 335
bigraded, 107
fibrewise, 356
leafwise, 355
of a ring, 336
of exterior forms, 353
of sheaves, 353
tangential, 355
de Rham theorem, 353
abstract, 344
density, 22
density-dual bundle, 131
density-dual vector bundle, 131
graded, 131
derivation, 326
contact, 66
projectable, 67
vertical, 67
graded, 104

contact, 122
nilpotent, 124
right, 124
derivation module, 326
graded, 105
differential
covariant, 33
vertical, 45
exterior, 21
total, 59
graded, 119
vertical, 59
graded, 119
differential calculus, 334
Chevalley–Eilenberg, 336
minimal, 336
leafwise, 355
minimal, 335
differential equation, 51
associated to a differential
operator, 53
formally integrable, 52
regular, 52
differential form, 57
graded, 118
differential ideal, 19
differential operator
as a morphism, 53
as a section, 52
degenerate, 155
graded, 104
of standard form, 53
on a module, 325
reducible, 157
Dirac Lagrangian, 259
Dirac operator, 250
graded, 258
total, 257
direct image of a sheaf, 346
direct limit, 322
direct sequence, 322
direct sum
of complexes, 328
of modules, 320
direct system of modules, 322
directed set, 322

distance function, 231
distribution, 19
  horizontal, 30
  involutive, 19
domain, 10
dual module, 320
dual vector bundle, 13

Ehresmann connection, 32
energy-momentum conservation law, 79
energy-momentum current, 80
  Hamiltonian, 306
  of gravity, 237
energy-momentum tensor, 79
  canonical, 80
  of gravity, 237
  metric, 192
equivalent G-bundle atlases, 169
equivalent G-bundles, 169
equivalent bundle atlases, 9
equivariant
  automorphism, 181
  connection, 177
  function, 182
Euler characteristic, 278
Euler class, 277
Euler–Lagrange equation, 64
  second order, 73
Euler–Lagrange operator, 63
  second order, 73
Euler–Lagrange–Cartan operator, 76
Euler–Lagrange-type operator, 63
even element, 100
even morphism, 101
exact sequence
  of modules, 321
    short, 321
    split, 322
  of vector bundles, 14
    short, 14
    split, 14
extension of a group, 332
exterior algebra, 324
  bigraded, 101
exterior bundle, 14

exterior differential, 21
exterior form, 21
  basic, 22
  graded, 114
  horizontal, 22
exterior product, 21
  graded, 101
  of vector bundles, 14

factor
  algebra, 319
  bundle, 14
  complex, 328
  module, 320
  sheaf, 339
fibration, 7
fibre, 7
fibre bundle, 8
fibre coordinates, 10
fibred coordinates, 8
fibred manifold, 7
fibrewise exterior differential, 357
fibrewise form, 356
fibrewise morphism, 11
field, 319
field system algebra, 120
field-antifield theory, 150
finite exactness, 93
first Noether theorem, 68
  for a graded Lagrangian, 124
first variational formula, 67
  for a graded Lagrangian, 123
flow, 18
foliated manifold, 20
foliation, 20
  horizontal, 40
  simple, 21
Frölicher–Nijenhuis bracket, 24
Fréchet ring, 350
frame, 12
  holonomic, 6
frame field, 217

gauge
  algebra, 175
  algebra bundle, 175

group, 182
operator, 146
  higher stage, 148
parameters, 70
potential, 178
transformation, 181
  infinitesimal, 182
gauge natural bundle, 219
gauge symmetry, 70
  k-stage, 146
    complete, 146
  Abelian, 150
  algebraically closed, 149
  complete, 144
  first-stage, 146
  generalized, 73
  of a graded Lagrangian, 143
  reducible, 73
Gel'fand topology, 349
general covariant transformation, 216
  infinitesimal, 216
generalized vector field, 67
  graded, 122
generating basis, 118
geodesic, 223
geodesic equation, 223
geometric space, 346
ghost, 140
Goldstone field, 213
graded
  algebra, 334
  bimodule, 100
  commutative algebra, 334
  commutative ring, 101
    Banach, 101
    real, 101
  connection, 112
  derivation, 104
    of a field system algebra, 124
  derivation module, 105
  differential form, 118
  differential operator, 104
  exterior differential, 114
  exterior form, 114
  exterior product, 101
  function, 108

interior product, 106
Leibniz rule, 334
manifold, 107
  composite, 116
  simple, 109
module, 100
  free, 100
  dual, 101
  morphism, 101
    even, 101
    odd, 101
  ring, 100
  vector field, 111
    generalized, 122
  vector space, 100
    (n, m)-dimensional, 100
graded-homogeneous element, 100
grading automorphism, 99
Grassmann algebra, 101
Grassmann manifold, 264
gravitational field, 225
Grothendieck's topology, 350
group bundle, 182
group module, 330
group ring, 330

Hamilton operator, 297
Hamilton–De Donder equation, 77
Hamiltonian connection, 294
Hamiltonian form, 295
  associated to a Lagrangian, 299
    weakly, 299
  constrained, 302
  vertical, 317
Hamiltonian map, 296
Heimholtz condition, 64
Higgs field, 200
holonomic
  atlas, 15
    of the frame bundle, 218
  automorphisms, 218
  coframe, 6
  coordinates, 6
  frame, 6
holonomy group, 266
  abstract, 266

Index   377

at a point, 266
restricted, 266
    abstract, 266
    at a point, 266
homogeneous space, 166
homology, 327
homology regularity condition, 138
homotopy operator, 328
horizontal
    distribution, 30
    foliation, 40
    form, 22
        graded, 119
    lift
        of a path, 32
        of a vector field, 31
    projection, 51
    splitting, 30
        canonical, 32
    vector field, 30

ideal, 319
    maximal, 319
    of nilpotents, 102
    prime, 319
    proper, 319
image of a sheaf
    direct, 346
    inverse, 346
imbedded submanifold, 7
imbedding, 7
immersion, 6
induced coordinates, 15
induced representation, 204
inductive limit, 324
infinitesimal generator, 18
infinitesimal transformation
    of a Lagrangian system, 66
        Grassmann-graded, 121
integral curve, 18
integral manifold, 20
    maximal, 20
integral section of a connection, 33
interior product, 22
    graded, 106
    of vector bundles, 13

inverse image of a sheaf, 346
inverse sequence, 324

Jacobi field, 91
jet
    first order, 26
    higher order, 46
    infinite order, 55
    of submanifolds, 286
    second order, 29
jet bundle, 27
    affine, 27
jet coordinates, 26
jet manifold, 26
    higher order, 46
    infinite order, 55
        graded, 118
    of submanifolds, 286
    repeated, 28
    second order, 29
    sesquiholonomic, 28
jet prolongation
    functor, 48
    of a differential equation, 51
    of a morphism, 28
        higher order, 48
    of a section, 27
        higher order, 48
        second order, 29
    of a structure group action, 176
    of a vector field, 28
        higher order, 49
juxtaposition rule, 335

kernel
    of a bundle morphism, 11
    of a differential operator, 53
    of a vector bundle morphism, 13
Komar superpotential
    generalized, 238
Koszul–Tate complex, 138
    of a differential operator, 159
Koszul–Tate operator, 138
KT-BRST complex, 153
KT-BRST operator, 153

Lagrangian, 63
    almost regular, 75
    constrained, 303
    degenerate, 132
    first order, 73
    gauge invariant, 188
    graded, 120
    hyperregular, 75
    regular, 75
    semiregular, 75
    variationally trivial, 64
Lagrangian constraint space, 74
Lagrangian system, 63
    N-stage reducible, 137
    Abelian, 150
    finitely degenerate, 133
    finitely reducible, 136
    Grassmann-graded, 120
    irreducible, 136
    reducible, 136
leaf, 20
leafwise differential calculus, 355
leafwise exterior differential, 355
leafwise form, 355
left-invariant form, 168
    canonical, 168
Legendre bundle, 74
    homogeneous, 76
    vertical, 91
Legendre map, 74
    homogeneous, 76
    vertical, 91
Leibniz rule, 326
    for a connection, 326
            Grassmann-graded, 107
        graded, 334
        Grassmann-graded, 104
Lepage equivalent, 65
    of a graded Lagrangian, 121
Lepage form, 65
Levi–Civita connection, 38
Levi–Civita symbol, 22
Lie algebra
    left, 166
    right, 166
Lie algebra bundle, 174

Lie bracket, 18
Lie coalgebra, 168
Lie derivative
    graded, 106
    of a tangent-valued form, 24
    of an exterior form, 23
Lie superalgebra, 103
Lie superbracket, 103
lift of a bundle, 198
lift of a vector field
    canonical, 18
    functorial, 216
    horizontal, 31
linear derivative of an affine
    morphism, 17
linear frame bundle, 217
local diffeomerphism, 6
local ring, 319
local-ringed space, 346
locally finite cover, 342
loop, 266
Lorentz atlas, 225
Lorentz connection, 226
Lorentz structure, 224
Lorentz subbundle, 224

manifold, 5
    fibred, 7
master equation, 152
matrix group, 168
matter field, 185
metric bundle, 225
metric connection, 227
metricity condition, 227
module, 320
    dual, 320
    finitely generated, 321
    free, 321
    graded, 100
    locally free, 347
    of finite rank, 321
    over a Lie algebra, 333
    over a Lie superalgebra, 103
    projective, 321
morphism
    of fibre bundles, 11

of graded manifolds, 110
of presheaves, 339
of ringed spaces, 347
of sheaves, 339
Mourer–Cartan equation, 168
multi-index, 46
multisymplectic form, 295
multisymplectic Liouville form, 77

Nambu–Goto Lagrangian, 292
natural bundle, 216
Nijenhuis differential, 24
nilpotent derivation, 124
Noether conservation law, 79
Noether current, 79
Noether identities, 132
  complete, 133
  first stage, 134
    non-trivial, 134
    trivial, 134
  first-stage
    complete, 136
  higher-stage, 138
  non-trivial, 132
  of a differential operator, 154
  trivial, 132
Noether theorem
  first, 68
  second
    direct, 72
    inverse, 140
non-metricity tensor, 227
normal bundle to a foliation, 20

odd element, 100
odd morphism, 101
on-shell, 69
open map, 7

paracompact space, 342
parallel displacement, 266
partition of unity, 342
path, 32
pin group, 244
Poincaré lemma, 353
  algebraic, 92

relative, 357
Poincaré–Cartan form, 75
polysymplectic form, 294
polysymplectic manifold, 294
Pontryagin class, 274
  of a manifold, 277
preconnection, 288
presheaf, 337
  canonical, 337
principal
  automorphism, 181
    of a connection bundle, 183
    of an associated bundle, 185
  bundle, 172
    continuous, 172
    spinor, 249
  connection, 177
    associated, 186
    canonical, 180
    conjugate, 179
  vector field, 182
    vertical, 182
product connection, 34
proper cover, 10
proper extension of a Lagrangian, 153
proper Lorentz group, 223
proper Lorentz structure, 224
proper Lorentz subbundle, 224
proper map, 7
proper solution of the master
  equation, 153
pull-back
  bundle, 11
  connection, 31
  form, 22
  section, 12
  vertical-valued form, 25

quasi-equivalent equations, 77

rank of a morphism, 6
reduced structure, 198
reduced subbundle, 199
reducible connection, 34
regular element of $J^1 Z_Q$, 289
relative Poincaré lemma, 357

representation of a Lie algebra, 167
resolution, 328
  of a sheaf, 343
    fine, 345
    flabby, 345
restriction of a bundle, 12
Ricci tensor, 221
right derivation, 124
right structure constants, 167
right-invariant form, 168
ring, 319
  graded, 100
  local, 319
ringed space, 346

section, 8
  global, 8
  integral, 33
  local, 8
  of a jet bundle, 27
    integrable, 27
    zero-valued, 13
Serre–Swan theorem, 351
  for graded manifolds, 109
sheaf, 337
  acyclic, 343
  constant, 337
  fine, 345
  flabby, 344
  flasque, 344
  locally free, 347
    of constant rank, 347
    of finite type, 347
  of continuous functions, 337
  of derivations, 347
  of graded derivations, 110
  of modules, 347
  of smooth functions, 337
sheaf cohomology, 341
smooth manifold, 5
soldered curvature, 35
soldering form, 25
  basic, 25
soul map, 102
space-time decomposition, 230
spatial
  atlas, 230
  distribution, 229
    g-compatible, 231
    foliation, 232
    structure, 229
spectrum of a ring, 348
  maximal, 349
  real, 349
spin group, 244
spinor
  bundle, 249
    principal, 249
    universal, 253
  connection, 250
  field, 249
  metric, 246
  space, 244
  structure
    Dirac, 246
    pseudo-Riemannian, 247
    Riemannian, 248
    universal, 253
spinor Lorentz group, 245
splitting domain, 108
splitting principle, 273
spontaneous symmetry breaking, 198
stalk, 337
Stiefel–Whitney class, 278
strength, 178
  canonical, 180
  form, 181
  of a gauge field, 179
  of a linear connection, 186
structure group, 169
  action, 172
  reduction, 198
structure module
  of a sheaf, 337
  of a vector bundle, 13
structure ring of a graded manifold, 108
structure sheaf
  of a graded manifold, 108
  of a ringed space, 346
  of a vector bundle, 351
subbundle, 11

submanifold, 7
submersion, 6
    continuous, 9
superdeterminant, 103
supergroup, 103
supermatrix, 102
    even, 102
    odd, 103
superpotential, 82
superspace, 102
supersymmetry, 123
supertrace, 103
supertransposition, 103
supervector space, 102
symmetry, 69
    classical, 69
    exact, 69
    gauge, 70
    generalized, 69
    variational, 68
symmetry current, 69
    Hamiltonian, 305

tangent bundle, 6
    affine, 17
    to a foliation, 20
    vertical, 15
tangent morphism, 6
    vertical, 15
tangent prolongation
    of a group action, 167
    of a structure group action, 174
tangent-valued form, 23
    canonical, 23
    horizontal, 24
        projectable, 25
tangent-valued Liouville form, 294
tensor algebra, 324
tensor bundle, 15
tensor product
    of Abelian groups, 321
    of commutative algebras, 321
    of complexes, 329
    of graded modules, 100
    of modules, 320
    of vector bundles, 14

    topological, 350
tensor product connection, 36
tetrad
    bundle, 225
    coframe, 226
    field, 225
    form, 226
        time-like, 230
    frame, 225
    function, 226
Todd class, 274
topological charge, 265
topological invariant, 348
torsion form, 35
    of a world connection, 37
    soldering, 221
total derivative, 27
    graded, 119
    higher order, 47
    infinite order, 59
total space, 7
transgression formula, 279
    on a base, 279
transition functions, 8
    G-valued, 169
trivial extension of a sheaf, 347
trivialization chart, 9
trivialization morphism, 9
tubular neighborhood, 20
typical fibre, 8

Utiyama theorem, 190

variation equation, 91
variational
    bicomplex, 62
        graded, 119
    complex, 62
        graded, 119
        short, 66
    derivative, 63
    formula, 64
    operator, 62
        graded, 119
    symmetry, 68
        classical, 69

of a graded Lagrangian, 123
vector bundle, 12
  characteristic, 109
  dual, 13
  graded, 131
vector field, 18
  complete, 18
  fundamental, 173
  generalized, 67
  graded, 111
  holonomic, 223
  horizontal, 30
    standard, 222
  integrable, 49
  left-invariant, 166
  parallel, 222
  principal, 182
  projectable, 18
    on a jet manifold, 49
  right-invariant, 166
  subordinate to a distribution, 19
  vertical, 18
vector space, 320
  graded, 100
vector-valued form, 26
vertical automorphism, 11
vertical extension
  of a Hamiltonian form, 317
  of an exterior form, 90
vertical splitting, 16
  of a vector bundle, 16
  of an affine bundle, 17
vertical-valued form, 25

weak conservation law, 69
Weil homomorphism, 265
Whitney sum
  of vector and affine bundles, 17
  of vector bundles, 13
world connection, 37
  affine, 238
  linear, 219
  on a tensor bundle, 37
  on the cotangent bundle, 37
  symmetric, 37
world manifold, 61

flat, 221
in gravitation theory, 215
parallelizable, 221
world metric, 38
  pseudo-Riemannian, 225
  Riemannian, 229
    g-compatible, 231
world spinor, 253

Yang–Mills
  equations, 190
  gauge theory, 190
  Lagrangian, 190
    graded, 197
  supergauge theory, 196

Zariski topology, 349